George Em Karniadakis (Ed.)
Handbook of Fractional Calculus with Applications

Handbook of Fractional Calculus with Applications

Edited by *J. A. Tenreiro Machado*

Volume 1: Theory
Anatoly Kochubei, Yuri Luchko (Eds.), 2019
ISBN 978-3-11-057081-6, e-ISBN (PDF) 978-3-11-057162-2,
e-ISBN (EPUB) 978-3-11-057063-2

Volume 2: Fractional Differential Equations
Anatoly Kochubei, Yuri Luchko (Eds.), 2019
ISBN 978-3-11-057082-3, e-ISBN (PDF) 978-3-11-057166-0,
e-ISBN (EPUB) 978-3-11-057105-9

Volume 4: Applications in Physics, Part A
Vasily E. Tarasov (Ed.), 2019
ISBN 978-3-11-057088-5, e-ISBN (PDF) 978-3-11-057170-7,
e-ISBN (EPUB) 978-3-11-057100-4

Volume 5: Applications in Physics, Part B
Vasily E. Tarasov (Ed.), 2019
ISBN 978-3-11-057089-2, e-ISBN (PDF) 978-3-11-057172-1,
e-ISBN (EPUB) 978-3-11-057099-1

Volume 6: Applications in Control
Ivo Petráš (Ed.), 2019
ISBN 978-3-11-057090-8, e-ISBN (PDF) 978-3-11-057174-5,
e-ISBN (EPUB) 978-3-11-057093-9

Volume 7: Applications in Engineering, Life and Social Sciences, Part A
Dumitru Baleanu, António Mendes Lopes (Eds.), 2019
ISBN 978-3-11-057091-5, e-ISBN (PDF) 978-3-11-057190-5,
e-ISBN (EPUB) 978-3-11-057096-0

Volume 8: Applications in Engineering, Life and Social Sciences, Part B
Dumitru Baleanu, António Mendes Lopes (Eds.), 2019
ISBN 978-3-11-057092-2, e-ISBN (PDF) 978-3-11-057192-9,
e-ISBN (EPUB) 978-3-11-057107-3

Preface

Fractional calculus (FC) originated in 1695, nearly at the same time as conventional calculus. However, FC attracted limited attention and remained a pure mathematical exercise in spite of the contributions of important mathematicians, physicists, and engineers. FC has seen a rapid development during the last few decades, both in mathematics and in applied sciences, and today it is recognized as a very effective tool for describing complex phenomena involving long-range memory effects and non-locality. A large number of research papers and books devoted to this subject have been published, and presently several specialized conferences and workshops are organized each year. The popularity of FC across all fields of science is due to its successful application in mathematical models, namely, in the form of FC operators and fractional integrals and especially differential equations. Presently, we are witnessing considerable progress of both the theoretical aspects and the applications of FC in areas such as physics, engineering, biology, medicine, economy, and finance.

The popularity of FC has attracted many researchers from all over the world and there is a demand for works covering all areas of science and engineering in a systematic and rigorous form. In fact, the literature devoted to FC and its applications is very extensive, but readers are often confronted with great heterogeneity and, in some cases, with inaccurate information. The *Handbook of fractional calculus with applications* (HFCA) intends to fill that gap and provides the readers with a solid and systematic treatment of the main aspects and applications of FC. Motivated by these ideas, the editors of the volumes involved a team of internationally recognized experts for a joint publishing project offering a survey of their own and other important results in their fields of research. As a result of these joint efforts, a modern encyclopedia of FC and its applications, reflecting present-day scientific knowledge, is now available with the HFCA. This work is distributed by several distinct volumes, each one developed under the supervision of its editors.

We have organized this volume into different thematic units, starting with time-fractional PDEs, proceeding with space- and space-time fractional PDEs, and concluding with singularities and fast solvers.

Part 1. Time-fractional derivatives

Chapter 1 describes many different numerical methods to compute fractional integrals and fractional derivatives, with emphasis placed on Caputo operators. It also covers several solvers for time-fractional differential equations. Chapter 2 introduces various approximation formulas for time-fractional problems, which can be theoretically proven. It also provides numerical schemes for the time-fractional subdiffusion equation and the diffusion-wave equation along with corresponding theoretical results.

https://doi.org/10.1515/9783110571684-201

Part 2. Space- and space-time fractional derivatives

Chapter 3 introduces six types of fractional derivatives, both for time- and space-fractional problems, and provides a comprehensive list of high-order discretization of these derivatives. It also provides examples and stability analysis of time- and space-fractional partial differential equations (PDEs). Chapter 4 analyzes two types of spectral methods for fractional differential equations. The first spectral method employs the classical orthogonal polynomials and the Galerkin formulation and is applied to space-time fractional PDEs. The second method employs Muntz polynomials and the weighted Galerkin formulation and is applied to time-fractional PDEs. Chapter 5 presents properties of the generalized Jacobi functions (GJFs) and their application to discretizing accurately space-time fractional PDEs using a Petrov–Galerkin formulation. The use of GJFs enables the resolution of singularities in the solution, which are inherent in fractional PDEs so that high-order numerical accuracy is achieved. Chapter 6 presents both spectral and spectral element methods using Galerkin and collocation formulations. It includes treatments of space-time fractional advection–diffusion as well as fractional Laplacians in multi-dimensions. It also includes hybrid formulations where finite differences are employed in time and spectral or spectral elements for spatial discretization. Chapter 7 focuses on discontinuous Galerkin methods for fractional advection–diffusion equations up to two dimensions. The high-order derivative is decomposed and the equation is recast into a first-order system. Then, a corresponding numerical scheme is presented and its stability and convergence are analyzed. Moreover, some algorithms in two dimensions are presented.

Chapter 8 presents both finite difference and finite element numerical methods for time-space fractional PDEs. In particular, finite difference methods are presented for the one-dimensional time-space fractional advection–dispersion equation and time-space Caputo–Reisz fractional diffusion equation in two dimensions. Also, an unstructured mesh finite element method for the two-dimensional time-space Riesz fractional diffusion equation on an irregular convex domain is presented. Finally, a two-dimensional time-space fractional diffusion equation based on the fractional Laplacian operator is considered. Chapter 9 deals with meshless methods for PDEs with fractional Laplacian in multi-dimensions. It compares two existing radial basis collocation methods on fractional Poisson problems with zero non-local boundary conditions. Numerical results show that the two methods have comparable solution accuracy and same-order time complexity, but different flexibility. Chapter 10 reviews the Lagrangian solvers developed in the last two decades for fractional differential equations using the proper dynamics of random walkers. For the vector fractional differential equations, a multi-scaling compound Poisson process can track the trajectory of particles moving along arbitrary directions with direction-dependent scaling rates. Different random walk particle tracking methods are presented to track particles whose mechanical dispersion follows streamlines.

Part 3. Singularities

Chapter 11 demonstrates by simple examples that singularities are very common in the solutions of fractional differential equations on bounded domains. These singularities reduce substantially the rate of convergence of numerical methods. The chapter reviews four main classes of specific methods that deal with singularities in the solutions of fractional differential equations.

Part 4. Fast solvers

Chapter 12 addresses the issue of computational complexity of the numerical solution of space-fractional PDEs. The non-local nature of fractional differential operators leads to dense or full stiffness matrices with complex structures. This chapter addresses the computational issues of fractional PDEs and outlines some of the recent developments of fast and accurate numerical methods and also discusses possible future directions in the field. Chapter 13 focuses on how to perform fast computation of the Mittag-Leffler function. The series representation is not suitable for computation and hence alternative approaches are presented based on the numerical inversion of the Laplace transform, e. g., a technique known as the optimal parabolic contour. Applications to the evaluation of derivatives of the Mittag-Leffler function and to matrix arguments are also presented.

My special thanks go to the authors of individual chapters that are excellent surveys of selected classical and new results in several important fields of FC. The editors believe that the HFCA will represent a valuable and reliable reference work for all scholars and professionals willing to develop research in the challenging and timely scientific area.

George Em Karniadakis

Contents

Kai Diethelm

Fundamental approaches for the numerical handling of fractional operators and time-fractional differential equations

Abstract: This article describes fundamental approaches for the numerical handling of problems arising in fractional calculus. This includes, in particular, methods for approximately computing fractional integrals and fractional derivatives, where the emphasis is placed on Caputo operators, as well as solvers for the associated differential and integral equations.

Keywords: Fractional differential equation, Caputo derivative, fractional integral, fractional derivative, numerical method

MSC 2010: Primary 65D25, Secondary 65D30, 65L05, 65L06, 65L60, 65R20

1 Introduction and statement of the problem

This chapter is devoted to a description of classical and modern approaches for the numerical computation of fractional integrals and derivatives. Moreover, these numerical methods will be applied to construct computational solution methods for associated differential and integral equations.

In this context the chapter will mainly concentrate on approximation methods for Riemann–Liouville integrals and for derivatives of the Caputo type.

In the approximation theory of classical (integer-order) differential and integral operators, one usually analyzes the error of such numerical methods under the assumption that the functions to be differentiated or integrated are in the set $C^k[a,b]$ with some, often large, $k \in \mathbb{N}$ and some interval $[a,b]$ that contains the region of interest. Many authors follow this path in the numerical approximation of fractional operators too, but in fact this is often not sufficient since most mathematical models of fractional order are known to have solutions whose second or even already first derivative has a singularity at one end point of the relevant interval; see, e. g., [9, Theorems 9

Acknowledgement: The preparation of this article was supported by the German Federal Ministry for Education and Research (BMBF) under Grant No. 01IS17096A. While writing the chapter, the author was affiliated to AG Numerik of Institut Computational Mathematics at Technische Universität Braunschweig.

Kai Diethelm, GNS Gesellschaft für numerische Simulation mbH, Am Gaußberg 2, 38114 Braunschweig, Germany; and Fakultät ANG, Hochschule Würzburg-Schweinfurt, Ignaz-Schön-Str. 11, 97421 Schweinfurt, Germany, e-mail: diethelm@gns-mbh.com

https://doi.org/10.1515/9783110571684-001

and 10], [8, § 6.4], or [2, 7, 23]. Therefore, it is usually desirable to have error bounds for the numerical methods under weaker assumptions, in particular admitting functions with unbounded derivatives. Unfortunately, such results are not always readily available.

In the theory of fractional calculus (see, e. g., [8, Chapters 2 and 3]), it is useful to introduce the fractional integrals first and the fractional derivatives afterwards. Our presentation for the corresponding numerical methods will follow this pattern.

It is important to keep in mind that there is a high degree of inconsistency of the nomenclature throughout the literature in this field. In particular, it is often possible to generalize approaches for the numerical handling of integer-order operators to the fractional setting in many different ways, and for each of the different outcomes of such a construction, it seems justified to name it after the underlying classical scheme; thus, two algorithms with the same name need not necessarily be identical. On the other hand, given a specific fractional numerical method, there are often multiple ways to derive it, and this may lead to different names being assigned to them; therefore, two algorithms with different names may actually coincide with each other.

Without loss of generality, in this chapter we shall assume that the starting point for all fractional operators is located at the origin.

2 Approximations for Riemann–Liouville integrals

Regarding the approximation of Riemann–Liouville integrals, this chapter will focus on six different approaches. The first four methods, discussed in Sections 2.1, 2.2, and 2.3, are classical. They have been in use for many years, and their properties are very well understood. The other two, presented in Sections 2.4 and 2.5, are much more recent, and they seem to have some very attractive properties, but at present their analysis is not fully complete yet.

2.1 Product rectangle and trapezoidal methods

Probably the simplest classical way of numerically approximating the Riemann–Liouville integral of a given function are the right and left *product rectangle rules* [11, 12]: Given a function $f : [0, b] \to \mathbb{R}$ whose Riemann–Liouville integral $I^\alpha f$ on the interval $[0, b]$ is to be determined, one chooses $n + 1$ grid points $0 = x_0 < x_1 < \cdots < x_{n-1} < x_n = b$, defines the function $f_n : [0, b] \to \mathbb{R}$ via the piece-wise constant approach

$$f_n(x) := f(x_j) \quad \text{for } x \in [x_j, x_{j+1}) \tag{1}$$

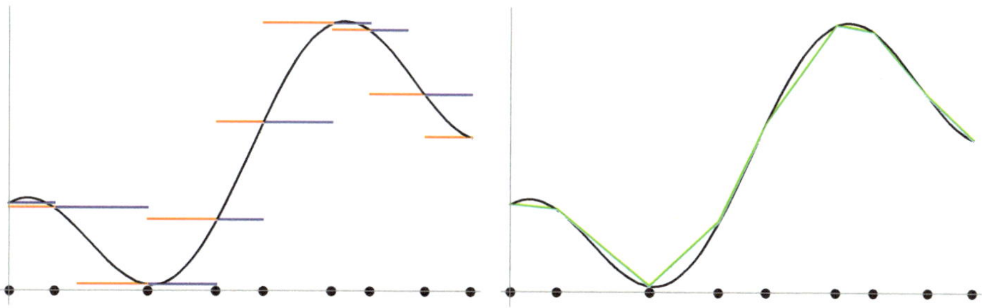

Figure 1: Left: Example graphs for a given function f (black) and functions f_n for the left product rectangle rule (blue) and right product rectangle rule (orange). Right: Example graphs for a given function f (black) and functions f_n for the product trapezoidal rule (green).

(see the blue curve in the left part of Figure 1), and uses $I^\alpha f_n$ as an approximation for $I^\alpha f$. This yields the left product rectangle rule, or *forward Euler rule*,

$$\begin{aligned} I^\alpha f(x_j) \approx I^\alpha_{n,\mathrm{LRe}} f(x_j) &:= I^\alpha f_n(x_j) \\ &= \frac{1}{\Gamma(\alpha)} \sum_{k=0}^{j-1} f(x_k) \int_{x_k}^{x_{k+1}} (x_j - s)^{\alpha-1}\,\mathrm{d}s, \end{aligned} \tag{2}$$

for $j = 1, 2, \dots, n$. Clearly, the integral in the rightmost expression can easily be computed explicitly, which yields the representation

$$I^\alpha_{n,\mathrm{LRe}} f(x_j) = \sum_{k=0}^{j-1} f(x_k) a^{\mathrm{LRe}}_{j,k} \tag{3a}$$

with

$$a^{\mathrm{LRe}}_{j,k} = \frac{1}{\Gamma(\alpha+1)}\big((x_j - x_k)^\alpha - (x_j - x_{k+1})^\alpha\big). \tag{3b}$$

One can see that the weights $a^{\mathrm{LRe}}_{j,k}$ are independent of n, but they do depend on the order α of the Riemann–Liouville integral; however, for the sake of simplicity, the notation does not reflect the latter fact. In the particularly important special case of an equispaced grid $\{x_j\}$, i. e., in the case $x_j = jh$ with $h = b/n$, the relations (3b) can be simplified to

$$a^{\mathrm{LRe}}_{j,k} = \frac{h^\alpha}{\Gamma(\alpha+1)}\big((j-k)^\alpha - (j-k-1)^\alpha\big). \tag{3c}$$

A particular point of interest in the representation (3c) is that the weights $a^{\mathrm{LRe}}_{j,k}$ actually do not show a dependency on j and k individually, but only on the difference $j-k$, i. e., they preserve the convolution structure that is present in the kernel of the Riemann–Liouville integral operator. In a concrete implementation, this feature can be exploited to reduce the memory requirements and the computational cost.

The *backward Euler method*, or right product rectangle rule, works in a very similar way. The only difference is that on each subinterval of the discretization, we use the value of f at the right end point instead of the left end point to define the approximating function, i. e., we replace equation (1) by

$$f_n(x) := f(x_{j+1}) \quad \text{for } x \in [x_j, x_{j+1}) \tag{4}$$

(see the orange curve in the left part of Figure 1). This yields the approximation formula

$$I^\alpha f(x_j) \approx I^\alpha_{n,\mathrm{RRe}} f(x_j) := \sum_{k=1}^{j} f(x_k) a_{j,k}^{\mathrm{RRe}} \tag{5a}$$

for $j = 1, 2, \ldots, n$, with

$$a_{j,k}^{\mathrm{RRe}} = a_{j,k-1}^{\mathrm{LRe}}. \tag{5b}$$

An approach closely related to the two rectangular methods is the *product trapezoidal rule* [11, 12]. Here, we replace the approximation (1) by the piece-wise linear function

$$f_n(x) := \frac{1}{x_{j+1} - x_j}((x - x_j) f(x_{j+1}) + (x_{j+1} - x) f(x_j)) \quad \text{for } x \in [x_j, x_{j+1}) \tag{6}$$

as indicated in the green curve in the right part of Figure 1. This yields the approximation formula

$$I^\alpha f(x_j) \approx I^\alpha_{n,\mathrm{Tr}} f(x_j) := \sum_{k=0}^{j} f(x_k) a_{j,k}^{\mathrm{Tr}} \tag{7a}$$

for $j = 1, 2, \ldots, n$, with

$$a_{j,0}^{\mathrm{Tr}} = \frac{1}{\Gamma(\alpha+2)} \cdot \frac{(x_j - x_1)^{\alpha+1} + x_j^\alpha [(\alpha+1)x_1 - x_j]}{x_1}, \tag{7b}$$

$$\begin{aligned} a_{j,k}^{\mathrm{Tr}} = \frac{1}{\Gamma(\alpha+2)} \Bigg(& \frac{(x_j - x_{k-1})^{\alpha+1} + (x_j - x_k)^\alpha [\alpha(x_{k-1} - x_k) + x_{k-1} - x_j]}{x_k - x_{k-1}} \\ & + \frac{(x_j - x_{k+1})^{\alpha+1} + (x_j - x_k)^\alpha [\alpha(x_k - x_{k+1}) + x_{k+1} - x_j]}{x_{k+1} - x_k} \Bigg) \\ & (1 \le k \le j-1), \end{aligned} \tag{7c}$$

$$a_{j,j}^{\mathrm{Tr}} = \frac{1}{\Gamma(\alpha+2)} (x_j - x_{j-1})^\alpha. \tag{7d}$$

As above, we note that these formulas can be simplified in the case of an equidistant mesh, where one finds

$$a_{j,k}^{\mathrm{Tr}} = \frac{h^\alpha}{\Gamma(\alpha+2)} \times \begin{cases} (j-1)^{\alpha+1} - (j-1-\alpha) j^\alpha & \text{for } k = 0, \\ ((j-k+1)^{\alpha+1} - 2(j-k)^{\alpha+1} + (j-k-1)^{\alpha+1}) & \text{for } 1 \le k < j, \\ 1 & \text{for } k = j. \end{cases} \tag{7e}$$

All these methods admit a relatively simple error estimate. Specifically, Figure 1 already indicates that the trapezoidal rule may be expected to provide a much better approximation than the rectangle rules; the following statement, taken from [12, Theorems 2.4 and 2.5], confirms this expectation.

Theorem 1. *Given an equidistant mesh on the interval* $[0, b]$, *the following error bounds hold for the product rectangle and trapezoidal formulas, respectively:*

(a) *Whenever* $f \in C^1[0, b]$,

$$|I^\alpha f(x_j) - I^\alpha_{n,\mathrm{LRe}} f(x_j)| \leq \frac{1}{\Gamma(\alpha+1)} x_j^\alpha h \sup_{0 \leq x \leq x_j} |f'(x)| = O(h)$$

uniformly on $[0, b]$.

(b) *For* $f(x) = x^p$ *with some* $p \in (0, 1)$,

$$|I^\alpha f(x_k) - I^\alpha_{n,\mathrm{LRe}} f(x_j)| \leq C_{\alpha,p} x_j^{\alpha+p-1} h$$

with some constant $C_{\alpha,p}$ *that depends on* α *and* p *but not on* h, k, *or* f.

(c) *Whenever* $f \in C^2[0, b]$,

$$|I^\alpha f(x_j) - I^\alpha_{n,\mathrm{Tr}} f(x_j)| \leq C_\alpha \sup_{0 \leq x \leq x_j} |f''(x)| x_j^\alpha h^2 = O(h^2)$$

uniformly on $[0, b]$, *where* C_α *is a constant that depends only on* α.

(d) *Whenever* $f \in C^1[0, b]$ *and* f' *satisfies a Lipschitz condition of order* $\mu \in (0, 1)$ *on* $[0, b]$,

$$|I^\alpha f(x_j) - I^\alpha_{n,\mathrm{Tr}} f(x_j)| \leq C_{\alpha,\mu} L_\mu(f') x_j^\alpha h^{1+\mu} = O(h^{1+\mu})$$

uniformly on $[0, b]$, *where* $C_{\alpha,\mu}$ *is a constant that depends only on* α *and* μ *and where* $L_\mu(f')$ *denotes the Lipschitz constant of order* μ *of the function* f'.

(e) *For* $f(x) = x^p$ *with some* $p \in (0, 2)$, $p \neq 1$,

$$|I^\alpha f(x_k) - I^\alpha_{n,\mathrm{LRe}} f(x_j)| \leq C_{\alpha,p} x_j^{\alpha+p-\rho} h^\rho$$

with $\rho = \min(2, p+1)$ *and with some constant* $C_{\alpha,p}$ *that depends on* α *and* p *but not on* h, k, *or* f.

In practical applications, it is frequently required to (approximately) compute the value of $I^\alpha f(x)$ not only for a single value of x but for all $x \in \{x_1, x_2, \ldots, x_n\}$. The methods described above admit to achieve this in an obvious manner. Since the computational cost of the jth step in this process is $O(j)$, the execution of the complete procedure can be done in $O(n^2)$ operations. In the classical case $\alpha = 1$ one could have rewritten the algorithms in a way that would have led to an $O(1)$ complexity in each individual step, and thus to an $O(n)$ complexity overall, but the approach does not directly admit that in the fractional case $\alpha \notin \mathbb{N}$. This reflects the non-locality of the fractional operators. However, the ideas of [14] and [13] indicate ways in which certain nodes may be left out of the numerical computation of the fractional integrals in such a way that the cost of each step can be reduced to $O(\log j)$, thus producing a computational cost of $O(n \log n)$, which is only marginally more expensive than in the classical case, for the complete process.

2.2 The G1 method

A different but also quite evident approach for numerically computing Riemann–Liouville integrals is the so-called *G1 method*, which was already discussed in very early works on numerical fractional calculus [31, § 8.2]. Starting from the definition of the Riemann–Liouville operator,

$$I^\alpha f(x) = \frac{1}{\Gamma(\alpha)} \int_0^x (x-t)^{\alpha-1} f(t)\, dt,$$

one recalls that it can be written, if f is continuous, in the so-called Grünwald–Letnikov form as the limit

$$I^\alpha f(x) = \lim_{n\to\infty} \frac{(x/n)^\alpha}{\Gamma(\alpha)} \sum_{k=0}^{n} \frac{\Gamma(k+\alpha)}{\Gamma(k+1)} f\left(x - k\frac{x}{n}\right);$$

cf., e. g., [8, Theorem 2.26]. Evidently, the limit does not change if one terminates the summation already at $k = n-1$, and so

$$I^\alpha f(x) = \lim_{n\to\infty} \frac{(x/n)^\alpha}{\Gamma(\alpha)} \sum_{k=0}^{n-1} \frac{\Gamma(k+\alpha)}{\Gamma(k+1)} f\left(x - k\frac{x}{n}\right).$$

The G1 method is then obtained from this formula by choosing a finite value for n instead of performing the limit operation $n \to \infty$, viz.

$$I^\alpha f(x) \approx I^\alpha_{n,\mathrm{G1}} f(x) := \frac{h^\alpha}{\Gamma(\alpha)} \sum_{k=0}^{n-1} \frac{\Gamma(k+\alpha)}{\Gamma(k+1)} f(x-kh), \tag{8}$$

where $h = x/n$.

Using standard methods from approximation theory essentially in the same way as in [8, Proof of Theorem 2.26], one can derive an error bound for this approach.

Theorem 2. *If $f \in C^1[0,b]$, then the G1 algorithm for the approximate computation of $I^\alpha f$ converges as $O(h)$, where $h = b/n$.*

We thus conclude that the G1 method does not converge very fast; however, it has a certain charm due to the fact that its weights satisfy a recurrence relation, namely,

$$\frac{\Gamma(k+\alpha)}{\Gamma(k+1)} = \frac{k+\alpha-1}{k}\frac{\Gamma(k+\alpha-1)}{\Gamma(k)}.$$

This identity can be exploited to rewrite the formula in the convenient form

$$I^{\alpha}_{n,\mathrm{G1}} f(x) = h^{\alpha}\Bigg(\Bigg(\Bigg(\cdots\Bigg(\Bigg(f(h)\frac{n+\alpha-2}{n-1} + f(2h)\Bigg)\frac{n+\alpha-3}{n-2} + f(3h)\Bigg)\cdots\Bigg)\frac{1+\alpha}{2} + f(x-h)\Bigg)\frac{\alpha}{1} + f(x)\Bigg),$$

which eliminates the necessity to evaluate any Gamma functions; see [31, equation (8.22)].

Like the methods of Section 2.1, the G1 method also requires $O(n^2)$ operations to compute the approximations to $I^{\alpha} f(x_j)$ for $j = 1, 2, \ldots, n$.

2.3 Fractional backward differentiation methods

A completely different, and much more sophisticated, class of methods was derived by Lubich [22, 24, 25, 26, 27] in the 1980s. The approach is based on a generalization of the classical linear multi-step methods for the solution of differential equations. In principle, the basic idea can be applied to any linear multi-step method [17]; in practice the backward differentiation formulas (BDFs) are by far the most significant subclass, and therefore they will be the only ones discussed explicitly here. Our presentation will essentially follow the path laid out by Garrappa [15].

Since the function F with

$$F(x) := y_0 + \int_0^x f(t)\,dt$$

is the solution to the initial value problem

$$y'(x) = f(x), \quad y(0) = y_0,$$

it is clear that methods for solving initial value problems of this kind can be employed to numerically approximate such integrals. Assuming an equidistant grid $x_j = jh$ with some fixed step size $h > 0$, the k-step BDF for this initial value problem takes the form

$$\sum_{j=0}^{k} \rho_j y_{n-j} = h\sigma_0 f(x_n), \quad n = k, k+1, \ldots; \tag{9}$$

the solution approximations at the points $x_1, x_2, \dots, x_{k-1}$ need to be found by a different method. The coefficients $\rho_1, \rho_2, \dots, \rho_k$ and σ_0 are chosen so that the order of the associated error term is maximized. Note that these formulas are only used for $k = 1, 2, \dots, 6$ because they become unstable for $k \geq 7$.

One can then show that it is possible to reformulate the set of equations (9) in a way that allows to find an explicit expression for y_n, viz.

$$y_n = h \sum_{j=0}^{n} \omega_{n-j} f(x_j), \quad n \geq k.$$

This expression is known as a convolution quadrature formula for the integral $I^1 f(x)$. The so-called convolution weights ω_j can be found from the coefficients of the BDF itself; specifically, one defines the first characteristic polynomial $\rho(z) = \rho_0 z^k + \rho_1 z^{k-1} + \dots + \rho_k$ and the second characteristic polynomial $\sigma(z) = \sigma_0 z^k$ and defines the generating function

$$\delta(\xi) = \frac{\rho(1/\xi)}{\sigma(1/\xi)}. \tag{10}$$

It then turns out that the ω_j are just the coefficients of the formal power series expansion of δ^{-1}, i. e.,

$$\frac{1}{\delta(\xi)} = \sum_{j=0}^{\infty} \omega_j \xi^j.$$

In order to extend this concept to a *convolution quadrature formula* for the Riemann–Liouville integral of order α, it suffices to simply consider the expression

$$I^\alpha f(x_n) \approx I^\alpha_{n,\mathrm{CQ}} f(x_n) := h^\alpha \sum_{k=0}^{n} \omega_{n-j,\alpha} f(x_j),$$

where the convolution quadrature weights $\omega_{n-j,\alpha}$ are given as the coefficients of the formal power series expansion

$$(\delta(\xi))^{-\alpha} = \sum_{j=0}^{\infty} \omega_{j,\alpha} \xi^j.$$

For this approach, a convergence estimate is also readily available [28, Theorem 2.1].

Theorem 3. *Consider the convolution quadrature method based on the k-step BDF with some $k \in \{1, 2, \dots, 6\}$ and assume that $f \in C^k[0, b]$. Then, there exists a constant C independent of h such that*

$$|I^\alpha f(x_n) - I^\alpha_{n,\mathrm{CQ}} f(x_n)| \leq C x_n^{\alpha-1-k} h^k.$$

In a direct implementation of the scheme, these algorithms also need $O(n^2)$ operations to compute the approximations to $I^\alpha f(x_j)$ for $j = 1, 2, \ldots, n$. However, it is known [16] that this can be reduced to $O(n \log n)$ if the fast Fourier transform technique is employed to perform the summations.

2.4 Methods based on alternative representations of the Riemann–Liouville operator

Two approaches for numerically computing fractional integrals that completely differ from the above discussed integrals have been proposed by Li [20] and by Jiang et al. [18], respectively; the latter idea has been picked up again and investigated in more detail by Zeng et al. [38] and Baffet [1]. Both these approaches are based on the following alternative representation for the Riemann–Liouville integral operator I^α.

Theorem 4. *Let $f \in C[0, b]$, $\alpha > 0$, and $x \in [0, b]$. Then,*

$$I^\alpha f(x) = \frac{1}{\Gamma(\alpha)\Gamma(1-\alpha)} \int_0^\infty g(\xi, x)\xi^{-\alpha}\, d\xi,$$

where

$$g(\xi, x) = \int_0^x \exp(-\xi(x-t)) f(t)\, dt.$$

Clearly, the function g in Theorem 4 is independent of α. Thus, if $I^\alpha f$ is to be calculated for multiple different values of α, then g needs to be determined only once; it is just the integration of $g(\cdot, x)(\cdot)^{-\alpha}$ that must be executed repeatedly.

Using the notation

$$\tilde{g}(\xi, x, h) = \int_x^{x+h} \exp(-\xi(x+h-t)) f(t)\, dt,$$

the definition of the function g given in Theorem 4 immediately implies that

$$g(\xi, x+h) = \exp(-\xi h) g(\xi, x) + \tilde{g}(\xi, x, h) \tag{11}$$

holds for all $\xi \geq 0$ if $f \in C[0, b]$, $\alpha > 0$, $h > 0$, and $x, x+h \in [0, b]$. Equation (11) in conjunction with the representation of Theorem 4 allows to establish a useful way of computing $I^\alpha f(x+h)$ for some $h > 0$ if $I^\alpha f(x)$ is already known. Specifically, while a direct application of the definition of I^α would require a re-computation of the entire integral from 0 to x, and thus a re-evaluation the complete history of f; this can be avoided by invoking the auxiliary function g. In fact, assuming that we have already calculated $g(\cdot, x)$, equation (11) implies that the computation of $g(\cdot, x+h)$ requires only the multiplication of $g(\cdot, x)$ by a constant and the addition of a locally defined integral, i. e., an integral over the short interval $[x, x+h]$ (whose length is fixed), followed by the evaluation of the integral $\int_0^\infty g(\xi, x+h)\xi^{-\alpha}\,d\xi$ by a suitable method. The crucial point in this procedure is then to find a fixed numerical formula for evaluating this latter integral that can be used for all x. If that can be achieved, then the process of calculating $I^\alpha f(x+h)$ on the basis of the knowledge of $I^\alpha f(x)$ can be performed in $O(1)$ operations, and so the approximate calculation of $I^\alpha f(x_j)$ for $j = 1, 2, \ldots, n$, with $x_j = jh$, does not need more than the $O(n)$ operations that one would also have to apply in the classical case $\alpha = 1$.

To this end, Li [20] has suggested to prescribe a tolerance $\epsilon > 0$ for the evaluation of this integral which is required to hold for all $x \geq h$ with the given step size h. It is then possible to develop a Q-point quadrature formula for $\int_0^\infty g(\xi, x)\xi^{-\alpha}\,d\xi$ that satisfies this accuracy requirement [20, § 3]. Essentially, the idea behind the construction of the method is to truncate the integration interval at a suitable point (that, depending on ϵ, can always be found in view of the decay behavior of the integrand), to subdivide the remaining interval into K carefully chosen subintervals, and to apply a P-point Gaussian quadrature formula on each subinterval, where K and P are appropriately chosen fixed numbers, so that the total number of quadrature nodes is $Q = KP$. In this way, one obtains a procedure that allows the computation of $I^\alpha f(x_j)$ with $x_j = jh$ for $j = 1, 2, \ldots, n$ in $O(nQ)$ operations and with a storage requirement of $O(Q)$ rather than $O(n)$ as it was necessary in the methods from the previous subsections.

In practice, the values ϵ and h will usually be fixed in advance, and then Q is a constant, so the complexity of this approach is $O(n)$. However, one may also vary these two parameters. In this case it can be shown that $Q = O((|\log \epsilon| + |\log h|)^2)$ as $\epsilon \to 0$ and $h \to 0$, which still yields a satisfactory computational cost.

It should be kept in mind here that Li's construction is, in a certain sense, nonuniform with respect to the order α of the integral; specifically, to achieve the required accuracy, one finds that the number P of quadrature points to be used on each subinterval is of the order

$$P = O((1-\alpha)^{-1}(|\log(\alpha\epsilon)| + |\log h|)^2),$$

so that many points are needed if α is close to 1. Nevertheless, the procedure appears to have the potential to be very useful in many situations where $0 < \alpha < 1$ and α is not too close to 1.

A related but slightly different approach that does not have this latter disadvantage has been discussed by Jiang et al. [18], Zeng et al. [38], and Baffet [1]. Their basic idea is also to split the integral $I^\alpha f(x)$ into a local part and a history part according to

$$I^\alpha f(x+h) = \frac{1}{\Gamma(\alpha)} \int_0^h (h-s)^{\alpha-1} f(x+s)\,ds + \int_0^x w_h(x-s) f(s)\,ds, \tag{12}$$

where

$$w_h(z) = \frac{1}{\Gamma(\alpha)}(z+h)^{\alpha-1}.$$

When h is kept fixed, the local part – i. e. the first integral on the right-hand side of equation (12) – can be approximated with a given (fixed) quadrature method at a computational cost of $O(1)$ operations. For the history part that forms the second integral on the right-hand side of equation (12), one notes that the integral that defines this term is actually the Laplace convolution of the kernel w_h and the integrand function f. It is therefore natural to attempt to approximate the kernel w_h with a sum of exponentials, viz.

$$w_h(x) \approx S(x) := \sum_{p=1}^{P} b_p \exp(-a_p x). \tag{13}$$

Assuming for the moment that the coefficients a_p and b_p ($p = 1, 2, \ldots, P$) have been determined in a way that implies that the difference between w_h and S is sufficiently small for all x in the region of interest, one then replaces w_h in the above convolution integral by S, which leads to an approximation for the history term having the form

$$(w_h * f)(x) \approx (S * f)(x) = \sum_{p=1}^{P} b_p \psi_p(x) \quad \text{with } \psi_p(x) = \int_0^x \exp(-a_p(x-s)) f(s)\,ds$$

that needs to be computed for all relevant values of x with a low computational cost. To this end, one notices that each ψ_p can be characterized as the unique solution to the linear inhomogeneous first-order initial value problem

$$\psi_p'(x) = -a_p \psi_p(x) + f(x), \quad \psi_p(0) = 0.$$

This representation indicates that, once the $\psi_p(x)$ ($p = 1, 2, \ldots, P$) are known for a certain x, they may be approximately computed at the point $x + h$ by a standard solver for first-order ordinary differential equations, which also requires only an $O(1)$ effort. It should be noted that typically one has to expect some of the coefficients a_p to be large in modulus and positive [1, § 2.2]. Therefore it is strongly advised to use an A-stable scheme for this step of the process, but of course this does not have an influence on the computational cost.

Thus, to complete the description of the approximation algorithm for the Riemann–Liouville integral $I^\alpha f$, the remaining task is the development of a strategy to find a suitable value for P and to compute appropriate values a_p and b_p in the representation (13). The basic idea in this context [1, § 4] is similar to the approach of Li discussed above in the sense that it also employs a numerical quadrature; specifically, one defines

$$q_0 := 0 \quad \text{and} \quad q_k := \frac{2^{k-1}}{b} \quad (k = 1, 2, \ldots, K+1),$$

where we assume that the integral is to be computed on the interval $[0, b]$. Starting from the representation

$$
\begin{aligned}
\frac{1}{\Gamma(\alpha)} x^{\alpha-1} &= \frac{\sin \pi\alpha}{\pi} \int_0^\infty s^{-\alpha} \exp(-sx)\, ds \\
&= \frac{\sin \pi\alpha}{\pi} \left(\sum_{k=0}^{K} \int_{q_k}^{q_{k+1}} s^{-\alpha} \exp(-sx)\, ds + \int_{q_{K+1}}^{\infty} s^{-\alpha} \exp(-sx)\, ds \right),
\end{aligned}
$$

one selects the value K sufficiently large so that the integral over $[q_{K+1}, \infty)$ is below an acceptable threshold and can be neglected. On each of the remaining subintervals, one uses an appropriately transformed J-point Gauss–Legendre quadrature formula (for $k = 1, 2, \ldots, K$) or a J-point Gauss–Jacobi quadrature with the weight function $s^{-\alpha}$ (for $k = 0$). This yields an approximation formula that can be rearranged to obtain the required form (13) with, in particular, $P = (K + 1)J$. The associated error analysis may be found, e. g., in [1, 18].

A quite similar method, using a different quadrature approximation for the history part of (12) that seems to allow the use of a smaller value of P, has been proposed and investigated by McLean [30].

All methods mentioned in this subsection are based on a two-step approach. The first step consists of defining some accuracy threshold ϵ, say, and replacing the integral in question itself by an expression that differs from the exact value by no more than this threshold. This is followed by a second step that constructs an approximation process depending on some parameter. As this parameter converges towards a suitable value, the associated approximation result then converges to the ϵ-perturbed integral. As such, the procedures of this subsection can be subsumed under the theory of *approximate approximations* as discussed, e. g., in [29].

2.5 Spectral methods

All the methods introduced so far can be considered as local methods in the sense that they can be used in a quite natural way to compute the approximations for $I^\alpha f(x_j)$, $j = 1, 2, \ldots, N$, with some N defined a priori, where at the jth step at least some of the information obtained in the previous steps can be re-used, so that only a local set of information needs to be newly computed (even though the way in which the “old” information is exploited is not necessarily local in nature). A significantly different approach is based on the fractional integral’s property to be a non-local operator. This gives rise to the idea of attempting a truly non-local approximation approach, the so-called *spectral methods* [37], i. e., methods based on an expansion of the solution in terms of eigenfunctions of a certain operator.

For $0 < \alpha < 1$, the basic idea is to recall the classical Jacobi polynomials with parameters $-\alpha/2$ and $\alpha/2$, i. e., the polynomials $P_k^{(-\alpha/2,\alpha/2)}$ of degree k that are orthogonal with respect to the weight function $(1-x)^{-\alpha/2}(1+x)^{\alpha/2}$ over the interval $[-1,1]$; cf., e. g., [34]. Assuming that the value to be approximately computed is $I^\alpha f(b)$ with some $b > 0$ (the underlying idea being that the fractional integral of f is to be computed for one value of its argument only), one defines the basis functions

$$\mathcal{P}_k^{(\alpha)}(x) := \left(\frac{2x}{b}\right)^{\alpha/2} P_{k-1}^{(-\alpha/2,\alpha/2)}\left(-1+\frac{2x}{b}\right)$$

for $x \in [0,b]$ and $k \in \mathbb{N}$.

The goal of the method is to find an approximation to $I^\alpha f(b)$ in the form

$$F_n(b) = \sum_{k=1}^{n} p_k \mathcal{P}_k^{(\alpha)}(b) \tag{14}$$

with suitable coefficients p_k. The determination of the coefficients is based on a Petrov–Galerkin approach, i. e., one chooses a set $\{\mathcal{Q}_k^{(\alpha)} : k = 1,2,\ldots,n\}$ of test functions and demands that all these test functions be orthogonal to the residual of the approximation F_n with respect to the standard L^2 inner product on the interval $[0,b]$, viz.

$$\int_0^b ({}^{\mathrm{C}}D^\alpha F_n(t) - f(t))\mathcal{Q}_j^{(\alpha)}(t)\,\mathrm{d}t = 0,$$

which is equivalent to requiring that

$$\sum_{k=1}^{n} p_k \int_0^b {}^{\mathrm{C}}D^\alpha \mathcal{P}_k^{(\alpha)}(t)\mathcal{Q}_j^{(\alpha)}(t)\,\mathrm{d}t = \int_0^b f(t)\mathcal{Q}_j^{(\alpha)}(t)\,\mathrm{d}t, \quad j = 1,2,\ldots,n. \tag{15}$$

According to the theory of fractional Sturm–Liouville eigenproblems [36], a suitable choice of the test functions is given by

$$\mathcal{Q}_k^{(\alpha)}(x) := \left(\frac{2(b-x)}{b}\right)^{\alpha/2} P_{k-1}^{(\alpha/2,-\alpha/2)}\left(-1+\frac{2x}{b}\right)$$

since these functions have the useful property

$${}^{\mathrm{C}}D_{b-}^{\alpha/2}\mathcal{Q}_k^{(\alpha)}(x) = {}^{\mathrm{C}}D^{\alpha/2}\mathcal{P}_k^{(\alpha)}(x).$$

In order to compute the approximation F_n, i. e., in order to find the coefficients p_k, it is necessary to solve the linear system (15), and to this end one may recall from [21] the identity

$$\int_0^b {}^{\mathrm{C}}D^\alpha \mathcal{P}_k^{(\alpha)}(t)\mathcal{Q}_j^{(\alpha)}(t)\,\mathrm{d}t = \int_0^b {}^{\mathrm{C}}D^{\alpha/2}\mathcal{P}_k^{(\alpha)}(t) \cdot {}^{\mathrm{C}}D_{b-}^{\alpha/2}\mathcal{Q}_j^{(\alpha)}(t)\,\mathrm{d}t,$$

which can be simplified to

$$\int_0^b {}^{\mathrm{C}}D^{\alpha}\mathcal{P}_k^{(\alpha)}(t)\mathcal{Q}_j^{(\alpha)}(t)\,\mathrm{d}t = \left(\frac{2}{b}\right)^{\alpha-1}\left(\frac{\Gamma(k+\alpha/2)}{\Gamma(k)}\right)^2 \frac{2}{2k-1}\delta_{jk}$$

so that the equation system (15) possesses a diagonal coefficient matrix. Therefore, its solution can be immediately computed as

$$I^{\alpha}f(b) \approx F_n(b) = \left(\frac{b}{2}\right)^{\alpha-1}\sum_{k=1}^{n}\left(\frac{\Gamma(k)}{\Gamma(k+\alpha/2)}\right)^2\frac{2k-1}{2}\int_0^b f(t)\mathcal{Q}_k^{(\alpha)}(t)\,\mathrm{d}t\cdot\mathcal{P}_k^{(\alpha)}(b). \tag{16}$$

Zayernouri and Karniadakis [37] report the results of this approximation method for various special examples, including cases where f is smooth and $I^{\alpha}f$ is non-smooth or vice versa; in all these experiments they observed rapid (typically exponential) convergence behavior. However, their examples are such that the integrals $\int_0^b f(t)\mathcal{Q}_k^{(\alpha)}(t)\,\mathrm{d}t$ arising in equation (16) for the desired approximate solution can be computed exactly. In general, this will not be the case, and then one has to resort to a numerical integration method whose application introduces an additional error component that needs to be taken into account in a complete error analysis.

3 Approximations for Caputo differential operators

We now leave the discussion of numerical fractional integration and turn our attention towards the question of numerical fractional differentiation. In this context we mostly concentrate on the approximation of fractional differential operators of the Caputo type [8, Chapter 3].

3.1 Grünwald–Letnikov and related methods

The most straightforward idea consists of rewriting the differential operator in the form of a fractional-order differential quotient (the so-called Grünwald–Letnikov form; cf. [8, Lemma 3.6]) and then by replacing this differential quotient by a corresponding difference quotient with a step size $h > 0$. This leads to the so-called *Grünwald–Letnikov methods*. Numerous variants are already discussed in the classical book of Oldham and Spanier [31, § 8.2]; a very popular choice among these discretizations of the fractional derivative is the so-called *G1 algorithm* for the Riemann–Liouville derivative,

$${}^{\mathrm{RL}}D^{\alpha}f(x) \approx {}^{\mathrm{RL}}D^{\alpha}_{n,\mathrm{G1}}f(x) := \frac{h^{-\alpha}}{\Gamma(-\alpha)}\sum_{k=0}^{n-1}\frac{\Gamma(k-\alpha)}{\Gamma(k+1)}f(x-kh), \tag{17}$$

where $h = x/n$. A comparison with the G1 algorithm for fractional integrals described in Section 2.2 reveals a direct correspondence between this new formula and the latter: A simple formal replacement of the value α in the definition of the G1 method for fractional integrals, viz. equation (8), by the value $-\alpha$ leads to the formula for the G1 method for Riemann–Liouville derivatives shown in equation (17).

Using the well-known relation between Riemann–Liouville and Caputo differential operators [8, Lemma 3.4], the formula from equation (17) can easily be modified to obtain a corresponding approximation method for Caputo operators, namely,

$$\begin{aligned} {}^{\mathrm{C}}D^{\alpha} f(x) &\approx {}^{\mathrm{C}}D^{\alpha}_{n,\mathrm{G1}} f(x) \\ &:= \frac{h^{-\alpha}}{\Gamma(-\alpha)} \sum_{k=0}^{n-1} \frac{\Gamma(k-\alpha)}{\Gamma(k+1)} f(x-kh) - \sum_{k=0}^{\lceil\alpha\rceil-1} \frac{x^{k-\alpha}}{\Gamma(k+1-\alpha)} f^{(k)}(0), \end{aligned} \tag{18}$$

where once again $h = x/n$.

A similar idea is to consider a direct discretization of the integral representation of the Caputo operator,

$${}^{\mathrm{C}}D^{\alpha} f(x) = I^{1-\alpha} f'(x),$$

where, for the sake of simplicity, we have momentarily restricted ourselves to the practically most important special case, $0 < \alpha < 1$; cf., e. g., [8, Chapter 3]. In contrast to the approach of the G1 method, one can now use an arbitrary – not necessarily uniform – mesh $0 = x_0 < x_1 < \cdots < x_n = b$ on the interval $[0, b]$ and employ, for $0 < \alpha < 1$, the straightforward first-order difference

$$f'(x) \approx \frac{f(x_{j+1}) - f(x_j)}{x_{j+1} - x_j} \qquad \text{for } x \in (x_j, x_{j+1})$$

to approximate the first derivative of the function in question inside the integral operator. This yields the so-called *L1 algorithm*

$${}^{\mathrm{C}}D^{\alpha} f(x_n) \approx {}^{\mathrm{C}}D^{\alpha}_{\mathrm{L1}} f(x_n) := \frac{1}{\Gamma(2-\alpha)} \sum_{j=0}^{n-1} w^{\alpha,\mathrm{L1}}_{n-j-1,n} \big(f(x_{n-j}) - f(x_{n-j-1})\big), \tag{19a}$$

where

$$w^{\alpha,\mathrm{L1}}_{\mu,n} = \frac{(x_n - x_\mu)^{1-\alpha} - (x_n - x_{\mu+1})^{1-\alpha}}{x_{n-\mu} - x_{n-\mu-1}}; \tag{19b}$$

cf., e. g., [35, Appendix A]. In the special case of an equidistant partitioning, i. e., when $x_j = jh$ with some fixed h, the representation of equation (19b) reduces to

$$w^{\alpha,\mathrm{L1}}_{\mu,n} = h^{-\alpha}\big((n-\mu)^{1-\alpha} - (n-\mu-1)^{1-\alpha}\big), \tag{19c}$$

which is already implicitly contained in [31, equation (8.2.6)] and for which the alternative representation

$$ {}^{\mathrm{C}}D^{\alpha}_{\mathrm{L1}}f(x_n) = \frac{1}{\Gamma(2-\alpha)} \sum_{j=0}^{n} \widetilde{w}^{\alpha,\mathrm{L1}}_{j,n} (f(x_{n-j}) - f(0)) \tag{20a} $$

with

$$ \widetilde{w}^{\alpha,\mathrm{L1}}_{j,n} = h^{-\alpha} \times \begin{cases} 1 & \text{for } j = 0, \\ (j-1)^{1-\alpha} - 2j^{1-\alpha} + (j+1)^{1-\alpha} & \text{for } j = 1, 2, \ldots, n-1, \\ (1-\alpha)j^{-\alpha} + (j-1)^{1-\alpha} - j^{1-\alpha} & \text{for } j = n \end{cases} \tag{20b} $$

has been developed in [4, 5].

Both methods are very easy to implement and can be used in an immediately obvious manner to construct associated solvers for fractional differential equations, but they tend to converge rather slowly; specifically, for sufficiently well-behaved functions f and a uniform grid with mesh spacing h, the G1 method converges as $O(h)$, and the L1 method for $0 < \alpha < 1$ behaves as $O(h^{2-\alpha})$; cf., e. g., [19, pp. 41 and 44]. Therefore they are of minor interest in the numerical treatment of ordinary differential equations of fractional order. However, they have proven to be useful building blocks for certain algorithms for solving partial differential equations.

A disadvantage of the L1 method is the fact that it is restricted to $0 < \alpha < 1$. A modification can, however, be applied to construct a scheme suitable for the case $1 < \alpha < 2$; it leads to the so-called *L2 method*. For this approach, one can start [19, Sections 2.2.2 and 2.3.1] from the representation

$$ {}^{\mathrm{C}}D^{\alpha}f(x) = I^{2-\alpha}f''(x) = \frac{1}{\Gamma(2-\alpha)} \sum_{k=0}^{n-1} \int_{x_k}^{x_{k+1}} t^{1-\alpha} f''(x-t)\,\mathrm{d}t $$

for the Caputo operator, where the grid points have to satisfy the relation $x_j = jh$ with $h = x/n$. One then employs the central difference approximation

$$ f''(x-t) = f''(x_n - t) \approx \frac{1}{h^2}(f(x_n - x_{k+1}) - 2f(x_n - x_k) + f(x_n - x_{k-1})) \tag{21} $$

for $t \in [x_k, x_{k+1}]$ and obtains the final L2 approximation scheme

$$ {}^{\mathrm{C}}D^{\alpha}f(x_n) \approx {}^{\mathrm{C}}D^{\alpha}_{\mathrm{L2}}f(x_n) := \frac{h^{-\alpha}}{\Gamma(3-\alpha)} \sum_{k=-1}^{n} w^{\alpha,\mathrm{L2}}_{k,n} f(x_{n-k}) $$

with

$$ w^{\alpha,\mathrm{L2}}_{k,n} = \begin{cases} 1 & \text{for } k = -1, \\ 2^{2-\alpha} - 3 & \text{for } k = 0, \\ (k+2)^{2-\alpha} - 3(k+1)^{2-\alpha} + 3k^{2-\alpha} - (k-1)^{2-\alpha} & \text{for } 1 \le k \le n-2, \\ -2n^{2-\alpha} + 3(n-1)^{2-\alpha} - (n-2)^{2-\alpha} & \text{for } k = n-1, \\ n^{2-\alpha} - (n-1)^{2-\alpha} & \text{for } k = n. \end{cases} $$

Note that this method requires the evaluation for f at the point $x_{n+1} = (n + 1)h$, which is located outside the interval $[0, x]$.

Also, one can observe that the central difference given in equation (21) is actually symmetric with respect to one of the end points of the associated subinterval $[x_k, x_{k+1}]$ and not about its mid point. If this is undesired, one may instead use the alternative

$$f''(x - t) = f''(x_n - t) \approx \frac{1}{h^2}(f(x_{n-k-2}) - f(x_{n-k-1}) + f(x_{n-k+1}) - f(x_{n-k}))$$

on this subinterval, which gives rise to the *L2C method*

$$^{\mathrm{C}}D^{\alpha}f(x_n) \approx {}^{\mathrm{C}}D^{\alpha}_{\mathrm{L2C}}f(x_n) := \frac{h^{-\alpha}}{2\Gamma(3-\alpha)} \sum_{k=-1}^{n} w_{k,n}^{\alpha,\mathrm{L2}} f(x_{n-k})$$

with

$$w_{k,n}^{\alpha,\mathrm{L2C}} = \begin{cases} 1 & \text{for } k = -1, \\ 2^{2-\alpha} - 2 & \text{for } k = 0, \\ 3^{2-\alpha} - 2^{2-\alpha} & \text{for } k = 1, \\ (k+2)^{2-\alpha} - 2(k+1)^{2-\alpha} + 2(k-1)^{2-\alpha} - (k-2)^{2-\alpha} & \text{for } 2 \le k \le n-2, \\ -n^{2-\alpha} - (n-3)^{2-\alpha} + 2(n-2)^{2-\alpha} & \text{for } k = n-1, \\ -n^{2-\alpha} + 2(n-1)^{2-\alpha} - (n-2)^{2-\alpha} & \text{for } k = n, \\ n^{2-\alpha} - (n-1)^{2-\alpha} & \text{for } k = n+1. \end{cases}$$

Like the L2 method, the L2C method also requires the evaluation of f outside the interval $[0, x]$; in fact one has to compute $f((n + 1)h)$ and $f(-h)$. Both the L2 and the L2C method exhibit a $O(h^{3-\alpha})$ convergence behavior for $1 < \alpha < 2$ if f is sufficiently well behaved; the constants implicitly contained in the O-terms seem to be smaller for the L2 method in the case $1 < \alpha < 1.5$ and for the L2C method if $1.5 < \alpha < 2$.

In fact, in the limit case $\alpha \to 1$, the L2 method reduces to the first-order backward difference, and the L2C method becomes the centered difference of the first order; for $\alpha \to 2$ the L2 method corresponds to the classical second-order central difference.

3.2 Fractional backward differentiation formulas

In a spirit similar to that underlying the derivation of the fractional BDFs for the Riemann–Liouville integral discussed in Section 2.3 and using an equispaced grid $x_j = jh$ $(j = 0, 1, 2, \ldots, n)$ as well, one can also construct corresponding *fractional backward differentiation formulas* for fractional derivatives. This can be simply done by formally replacing the value α in the construction of Section 2.3 by $-\alpha$. In this way, one obtains the approximation methods for the Riemann–Liouville derivatives of order α. If one is interested in approximating Caputo derivatives, one has to incorporate the standard relation between the two types of derivative. This formally yields the convolution formula

$$\begin{aligned} {}^{\mathrm{C}}D^{\alpha}f(x_n) &\approx {}^{\mathrm{C}}D^{\alpha}_{n,\mathrm{CQ}}f(x_n) \\ &:= h^{-\alpha}\sum_{k=0}^{n}\omega_{n-j,-\alpha}f(x_j) - \sum_{k=0}^{\lceil\alpha\rceil-1}\frac{x_n^{k-\alpha}}{\Gamma(k+1-\alpha)}f^{(k)}(0), \end{aligned} \tag{22}$$

where the convolution weights $\omega_{n-j,-\alpha}$ are given as the coefficients of the formal power series expansion

$$(\delta(\xi))^{\alpha} = \sum_{j=0}^{\infty}\omega_{j,-\alpha}\xi^{j}$$

and where δ is, as above, the generating function of the underlying classical BDF as given in equation (10).

If the function in question is sufficiently smooth, then these methods retain the convergence orders of the classical backward differentiation methods on which they are based.

4 Numerical solution of fractional differential equations

4.1 General statements

Numerical methods for solving initial value problems for ordinary differential equations of fractional order, i. e., equations of the form

$$ {}^{\mathrm{C}}D^{\alpha}y(x) = f(x, y(x)), \quad y^{(k)}(0) = y_0^{(k)} \quad (k = 0, 1, \ldots, \lceil\alpha\rceil - 1), \tag{23}$$

or for semi-discretizing time-fractional partial differential equations with respect to the time variable, are usually based on

(a) a direct discretization of the fractional derivative using, e. g., one of the methods described in Section 3;

(b) an application of a Riemann–Liouville integral operator (with respect to the time variable in the case of a time-fractional partial differential equation), together with an incorporation of the initial condition(s), thus rewriting the equation in the form of a Volterra integral equation [8, Lemma 6.2], followed by a discretization of the integral operator by, e. g., one of the schemes mentioned in Section 2; or

(c) a generalization of some other technique for classical (first-order) differential equations to the fractional setting. This latter approach includes, in principle, also the class of *Runge–Kutta methods* that are very popular in the classical case.

As far as the Runge–Kutta methods mentioned in item (c) are concerned, it is rather difficult to carry over the underlying approach to the fractional situation and implement it there [23] because the conditions that need to be satisfied to obtain a certain convergence order tend to be very complicated, especially if α is close to 0. Therefore, the application of fractional Runge–Kutta methods in practice is quite uncommon.

The practical application of the other approaches, especially with respect to those mentioned in items (a) and (b), is relatively straightforward and does not present serious difficulties. The analysis of these methods with respect to their convergence and stability properties, however, is a topic of great significance that has been studied intensively. Details on these questions will be described in a separate chapter in this handbook [33]. Here we shall only mention a few special points.

The investigation of the properties of the numerical methods is strongly influenced by the fact that initial value problems for fractional differential equations – in stark contrast to their counterparts with classical differential equations of integer order – usually have solutions that exhibit certain singularities in a neighborhood of the starting point [8, § 6.4]. It is therefore not reasonable [32] to perform this analysis under the assumption of smooth solutions; instead, as exemplified in [12], the nature of the singularities has to be taken into account.

4.2 Fractional backward differentiation formulas

A common outcome of such a full error analysis is the observation that the convergence is relatively slow. This is due to the fact that the construction of most straightforward numerical methods follows classical principles and thus can only poorly handle the singular behavior of the exact solutions. To make up for this undesired property, it is possible to derive modifications of the algorithms in question that can theoretically provide exact results for the lower-order (i. e., for the most relevant) components of the analytical solutions, thus eliminating the most disturbing parts of the error and leaving only a rapidly convergent part. This idea has been most thoroughly developed and discussed in the context of the p-step *fractional backward differentiation formulas* [22, 24, 25]. In this case, this modification leads to the final formula

$$y(nh) \approx y_n := \sum_{k=0}^{\lceil\alpha\rceil-1} \frac{y_0^{(k)}}{k!}(jh)^k + h^\alpha\left(\sum_{j=0}^{n} \omega_{n-j,\alpha} f(jh, y_j) + \sum_{j=0}^{s} w_{n,j}^{(\alpha)} f(jh, y_j)\right) \tag{24}$$

for the initial value problem (23) with a given step size $h > 0$. Here, the first sum in the parentheses on the right-hand side is the convolution quadrature encountered above, and the second sum is the so-called *starting quadrature*. In this starting quadrature, $s + 1$ is the number of elements of the set

$$A_{p,\alpha} := \{k + j\alpha : j, k \in \mathbb{N}_0\} \cap [0, p-1], \tag{25}$$

and the coefficients $w_{n,j}^{(\alpha)}$ are characterized as being the solution to the $(s+1)$-dimensional linear equation system

$$\sum_{j=0}^{s} w_{n,j}^{(\alpha)} j^{\gamma} = \frac{\Gamma(\gamma+1)}{\Gamma(\gamma+\alpha+1)} n^{\gamma+\alpha} - \sum_{j=0}^{n} \omega_{n-j,\alpha} j^{\gamma}, \quad \gamma \in A_{p,\alpha}, \tag{26}$$

whose coefficient matrix has a generalized Vandermonde structure which asserts that the system (26) has a unique solution. This term is introduced into equation (24) in order to account for the non-smooth behavior of the differential equation's solution in a way that retains the original formula's convergence order.

Thus, as a consequence of this construction, the convergence and stability properties of the fractional-order method are essentially identical to those of the underlying classical formula [24].

A slight disadvantage of this approach is that its extension to the scenario of a non-equispaced grid is rather difficult. More significant in the practical application of such formulas is the observation [10] that the equation system (26) depends on the order α in a very involved manner. In particular, it is likely to be severely ill-conditioned for certain values of α, and therefore the effective computation of the values $w_{n,j}^{(\alpha)}$ required in the definition of the starting quadrature is likely to suffer from a lack of accuracy that, in turn, has adverse effects on the overall quality of the final result. Nevertheless, as discussed in [10], there also exist values of α for which the condition number of the equation system is still within a manageable range so that these methods may be successfully employed.

4.3 The Adams predictor–corrector method

Another method that has gained much popularity over the last two decades is the predictor–corrector technique known as the *fractional Adams method* [11, 12]. It is based on rewriting the initial value problem (23) in the equivalent form

$$y(x) = \sum_{k=0}^{\lceil\alpha\rceil-1} \frac{y_0^{(k)}}{k!} x^k + I^{\alpha}[f(\cdot, y(\cdot))](x) \tag{27}$$

and applying the left product rectangle rule (i. e., the forward Euler scheme) from Section 2.1 at $x = x_n$ to obtain a preliminary approximation, the so-called predictor, for $y(x_n)$. This is then improved in a corrector step based on the product trapezoidal scheme. If desired, this correction can be repeated m times with an appropriately chosen value for m to improve the convergence behavior [6]. This leads to the overall scheme

$$y_{k,0} = \sum_{j=0}^{\lceil\alpha\rceil-1} \frac{x_k^j}{j!} y_0^{(j)} + \sum_{j=0}^{k-1} a_{kj}^{\mathrm{LRe}} f(x_j, y_j), \tag{28a}$$

$$y_{k,\mu} = \sum_{j=0}^{\lceil\alpha\rceil-1} \frac{x_k^j}{j!} y_0^{(j)} + \sum_{j=0}^{k-1} a_{kj}^{\mathrm{Tr}} f(x_j, y_j) + a_{kk}^{\mathrm{Tr}} f(x_k, y_{k,\mu-1}) \quad (\mu = 1, 2, \ldots, m), \tag{28b}$$

$$y_k = y_{k,m} \tag{28c}$$

for $k = 1, 2, \ldots, n$, where y_k is then used as the final approximation to $y(x_k)$. In equation (28), the coefficients are those given in equations (3b) and (7b), (7c) and (7d), respectively.

Numerous techniques to improve the performance of this method have been proposed, including

- the use of the product trapezoidal method for the history part of the predictor step [3] which reduces the computational cost by almost 50 % without adverse effects on the accuracy; and
- the use of a graded (nested) mesh, reducing the computational cost of the overall method with n grid points from $O(n^2)$ to $O(n \log n)$ [13, 14].

Bibliography

[1] D. Baffet, A Gauss–Jacobi kernel compression scheme for fractional differential equations, *J. Sci. Comput.*, to appear, https://doi.org/10.1007/s10915-018-0848-x

[2] H. Brunner, A. Pedas, and G. Vainikko, The piecewise polynomial collocation method for nonlinear weakly singular Volterra equations, *Math. Comput.*, **68** (1999), 1079–1095.

[3] W. H. Deng, Numerical algorithm for the time fractional Fokker–Planck equation, *J. Comput. Phys.*, **227** (2007), 1510–1522.

[4] K. Diethelm, An algorithm for the numerical solution of differential equations of fractional order, *Electron. Trans. Numer. Anal.*, **5** (1997), 1–6.

[5] K. Diethelm, Generalized compound quadrature formulae for finite-part integrals, *IMA J. Numer. Anal.*, **17** (1997), 479–493.

[6] K. Diethelm, Efficient solution of multi-term fractional differential equations using P(EC)^{m}E methods, *Computing*, **71** (2003), 305–319.

[7] K. Diethelm, Smoothness properties of solutions of Caputo-type fractional differential equations, *Fract. Calc. Appl. Anal.*, **10** (2007), 151–160.

[8] K. Diethelm, *The Analysis of Fractional Differential Equations*, Springer, Berlin, 2010.

[9] K. Diethelm, General theory of Caputo-type fractional differential equations, in J. A. T. Machado (ed.), *Handbook of Fractional Calculus with Applications*, de Gruyter, Berlin, 2019.

[10] K. Diethelm, J. M. Ford, N. J. Ford, and M. Weilbeer, Pitfalls in fast numerical solvers for fractional differential equations, *J. Comput. Appl. Math.*, **186** (2006), 482–503.

[11] K. Diethelm, N. J. Ford, and A. D. Freed, A predictor-corrector approach for the numerical solution of fractional differential equations, *Nonlinear Dyn.*, **29** (2002), 3–22.

[12] K. Diethelm, N. J. Ford, and A. D. Freed, Detailed error analysis for a fractional Adams method, *Numer. Algorithms*, **36** (2004), 31–52.

[13] K. Diethelm and A. D. Freed, An efficient algorithm for the evaluation of convolution integrals, *Comput. Math. Appl.*, **51** (2006), 51–72.

[14] N. J. Ford and A. C. Simpson, The numerical solution of fractional differential equations: Speed versus accuracy, *Numer. Algorithms*, **26** (2001), 333–346.

[15] R. Garrappa, Numerical solution of fractional differential equations: A survey and a software tutorial, *Mathematics*, **6** (2018), 16.
[16] E. Hairer, C. Lubich, and M. Schlichte, Fast numerical solution of nonlinear Volterra convolution equations, *SIAM J. Sci. Stat. Comput.*, **6** (1985), 532–541.
[17] E. Hairer, S. P. Nørsett, and G. Wanner, *Solving Ordinary Differential Equations. I: Nonstiff Problems*, 2nd revised ed., 3rd corrected printing, Springer, Berlin, 2010.
[18] S. Jiang, J. Zhang, Q. Zhang, and Z. Zhang, Fast evaluation of the Caputo fractional derivative and its application to fractional diffusion equations, *Commun. Comput. Phys.*, **21** (2017), 650–678.
[19] C. P. Li and F. H. Zeng, *Numerical Methods for Fractional Calculus*, CRC Press, Boca Raton, 2015.
[20] J.-R. Li, A fast time stepping method for evaluating fractional integrals, *SIAM J. Sci. Comput.*, **31** (2010), 4696–4714.
[21] X. Li and C. Xu, A space-time spectral method for the time fractional diffusion equation, *SIAM J. Numer. Anal.*, **47** (2009), 2108–2131.
[22] C. Lubich, On the stability of linear multistep methods for Volterra convolution equations, *IMA J. Numer. Anal.*, **3** (1983), 439–465.
[23] C. Lubich, Runge–Kutta theory for Volterra and Abel integral equations of the second kind, *Math. Comput.*, **41** (1983), 87–102.
[24] C. Lubich, Fractional linear multistep methods for Abel–Volterra integral equations of the second kind, *Math. Comput.*, **45** (1985), 463–469.
[25] C. Lubich, Discretized fractional calculus, *SIAM J. Math. Anal.*, **17** (1986), 704–719.
[26] C. Lubich, Convolution quadrature and discretized operational calculus. I, *Numer. Math.*, **52** (1988), 129–145.
[27] C. Lubich, Convolution quadrature and discretized operational calculus. II, *Numer. Math.*, **52** (1988), 413–425.
[28] C. Lubich, Convolution quadrature revisited, *BIT Numer. Math.*, **44** (2004), 503–514.
[29] V. Maz'ya and G. Schmidt, *Approximate Approximations*, Amer. Math. Soc., Providence, 2007.
[30] W. McLean, Exponential sum approximation for $t^{-\beta}$, in J. Dick, F. Y. Kuo, and H. Wozniakowski (eds.), *Contemporary Computational Mathematics – A Celebration of the 80th Birthday of Ian Sloan*, pp. 911–930, Springer International, Cham, 2018.
[31] K. B. Oldham and J. Spanier, *The Fractional Calculus*, Academic Press, New York, 1974.
[32] M. Stynes, Too much regularity may force too much uniqueness, *Fract. Calc. Appl. Anal.*, **19** (2016), 1554–1562.
[33] Z. Z. Sun and X. Zhao, Time-fractional derivatives, in J. A. T. Machado (ed.), *Handbook of Fractional Calculus with Applications*, de Gruyter, Berlin, 2019.
[34] G. Szegő, *Orthogonal Polynomials*, 4th edn., Amer. Math. Soc., Providence, 1975.
[35] S. B. Yuste and J. Quintana-Murillo, A finite difference method with non-uniform timesteps for fractional diffusion equations, *Comput. Phys. Commun.*, **183** (2012), 2594–2600. corrected version: arXiv:1109.6622v3 [math.NA], 2013.
[36] M. Zayernouri and G. E. Karniadakis, Fractional Sturm–Liouville eigen-problems: Theory and numerical approximations, *J. Comput. Phys.*, **252** (2013), 495–517.
[37] M. Zayernouri and G. E. Karniadakis, Exponentially accurate spectral and spectral element methods for fractional ODEs, *J. Comput. Phys.*, **257** (2014), 460–480.
[38] F. H. Zeng, I. Turner, and K. Burrage, A stable fast time-stepping method for fractional integral and derivative operators, *J. Sci. Comput.*, **77** (2018), 283–307.

Xuan Zhao and Zhi-Zhong Sun

Time-fractional derivatives

Abstract: The definitions of the fractional integral, Grünwald–Letnikov fractional derivative, Riemann–Liouville fractional derivative, and Caputo fractional derivative are presented. The numerical approximations for the Riemann–Liouville fractional derivative based on the shifted Grünwald–Letnikov formula are provided. The $L1$ interpolation approximation and $L2\text{-}1_\sigma$ interpolation approximation for the Caputo fractional derivative are given. The finite difference methods based on Grünwald–Letnikov formula, $L1$ formula and $L2\text{-}1_\sigma$ formula for the fractional ordinary equation are derived. Four finite difference schemes based on the first-order Grünwald–Letnikov formula, the second-order shifted Grünwald–Letnikov formula, the $L1$ formula and the $L2\text{-}1_\sigma$ formula are constructed for the time-fractional subdiffusion equations. Two difference schemes by using the $L1$ formula and the $L2\text{-}1_\sigma$ formula are developed for the time-fractional diffusion-wave equations. For each scheme, the convergence result is given.

Keywords: Grünwald–Letnikov fractional derivative, Riemann–Liouville fractional derivative, Caputo fractional derivative, shifted Grünwald–Letnikov formula, $L1$ formula, $L2\text{-}1_\sigma$ formula, fractional ordinary equation, time-fractional subdiffusion equation, time-fractional diffusion-wave equation, finite difference scheme, convergence

MSC 2010: 26A33, 65L12, 65L20, 65L70, 65M06, 65M12, 65M15

1 Introduction

Differential equations of fractional order have proven to be valuable tools in the modeling of many phenomena in various fields of science and engineering [16, 30]. Fractional subdiffusion equations and diffusion-wave equations are two classes of important time-fractional partial differential equations, which are obtained by replacing the time derivative in the ordinary diffusion and wave equation by a fractional derivative of order α with $0 < \alpha < 1$ and $1 < \alpha < 2$, respectively. They have been widely applied in the modeling of anomalous diffusive and subdiffusive systems, the description of fractional random walk, the unification of diffusion and wave propagation phenomena, etc. [2, 28, 36].

Xuan Zhao, Jiangsu Provincial Key Laboratory of Networked Collective Intelligence, School of Mathematics, Southeast University, Nanjing 210096, China, e-mail: xuanzhao11@seu.edu.cn
Zhi-Zhong Sun, School of Mathematics, Southeast University, Nanjing 210096, China, e-mail: zzsun@seu.edu.cn

https://doi.org/10.1515/9783110571684-002

The presence of the integral in the non-integer-order derivatives makes the problem global. Unlike the classical case, one requires information about all the previous time layers, when numerically approximating a time-fractional diffusion equation for a certain time layer. For that reason algorithms for solving the time-fractional diffusion equations are rather time consuming, even in the one-dimensional case. In this regard constructing stable and high order accurate difference schemes is a very important task.

In fact, the weakly singular partial integro-differential equations have been studied numerically for many years. We should point out that some of these equations can be viewed as fractional partial differential equations. For example, Lopez-Marcos [24], Lubich [25], and Tang [42] analyzed finite difference schemes for a partial integro-differential equation, which is equivalent to the fractional diffusion-wave equation under some assumptions on the exact solution.

In the beginning of the numerical investigation for the time-fractional diffusion equation, most of the pioneers [48, 49] designed the numerical methods based on the first-order Grünwald–Letnikov formula and L1 formula. Due to the fundamental property of the global dependence for the time-fractional problems, all previous solutions have to be saved in order to compute the solution at the current time level, which makes the storage very expensive if low-order methods are employed. Later on, researchers focused on designing the high-order approximations for the time-fractional derivatives. The main work consists of two ways to numerically discretize the time-fractional derivatives in finite difference methods. One class of approximations is along the line of the Grünwald–Letnikov formula, the other is to approximate the functions under the integration with different interpolating polynomials. In the current handbook, we restrict our introduction to the approximation formulas which are theoretically proven and widely adopted to solve the time-fractional problems.

The organization of this chapter is as follows. In Section 2, we introduce the definitions of the fractional derivatives usually used in the time-fractional problems. Moreover, the relations of the fractional integral and the fractional derivatives are presented. In Section 3, we show the first-order and second-order approximations for the Riemann–Liouville fractional derivative based on the Grünwald–Letnikov formula. The $L1$ formula and $L2\text{-}1_\sigma$ formula for the approximation of the Caputo derivative are given in Section 4. In Section 5, the approximation formulas from Section 2 and Section 3 are utilized for solving the fractional ordinary differential equations. The numerical schemes for the time-fractional subdiffusion equation and the diffusion-wave equation are provided and the theoretical results are proven in Section 6 and Section 7, respectively. In the last section concluding remarks and some comments on the fractional high-dimensional problems and the other problems are provided.

2 Definitions and properties of time-fractional derivatives

As is well known, there are different definitions of fractional derivatives. In this section, we just show the definitions usually used in the time-fractional problems.

2.1 Fractional integral

Definition 1. Let α be a positive real number. The left and right fractional integrals of the function $f(t)$ are respectively defined as follows:

$$I_{a+}^{\alpha} f(t) := \frac{1}{\Gamma(\alpha)} \int_a^t \frac{f(s)}{(t-s)^{1-\alpha}} \mathrm{d}s,$$

$$I_{a-}^{\alpha} f(t) := \frac{1}{\Gamma(\alpha)} \int_t^a \frac{f(s)}{(s-t)^{1-\alpha}} \mathrm{d}s,$$

where $\Gamma(z)$ is *the Gamma* function defined by

$$\Gamma(z) = \int_0^{\infty} \mathrm{e}^{-s} s^{z-1} \mathrm{d}s, \quad \mathrm{Re}(z) > 0.$$

2.2 Grünwald–Letnikov fractional derivative

Definition 2. Let α be a positive real number, $n-1 \le \alpha < n$, $n \in \mathbb{Z}^+$. The Grünwald–Letnikov fractional derivative for $f(t)$ of order α is defined as

$${}^{\mathrm{GL}}D_a^{\alpha} f(t) = \lim_{h \to 0} h^{-\alpha} \sum_{j=0}^{[(t-a)/h]} (-1)^j \binom{\alpha}{j} f(t-jh),$$

where $\binom{\alpha}{j}$ denotes the binomial coefficient

$$\binom{\alpha}{j} = \frac{\alpha(\alpha-1)\cdots(\alpha-j+1)}{j!}.$$

2.3 Riemann–Liouville fractional derivative

Definition 3. Let α be a positive real number, $n-1 \le \alpha < n$, $n \in \mathbb{Z}^+$. The Riemann–Liouville fractional derivative for $f(t)$ of order α is defined as

$${}^{\mathrm{RL}}D_a^{\alpha} f(t) = \frac{\mathrm{d}^n}{\mathrm{d}t^n} \left(\frac{1}{\Gamma(n-\alpha)} \int_a^t \frac{f(s)\mathrm{d}s}{(t-s)^{\alpha-n+1}} \right).$$

2.4 Caputo fractional derivative

Definition 4. Let α be a positive real number, $n-1 < \alpha \le n$, $n \in \mathbb{Z}^+$. The Caputo fractional derivative for $f(t)$ of order α is defined as

$$ {}^{C}D_a^{\alpha} f(t) = \frac{1}{\Gamma(n-\alpha)} \int_a^t \frac{f^{(n)}(s)\mathrm{d}s}{(t-s)^{\alpha-n+1}}. $$

From the above definitions, we have the relations of the fractional integral and the fractional derivatives in the following. It is obvious that

$$ {}^{\mathrm{RL}}D_a^{\alpha} f(t) = \frac{\mathrm{d}^n}{\mathrm{d}t^n}[I_{a+}^{(n-\alpha)} f(t)], \quad {}^{C}D_a^{\alpha} f(t) = I_{a+}^{(n-\alpha)} f^{(n)}(t). $$

When $f^{(k)}(\tau)$, $k = 0, 1, 2, \ldots, n$, is continuous on $[a, t]$, $n = \lceil \alpha \rceil$, it can be proven that

$$ {}^{\mathrm{GL}}D_a^{\alpha} f(t) = \sum_{j=0}^{n-1} \frac{f^{(j)}(a)(t-a)^{j-\alpha}}{\Gamma(1+j-\alpha)} + {}^{C}D_a^{\alpha} f(t). $$

When $f^{(k)}(\tau)$, $k = 0, 1, 2, \ldots, n$, is continuous on $[a, t]$ and $f^{(n)}(\tau)$ is integrable on $[a, t]$, the Riemann–Liouville fractional derivative of $f(t)$ with order α is equivalent to the Grünwald–Letnikov fractional derivative of $f(t)$ with order α. When the function $f(t)$ satisfies

$$ f^{(j)}(a) = 0, \quad j = 0, 1, \ldots, \lfloor \alpha \rfloor, $$

the Riemann–Liouville fractional derivative of $f(t)$ with order α is equivalent to the Caputo fractional derivative of $f(t)$ with order α.

3 The numerical approximations for the Riemann–Liouville fractional derivative

In this section, we present the numerical approximations of ${}^{\mathrm{RL}}D_{-\infty}^{\alpha} f(t)$, where $0 < \alpha \le 2$, based on the Grünwald–Letnikov formula. The first-order formula and the second-order formula will be given in detail in order to solve the time-fractional differential equations.

Define the shifted Grünwald–Letnikov formula as follows:

$$ A_{\tau,p}^{\alpha} f(t) = \tau^{-\alpha} \sum_{k=0}^{\infty} g_k^{(\alpha)} f(t-(k-p)\tau), \tag{1} $$

where p is an integer,

$$ g_k^{(\alpha)} = (-1)^k \binom{\alpha}{k}. $$

When $p = 0$, equation (1) is the standard Grünwald–Letnikov formula. The coefficients $\{g_k^{(\alpha)}\}$ are the coefficients of the power series of the function $(1-z)^\alpha$; we have

$$(1-z)^\alpha = \sum_{k=0}^{\infty} (-1)^k \binom{\alpha}{k} z^k = \sum_{k=0}^{\infty} g_k^{(\alpha)} z^k, \quad -1 < z \leq 1.$$

They have the following recursive relationship:

$$g_0^{(\alpha)} = 1, \quad g_k^{(\alpha)} = \left(1 - \frac{\alpha+1}{k}\right) g_{k-1}^{(\alpha)}, \quad k = 1, 2, \ldots.$$

Lemma 1. *The coefficients* $\{g_k^{(\alpha)}\}$ *in equation* (1) *satisfy the following:*

(I) *when* $\alpha = 0$,

$$g_0^{(\alpha)} = 1, \quad g_1^{(\alpha)} = g_2^{(\alpha)} = \cdots = 0;$$

(II) *when* $0 < \alpha < 1$,

$$g_0^{(\alpha)} = 1, \quad g_1^{(\alpha)} = -\alpha, \quad g_2^{(\alpha)} < g_3^{(\alpha)} < \cdots < 0,$$
$$\sum_{k=0}^{\infty} g_k^{(\alpha)} = 0, \quad \sum_{k=0}^{m} g_k^{(\alpha)} > 0, \quad m \geq 1;$$

(III) *when* $\alpha = 1$,

$$g_0^{(\alpha)} = 1, \quad g_1^{(\alpha)} = -1, \quad g_2^{(\alpha)} = g_3^{(\alpha)} = \cdots = 0;$$

(IV) *when* $1 < \alpha < 2$,

$$g_0^{(\alpha)} = 1, \quad g_1^{(\alpha)} = -\alpha, \quad g_2^{(\alpha)} > g_3^{(\alpha)} > \cdots > 0,$$
$$\sum_{k=0}^{\infty} g_k^{(\alpha)} = 0, \quad \sum_{k=0}^{m} g_k^{(\alpha)} < 0, \quad m \geq 1;$$

(V) *when* $\alpha = 2$,

$$g_0^{(\alpha)} = 1, \quad g_1^{(\alpha)} = -2, \quad g_2^{(\alpha)} = 1, \quad g_3^{(\alpha)} = g_4^{(\alpha)} = \cdots = 0.$$

Define

$$\mathscr{C}^{n+\alpha}(\mathcal{R}) = \left\{ f \mid f \in L^1(\mathcal{R}), \int_{-\infty}^{\infty} (1 + |\omega|)^{n+\alpha} |F(\omega)| \mathrm{d}\omega < \infty \right\},$$

where $F(\omega) = \int_{-\infty}^{\infty} f(t) e^{i\omega t} \mathrm{d}t$ is the Fourier transform of $f(t)$.

Tuan and Gorenflo [44] showed the asymptotic expansion of the standard Grünwald–Letnikov formula. Tadjeran et al. [41] presented the asymptotical expansion of the shifted Grünwald–Letnikov formula. In [27], the authors gave the case of $n = 1$.

Theorem 1. *Let* $f \in \mathscr{C}^{n+\alpha}(\mathcal{R})$. *Then*

$$A^{\alpha}_{\tau,p}f(t) = {}^{\mathrm{RL}}D^{\alpha}_{-\infty}f(t) + \sum_{l=1}^{n-1} c_l^{\alpha,p}\,{}^{\mathrm{RL}}D^{\alpha+l}_{-\infty}f(t)\tau^l + O(\tau^n)$$

for $t \in \mathcal{R}$, *where* $\{c_l^{\alpha,p}\}$ *are the coefficients of the power series of the function* $W_{\alpha,p}(z) = (\frac{1-e^{-z}}{z})^{\alpha}e^{pz}$, *that is,*

$$W_{\alpha,p}(z) = \sum_{l=0}^{\infty} c_l^{\alpha,p} z^l = 1 + c_1^{\alpha,p}z + c_2^{\alpha,p}z^2 + c_3^{\alpha,p}z^3 + O(|z|^4),$$

where

$$c_1^{\alpha,p} = p - \frac{\alpha}{2}, \quad c_2^{\alpha,p} = \frac{p^2}{2} - \frac{\alpha p}{2} + \frac{\alpha(3\alpha+1)}{24},$$
$$c_3^{\alpha,p} = \frac{p^3}{6} - \frac{\alpha p^2}{4} + \frac{\alpha(3\alpha+1)p}{24} - \frac{\alpha^2(\alpha+1)}{48}.$$

3.1 The first-order approximation

Theorem 2 ([43]). *Let* $f \in \mathscr{C}^{1+\alpha}(\mathcal{R})$. *Then*

$$A^{\alpha}_{\tau,p}f(t) = {}^{\mathrm{RL}}D^{\alpha}_{-\infty}f(t) + O(\tau)$$

for $t \in \mathcal{R}$.

3.2 The second-order approximation

Theorem 3 ([43]). *Let* $f \in \mathscr{C}^{2+\alpha}(\mathcal{R})$ *and* $p \neq q$. *Then*

$$\lambda_1 A^{\alpha}_{\tau,p}f(t) + \lambda_2 A^{\alpha}_{\tau,q}f(t) = {}^{\mathrm{RL}}D^{\alpha}_{-\infty}f(t) + O(\tau^2)$$

for $t \in \mathcal{R}$, *where*

$$\lambda_1 = \frac{\alpha - 2q}{2(p-q)}, \quad \lambda_2 = \frac{2p - \alpha}{2(p-q)}.$$

Corollary 1. *When* $\alpha \in (0,1)$, *take* $(p,q) = (0,-1)$. *Then* $\lambda_1 = 1 + \frac{\alpha}{2}$, $\lambda_2 = -\frac{\alpha}{2}$. *This formula is used in solving the time-fractional problem* [45].

3.3 The third-order approximation

Theorem 4 ([43]). *Let* $f \in \mathscr{C}^{3+\alpha}(\mathcal{R})$, *and* p, q, r *are not equal to each other. Then*

$$\lambda_1 A^{\alpha}_{\tau,p}f(t) + \lambda_2 A^{\alpha}_{\tau,q}f(t) + \lambda_3 A^{\alpha}_{\tau,r}f(t) = {}^{\mathrm{RL}}D^{\alpha}_{-\infty}f(t) + O(\tau^3)$$

for $t \in \mathcal{R}$, *where*

$$\lambda_1 = \frac{12qr - (6q + 6r + 1)\alpha + 3\alpha^2}{12(qr - pq - pr + p^2)},$$

$$\lambda_2 = \frac{12pr - (6p + 6r + 1)\alpha + 3\alpha^2}{12(pr - pq - qr + q^2)},$$

$$\lambda_3 = \frac{12pq - (6p + 6q + 1)\alpha + 3\alpha^2}{12(pq - pr - qr + r^2)}.$$

Corollary 2. *When* $\alpha \in (0,1)$, *take* $(p,q,r) = (0,-1,-2)$. *Then*

$$\lambda_1 = \frac{24 + 17\alpha + 3\alpha^2}{24}, \quad \lambda_2 = -\frac{11\alpha + 3\alpha^2}{12}, \quad \lambda_3 = \frac{5\alpha + 3\alpha^2}{24}.$$

The formula is used in solving the time-fractional problem [17, 18].

4 The numerical approximations of the Caputo fractional derivative

In this section, the $L1$ formula and $L2\text{-}1_\sigma$ formula are introduced for the approximations of the Caputo fractional derivative.

4.1 $L1$ interpolation approximation

Denote $t_n = n\tau$. For the Caputo fractional derivative with order α $(0 < \alpha < 1)$

$${}^{C}D_0^\alpha f(t) = \frac{1}{\Gamma(1-\alpha)} \int_0^t \frac{f'(s)}{(t-s)^\alpha}\, ds,$$

the $L1$ approximation is

$$D_\tau^\alpha f(t_n) := \frac{\tau^{-\alpha}}{\Gamma(2-\alpha)} \left[a_0^{(\alpha)} f(t_n) - \sum_{k=1}^{n-1} (a_{n-k-1}^{(\alpha)} - a_{n-k}^{(\alpha)}) f(t_k) - a_{n-1}^{(\alpha)} f(t_0) \right].$$

The coefficients $\{a_l^{(\alpha)}\}$ satisfy the following.

Lemma 2. *Let* $\alpha \in (0,1)$, $a_l^{(\alpha)} = (l+1)^{1-\alpha} - l^{1-\alpha}$, $l \geq 0$, $l = 0, 1, 2, \ldots$. *Then*

(I) $1 = a_0^{(\alpha)} > a_1^{(\alpha)} > a_2^{(\alpha)} > \cdots > a_l^{(\alpha)} > 0$; $a_l^{(\alpha)} \to 0$ *when* $l \to \infty$;

(II) $(1-\alpha)l^{-\alpha} < a_{l-1}^{(\alpha)} < (1-\alpha)(l-1)^{-\alpha}$, $l \geq 1$.

The authors of [21, 22, 23, 40] theoretically proved that the $L1$ formula keeps the order of $2 - \alpha$ for $\alpha \in (0,1)$. Here we present the truncation error of the $L1$ formula in the following.

Theorem 5 ([40]). *Let $f(t) \in C^2[t_0, t_n]$, $\alpha \in (0,1)$. Then*

$$
\begin{aligned}
&|{}^C D_0^\alpha f(t)|_{t_n} - D_\tau^\alpha f(t_n)| \\
&\quad \le \frac{1}{2\Gamma(1-\alpha)}\left[\frac{1}{4} + \frac{\alpha}{(1-\alpha)(2-\alpha)}\right] \max_{0\le t\le t_n} |f''(t)|\, \tau^{2-\alpha}.
\end{aligned}
$$

For $\alpha \in (1,2)$, the approximation error is proven as follows.

Theorem 6 ([40]). *Let $f(t) \in C^3[t_0, t_n]$, $\alpha \in (1,2)$. Then*

$$
\begin{aligned}
&\frac{1}{2}[{}^C D_0^\alpha f(t)|_{t=t_n} + {}^C D_0^\alpha f(t)|_{t=t_{n-1}}] \\
&\quad = \frac{\tau^{1-\alpha}}{\Gamma(3-\alpha)}\left[b_0^{(\alpha)} \delta_t f^{n-\frac{1}{2}} - \sum_{k=1}^{n-1} (b_{n-k-1}^{(\alpha)} - b_{n-k}^{(\alpha)}) \delta_t f^{k-\frac{1}{2}} - b_{n-1}^{(\alpha)} f'(t_0)\right] \\
&\quad\quad + \hat{R}^{n-\frac{1}{2}},
\end{aligned}
$$

where $b_l^{(\alpha)} = a_l^{(\alpha-1)} = (l+1)^{2-\alpha} - l^{2-\alpha}$, $l = 0, 1, 2, \ldots$, and

$$
|\hat{R}^{n-\frac{1}{2}}| \le \left\{\frac{1}{6\Gamma(3-\alpha)} + \frac{1}{2\Gamma(2-\alpha)}\left[\frac{1}{4} + \frac{\alpha-1}{(2-\alpha)(3-\alpha)}\right]\right\} \max_{0\le t\le t_n} |f'''(t)|\, \tau^{3-\alpha}.
$$

4.2 The $L2\text{-}1_\sigma$ interpolation approximation

Similarly as the $L1$ interpolation, Alikhanov [1] constructed the $L2\text{-}1_\sigma$ formula, which achieves the order of $3-\alpha$.

Let $0 < \alpha < 1$, $\sigma = 1 - \frac{\alpha}{2}$, $t_{n-1+\sigma} = t_{n-1} + \sigma\tau$. Define

$$
\begin{aligned}
\hat{a}_0^{(\alpha)} &= \sigma^{1-\alpha}, \quad \hat{a}_l^{(\alpha)} = (l+\sigma)^{1-\alpha} - (l-1+\sigma)^{1-\alpha}, \quad l \ge 1, \\
\hat{b}_l^{(\alpha)} &= \frac{1}{2-\alpha}[(l+\sigma)^{2-\alpha} - (l-1+\sigma)^{2-\alpha}] - \frac{1}{2}[(l+\sigma)^{1-\alpha} + (l-1+\sigma)^{1-\alpha}], \quad l \ge 1.
\end{aligned}
$$

When $n = 1$, denote

$$
c_0^{(n,\alpha)} = \hat{a}_0^{(\alpha)}. \tag{2}
$$

When $n \ge 2$, denote

$$
c_k^{(n,\alpha)} = \begin{cases} \hat{a}_0^{(\alpha)} + \hat{b}_1^{(\alpha)}, & k = 0, \\ \hat{a}_k^{(\alpha)} + \hat{b}_{k+1}^{(\alpha)} - \hat{b}_k^{(\alpha)}, & 1 \le k \le n-2, \\ \hat{a}_k^{(\alpha)} - \hat{b}_k^{(\alpha)}, & k = n-1. \end{cases} \tag{3}
$$

Without confusion, we omit the superscript α in $a_k^{(n,\alpha)}$, $b_k^{(n,\alpha)}$, and $c_k^{(n,\alpha)}$.

The $L2\text{-}1_\sigma$ formula is obtained as follows:

$$
\Delta_\tau^\alpha f(t)|_{t=t_{n-1+\sigma}} = \frac{\tau^{-\alpha}}{\Gamma(2-\alpha)} \sum_{k=0}^{n-1} c_k^{(n)} [f(t_{n-k}) - f(t_{n-k-1})].
$$

Lemma 3 ([1]). *Let* α $(0 < \alpha < 1)$, $\sigma = 1 - \frac{\alpha}{2}$, $c_k^{(n)}$ $(0 \le k \le n-1, n \ge 1)$, *defined by* (2)–(3). *Then we have*

$$c_k^{(n)} > \frac{1-\alpha}{2}(k+\sigma)^{-\alpha},$$
$$c_0^{(n)} > c_1^{(n)} > c_2^{(n)} > \cdots > c_{n-2}^{(n)} > c_{n-1}^{(n)},$$
$$(2\sigma - 1)c_0^{(n)} - \sigma c_1^{(n)} > 0.$$

Theorem 7 ([1]). *Let* $f(t) \in C^3[t_0, t_n]$, $0 < \alpha < 1$, $\sigma = 1 - \frac{\alpha}{2}$. *Then*

$$\left|{}^C D_0^\alpha f(t)|_{t=t_{n-1+\sigma}} - \Delta_\tau^\alpha f(t)|_{t=t_{n-1+\sigma}}\right| \le \frac{(4\sigma-1)\sigma^{-\alpha}}{12\Gamma(2-\alpha)} \max_{0\le t\le t_n} |f'''(t)| \tau^{3-\alpha}.$$

Remark 1. The parameter $\sigma = 1 - \frac{\alpha}{2}$ is determined by the following equation:

$$\int_{t_{n-1}}^{t_{n-1+\sigma}} \frac{\xi - t_{n-\frac{1}{2}}}{(t_{n-1+\sigma} - \xi)^\alpha} d\xi = 0.$$

5 Finite difference method for the fractional ordinary equation

In this section, we construct the finite difference schemes for the fractional initial/boundary value problems with $0 < \alpha < 1$ and $1 < \alpha < 2$ by applying the Grünwald–Letnikov formula, the $L1$ formula, and the $L2\text{-}1_\sigma$ formula, respectively. The results of the stability and the convergence for the schemes are presented [39].

5.1 Schemes based on the Grünwald–Letnikov formula

Problem 1. Solve the following initial value problem with $\alpha \in (0, 1)$:

$$\begin{cases} {}^{\mathrm{RL}}D_0^\alpha y(t) = f(t), & 0 < t \le T, \\ y(0) = 0. \end{cases} \tag{4}$$

Define the function

$$\hat{u}(t) = \begin{cases} 0, & t < 0, \\ y(t), & 0 \le t \le T, \\ v(t), & T < t < 2T, \\ 0, & t \ge 2T, \end{cases}$$

where $v(t)$ is a smooth function which satisfies $v^{(k)}(T) = y^{(k)}(T)$, $v^{(k)}(2T) = 0$, $k = 0, 1, 2$. The introduction of the function $v(t)$ is just for the theoretical analysis. It is never used for the computation. Suppose $\hat{u} \in \mathscr{C}^{1+\alpha}(\mathcal{R})$. According to Theorem 2, there exists a constant c_1 such that

$$\left|{}^{\mathrm{RL}}D_0^{\alpha}y(t)|_{t=t_n} - \tau^{-\alpha}\sum_{k=0}^{n} g_k^{(\alpha)} y(t_{n-k})\right| \le c_1\tau, \quad 1 \le n \le N.$$

The finite difference scheme for the problem (4) is listed as follows:

$$\begin{cases} \tau^{-\alpha}\sum\limits_{k=0}^{n} g_k^{(\alpha)} y^{n-k} = f(t_n), & 1 \le n \le N, \\ y^0 = 0. \end{cases} \tag{5}$$

Theorem 8. *Let $\{y^n \mid n = 0,1,2,\ldots,N\}$ be the solution of the difference scheme (5). Then*

$$|y^k| \le \frac{5}{(1-\alpha)2^{\alpha}} t_k^{\alpha} \max_{1\le m\le k} |f(t_m)|, \quad 1 \le k \le N.$$

Theorem 9. *Let $\{y(t_n) \mid n = 0,1,2,\ldots,N\}$ be the solution of the fractional initial problem (4) and let $\{y^n \mid n = 0,1,2,\ldots,N\}$ be the solution of the difference scheme (5). Denote*

$$e^n = y(t_n) - y^n, \quad n = 0,1,2,\ldots,N.$$

Then

$$|e^n| \le \frac{5c_1}{(1-\alpha)2^{\alpha}} T^{\alpha}\tau, \quad 1 \le n \le N.$$

Problem 2. Solve the following initial value problem with $\alpha \in (1,2)$:

$$\begin{cases} {}^{\mathrm{RL}}D_0^{\alpha}y(t) = f(t), & 0 < t \le T, \\ y(0) = 0, \quad y'(0) = 0. \end{cases} \tag{6}$$

Define the function

$$\hat{u}(t) = \begin{cases} 0, & t < 0, \\ y(t), & 0 \le t \le T, \\ v(t), & T < t < 2T, \\ 0, & t \ge 2T, \end{cases}$$

where $v(t)$ is a smooth function which satisfies $v^{(k)}(T) = y^{(k)}(T)$, $v^{(k)}(2T) = 0$, $k = 0,1,2,3$. Suppose $\hat{u} \in \mathscr{C}^{1+\alpha}(\mathcal{R})$.

Let

$$z(t) = y'(t), \quad \beta = \alpha - 1.$$

Then

$${}^{\mathrm{RL}}D_0^{\alpha}y(t) = {}^{\mathrm{RL}}D_0^{\beta}z(t).$$

Denote

$$Y^n = y(t_n), \quad Z^n = z(t_n), \quad 0 \le n \le N,$$
$$Y^{n-\frac{1}{2}} = \frac{1}{2}(Y^n + Y^{n-1}), \quad \delta_t Y^{n-\frac{1}{2}} = \frac{1}{\tau}(Y^n - Y^{n-1}).$$

According to Theorem 2, we have

$${}^{\mathrm{RL}}D_0^{\beta} z(t)|_{t_n} = \tau^{-\beta} \sum_{k=0}^{n} g_k^{(\beta)} z(t_{n-k}) = O(\tau), \quad 1 \le n \le N.$$

Consequently,

$$\begin{aligned}
&\frac{1}{2}[{}^{\mathrm{RL}}D_0^{\alpha} y(t)|_{t_{n-1}} + {}^{\mathrm{RL}}D_0^{\alpha} y(t)|_{t_n}] \\
&= \frac{1}{2}[{}^{\mathrm{RL}}D_0^{\beta} z(t)|_{t_{n-1}} + {}^{\mathrm{RL}}D_0^{\beta} z(t)|_{t_n}] \\
&= \frac{1}{2}\left[\tau^{-\beta} \sum_{k=0}^{n-1} g_k^{(\beta)} z(t_{n-1-k}) + \tau^{-\beta} \sum_{k=0}^{n} g_k^{(\beta)} z(t_{n-k})\right] + O(\tau) \\
&= \tau^{-\beta} \sum_{k=0}^{n-1} g_k^{(\beta)} \frac{Z^{n-1-k} + Z^{n-k}}{2} + O(\tau) \\
&= \tau^{1-\alpha} \sum_{k=0}^{n-1} g_k^{(\alpha-1)} \delta_t Y^{n-k-\frac{1}{2}} + O(\tau).
\end{aligned}$$

The difference scheme for the problem (6) is listed as follows:

$$\begin{cases} \tau^{1-\alpha} \sum_{k=0}^{n-1} g_k^{(\alpha-1)} \delta_t y^{n-k-\frac{1}{2}} = f^{n-\frac{1}{2}}, & 1 \le n \le N, \\ y^0 = 0. \end{cases} \tag{7}$$

Theorem 10. *Let $\{y^n \mid n = 0, 1, 2, \ldots, N\}$ be the solution of the scheme (7). Then we have*

$$|\delta_t y^{k-\frac{1}{2}}| \le \frac{5}{(2-\alpha)2^{\alpha-1}} t_k^{\alpha-1} \max_{1 \le m \le k} |f^{m-\frac{1}{2}}|, \quad 1 \le k \le N,$$

and

$$|y^k| \le \frac{5}{(2-\alpha)2^{\alpha-1}} t_k^{\alpha} \max_{1 \le m \le k} |f^{m-\frac{1}{2}}|, \quad 1 \le k \le N.$$

Theorem 11. *Let $\{y(t_n) \mid n = 0, 1, 2, \ldots, N\}$ be the solution of the fractional initial problem (6) and let $\{y^n \mid n = 0, 1, 2, \ldots, N\}$ be the solution of the difference scheme (7). Denote*

$$e^n = y(t_n) - y_n, \quad n = 0, 1, 2, \ldots, N.$$

Then there exists a constant c_2 such that

$$|e^n| \le \frac{5c_2}{(2-\alpha)2^{\alpha-1}} T^{\alpha} \tau, \quad 1 \le n \le N.$$

Problem 3. Solve the following boundary value problem with $\alpha \in (1,2)$:

$$\begin{cases} {}_0\mathbf{D}_t^{\alpha} y(t) = f(t), & 0 < t < T, \\ y(0) = 0, \quad y(T) = B. \end{cases} \tag{8}$$

Define the function

$$\hat{u}(t) = \begin{cases} 0, & t < 0, \\ y(t), & 0 \le t \le T, \\ v(t), & T < t \le 2T, \\ 0, & t > 2T, \end{cases}$$

where $v(t)$ is a smooth function which satisfies $v^{(k)}(T) = y^{(k)}(T)$, $v^{(k)}(2T) = 0$, $k = 0,1,2,3$. Suppose $\hat{u} \in \mathscr{C}^{1+\alpha}(\mathcal{R})$.

The difference scheme for (8) is constructed as follows:

$$\begin{cases} \tau^{-\alpha} \sum\limits_{k=0}^{n+1} g_k^{(\alpha)} y^{n-k+1} = f(t_n), & 1 \le n \le N-1, \\ y^0 = 0, \quad y^N = B. \end{cases} \tag{9}$$

Theorem 12. *Let* $\{y^n \mid n = 0,1,2,\ldots,N\}$ *be the solution for the difference scheme*

$$\begin{cases} \tau^{-\alpha} \sum\limits_{k=0}^{n+1} g_k^{(\alpha)} y^{n-k+1} = f(t_n), & 1 \le n \le N-1, \\ y^0 = 0, \quad y^N = 0. \end{cases}$$

Then

$$\|y\|_\infty \le \frac{45}{(\alpha-1)(2-\alpha)(3-\alpha)} \left(\frac{T}{4}\right)^{\alpha} \|f\|_\infty,$$

where

$$\|y\|_\infty = \max_{1 \le n \le N-1} |y^n|, \qquad \|f\|_\infty = \max_{1 \le n \le N-1} |f(t_n)|.$$

Theorem 13. *Let* $\{y(t_n) \mid n = 0,1,\ldots,N\}$ *be the solution of the fractional initial boundary problem* (8) *and let* $\{y^n \mid n = 0,1,\ldots,N\}$ *be the solution of the difference scheme* (9). *Denote*

$$e^n = y(t_n) - y^n, \quad n = 0,1,\ldots,N.$$

Then there exists a constant c_3 *such that*

$$\|e\|_\infty \le \frac{45}{(\alpha-1)(2-\alpha)(3-\alpha)} \left(\frac{T}{4}\right)^{\alpha} c_3 \tau.$$

5.2 Schemes based on the $L1$ formula

Problem 4. Solve the following initial value problem with $\alpha \in (0,1)$:

$$\begin{cases} {}^C D_0^\alpha y(t) = f(t), & 0 < t \le T, \\ y(0) = A. \end{cases} \tag{10}$$

The finite difference scheme for the problem (10) is listed as follows:

$$\begin{cases} \dfrac{\tau^{-\alpha}}{\Gamma(2-\alpha)} \left[a_0^{(\alpha)} y^n - \sum_{k=1}^{n-1} (a_{n-k-1}^{(\alpha)} - a_{n-k}^{(\alpha)}) y^k - a_{n-1}^{(\alpha)} y^0 \right] = f(t_n), & 1 \le n \le N, \\ y^0 = A. \end{cases} \tag{11}$$

Theorem 14. *Let $\{y^n \mid n = 0,1,2,\ldots,N\}$ be the solution of the difference scheme* (11). *Then*

$$|y^n| \le |y^0| + t_n^\alpha \Gamma(1-\alpha) \max_{1 \le l \le n} |f(t_l)|, \quad 1 \le n \le N.$$

Theorem 15. *Let $\{y(t_n) \mid n = 0,1,2,\ldots,N\}$ be the solution of the fractional initial value problem* (10) *and let $\{y^n \mid n = 0,1,2,\ldots,N\}$ be the solution of the difference scheme* (11). *Denote*

$$e^n = y(t_n) - y^n, \quad n = 0,1,2,\ldots,N.$$

Then there exists a constant c_4 such that

$$|e^n| \le c_4 T^\alpha \Gamma(1-\alpha) \tau^{2-\alpha}, \quad 1 \le n \le N.$$

Problem 5. Solve the following initial value problem with $\alpha \in (1,2)$:

$$\begin{cases} {}^C D_0^\alpha y(t) = f(t), & 0 < t \le T, \\ y(0) = A, \quad y'(0) = B. \end{cases} \tag{12}$$

The difference scheme for the problem (12) is listed as follows:

$$\begin{cases} \dfrac{\tau^{1-\alpha}}{\Gamma(3-\alpha)} \left[b_0^{(\alpha)} \delta_t y^{n-\frac{1}{2}} - \sum_{k=1}^{n-1} (b_{n-k-1}^{(\alpha)} - b_{n-k}^{(\alpha)}) \delta_t y^{k-\frac{1}{2}} - b_{n-1}^{(\alpha)} B \right] = f^{n-\frac{1}{2}}, & 1 \le n \le N, \\ y^0 = A, \end{cases} \tag{13}$$

where $f^{n-\frac{1}{2}} = \frac{1}{2}[f(t_n) + f(t_{n-1})]$.

Theorem 16. *Let $\{y^n \mid n = 0,1,2,\ldots,N\}$ be the solution of the difference scheme* (13). *Then we have*

$$|y^n| \le |A| + T\Big(|B| + t_n^{\alpha-1} \Gamma(2-\alpha) \max_{1 \le l \le n} |f^{l-\frac{1}{2}}|\Big), \quad 1 \le n \le N.$$

Theorem 17. *Let $\{y(t_n) \mid n = 0, 1, 2, \ldots, N\}$ be the solution of the fractional initial problem* (12) *and let $\{y^n \mid n = 0, 1, 2, \ldots, N\}$ be the solution of the difference scheme* (13). *Denote*

$$e^n = y(t_n) - y^n, \quad n = 0, 1, 2, \ldots, N.$$

Then we have

$$|e^n| \le c_5 T^\alpha \Gamma(2-\alpha)\tau^{3-\alpha}, \quad 1 \le n \le N,$$

where c_5 is a constant.

5.3 Schemes based on the $L2\text{-}1_\sigma$ formula

Consider Problem 4. The following scheme based on the $L2\text{-}1_\sigma$ formula can be constructed:

$$\begin{cases} \dfrac{\tau^{-\alpha}}{\Gamma(2-\alpha)} \displaystyle\sum_{k=0}^{n-1} c_k^{(n)}(y^{n-k} - y^{n-k-1}) = f(t_{n-1+\sigma}), & n = 1, 2, \ldots, N, \\ y^0 = A. \end{cases} \tag{14}$$

Theorem 18. *Let $\{y^n \mid n = 0, 1, \ldots, N\}$ be the solution of the difference scheme*

$$\begin{cases} \dfrac{\tau^{-\alpha}}{\Gamma(2-\alpha)} \displaystyle\sum_{k=0}^{n-1} c_k^{(n)}(y^{n-k} - y^{n-k-1}) = g^n, & n = 1, 2, \ldots, N, \\ y^0 = A. \end{cases}$$

Then we have

$$|y^k| \le |y^0| + 2t_k^\alpha \Gamma(1-\alpha) \max_{1\le m\le k} |g^m|, \quad k = 1, 2, \ldots, N.$$

Theorem 19. *Let $\{y(t_n) \mid n = 0, 1, \ldots, N\}$ be the solution of the fractional initial value problem* (10) *and let $\{y^n \mid n = 0, 1, \ldots, N\}$ be the solution of the difference scheme* (14). *Denote*

$$e^n = y(t_n) - y^n, \quad n = 0, 1, 2, \ldots, N.$$

Then there exists a constant c_6 such that

$$|e^n| \le 2c_6 T^\alpha \Gamma(1-\alpha)\tau^{3-\alpha}, \quad n = 1, 2, \ldots, N.$$

Consider the problem (12) and let $z(t) = y'(t)$. Then we obtain the following initial value problem for $z(t)$:

$$\begin{cases} {}^C D_0^{\alpha-1} z(t) = f(t), & 0 < t \le T, \\ z(0) = B. \end{cases} \tag{15}$$

The problem (15) has the same form as Problem 4. Using the $L2\text{-}1_\sigma$ formula, we can obtain the approximate value z^n for $z(t_n)$, $n = 0, 1, 2, \ldots, N$. With the help of the trapezoidal rule or Simpson's rule, from

$$y(t_n) = A + \int_{t_0}^{t_n} z(\eta)\mathrm{d}\eta,$$

it is easy to get the approximate value y^n of $y(t_n)$, $n = 0, 1, 2, \ldots, N$.

6 The finite difference methods for the time-fractional subdiffusion equation

In this section, we construct the finite difference schemes for the time-fractional subdiffusion equation applying the first-order Grünwald–Letnikov formula, the second-order shifted Grünwald–Letnikov formula, the $L1$ formula, and the $L2\text{-}1_\sigma$ formula, respectively. The results for the stability and the convergence of the schemes are presented.

6.1 The first-order Grünwald–Letnikov formula

Consider the following time-fractional subdiffusion equation:

$$\begin{cases} {}^C D_0^\alpha u(x,t) = \dfrac{\partial^2 u}{\partial x^2}(x,t) + f(x,t), & x \in (0,L),\ t \in (0,T], \\ u(x,0) = 0, & x \in (0,L), \\ u(0,t) = \mu(t), \quad u(L,t) = \nu(t), & t \in [0,T], \end{cases} \tag{16}$$

where $\alpha \in (0,1)$, f, μ, ν are known functions, $\mu(0) = \nu(0) = 0$, and

$${}^C D_0^\alpha u(x,t) = \frac{1}{\Gamma(1-\alpha)} \int_0^t \frac{u_s(x,s)}{(t-s)^\alpha} \mathrm{d}s.$$

Take two positive integers M and N. Let $h = L/M$, $\tau = T/N$. Denote $x_i = ih$, $t_n = n\tau$, $\Omega_h = \{x_i \mid 0 \le i \le M\}$, $\Omega_\tau = \{t_n \mid 0 \le n \le N\}$, $\Omega_{h\tau} = \Omega_h \times \Omega_\tau$. If $v = \{v_i^n \mid 0 \le i \le M, 0 \le n \le N\}$ is a grid function defined on $\Omega_{h\tau}$, introduce the following notations:

$$\delta_x v_{i-\frac{1}{2}}^n = \frac{1}{h}(v_i^n - v_{i-1}^n), \quad \delta_x^2 v_i^n = \frac{1}{h^2}(v_{i-1}^n - 2v_i^n + v_{i+1}^n),$$
$$v_i^{n-\frac{1}{2}} = \frac{1}{2}(v_i^n + v_i^{n-1}), \quad \delta_t v_i^{n-\frac{1}{2}} = \frac{1}{\tau}(v_i^n - v_i^{n-1}).$$

Define the grid function space

$$\mathcal{U}_h = \{u \mid u = (u_0, u_1, \ldots, u_M)\}, \quad \mathring{\mathcal{U}}_h = \{u \mid u \in \mathcal{U}_h, u_0 = u_M = 0\}.$$

For $u \in \mathring{\mathcal{U}}_h$, denote

$$\|u\| = \sqrt{h \sum_{i=1}^{M-1} u_i^2}, \quad \|u\|_\infty = \max_{1 \le i \le M-1} |u_i|.$$

Define the grid functions

$$U_i^n = u(x_i, t_n), \quad f_i^n = f(x_i, t_n), \quad 0 \le i \le M,\ 0 \le n \le N.$$

For $x \in [0, L]$, define the function

$$\hat{u}(t) = \begin{cases} 0, & t < 0, \\ u(x,t), & 0 \le t \le T, \\ v(t), & T < t < 2T, \\ 0, & t \ge 2T, \end{cases}$$

where $v(t)$ is a smooth function which satisfies $v^{(k)}(t)|_{t=T} = \frac{\partial^k u(x,t)}{\partial t^k}|_{t=T}$, $v^{(k)}(2T) = 0$, $k = 0, 1, 2$. Suppose $\hat{u} \in \mathscr{C}^{1+\alpha}(\mathcal{R})$.

The difference scheme for the problem (16) is presented as follows:

$$\begin{cases} \tau^{-\alpha} \sum_{k=0}^{n} g_k^{(\alpha)} u_i^{n-k} = \delta_x^2 u_i^n + f_i^n, & 1 \le i \le M-1, 1 \le n \le N, \\ u_i^0 = 0, & 1 \le i \le M-1, \\ u_0^n = \mu(t_n), \quad u_M^n = v(t_n), & 0 \le n \le N. \end{cases} \tag{17}$$

Theorem 20. *The difference scheme* (17) *is uniquely solvable.*

Theorem 21. *Let* $\{v_i^n \mid 0 \le i \le M, 0 \le n \le N\}$ *be the solution of the difference scheme*

$$\begin{cases} \tau^{-\alpha} \sum_{k=0}^{n} g_k^{(\alpha)} v_i^{n-k} = \delta_x^2 v_i^n + f_i^n, & 1 \le i \le M-1,\ 1 \le n \le N, \\ v_i^0 = \varphi(x_i), & 1 \le i \le M-1, \\ v_0^n = 0, \quad v_M^n = 0, & 0 \le n \le N. \end{cases}$$

Then we have

$$\|v^n\|_\infty \le \frac{5}{1-\alpha} \|v^0\|_\infty + \frac{5}{(1-\alpha)2^\alpha} t_n^\alpha \max_{1 \le m \le n} \|f^m\|_\infty, \quad 1 \le n \le N,$$

where $\|f^m\|_\infty = \max_{1 \le i \le M-1} |f_i^m|$.

The above lemma indicates that the difference scheme (17) is unconditionally stable to the initial value φ and the source term f.

Theorem 22. *Let $\{U_i^n \mid 0 \le i \le M, 0 \le n \le N\}$ be the solution of the fractional subdiffusion problem* (16) *and let $\{u_i^n \mid 0 \le i \le M, 0 \le n \le N\}$ be the solution of the difference scheme* (17). *Denote*

$$e_i^n = U_i^n - u_i^n, \quad 0 \le i \le M,\ 0 \le n \le N.$$

Then we have

$$\|e^n\|_\infty \le \frac{5}{(1-\alpha)2^\alpha} T^\alpha c_7(\tau + h^2), \quad 1 \le n \le N,$$

where c_7 is a constant.

6.2 The second-order shifted Grünwald–Letnikov formula

We consider the problem (16). Denote

$$w_0^{(\alpha)} = \left(1 + \frac{\alpha}{2}\right) g_0^{(\alpha)},$$
$$w_k^{(\alpha)} = \left(1 + \frac{\alpha}{2}\right) g_k^{(\alpha)} - \frac{\alpha}{2} g_{k-1}^{(\alpha)}, \quad k \ge 1.$$

The following difference scheme is constructed for the problem (16):

$$\begin{cases} \tau^{-\alpha} \sum\limits_{k=0}^{n} w_k^{(\alpha)} u_i^{n-k} = \delta_x^2 u_i^n + f_i^n, & 1 \le i \le M-1,\ 1 \le n \le N, \\ u_i^0 = 0, & 1 \le i \le M-1, \\ u_0^n = \psi_1(t_n), \quad u_M^n = \psi_2(t_n), & 0 \le n \le N. \end{cases} \tag{18}$$

Theorem 23. *The difference scheme* (18) *is uniquely solvable.*

Theorem 24. *Let $\{U_i^n \mid 0 \le i \le M, 0 \le n \le N\}$ be the solution of the fractional subdiffusion problem* (16) *and let $\{u_i^n \mid 0 \le i \le M, 0 \le n \le N\}$ be the solution of the difference scheme* (18). *Denote*

$$e_i^n = U_i^n - u_i^n, \quad 0 \le i \le M,\ 0 \le n \le N.$$

Then we have

$$\tau \sum_{n=1}^{N} \|e^n\|_\infty \le \frac{\sqrt{6}}{12} L^2 T c_8(\tau^2 + h^2),$$

where c_8 is a constant.

6.3 The $L1$ formula

Consider the following time-fractional subdiffusion equation:

$$\begin{cases} {}^C D_0^\alpha u(x,t) = \dfrac{\partial^2 u}{\partial x^2}(x,t) + f(x,t), & x \in (0,L),\ t \in (0,T], \\ u(x,0) = \varphi(x), & x \in (0,L), \\ u(0,t) = \mu(t), \quad u(L,t) = \nu(t), & t \in [0,T], \end{cases} \tag{19}$$

where $\alpha \in (0,1)$, f, φ, μ, ν are known functions, and $\varphi(0) = \mu(0)$, $\varphi(L) = \nu(0)$.

The following difference scheme is established for the problem (19):

$$\begin{cases} \dfrac{\tau^{-\alpha}}{\Gamma(2-\alpha)}\left[a_0 u_i^n - \sum_{k=1}^{n-1}(a_{n-k-1} - a_{n-k})u_i^k - a_{n-1}u_i^0\right] = \delta_x^2 u_i^n + f_i^n, \\ \qquad\qquad 1 \le i \le M-1,\ 1 \le n \le N, \\ u_i^0 = \varphi(x_i), \qquad 1 \le i \le M-1, \\ u_0^n = \mu(t_n), \quad u_M^n = \nu(t_n), \qquad 0 \le n \le N. \end{cases} \tag{20}$$

Theorem 25. *The difference scheme* (20) *is uniquely solvable.*

Theorem 26. *Let* $\{v_i^n \mid 0 \le i \le M, 0 \le n \le N\}$ *be the solution of the difference scheme*

$$\begin{cases} \dfrac{\tau^{-\alpha}}{\Gamma(2-\alpha)}\left[a_0 v_i^n - \sum_{k=1}^{n-1}(a_{n-k-1} - a_{n-k})v_i^k - a_{n-1}v_i^0\right] = \delta_x^2 v_i^n + f_i^n, \\ \qquad\qquad 1 \le i \le M-1,\ 1 \le n \le N, \\ v_i^0 = \varphi(x_i), \qquad 1 \le i \le M-1, \\ v_0^n = 0, \quad v_M^n = 0, \qquad 0 \le n \le N. \end{cases}$$

Then we have

$$\|v^n\|_\infty \le \|v^0\|_\infty + t_n^\alpha \Gamma(1-\alpha) \max_{1 \le m \le n} \|f^m\|_\infty, \quad 1 \le n \le N,$$

where

$$\|f^m\|_\infty = \max_{1 \le i \le M-1} |f_i^m|.$$

Theorem 27. *Let* $\{U_i^n \mid 0 \le i \le M, 0 \le n \le N\}$ *be the solution of the fractional subdiffusion problem* (19) *and let* $\{u_i^n \mid 0 \le i \le M, 0 \le n \le N\}$ *be the solution of the difference scheme* (20). *Denote*

$$e_i^n = U_i^n - u_i^n, \quad 0 \le i \le M,\ 0 \le n \le N.$$

Then we have

$$\|e^n\|_\infty \le c_9 T^\alpha \Gamma(1-\alpha)(\tau^{2-\alpha} + h^2), \quad 1 \le n \le N,$$

where c_9 *is a constant.*

6.4 The $L2\text{-}1_\sigma$ formula

We discretize the problem (19) using the $L2\text{-}1_\sigma$ formula for the time-fractional derivative.

The following difference scheme is derived for the problem (19):

$$\begin{cases} \dfrac{\tau^{-\alpha}}{\Gamma(2-\alpha)} \sum\limits_{k=0}^{n-1} c_k^{(n)} (u_i^{n-k} - u_i^{n-k-1}) = \sigma \delta_x^2 u_i^n + (1-\sigma)\delta_x^2 u_i^{n-1} + f_i^{n-1+\sigma}, \\ \qquad\qquad\qquad\qquad\qquad 1 \le i \le M-1,\ 1 \le n \le N, \\ u_i^0 = \varphi(x_i), \qquad\qquad\qquad\quad 1 \le i \le M-1, \\ u_0^n = \mu(t_n), \quad u_M^n = \nu(t_n), \qquad 0 \le n \le N. \end{cases} \tag{21}$$

Theorem 28. *The difference scheme (21) is uniquely solvable.*

Theorem 29. *Let $\{v_i^n \mid 0 \le i \le M, 0 \le n \le N\}$ be the solution of the difference scheme*

$$\begin{cases} \dfrac{\tau^{-\alpha}}{\Gamma(2-\alpha)} \sum\limits_{k=0}^{n-1} c_k^{(n)} (v_i^{n-k} - v_i^{n-k-1}) = \sigma \delta_x^2 v_i^n + (1-\sigma)\delta_x^2 v_i^{n-1} + f_i^{n-1+\sigma}, \\ \qquad\qquad\qquad\qquad\qquad 1 \le i \le M-1,\ 1 \le n \le N, \\ v_i^0 = \varphi(x_i), \qquad\qquad\qquad\quad 1 \le i \le M-1, \\ v_0^n = 0, \quad v_M^n = 0, \qquad\qquad 0 \le n \le N. \end{cases}$$

Then we have

$$\|v^n\|^2 \le \|v^0\|^2 + \frac{L^2}{6} t_n^\alpha \sigma(1-\alpha) \max_{1 \le m \le n} \|f^{m-1+\sigma}\|^2, \quad 1 \le n \le N.$$

Theorem 30. *Let $\{U_i^n \mid 0 \le i \le M, 0 \le n \le N\}$ be the solution of the fractional subdiffusion problem (19) and let $\{u_i^n \mid 0 \le i \le M, 0 \le n \le N\}$ be the solution of the difference scheme (21). Denote*

$$e_i^n = U_i^n - u_i^n, \quad 0 \le i \le M,\ 0 \le n \le N.$$

Then there exists a constant c_{10} such that

$$\|e^n\| \le c_{10}(\tau^2 + h^2), \quad 1 \le n \le N.$$

7 The finite difference methods for the time-fractional diffusion-wave equation

In this section, we construct the finite difference schemes for the time-fractional diffusion-wave equation applying the $L1$ formula and the $L2\text{-}1_\sigma$ formula, respectively. The results of the stability and the convergence for the schemes are presented.

7.1 The $L1$ formula

Consider the following time-fractional diffusion-wave equation:

$$\begin{cases} {}^C D_0^\alpha u(x,t) = \dfrac{\partial^2 u}{\partial x^2}(x,t) + f(x,t), & x \in (0,L),\ t \in (0,T], \\ u(x,0) = \phi(x), \quad u_t(x,0) = \psi(x), & x \in (0,L), \\ u(0,t) = \mu(t), \quad u(L,t) = \nu(t), & t \in [0,T], \end{cases} \tag{22}$$

where $\alpha \in (1,2)$, f, ϕ, ψ, μ, ν are known functions, $\phi(0) = \mu(0)$, $\phi(L) = \nu(0)$, $\psi(0) = \mu'(0)$, $\psi(L) = \nu'(0)$, and

$$ {}^C D_0^\alpha u(x,t) = \frac{1}{\Gamma(2-\alpha)} \int_0^t \frac{u_{ss}(x,s)}{(t-s)^{\alpha-1}} \mathrm{d}s.$$

Denote

$$U_i^n = u(x_i, t_n), \quad f_i^n = f(x_i, t_n), \quad \psi_i = \psi(x_i), \quad 0 \le i \le M,\ 0 \le n \le N.$$

The following difference scheme can be derived for the problem (22):

$$\begin{cases} \dfrac{\tau^{1-\alpha}}{\Gamma(3-\alpha)} \left[b_0 \delta_t u_i^{n-\frac12} - \displaystyle\sum_{k=1}^{n-1} (b_{n-k-1} - b_{n-k}) \delta_t u_i^{k-\frac12} - b_{n-1}\psi_i \right] = \delta_x^2 u_i^{n-\frac12} + f_i^{n-\frac12}, \\ \qquad\qquad 1 \le i \le M-1,\ 1 \le n \le N, \\ u_i^0 = \phi(x_i), \qquad 1 \le i \le M-1, \\ u_0^n = \mu(t_n), \quad u_M^n = \nu(t_n), \qquad 0 \le n \le N. \end{cases} \tag{23}$$

Theorem 31. *The difference scheme (23) is uniquely solvable.*

Theorem 32. *Let $\{v_i^n \mid 0 \le i \le M, 0 \le n \le N\}$ be the solution of the difference scheme*

$$\begin{cases} \dfrac{\tau^{1-\alpha}}{\Gamma(3-\alpha)} \left[b_0 \delta_t v_i^{n-\frac12} - \displaystyle\sum_{k=1}^{n-1} (b_{n-k-1} - b_{n-k}) \delta_t v_i^{k-\frac12} - b_{n-1}\psi_i \right] = \delta_x^2 v_i^{n-\frac12} + f_i^{n-\frac12}, \\ \qquad\qquad 1 \le i \le M-1,\ 1 \le n \le N, \\ v_i^0 = \phi(x_i), \qquad 1 \le i \le M-1, \\ v_0^n = 0, \quad v_M^n = 0, \qquad 0 \le n \le N. \end{cases}$$

Then we have

$$\|\delta_x v^n\|^2 \le \|\delta_x v^0\|^2 + \frac{t_n^{2-\alpha}}{\Gamma(3-\alpha)} \|\psi\|^2 + t_n^{\alpha-1}\Gamma(2-\alpha)\tau \sum_{k=1}^n \|f^{k-\frac12}\|^2, \quad 1 \le n \le N,$$

where

$$\|\psi\|^2 = h \sum_{i=1}^{M-1} \psi_i^2, \quad \|f^{k-\frac12}\|^2 = h \sum_{i=1}^{M-1} (f_i^{k-\frac12})^2.$$

Theorem 33. *Let $\{U_i^n \mid 0 \le i \le M, 0 \le n \le N\}$ be the solution of the fractional diffusion-wave problem* (22) *and let $\{u_i^n \mid 0 \le i \le M, 0 \le n \le N\}$ be the solution of the difference scheme* (23)*. Denote*

$$e_i^n = U_i^n - u_i^n, \quad 0 \le i \le M,\ 0 \le n \le N.$$

Then we have

$$\|e^n\|_\infty \le \frac{1}{2} c_{11} L \sqrt{T^\alpha \Gamma(2-\alpha)} (\tau^{3-\alpha} + h^2), \quad 1 \le n \le N,$$

where c_{11} is a constant.

7.2 The $L2\text{-}1_\sigma$ formula

Consider (22). Let $v = u_t$. Then (22) is equivalent to [38]

$$\begin{cases} {}^C D_0^{\alpha-1} v(x,t) = \dfrac{\partial^2 u}{\partial x^2}(x,t) + f(x,t), & x \in (0,L),\ t \in (0,T], \\ \dfrac{\partial^3 u}{\partial x^2 \partial t}(x,t) = \dfrac{\partial^2 v}{\partial x^2}(x,t), & x \in (0,L),\ t \in (0,T], \\ u(x,0) = \phi(x), \quad v(x,0) = \psi(x), & x \in (0,L), \\ u(0,t) = \mu(t), \quad u(L,t) = \nu(t), & t \in [0,T]. \end{cases}$$

We construct the difference scheme for the above problem as follows:

$$\frac{\tau^{1-\alpha}}{\Gamma(3-\alpha)} \sum_{k=0}^{n-1} c_k^{(n,\alpha-1)} (v_i^{n-k} - v_i^{n-k-1}) = \sigma \delta_x^2 u_i^n + (1-\sigma)\delta_x^2 u_i^{n-1} + f_i^{n-1+\sigma},$$
$$1 \le i \le M-1,\ 1 \le n \le N, \tag{24}$$
$$\delta_t \delta_x^2 u_i^{\frac{1}{2}} = \delta_x^2 v_i^{\frac{1}{2}}, \quad 1 \le i \le M-1, \tag{25}$$
$$D_{\hat{t}} \delta_x^2 u_i^n = \sigma \delta_x^2 v_i^{n+1} + (1-\sigma)\delta_x^2 v_i^n, \quad 1 \le i \le M-1,\ 1 \le n \le N-1, \tag{26}$$
$$u_i^0 = \phi(x_i), \quad v_i^0 = \psi(x_i), \quad 1 \le i \le M-1, \tag{27}$$
$$u_0^n = \mu(t_n), \quad u_M^n = \psi(t_n), \quad 0 \le n \le N, \tag{28}$$
$$v_0^n = \mu'(t_n), \quad v_M^n = \nu'(t_n), \quad 0 \le n \le N, \tag{29}$$

where

$$D_{\hat{t}} u_i^n = \frac{1}{2\tau}[(2\sigma+1)u_i^{n+1} - 4\sigma u_i^n + (2\sigma-1)u_i^{n-1}], \quad 1 \le n \le N-1.$$

Equations (25) and (26) can be written as

$$\delta_x^2 u_i^1 = \delta_x^2 u_i^0 + \frac{\tau}{2}(\delta_x^2 v_i^1 + \delta_x^2 v_i^0), \quad 1 \le i \le M-1, \tag{30}$$

$$\delta_x^2 u_i^{n+1} = \frac{1}{2\sigma+1}[4\sigma\delta_x^2 u_i^n - (2\sigma-1)\delta_x^2 u_i^{n-1}] + \frac{2\tau}{2\sigma+1}[\sigma\delta_x^2 v_i^{n+1} + (1-\sigma)\delta_x^2 v_i^n], \quad 1 \le i \le M-1,\ 1 \le n \le N-1. \tag{31}$$

Inserting (30) into (24) and combining with (29) for $n = 1$, we obtain a tridiagonal linear system about $\{v_i^1 \mid 1 \le i \le M-1\}$. Then from (30) and (28), we obtain a tridiagonal linear system about $\{u_i^1 \mid 1 \le i \le M-1\}$.

If, for $0 \le k \le n$, $\{u_i^k \mid 1 \le i \le M-1\}$ and $\{v_i^k \mid 1 \le i \le M-1\}$ have been obtained, then inserting (31) into (24) yields a tridiagonal linear system about $\{v_i^{n+1} \mid 1 \le i \le M-1\}$. After $\{v_i^{n+1} \mid 1 \le i \le M-1\}$ has been solved, inserting it into (31) yields a tridiagonal linear system about $\{u_i^{n+1} \mid 1 \le i \le M-1\}$. The double-sweep methods may be used to solve these tridiagonal linear systems.

Theorem 34. *Suppose* $\{w_i^n \mid 0 \le i \le M,\ 0 \le n \le N\}$ *and* $\{z_i^n \mid 0 \le i \le M,\ 0 \le n \le N\}$ *satisfy*

$$\frac{\tau^{1-\alpha}}{\Gamma(3-\alpha)} \sum_{k=0}^{n-1} c_k^{(n,\alpha-1)}(z_i^{n-k} - z_i^{n-k-1}) = \sigma\delta_x^2 w_i^n + (1-\sigma)\delta_x^2 w_i^{n-1} + f_i^{n-1+\sigma},$$

$$1 \le i \le M-1,\ 1 \le n \le N,$$

$$\delta_t\delta_x^2 w_i^{\frac{1}{2}} = \delta_x^2 z_i^{\frac{1}{2}} + g_i^{\frac{1}{2}}, \quad 1 \le i \le M-1,$$

$$D_{\hat t}\delta_x^2 w_i^n = \sigma\delta_x^2 z_i^{n+1} + (1-\sigma)\delta_x^2 z_i^n + g_i^{n+\sigma}, \quad 1 \le i \le M-1,\ 1 \le n \le N-1,$$

$$w_i^0 = \phi(x_i), \quad z_i^0 = \psi(x_i), \quad 1 \le i \le M-1,$$

$$w_0^n = 0, \quad w_M^n = 0, \quad z_0^n = 0, \quad z_M^n = 0, \quad 0 \le n \le N,$$

where $\phi(x_i) = 0$, $\psi(x_i) = 0$ *for* $i = 0, M$. *Then we have*

$$\|\delta_x w^n\|^2 \le c_{12}\exp\left(\frac{2\sigma L^2}{3}T\right)G_n, \quad \tau\sum_{k=1}^{n}\|z^k\|^2 \le c_{13}G_n, \quad 0 \le n \le N,$$

where c_{12} *and* c_{13} *are two constants and*

$$G_n = \|\delta_x w^0\|^2 + \|\delta_x^2 w^0\|^2 + \|z^0\|^2 + \|\delta_x z^0\|^2 + \|f^\sigma\|^2 + \|g^{\frac{1}{2}}\|^2 + \tau\sum_{k=1}^{n-1}\|f^{k+\sigma}\|^2 + \tau\sum_{k=1}^{n-1}\|g^{k+\sigma}\|^2.$$

From Theorem 34, we can obtain the stability of the difference scheme.

Theorem 35. *The solution of the difference scheme* (24)–(29) *is unconditionally stable with respect to the initial values* ϕ, ψ *and the right-hand side function* f.

We have the following result on the convergence of the difference scheme.

Theorem 36. *Let* $\{U_i^k \mid 0 \le i \le M,\ 0 \le k \le N\}$ *be the solution of the problem* (22) *and let* $\{u_i^k \mid 0 \le i \le M,\ 0 \le k \le N\}$ *be the solution of the difference scheme* (24)–(29). *Denote*

$$e_i^k = U_i^k - u_i^k, \quad 0 \le i \le M,\ 0 \le k \le N.$$

Then there exists a constant c_{14} *such that*

$$\|\delta_x e^n\| \le c_{14}(\tau^2 + h^2), \quad 0 \le n \le N.$$

8 Concluding remarks

The above numerical schemes for the time-fractional subdiffusion equation and the diffusion-wave equation could be extended to solve the high-dimensional problems. With the help of the alternating direction implicit (ADI) method, the problem could be split into a series of one-dimensional problems to reduce the storage and the computational cost [5, 50, 51]. When solving the time-fractional problems, the compact finite difference schemes [4, 6, 10, 45] could also be used in order to increase the accuracy in space. The above time-fractional subdiffusion equation and the diffusion-wave equation with Neumann boundary conditions are also numerically solved and the stability results are theoretically guaranteed [31, 34, 52].

There are some works on the finite difference methods for the problems of multi-term fractional differential equations [32, 33, 8] and the distributed-order differential equations [7, 47, 9, 12, 11, 14, 13].

There exist some fractional multi-step methods [46] and high-order formulas resulted from the high-order interpolation of the function [3, 15, 20, 26, 53] for the approximation of the fractional derivatives. In [29, 19], the fast solvers are proposed for solving the linear system arising from the time-fractional partial differential equation. Considering that the solution of the initial boundary value problem of the fractional differential equations may have a weak singularity at the initial time, the graded mesh technique has recently been developed [37, 35].

Bibliography

[1] A. A. Alikhanov, A new difference scheme for the time fractional diffusion equation, *J. Comput. Phys.*, **280** (2015), 424–438.
[2] J. Bouchaud and A. Georges, Anomalous diffusion in disordered media: Statistical mechanisms, models and physical applications, *Phys. Rep.*, **195** (1990), 127–293.
[3] J. Cao and C. Xu, A high order schema for the numerical solution of the fractional ordinary differential equations, *J. Comput. Phys.*, **238** (2013), 154–168.
[4] M. R. Cui, Compact finite difference method for the fractional diffusion equation, *J. Comput. Phys.*, **228** (2009), 7792–7804.

[5] M. R. Cui, Compact alternating direction implicit method for two-dimensional time fractional diffusion equation, *J. Comput. Phys.*, **231** (2012), 2621–2633.

[6] R. Du, W. R. Cao, and Z. Z. Sun, A compact difference scheme for the fractional diffusion-wave equation, *Appl. Math. Model.*, **34** (2010), 2998–3007.

[7] N. J. Ford, M. L. Morgado, and M. Rebelo, A numerical method for the distributed order time-fractional diffusion equation, in: *ICFDA'14 Catania*, 23–25 June 2014, ISBN 978-1-4799-2590-2.

[8] G. H. Gao, A. A. Alikhanov, and Z. Z. Sun, The temporal second order difference schemes based on the interpolation approximation for solving the time multi-term and distributed-order fractional sub-diffusion equations, *J. Sci. Comput.*, **73** (2017), 93–121.

[9] G. H. Gao H. W. Sun, and Z. Z. Sun, Some high-order difference schemes for the distributed-order differential equations, *J. Comput. Phys.*, **298** (2015), 337–359.

[10] G. H. Gao and Z. Z. Sun, A compact finite difference scheme for the fractional sub-diffusion equations, *J. Comput. Phys.*, **230** (2011), 586–595.

[11] G. H. Gao and Z. Z. Sun, Two alternating direction implicit difference schemes with the extrapolation method for the two-dimensional distributed-Order differential equations, *Comput. Math. Appl.*, **69** (2015), 926–948.

[12] G. H. Gao and Z. Z. Sun, Two unconditionally stable and convergent difference schemes with the extrapolation method for the one-dimensional distributed-order differential equations, *Numer. Methods Partial Differ. Equ.*, **32** (2016), 591–615.

[13] G. H. Gao and Z. Z. Sun, Two alternating direction implicit difference schemes for two-dimensional distributed-order fractional diffusion equations, *J. Sci. Comput.*, **66** (2016), 1281–1312.

[14] G. H. Gao and Z. Z. Sun, Two difference schemes for solving the one-dimensional time distributed-order fractional wave equations, *Numer. Algorithms*, **74** (2017), 675–697.

[15] G. H. Gao, Z. Z. Sun, and H. W. Zhang, A new fractional numerical differentiation formula to approximate the Caputo fractional derivative and its applications, *J. Comput. Phys.*, **259** (2014), 33–50.

[16] R. Hilfer (ed.), *Applications of Fractional Calculus in Physics*, World Scientific, Singapore, 2000.

[17] C. C. Ji and Z. Z. Sun, A high-order compact finite difference scheme for the fractional sub-diffusion equation, *J. Sci. Comput.*, **64** (2015), 959–985.

[18] C. C. Ji and Z. Z. Sun, The high-order compact numerical algorithms for the two-dimensional fractional sub-diffusion equation, *Appl. Math. Comput.*, **269** (2015), 775–791.

[19] R. H. Ke, M. K. Ng, and H. W. Sun, A fast direct method for block triangular Toeplitz-like with tri-diagonal block systems from time-fractional partial differential equation, *J. Comput. Phys.*, **303** (2015), 203–211.

[20] C. P. Li, A. Chen, and J. Ye, Numerical approaches to fractional calculus and fractional ordinary differential equation, *J. Comput. Phys.*, **230** (2011), 3352–3368.

[21] M. Li, X. T. Xiong, and Y. J. Wang, A numerical evaluation and regularization of Caputo fractional derivatives, *J. Phys. Conf. Ser.*, **290** (2011), 012011.

[22] Y. Lin, X. Li, and C. Xu, Finite difference/spectral approximations for the fractional cable equation, *Math. Comput.*, **80** (2011), 1369–1396.

[23] Y. Lin and C. Xu, Finite difference/spectral approximations for the time-fractional diffusion equation, *J. Comput. Phys.*, **225** (2007), 1533–1552.

[24] J. C. Lopez-Marcos, A difference scheme for a nonlinear partial integrodifferential equation, *SIAM J. Numer. Anal.*, **27** (1990), 20–31.

[25] C. Lubich, Discretized fractional calculus, *SIAM J. Math. Anal.*, **17** (1986), 704–719.

[26] V. Lynch, B. Carreras, D. Castillo-Negrete, K. Ferreira-Mejias, and H. Hicks, Numerical methods for the solution of partial differential equations of fractional order, *J. Comput. Phys.*, **192** (2003), 406–421.

[27] M. M. Meerschaert and C. Tadjeran, Finite difference approximations for fractional advection-dispersion flow equations, *J. Comput. Appl. Math.*, **172** (2004), 65–77.

[28] R. Metzler and J. Klafter, The random walks guide to anomalous diffusion: A fractional dynamics approach, *Phys. Rep.*, **339** (2000), 1–77.

[29] H. K. Pang and H. W. Sun, Fast numerical contour integral method for fraction diffusion equation, *J. Sci. Comput.*, **66** (2016), 41–66.

[30] I. Podlubny, *Fractional Differential Equations*, Academic Press, New York, 1999.

[31] J. C. Ren and Z. Z. Sun, Numerical algorithm with high spatial accuracy for the fractional diffusion-wave equation with Neumann boundary conditions, *J. Sci. Comput.*, **56** (2013), 381–408.

[32] J. C. Ren and Z. Z. Sun, Efficient and stable numerical methods for multi-term time fractional sub-diffusion equations, *East Asian J. Appl. Math.*, **4** (2014), 242–266.

[33] J. C. Ren and Z. Z. Sun, Efficient numerical approximation of the multi-term time fractional diffusion-wave equations, *East Asian J. Appl. Math.*, **5** (2015), 1–28.

[34] J. C. Ren, Z. Z. Sun, and X. Zhao, Compact difference scheme for the fractional sub-diffusion equation with Neumann boundary conditions, *J. Comput. Phys.*, **232** (2013), 456–467.

[35] J. Y. Shen, Z. Z. Sun , and R. Du, Fast finite difference schemes for the time-fractional diffusion equation with a weak singularity at the initial time, *East Asian J. Appl. Math.*, **8** (2018), 834–858.

[36] T. H. Solomon, E. R. Weeks, and H. L. Swinney, Observations of anomalous diffusion and Lévy flights in a 2-dimensional rotating flow, *Phys. Rev. Lett.*, **71** (1993), 3975–3979.

[37] M. Stynes, E. O'riordan, and J. L. Gracia, Error analysis of a finite difference method on graded meshes for a time-fractional diffusion equation, *SIAM J. Numer. Anal.*, **55** (2017), 1057–1079.

[38] H. Sun, Z. Z. Sun, and G. H. Gao, Some temporal second order difference schemes for fractional wave equations, *Numer. Methods Partial Differ. Equ.*, **32** (2016), 970–1001.

[39] Z. Z. Sun and G. H. Gao, *Numerical Methods for Fractional Differential Equations*, Science Press, 2015.

[40] Z. Z. Sun and X. Wu, A fully discrete difference scheme for a diffusion-wave system, *Appl. Numer. Math.*, **56** (2006), 193–209.

[41] C. Tadjeran, M. M. Meerschaert, and H. P. Scheffler, A second-order accurate numerical approximation for the fractional diffusion equation, *J. Comput. Phys.*, **213** (2006), 205–213.

[42] T. Tang, A finite difference scheme for partial integro-differential equations with a weakly singular kernel, *Appl. Numer. Math.*, **11** (1993), 309–319.

[43] W. Y. Tian, H. Zhou, and W. H. Deng, A class of second order difference approximations for solving space fractional diffusion equations, *Math. Comput.*, **84** (2015), 1703–1727.

[44] V. K. Tuan and R. Gorenflo, Extrapolation to the limit for numerical fractional differentiation, *Z. Angew. Math. Mech.*, **75** (1995), 646–648.

[45] Z. B. Wang and S. Vong, Compact difference schemes for the modified anomalous fractional sub-diffusion equation and the fractional diffusion-wave equation, *J. Comput. Phys.*, **277** (2014), 1–15.

[46] J. Y. Yang, J. F. Huang, D. M. Liang, and Y. F. Tang, Numerical solution of fractional diffusion-wave equation based on fractional multistep method, *Appl. Math. Model.*, **38** (2014), 3652–3661.

[47] H. Ye, F. Liu, V. Anh, and I. Turner, Numerical analysis for the time distributed-order and Riesz space fractional diffusions on bounded domains, *IMA J. Appl. Math.*, **80** (2015), 825–838.

[48] S. B. Yuste, Weighted average finite difference methods for fractional diffusion equations, *J. Comput. Phys.*, **216** (2006), 264–274.

[49] S. B. Yuste and L. Acedo, An explicit finite difference method and a new Von Neumann-type stability analysis for fractional diffusion-wave equations, *SIAM J. Numer. Anal.*, **42**(5) (2005), 1862–1874.
[50] Y. N. Zhang and Z. Z. Sun, Alternating direction implicit schemes for the two-dimensional fractional sub-diffusion equation, *J. Comput. Phys.*, **230** (2011), 8713–8728.
[51] Y. N. Zhang, Z. Z. Sun, and X. Zhao, Compact alternating direction implicit schemes for the two-dimensional fractional diffusion-wave equation, *SIAM J. Numer. Anal.*, **50** (2012), 1535–1555.
[52] X. Zhao and Z. Z. Sun, A box-type scheme for fractional sub-diffusion equation with Neumann boundary conditions, *J. Comput. Phys.*, **230** (2011), 6061–6074.
[53] X. Zhao, Z. Z. Sun, and G. Karniadakis, Second-order approximations for variable order fractional derivatives: Algorithms and applications, *J. Comput. Phys.*, **293** (2015), 184–200.

Hengfei Ding and Changpin Li

High-order finite difference methods for fractional partial differential equations

Abstract: More than six kinds of fractional derivatives have been described. Normally, the time-fractional derivatives appear in the Caputo or Riemann–Liouville sense. As for the space-fractional derivative, it is commonly defined as an operator inverse to the Riesz potential and referred to as the Riesz derivative. Here, we mainly focus on introducing numerical approximations for these fractional derivatives. Meanwhile, we give applications of these numerical approximation formulas in fractional partial differential equations.

Keywords: Caputo derivative, Riemann–Liouville derivative, Riesz derivative, numerical test

MSC 2010: 62M20, 62N02, 93D05, 93D20, 93B52

1 Introduction

In recent years, fractional calculus has received increasing attention due to its widespread applications in science and engineering [40, 43, 45]. For most fractional differential equations, to obtain the analytical solution is not easy or even impossible, so many researchers have to solve fractional differential equations by using various kinds of numerical methods. Different from the typical differential equations, even if we use lower-order methods for solving the fractional differential equations, we still need more calculations and storage space. If we use higher-order methods for fractional differential equations, the calculations and memory capacities cannot increase much. In this sense, the higher-order numerical methods for fractional calculus and fractional differential equations attract more and more interest.

To solve fractional differential equations numerically, the key issue is how to deal with the fractional derivatives. This chapter is organized as follows. In Section 2, we mainly provide some high-order numerical approaches to approximate fractional

Acknowledgement: The work was supported by the National Natural Science Foundation of China (Nos. 11561060 and 11671251) and the Natural Science Foundation of Gansu Province of China (No. 17JR5RE009).

Hengfei Ding, School of Mathematics and Statistics, Tianshui Normal University, Tianshui 741001, China, e-mail: dinghf05@163.com
Changpin Li, Department of Mathematics, Shanghai University, Shanghai 200444, China, e-mail: lcp@shu.edu.cn

https://doi.org/10.1515/9783110571684-003

derivatives. In Section 3, we briefly introduce a high-order finite difference method for fractional partial differential equations.

2 High-order numerical approximation formulas for the fractional derivatives

For the single-variable function $u(t)$ ($u(x)$) defined on $[0, T]$ ($[a, b]$), we define the uniform step sizes by $\tau = T/N$ ($h = (b-a)/M$). Here, N (M) is a positive integer. The grid points $t_k = k\tau$ ($x_j = jh$), $k = 0, 1, \ldots, N$ ($j = 0, 1, \ldots, M$). The numerical approximation of function $u(t_k)$ ($u(x_j)$) is denoted by u^k (u^j) or u_k (u_j) unless otherwise specified.

2.1 The Caputo derivative

2.1.1 Based on the difference quotient methods

The L1 formula

Considering the Caputo derivative ${}_C D_{0,t}^{\alpha} u(t)$ for $0 < \alpha < 1$ at point $t = t_n$, one has [25, 43]

$$\begin{aligned}\left[{}_C D_{0,t}^{\alpha} u(t)\right]_{t=t_n} &= \frac{1}{\Gamma(1-\alpha)} \sum_{k=0}^{n-1} \int_{t_k}^{t_{k+1}} (t_n - s)^{-\alpha} u'(s)\,\mathrm{d}s \\ &= \frac{1}{\Gamma(1-\alpha)} \sum_{k=0}^{n-1} \int_{t_k}^{t_{k+1}} (t_n - s)^{-\alpha} \left[\frac{u(t_{k+1}) - u(t_k)}{\tau} + \mathcal{O}(\tau)\right] \mathrm{d}s \\ &= \sum_{k=0}^{n-1} b_{n-k-1}\big(u(t_{k+1}) - u(t_k)\big) + \mathcal{O}(\tau^{2-\alpha}),\end{aligned}$$

where the weights are defined by

$$b_k = \frac{\tau^{-\alpha}}{\Gamma(2-\alpha)}\left[(k+1)^{1-\alpha} - k^{1-\alpha}\right], \quad k = 0, 1, \ldots, n-1.$$

The L1C formula

Using the central difference formula

$$u'(s) = \frac{u(t_{i+1}) - u(t_{i-1})}{2\tau} + \frac{u(t_{i+1}) - 2u(t_i) + u(t_{i-1})}{\tau^2}(s - t_i) + \mathcal{O}(\tau^2)$$

to deal with the derivative $u'(s)$, the so-called L1C formula for the Caputo derivative has the following form [28]:

$$\begin{aligned}[{}_CD^{\alpha}_{0,t}u(t)]_{t=t_k} &= \frac{\tau^{-\alpha}}{\Gamma(3-\alpha)}\sum_{i=0}^{k-1}\big[w_{1,k-i}\big(u(t_{i+1})-u(t_{i-1})\big)\\ &\quad + w_{2,k-i}\big(u(t_{i+1})-2u(t_i)+u(t_{i-1})\big)\big] + \mathcal{O}(\tau^{3-\alpha}),\end{aligned} \tag{1}$$

where

$$\begin{aligned}w_{1,k-i} &= \frac{2-\alpha}{2}\big[(k-i)^{1-\alpha}-(k-i-1)^{1-\alpha}\big],\\ w_{2,k-i} &= (k-i)^{2-\alpha}-(k-i-1)^{2-\alpha}-(2-\alpha)(k-i-1)^{1-\alpha}.\end{aligned}$$

Remark 1. In equation (1), if $i = 0$, then $u_{i-1} = u_{-1}$ is defined outside of $[0, T]$. In numerical calculation, we mainly use the neighboring function values to approximate u_{-1}, that is, $u_{-1} = u(0) - \tau u'(0) + \frac{\tau^2}{2}u''(0) + \mathcal{O}(\tau^3)$.

(1) When $u'(0) = u''(0) = 0$, then $u_{-1} = u_0 + \mathcal{O}(\tau^3)$, the convergence order of (1) is $\mathcal{O}(\tau^{3-\alpha})$.
(2) When $u'(0) = 0$, $u''(0) \neq 0$, then $u_{-1} = u_0 + \frac{\tau^2}{2}u''(0) + \mathcal{O}(\tau^3)$, the convergence order of (1) is $\mathcal{O}(\tau^2)$.
(3) When $u'(0) \neq 0$, then the convergence order is $\mathcal{O}(\tau)$.

The L2 formula

Note that, for the case $\alpha \in (1, 2)$, the Caputo derivative has the following form:

$$[{}_CD^{\alpha}_{0,t}u(t)]_{t=t_n} = \frac{1}{\Gamma(2-\alpha)}\sum_{k=0}^{n-1}\int_{t_k}^{t_{k+1}} s^{1-\alpha}u''(t_n-s)\,ds. \tag{2}$$

Using the approximation

$$u''(t_n-s) = \frac{u(t_n-t_{k+1})-2u(t_n-t_k)+u(t_n-t_{k-1})}{\tau^2} + \mathcal{O}(\tau^2)$$

on each interval $[t_k, t_{k+1}]$, we can obtain the following L2 method for the Caputo derivative [37, 51]:

$$[{}_CD^{\alpha}_{0,t}u(t)]_{t=t_n} = \sum_{k=0}^{n-1} c_k\big(u(t_{n-k-1})-2u(t_{n-k})+u(t_{n-k+1})\big) + \mathcal{O}(\tau^{3-\alpha}),$$

where

$$c_k = \frac{\tau^{-\alpha}}{\Gamma(3-\alpha)}\big[(k+1)^{2-\alpha}-k^{2-\alpha}\big], \quad k = 0, 1, \ldots, n-1.$$

The L2C formula

If we use the four-point difference scheme

$$u''(t_n - s) = \frac{u(t_n - t_{k+2}) - u(t_n - t_{k+1}) + u(t_n - t_{k-1}) - u(t_n - t_k)}{2\tau^2} + O(\tau^2)$$

for the second derivative $u''(t_n - s)$ on each interval $[t_k, t_{k+1}]$ in (2), then we can obtain the following L2C method for the Caputo derivative:

$$[{}_C D_{0,t}^{\alpha} u(t)]_{t=t_n} = \sum_{k=0}^{n-1} c_k \big(u(t_{n-k-2}) - u(t_{n-k-1}) + u(t_{n-k+1}) - u(t_{n-k})\big) + \mathcal{O}(\tau^{3-\alpha}).$$

The accuracy of the L2 method and the L2C method depends on α. Experiments show that the L2 method is more accurate than the L2C method for $1 < \alpha < 1.5$, and the reverse holds for $1.5 < \alpha < 2$. These two methods have almost similar results around $\alpha = 1.5$ [37].

2.1.2 Based on the polynomial interpolation methods

The linear interpolation formula

In fact, a piece-wise linear interpolation approximation for the integrand function of the Caputo derivative leads to a formula of the same form as the L1 formula.

The L1-2 formula

Based on the idea of the linear interpolation formula, a natural extension is to apply a higher-order interpolation instead of the linear interpolation to improve the numerical accuracy. In [21], Gao et al. used the linear interpolation approximation on the first small interval $[t_0, t_1]$ and applied the quadratic interpolation approximation on the remaining interval $[t_{k-1}, t_k]$ $(k \geq 2)$. They obtained a difference formula (called the *L*1-2 formula) for the Caputo ${}_C D_{0,t}^{\alpha} u(t)$ $(\alpha \in (0,1))$ derivative as follows:

$$[{}_C D_{0,t}^{\alpha} u(t)]_{t=t_k} = \frac{\tau^{-\alpha}}{\Gamma(2-\alpha)} \left[c_0^{(\alpha)} f(t_k) - \sum_{j=0}^{k-1} (c_{k-j-1}^{(\alpha)} - c_{k-j}^{(\alpha)}) f(t_j) - c_{k-1}^{(\alpha)} f(t_0) \right] + r^k,$$

where $r^k = \mathcal{O}(\tau^{2-\alpha})$ for $k = 1$ and $r^k = \mathcal{O}(\tau^{3-\alpha})$ for $k \geq 2$. Here, the coefficients $a_j^{(\alpha)}$, $b_j^{(\alpha)}$, and $c_j^{(\alpha)}$ are defined by

$$\begin{aligned}
a_j^{(\alpha)} &= (j+1)^{1-\alpha} - j^{1-\alpha}, \\
b_j^{(\alpha)} &= \frac{1}{2-\alpha}[(j+1)^{2-\alpha} - j^{2-\alpha}] - \frac{1}{2}[(j+1)^{1-\alpha} - j^{1-\alpha}], \\
c_j^{(\alpha)} &= \begin{cases} a_0^{(\alpha)} + b_0^{(\alpha)}, & j = 0, \\ a_j^{(\alpha)} + b_j^{(\alpha)} - b_{j-1}^{(\alpha)}, & 1 \leq j \leq k-2, \\ a_j^{(\alpha)} - b_{j-1}^{(\alpha)}, & j = k-1. \end{cases}
\end{aligned}$$

The L2-1_σ formula

In [1], Alikhanov constructed a $(3-\alpha)$th-order difference formula (called the L2-1_σ formula) for Caputo derivatives with $0 < \alpha < 1$ as follows:

$$[{}_CD^\alpha_{0,t}u(t)]_{t=t_{n+\sigma}} = \frac{\tau^{-\alpha}}{\Gamma(2-\alpha)}\sum_{j=0}^{n} c^{(\alpha,\sigma)}_{n-j}(u(t_{j+1}) - u(t_j)) + O(\tau^{3-\alpha}),$$

where

$$a^{(\alpha,\sigma)}_j = \begin{cases} \sigma^{1-\alpha}, & j=0,\\ (j+\sigma)^{1-\alpha} - (j-1+\sigma)^{1-\alpha}, & j \geq 1, \end{cases}$$

$$b^{(\alpha,\sigma)}_j = \frac{1}{2-\alpha}[(j+\sigma)^{2-\alpha} - (j-1+\sigma)^{2-\alpha}] - \frac{1}{2}[(j+\sigma)^{1-\alpha} - (j-1+\sigma)^{1-\alpha}],$$

$$c^{(\alpha,\sigma)}_j = \begin{cases} a^{(\alpha,\sigma)}_0, & n=0,\\ a^{(\alpha,\sigma)}_0 + b^{(\alpha,\sigma)}_1, & j=0, n\geq 1,\\ a^{(\alpha,\sigma)}_j + b^{(\alpha,\sigma)}_{j+1} - b^{(\alpha,\sigma)}_j, & 1 \leq j \leq n-1,\\ a^{(\alpha,\sigma)}_n - b^{(\alpha,\sigma)}_n, & j = n \geq 1. \end{cases}$$

However, when we use this formula to solve time-fractional differential equations, the resulting convergence order reduces to $O(\tau^2)$.

The linear and cubic interpolation formulas

Note that the Caputo derivative with order α $(0 < \alpha < 1)$ of the function $f(t)$ at grid point t_n is defined by

$$[{}_CD^\alpha_{0,t}f(t)]_{t=t_n} = \frac{1}{\Gamma(1-\alpha)}\sum_{j=1}^{n}\int_{t_{j-1}}^{t_j}(t_n - s)^{-\alpha}f'(s)ds.$$

On the first subinterval $[t_0, t_1]$, we use linear interpolation to approximate $f(t)$, thus

$$\begin{aligned}\frac{1}{\Gamma(1-\alpha)}\int_{t_0}^{t_1}(t_n - s)^{-\alpha}f'(s)ds &\approx \frac{f_1 - f_0}{\tau\Gamma(1-\alpha)}\int_{t_0}^{t_1}(t_n - s)^{-\alpha}ds\\ &= \frac{\tau^{-\alpha}a_{n-1}}{\Gamma(2-\alpha)}(f_1 - f_0),\end{aligned}$$

where $a_{n-1} = n^{1-\alpha} - (n-1)^{1-\alpha}$.

On the second subinterval $[t_1, t_2]$, we have the approximation as follows:

$$\frac{1}{\Gamma(1-\alpha)}\int_{t_1}^{t_2}(t_n - s)^{-\alpha}f'(s)ds \approx \frac{\tau^{-\alpha}}{\Gamma(2-\alpha)}[(a_{n-2} + b_{n-2})f_2 - (a_{n-2} + 2b_{n-2})f_1 + b_{n-2}f_0],$$

in which

$$a_{n-2} = (n-1)^{1-\alpha} - (n-2)^{1-\alpha}$$

and

$$b_{n-2} = \frac{(n-1)^{2-\alpha} - (n-2)^{2-\alpha}}{2-\alpha} - \frac{(n-1)^{1-\alpha} + (n-2)^{1-\alpha}}{2}.$$

For the other subdomains ($j \geq 3$), using the cubic interpolation function

$$p_3(t) = \sum_{l=0}^{3} f(t_{j-l}) \prod_{i=0, i\neq l}^{3} \frac{t - t_{j-i}}{t_{j-l} - t_{j-i}}$$

to approximate $f(t)$ yields

$$\begin{aligned}\frac{1}{\Gamma(1-\alpha)} \sum_{j=3}^{n} \int_{t_{j-1}}^{t_j} (t_n - s)^{-\alpha} f'(s) ds &\approx \frac{1}{\Gamma(1-\alpha)} \sum_{j=3}^{n} \int_{t_{j-1}}^{t_j} (t_n - s)^{-\alpha} p_3'(s) ds \\ &= \frac{\tau^{-\alpha}}{\Gamma(2-\alpha)} \sum_{j=3}^{n} [w_{1,n-j} f(t_j) + w_{2,n-j} f(t_{j-1}) + w_{3,n-j} f(t_{j-2}) + w_{4,n-j} f(t_{j-3})],\end{aligned}$$

where

$$\begin{cases} w_{1,n-j} = \frac{1}{6}[2(n-j+1)^{1-\alpha} - 11(n-j)^{1-\alpha}] - \frac{1}{2-\alpha}[2(n-j)^{2-\alpha} \\ \qquad - (n-j+1)^{2-\alpha}] - \frac{1}{(2-\alpha)(3-\alpha)}[(n-j)^{3-\alpha} - (n-j+1)^{3-\alpha}], \\ w_{2,n-j} = \frac{1}{2}[6(n-j)^{1-\alpha} + (n-j+1)^{1-\alpha}] + \frac{1}{2-\alpha}[5(n-j)^{2-\alpha} \\ \qquad - 2(n-j+1)^{2-\alpha}] + \frac{3}{(2-\alpha)(3-\alpha)}[(n-j)^{3-\alpha} - (n-j+1)^{3-\alpha}], \\ w_{3,n-j} = -\frac{1}{2}[3(n-j)^{1-\alpha} + 2(n-j+1)^{1-\alpha}] - \frac{1}{2-\alpha}[4(n-j)^{2-\alpha} \\ \qquad - (n-j+1)^{2-\alpha}] - \frac{3}{(2-\alpha)(3-\alpha)}[(n-j)^{3-\alpha} - (n-j+1)^{3-\alpha}], \\ w_{4,n-j} = \frac{1}{6}[2(n-j)^{1-\alpha} + (n-j+1)^{1-\alpha}] + \frac{1}{2-\alpha}(n-j)^{2-\alpha} \\ \qquad + \frac{1}{(2-\alpha)(3-\alpha)}[(n-j)^{3-\alpha} - (n-j+1)^{3-\alpha}], \quad 3 \leq j \leq n. \end{cases}$$

Finally, combining the above analysis, we can obtain a new numerical approximation formula for the Caputo derivative with order α ($0 < \alpha < 1$) as follows [2]:

$$
\begin{aligned}
[{}_C D_{0,t}^{\alpha} f(t)]_{t=t_n} &= \frac{1}{\Gamma(1-\alpha)} \int_0^{t_n} (t_n - s)^{-\alpha} f'(s) ds \\
&= \frac{1}{\Gamma(1-\alpha)} \left[\int_{t_0}^{t_1} (t_n - s)^{-\alpha} f'(s) ds + \int_{t_1}^{t_2} (t_n - s)^{-\alpha} f'(s) ds + \sum_{j=3}^{n} \int_{t_{j-1}}^{t_j} (t_n - s)^{-\alpha} f'(s) ds \right] \\
&= \frac{\tau^{-\alpha}}{\Gamma(2-\alpha)} \sum_{j=0}^{n} g_j f_{n-j} + r^n. \qquad (3)
\end{aligned}
$$

Here the coefficients have different values for different n.

For $n = 1$,

$$
\begin{cases}
g_0 = a_0, \\
g_1 = -a_0.
\end{cases}
$$

For $n = 2$,

$$
\begin{cases}
g_0 = a_0 + b_0, \\
g_1 = a_1 - a_0 - 2b_0, \\
g_2 = b_0 - a_1.
\end{cases}
$$

For $n = 3$,

$$
\begin{cases}
g_0 = w_{1,0}, \\
g_1 = w_{2,0} + a_1 + b_1, \\
g_2 = w_{3,0} + a_2 - a_1 - 2b_1, \\
g_3 = w_{4,0} - a_2 + b_1.
\end{cases}
$$

For $n = 4$,

$$
\begin{cases}
g_0 = w_{1,0}, \\
g_1 = w_{1,1} + w_{2,0}, \\
g_2 = w_{2,1} + w_{3,0} + a_2 + b_2, \\
g_3 = w_{3,1} + w_{4,0} + a_3 - a_2 - 2b_2, \\
g_4 = w_{4,1} - a_3 + b_2.
\end{cases}
$$

For $n = 5$,

$$
\begin{cases}
g_0 = w_{1,0}, \\
g_1 = w_{1,1} + w_{2,0}, \\
g_2 = w_{1,2} + w_{2,1} + w_{3,0}, \\
g_3 = w_{2,2} + w_{3,1} + w_{4,0} + a_3 + b_3, \\
g_4 = w_{3,2} + w_{4,1} + a_4 - a_3 - 2b_3, \\
g_5 = w_{4,2} - a_4 + b_3.
\end{cases}
$$

And for $n \geq 6$, the coefficients satisfy the following relations:

$$\begin{cases} g_0 = w_{1,0}, \\ g_1 = w_{1,1} + w_{2,0}, \\ g_2 = w_{1,2} + w_{2,1} + w_{3,0}, \\ g_j = w_{1,j} + w_{2,j-1} + w_{3,j-2} + w_{4,j-3}, \quad 3 \leq j \leq n-3, \\ g_{n-2} = a_{n-2} + b_{n-2} + w_{2,n-3} + w_{3,n-4} + w_{4,n-5}, \\ g_{n-1} = w_{3,n-3} + w_{4,n-4} + a_{n-1} - a_{n-2} - 2b_{n-2}, \\ g_n = w_{4,n-3} - a_{n-1} + b_{n-2}. \end{cases}$$

Theorem 1. *Assume that $f(t) \in C^4[0, t_n]$. For any α $(0 < \alpha < 1)$, the truncation errors $|r^n|$ defined in (3) satisfy*

$$(1) \quad |r^1| \leq c_1 \max_{t_0 \leq t \leq t_1} |f''(t)| \tau^{2-\alpha}, \quad c_1 > 0,\ n = 1,$$

$$(2) \quad |r^2| \leq c_2 \max_{t_0 \leq t \leq t_2} |f'''(t)| \tau^{3-\alpha}, \quad c_2 > 0,\ n = 2,$$

$$(3) \quad |r^n| \leq \frac{1}{\Gamma(1-\alpha)} \left\{ \frac{2\alpha}{3} \max_{t_0 \leq t \leq t_2} |f'''(t)| (t_n - t_2)^{-\alpha-1} \tau^4 + \left[\frac{1}{12} + \frac{3\alpha^2}{2(1-\alpha)(2-\alpha)} \right] \max_{t_0 \leq t \leq t_n} |f^{(4)}(t)| \tau^{4-\alpha} \right\}, \quad n \geq 3.$$

The rth ($4 \leq r$ is a positive integer) degree interpolation formula

At the grid point t_n, the Caputo derivative can be rewritten as

$$_C D_{0,t}^{\alpha} f(t)\big|_{t=t_n} = \frac{1}{\Gamma(1-\alpha)} \sum_{j=1}^{n} \int_{t_{j-1}}^{t_j} (t_n - s)^{-\alpha} f'(s) \mathrm{d}s.$$

Two different cases regarding the subinterval $[t_{j-1}, t_j]$ should be discussed separately.

(i) On the interval $[t_{j-1}, t_j]$, $n \geq j \geq r$, $N \geq n \geq r$.

By using the points $(t_{j-r}, f(t_{j-r}))$, $(t_{j-r+1}, f(t_{j-r+1}))$, $\ldots$, $(t_j, f(t_j))$, the Lagrange interpolation polynomial (on the interval $[t_{j-1}, t_j]$) is given by

$$p_r(t) = \sum_{i=0}^{r} f(t_{j-i}) \prod_{k=0,\ k \neq i}^{r} \frac{t - t_{j-k}}{t_{j-i} - t_{j-k}}, \quad t \in [t_{j-1}, t_j].$$

Denote

$$I_j[f(t)] = \frac{1}{\Gamma(1-\alpha)} \int_{t_{j-1}}^{t_j} (t_n - s)^{-\alpha} f'(s) \mathrm{d}s,$$

$$I_j[p_r(t)] = \frac{1}{\Gamma(1-\alpha)} \int_{t_{j-1}}^{t_j} (t_n - s)^{-\alpha} p_r'(s)\mathrm{d}s.$$

Using $p_r(t)$ to approximate $f(t)$ on the interval $[t_{j-1}, t_j]$, where $j \geq r$, we have

$$\begin{aligned}
\frac{1}{\Gamma(1-\alpha)} \int_{t_{j-1}}^{t_j} (t_n - s)^{-\alpha} f'(s)\mathrm{d}s &\approx I_j[p_r(t)] = \frac{1}{\Gamma(1-\alpha)} \int_{t_{j-1}}^{t_j} (t_n - s)^{-\alpha} p_r'(s)\mathrm{d}s \\
&= \frac{1}{\Gamma(1-\alpha)} \sum_{i=0}^{r} \frac{(-1)^i f(t_{j-i})}{i!(r-i)!\tau^r} \int_{t_{j-1}}^{t_j} (t_n - s)^{-\alpha} \left[\prod_{k=0,\, k\neq i}^{r} (s - t_{j-k}) \right]' \mathrm{d}s \\
&= \frac{1}{\Gamma(1-\alpha)} \left\{ \sum_{i=0}^{r} \frac{(-1)^{i+1} f(t_{j-i})}{i!(r-i)!\tau^r} \left[\sum_{l=1}^{r} \frac{(t_n - s)^{l-\alpha}}{p_l} \left(\prod_{k=0\, k\neq i}^{r} (s - t_{j-k}) \right)^{(l)} \right] \right\} \Bigg|_{t_{j-1}}^{t_j} \\
&= \frac{\tau^{-\alpha}}{\Gamma(1-\alpha)} \sum_{i=0}^{r} \omega_{i,n-j}^{r} f(t_{j-i}).
\end{aligned}$$

For the weight coefficients $\omega_{i,n-j}^r$, we give some more notations as follows:

$$a_k^s = s^{k-\alpha}, \quad b_k^s = (s+1)^{k-\alpha}, \quad p_k = \prod_{l=1}^{k} (l - \alpha), \quad k \geq 1.$$

Let

$$\alpha_{j,i}^k = \begin{cases} \phi_{j,i}^k & k \neq 0, \\ 1 & k = 0, \end{cases} \qquad \beta_{j,i}^k = \begin{cases} \psi_{j,i}^k & k \neq 0, \\ 1 & k = 0. \end{cases}$$

Here $\phi_{j,i}^k$ is the sum of products of all the different combinations of k elements in set $A_{j,i} = \{\bar{a} \mid \bar{a} \in [0, j-1], \bar{a} \neq i, \bar{a} \in \mathbb{Z}\}$; $\psi_{j,i}^k$ is the sum of products of all the different combinations of k elements in set $B_{j,i} = \{\bar{b} \mid \bar{b} \in [-1, j-2], \bar{b} \neq i-1, \bar{b} \in \mathbb{Z}\}$. Then, $\omega_{i,n-j}^r$ can be expressed as

$$\omega_{i,n-j}^r = \frac{(-1)^{i+1}}{i!(r-i)!} \sum_{l=1}^{r} \left[\frac{l!}{p_l} \left(\alpha_{r+1,i}^{r-l} a_l^{n-j} - \beta_{r+1,i}^{r-l} b_l^{n-j} \right) \right], \quad 0 \leq i \leq r-1.$$

(ii) On the interval $[t_{j-1}, t_j]$, $0 < j < r$, $1 \leq n \leq N$.

Because $j < r$, there are not enough points to construct an rth-degree polynomial. Similar to case (i), we use $I_j[p_j(t)]$ to approximate $I_j[f(t)]$. In particular, for $j = 1$ we use $I_1[p_1(t)]$ on the interval $[t_0, t_1]$, where the interpolation polynomial is a linear function.

Hence, combining cases (i) and (ii) yields the numerical approximation for the Caputo fractional derivatives of the following form [32]:

$$
\begin{aligned}
[{}_{C}D^{\alpha}_{0,t}f(t)]_{t=t_n} &= \frac{1}{\Gamma(1-\alpha)}\sum_{j=1}^{n}\int_{t_{j-1}}^{t_j}(t_n-t)^{-\alpha}f'(t)\mathrm{d}t \\
&\approx \sum_{j=1}^{r-1} I_j[p_j(t)] + \sum_{j=r}^{n} I_j[p_r(t)], \quad N \ge n \ge r.
\end{aligned}
$$

There are some other numerical schemes for Caputo derivatives [26, 36] available.

2.2 The Riemann–Liouville derivative

2.2.1 Based on the Caputo derivative approximation methods

Since the Riemann–Liouville derivative and the Caputo derivative have the relations [31, 45, 47]

$$
{}_{\mathrm{RL}}D^{\alpha}_{0,t}u(t) = {}_{C}D^{\alpha}_{0,t}u(t) + \frac{u(0)}{\Gamma(1-\alpha)}t^{-\alpha}, \quad 0 < \alpha < 1, .
$$

and

$$
{}_{\mathrm{RL}}D^{\alpha}_{0,t}u(t) = {}_{C}D^{\alpha}_{0,t}u(t) + \frac{u(0)}{\Gamma(1-\alpha)}t^{-\alpha} + \frac{u'(0)}{\Gamma(2-\alpha)}t^{1-\alpha}, \quad 1 < \alpha < 2,
$$

almost all the numerical methods for the Caputo derivative can be theoretically extended to the Riemann–Liouville case if $u(t)$ satisfies suitable smooth conditions. Here, we first list some algorithms that are often used in the simulation of the Riemann–Liouville derivative in fractional partial equations based on the above relations.

The L1 formula

We have

$$
[{}_{\mathrm{RL}}D^{\alpha}_{0,t}u(t)]_{t=t_n} = \frac{u(0)t_n^{-\alpha}}{\Gamma(1-\alpha)} + \sum_{k=0}^{n-1} b_{n-k-1}[u(t_{k+1}) - u(t_k)] + \mathcal{O}(\tau^{2-\alpha}),
$$

where

$$
b_k = \frac{\Delta t^{-\alpha}}{\Gamma(2-\alpha)}[(k+1)^{1-\alpha} - k^{1-\alpha}], \quad 0 < \alpha < 1.
$$

The L1C formula

We have

$$
\begin{aligned}
[{}_{\mathrm{RL}}D^{\alpha}_{0,t}u(t)]_{t=t_n} &= \frac{u(0)t_n^{-\alpha}}{\Gamma(1-\alpha)} + \frac{\tau^{-\alpha}}{\Gamma(3-\alpha)}\sum_{i-0}^{k-1}[w_{1,k-i}(f(t_{i+1}) - f(t_{i-1})) \\
&\quad + w_{2,k-i}(f(t_{i+1}) - 2f(t_i) + f(t_{i-1}))] + \mathcal{O}(\tau^{3-\alpha}),
\end{aligned}
$$

where $0 < \alpha < 1$ and the coefficients $w_{1,k-i}$ and $w_{2,k-i}$ are the same as those in equation (1).

The L2 formula

We have

$$\begin{aligned}[{}_{\mathrm{RL}}D_{0,t}^{\alpha}u(t)]_{t=t_n} &= \frac{u(0)t_n^{-\alpha}}{\Gamma(1-\alpha)} + \frac{u'(0)t_n^{1-\alpha}}{\Gamma(2-\alpha)} \\ &\quad + \frac{\Delta t^{-\alpha}}{\Gamma(3-\alpha)}\sum_{k=0}^{n-1} b_k(u_{n-k-1} - 2u_{n-k} + u_{n-k+1}) + \mathcal{O}(\tau^{3-\alpha}),\end{aligned}$$

where

$$b_k = (k+1)^{2-\alpha} - k^{2-\alpha}, \quad 1 < \alpha < 2.$$

The L2C formula

We have

$$\begin{aligned}[{}_{\mathrm{RL}}D_{0,t}^{\alpha}u(t)]_{t=t_n} &= \frac{u(0)t_n^{-\alpha}}{\Gamma(1-\alpha)} + \frac{u'(0)t_n^{1-\alpha}}{\Gamma(2-\alpha)} \\ &\quad + \frac{\Delta t^{-\alpha}}{\Gamma(3-\alpha)}\sum_{k=0}^{n-1} b_k(u_{n-k-2} - u_{n-k-1} + u_{n-k+1} - u_{n-k}) \\ &\quad + \mathcal{O}(\tau^{3-\alpha}),\end{aligned}$$

where

$$b_k = (k+1)^{2-\alpha} - k^{2-\alpha}, \quad 1 < \alpha < 2.$$

The centered difference and piece-wise linear interpolation formula

Observing that

$${}_{\mathrm{RL}}D_{0,t}^{\alpha}u(t) = D^2\,{}_{\mathrm{RL}}D_{0,t}^{-(2-\alpha)}u(t) \quad \text{for } \alpha \in (1,2),$$

we have

$$\begin{aligned}[{}_{\mathrm{RL}}D_{0,t}^{\alpha}u(t)]|_{t=t_n} &= \frac{1}{\tau^2}({}_{\mathrm{RL}}D_{0,t}^{-(2-\alpha)}u(t_{n+1}) - 2\,{}_{\mathrm{RL}}D_{0,t}^{-(2-\alpha)}u(t_n) \\ &\quad + {}_{\mathrm{RL}}D_{0,t}^{-(2-\alpha)}u(t_{n-1})) + \mathcal{O}(\tau^2).\end{aligned} \tag{4}$$

For the numerical approximation of ${}_{\mathrm{RL}}D_{0,t}^{-(2-\alpha)}u(t)$ at $t = t_n$, we denote the corresponding integral function $u(s)$ as $u_n(s)$. Suppose that $u_n(s) = \sum_{k=1}^{n} u_k\varphi_{k,n}(s)$, where $\varphi_{k,n}(s)$ is a piece-wise linear function defined by

$$\varphi_{k,n}(s) = \begin{cases} \frac{s-t_{k-1}}{\tau}, & s \in (t_{k-1}, t_k), \\ \frac{t_{k+1}-s}{\tau}, & s \in (t_k, t_{k+1}), \\ 0, & \text{otherwise}, \end{cases}$$

with $k \le n-1$ and $\varphi_{n,n}(s) = \frac{s-t_{n-1}}{\tau}$, $s \in (t_{n-1}, t_n)$. Then, we have

$$\left[{}_{\mathrm{RL}}D_{0,t}^{-(2-\alpha)}u(t)\right]\big|_{t=t_n} = \frac{\tau^{2-\alpha}}{\Gamma(4-\alpha)} \sum_{k=0}^{n} w_{k,n}u_k + \mathcal{O}(\tau^2), \tag{5}$$

where $w_{k,n} = (n-k+1)^{3-\alpha} - 2(n-k)^{3-\alpha} + (n-k-1)^{3-\alpha}$ for $k \le n$ and $w_{n,n} = 1$.

Thus, combining (4) with (5), we obtain [49]

$$\begin{aligned}\left[{}_{\mathrm{RL}}D_{0,t}^{\alpha}u(t)\right]_{t=t_n} = {}& \frac{\tau^{-\alpha}}{\Gamma(4-\alpha)}\left[\sum_{k=0}^{n+1} w_{k,n+1}u_k - 2\sum_{k=0}^{n} w_{k,n}u_k + \sum_{k=0}^{n-1} w_{k,n-1}u_k\right] \\ & + \mathcal{O}(\tau^2).\end{aligned}$$

2.2.2 Based on the generating function methods

The Lubich high-order formula

The Riemann–Liouville derivative ${}_{\mathrm{RL}}D_{0,t}^{\alpha}u(t)$ can be approximated by the Grünwald–Letnikov derivative under suitable regularity assumptions for $u(t)$. The basic starting point of the standard Grünwald–Letnikov formula is the following [45]:

$$\left[{}_{\mathrm{RL}}D_{0,t}^{\alpha}u(t)\right]_{t=t_n} = \frac{1}{\tau^{\alpha}} \sum_{j=0}^{n} \varpi_{1,j}^{(\alpha)} u(t_{n-j}) + \mathcal{O}(\tau), \tag{6}$$

where

$$\varpi_{1,\ell}^{(\alpha)} = (-1)^{\ell}\binom{\alpha}{\ell} = (-1)^{\ell}\frac{\Gamma(\alpha+1)}{\Gamma(\ell+1)\Gamma(\alpha-\ell+1)}, \quad \ell = 0,1,\dots,m,$$

are the binomial coefficients. They have the following recurrence relationships:

$$\varpi_{1,0}^{(\alpha)} = 1, \quad \varpi_{1,\ell}^{(\alpha)} = \left(1 - \frac{1+\alpha}{\ell}\right)\varpi_{1,\ell-1}^{(\alpha)}, \quad \ell = 1,2,\dots,m.$$

Obviously, $\varpi_{1,\ell}^{(\alpha)}$ $(\ell = 0,1,\dots,m)$ are just the first $m+1$ coefficients of the Taylor series of the expansion of the following function:

$$W_1^{(\alpha)}(z) = (1-z)^{\alpha} = \sum_{\ell=0}^{\infty}(-1)^{\ell}\binom{\alpha}{\ell}z^{\ell} = \sum_{\ell=0}^{\infty}\varpi_{1,\ell}^{(\alpha)}z^{\ell}.$$

Equation (6) is not suitable for the discretization of fractional partial differential equations when $\alpha \in (1,2)$ since it leads to an unstable numerical scheme, even for traditionally stable methods, such as implicit and Crank–Nicolson methods. The following shifted Grünwald–Letnikov formula can remedy this issue properly [38]:

$$\left[{}_{\mathrm{RL}}D_{0,t}^{\alpha}u(t)\right]_{t=t_n} = \frac{1}{\tau^{\alpha}} \sum_{j=0}^{n+p} \omega_{1,j}^{(\alpha)} u(t_{n-j+p}) + O(\tau). \tag{7}$$

Here, p is the shift and $p \in \mathbb{N}$. It is shown that the best performance of the above approximation comes from minimizing $|p - \alpha/2|$. If $\alpha = 2$, (7) coincides with the centered difference of the second-order derivative. If the shift is not chosen as integers, the numerical approximation (7) may have some superconvergent behaviors; see [19, 42, 64].

We can see that equation (6) has only first-order accuracy if $f(0+) = 0$. Therefore, to seek high-order numerical methods for the fractional derivatives is of great importance. In [35], Lubich firstly proposed numerical schemes of order two, three, four, five, and six, which are called fractional linear multi-step formulas. Here, it must be mentioned that the fractional linear multi-step method is different from the usual linear multi-step method. The former consists of varied steps. That is to say, the value of the mth step x_m depends on the preceding step values $x_0, x_1, \ldots, x_{m-1}$, which means the multi-step number is increasing, while the latter is of a fixed multi-step number. Actually, if the function $u(x)$ is sufficiently smooth and has many vanishing derivatives at 0, then the Lubich formula with order p ($p = 2 \sim 6$) has the following form:

$$\left[{}_{\mathrm{RL}}D_{0,t}^{\alpha} u(t)\right]_{t=t_n} = \frac{1}{\tau^\alpha} \sum_{j=0}^{n} \varpi_{p,j}^{(\alpha)} u(t_{n-j}) + \mathcal{O}(\tau^p), \quad p = 2 \sim 6,$$

where the weights $\{\varpi_{p,j}^{(\alpha)}\}$ can be obtained by the following generating functions:

$$W_p^{(\alpha)}(z) = \sum_{\ell=0}^{\infty} \varpi_{p,\ell}^{(\alpha)} z^\ell, \quad p = 2 \sim 6,$$

where

$$\begin{aligned}
W_2^{(\alpha)}(z) &= \left(\frac{3}{2} - 2z + \frac{1}{2}z^2\right)^\alpha, \\
W_3^{(\alpha)}(z) &= \left(\frac{11}{6} - 3z + \frac{3}{2}z^2 - \frac{1}{3}z^3\right)^\alpha, \\
W_4^{(\alpha)}(z) &= \left(\frac{25}{12} - 4z + 3z^2 - \frac{4}{3}z^3 + \frac{1}{4}z^4\right)^\alpha, \\
W_5^{(\alpha)}(z) &= \left(\frac{137}{60} - 5z + 5z^2 - \frac{10}{3}z^3 + \frac{5}{4}z^4 - \frac{1}{5}z^5\right)^\alpha, \\
W_6^{(\alpha)}(z) &= \left(\frac{147}{60} - 6z + \frac{15}{2}z^2 - \frac{20}{3}z^3 + \frac{15}{4}z^4 - \frac{6}{5}z^5 + \frac{1}{6}z^6\right)^\alpha.
\end{aligned}$$

From the above introduction, the key question is how to compute the coefficients $\varpi_{p,\ell}^{(\alpha)}$. As far as we know, there are three methods to compute these coefficients. One way is to use the fast Fourier transform method [45] and obtain

$$\varpi_{p,\ell}^{(\alpha)} = \frac{1}{2\pi} \int_0^{2\pi} W_p^{(\alpha)}(e^{i\theta}) e^{-i\ell\theta} \mathrm{d}\theta, \quad p = 2, 3, 4, 5, 6.$$

Obviously, it is not an easy task to obtain an explicit analytical expression of the coefficients $\varpi_{p,\ell}^{(\alpha)}$ according to the above formula if $\alpha \notin \mathbb{Z}$.

Another way of computing the coefficients $\varpi_{p,\ell}^{(\alpha)}$ is by the automatic differentiation technique [46], i. e.,

$$\varpi_{p,\ell}^{(\alpha)} = \frac{1}{\ell u_0} \sum_{j=0}^{\ell-1} [\alpha(\ell - j) - j] \varpi_{p,j}^{(\alpha)} u_{\ell-j}, \quad p = 2, \ldots, 6.$$

Here the values u_k $(k = 0, 1, \ldots)$ denote the Taylor expansion coefficients of the generating functions of the classical linear multi-step methods.

The third method is to use the expansion of series. These coefficients are not given in the form of recurrence relationships but explicit expressions [11, 27]. Such analytical expressions are useful for theoretical analysis, such as stability and convergence analysis.

The weighted and shifted Grünwald–Letnikov formula

The second-order numerical approximation on the finite interval $[t_0, T]$ can be developed through assembling the Grünwald–Letnikov formula with different weights and shifts [52]. Let $t_0 = -\infty$ and denote the first-order shifted Grünwald–Letnikov difference operator as

$$A_{1,p}^{\alpha} u(t) = \frac{1}{\tau^{\alpha}} \sum_{j=0}^{\infty} \varpi_{1,j}^{(\alpha)} u(t - (j - p)\tau).$$

Assume the second-order weighted and shifted Grünwald–Letnikov difference (WSGD) operator has the form

$$B_{1,p,q}^{\alpha} u(t) = a A_{1,p}^{\alpha} u(t) + b A_{1,q}^{\alpha} u(t).$$

Based on the Fourier transform method, we know that the difference operator $B_{1,p,q}^{\alpha}$ can reach second-order convergence for the Riemann–Liouville derivative by choosing $a + b = 1$ and $a(p - \alpha/2) + b(q - \alpha/2) = 0$. For the well-defined function $u(t)$ on the bounded interval $[t_0, T]$ with $u(t_0) = 0$, using the WSGD operator mentioned above, one has

$$\left[{}_{\mathrm{RL}}D_{t_0,t}^{\alpha} u(t)\right]_{t=t_n} = \widetilde{B}_{1,p,q}^{\alpha} u(t) + O(\tau^2),$$

where

$$\widetilde{B}_{1,p,q}^{\alpha} u(t) = a \widetilde{A}_{1,p}^{\alpha} u(t) + b \widetilde{A}_{1,q}^{\alpha} u(t),$$

$$\widetilde{A}_{1,p}^{\alpha} u(t) = \frac{1}{\tau^{\alpha}} \sum_{k=0}^{[\frac{t_n - t_0}{\tau}]+p} w_{1,k}^{(\alpha)} u(t_n - (k - p)\tau),$$

$$\widetilde{A}_{1,q}^{\alpha} u(t) = \frac{1}{\tau^{\alpha}} \sum_{k=0}^{[\frac{t_n - t_0}{\tau}]+q} w_{1,k}^{(\alpha)} u(t_n - (k - q)\tau).$$

It has been shown that the above numerical approximation leads to an unstable numerical scheme for fractional partial differential equations when the shift pair $(p,q) = (0,-1)$. When $(p,q) = (1,0)$ or $(1,-1)$, the corresponding numerical scheme is stable.

Similar to the idea of the WSGD operators, Tian et al. further proposed a discrete operator (called the 3-WSGD operator) for approximating the Riemann–Liouville fractional derivative with third-order accuracy [52]. However, the authors pointed out that it failed to obtain the stable numerical scheme for all $\alpha \in (1,2)$. Consequently, based on the WSGD operators and the Taylor expansions of the shifted Grünwald finite difference formula, Zhou et al. derived a novel third-order numerical approximation for the well-defined function $u(t)$ on the bounded interval $[t_0, T]$ with $u(t_0) = 0$ [66], i. e.,

$$\widetilde{B}^{\alpha}_{1,p,q}u(t) = \mathcal{C}({}_{\mathrm{RL}}D^{\alpha}_{t_0,t}u(t)) + O(\tau^3).$$

Here the finite difference operator $\mathcal{C}$ (also called CWSGD operator) is defined by

$$\mathcal{C} = 1 + c^{\alpha}_{p,q,2}\tau^2\delta_t^2,$$

in which $c^{\alpha}_{p,q,2} = \frac{\alpha-2q}{2(p-q)}a^{\alpha}_{p,2} + \frac{2p-\alpha}{2(p-q)}a^{\alpha}_{q,2}$, δ_t^2 is the centered difference operator, and $a^{\alpha}_{p,k}$ are the coefficients of the power series expansion of function

$$w_{\alpha,p}(z) = \left(\frac{1-e^{-z}}{z}\right)^{\alpha} e^{pz} = \sum_{k=0}^{\infty} a^{\alpha}_{p,k}z^k.$$

The appropriate choices of the pair (p,q) are $(1,0)$ and $(1,-1)$ since taking $(p,q) = (0,-1)$ will lead to an unstable numerical scheme for time-dependent problems.

Inspired by the shifted Grünwald–Letnikov operator and the Taylor expansion, Yu et al. [57] and Hao et al. [22] independently developed the following fourth-order quasi-compact scheme for the Riemann–Liouville derivative:

$$\mathcal{P}_x({}_{\mathrm{RL}}D^{\alpha}_{t_0,t}u(t)) = \mu_1\widetilde{A}^{\alpha}_{\tau,1}u(t) + \mu_0\widetilde{A}^{\alpha}_{\tau,0}u(t) + \mu_{-1}\widetilde{A}^{\alpha}_{\tau,-1}u(t) + O(\tau^4),$$

where $\mathcal{P}_x = 1 + b_2^{\alpha}\tau^2\delta_t^2$ is also called the CWSGD operator. The coefficients b_2^{α}, μ_1, μ_0, and μ_{-1} are the functions of α and

$$\begin{cases} \mu_1 = (\alpha+1)(\alpha+2)/12, \\ \mu_0 = -(\alpha-2)(\alpha+2)/12, \\ \mu_{-1} = (\alpha-2)(\alpha-1)/12, \\ b_2^{\alpha} = (4+\alpha-\alpha^2)/24. \end{cases}$$

The weighted and shifted Lubich second-order formula

It is known that the Grünwald–Letnikov formula coincides with the first-order Lubich method with first-order convergence. Hence, it is possible to derive other efficient numerical approximations for Riemann–Liouville derivatives with higher accuracy based on the combination of the Lubich formula with different weights and shifts by using other generating functions. Chen and Deng developed a class of high-accuracy numerical schemes with second-, third-, and fourth-order accuracy by adopting the generating function $W_2^{(\alpha)}(z) = (3/2 - 2z + z^2/2)^\alpha$ in this spirit [7]. Since this idea is similar to the case discussed by Tian et al., we omit the details and only list the corresponding approximations.

Denote

$$A_{2,p}^{\alpha}u(t) = \frac{1}{\tau^\alpha}\sum_{j=0}^{\infty}\omega_{2,j}^{(\alpha)}u(t-(j-p)\tau),$$
$$B_{2,p,q}^{\alpha}u(t) = w_p A_{2,p}^{\alpha}u(t) + w_q A_{2,q}^{\alpha}u(t).$$

Then the second-order approximation for the left Riemann–Liouville derivative is

$${}_{\mathrm{RL}}D_{-\infty,t}^{\alpha}u(t) = B_{2,p,q}^{\alpha}u(t) + \mathcal{O}(\tau^2),$$

where

$$w_p = \frac{q}{q-p}, \quad w_q = \frac{p}{p-q},$$

$p \neq q$, and p, q are integers.

Similarly, denote

$$B_{2,p,q,r,s}^{\alpha}u(t) = w_{p,q}B_{2,p,q}^{\alpha}u(t) + w_{r,s}B_{2,r,s}^{\alpha}u(t).$$

Then the third-order approximation for the left Riemann–Liouville derivative is

$${}_{\mathrm{RL}}D_{-\infty,t}^{\alpha}u(t) = B_{2,p,q,r,s}^{\alpha}u(t) + O(\tau^3),$$

where

$$w_{p,q} = \frac{3rs+2\alpha}{3(rs-pq)}, \quad w_{r,s} = \frac{3pq+2\alpha}{3(pq-rs)},$$

$pq \neq rs$, and p, q, r, s are integers.

Finally, we also define the operator

$$B_{2,p,q,r,s,\bar{p},\bar{q},\bar{r},\bar{s}}^{\alpha}u(t) = w_{p,q,r,s}B_{2,p,q,r,s}^{\alpha}u(t) + w_{\bar{p},\bar{q},\bar{r},\bar{s}}B_{2,\bar{p},\bar{q},\bar{r},\bar{s}}^{\alpha}u(t).$$

Then the fourth-order approximation for the left Riemann–Liouville derivative is

$${}_{\mathrm{RL}}D_{-\infty,t}^{\alpha}u(t) = B_{2,p,q,r,s,\bar{p},\bar{q},\bar{r},\bar{s}}^{\alpha}u(t) + O(\tau^4),$$

where

$$w_{p,q,r,s} = \frac{a_{p,q,r,s}\bar{b}_{\bar{p},\bar{q},\bar{r},\bar{s}}}{a_{p,q,r,s}\bar{b}_{\bar{p},\bar{q},\bar{r},\bar{s}} - \bar{a}_{\bar{p},\bar{q},\bar{r},\bar{s}}b_{p,q,r,s}}, \quad w_{\bar{p},\bar{q},\bar{r},\bar{s}} = \frac{\bar{a}_{\bar{p},\bar{q},\bar{r},\bar{s}}b_{p,q,r,s}}{\bar{a}_{\bar{p},\bar{q},\bar{r},\bar{s}}b_{p,q,r,s} - a_{p,q,r,s}\bar{b}_{\bar{p},\bar{q},\bar{r},\bar{s}}},$$

with

$$\begin{aligned} a_{p,q,r,s} &= rs - pq, \\ b_{p,q,r,s} &= 6pqrs(r+s-p-q) + 4\alpha[rs(r+s) - pq(p+q)] + 9\alpha(rs-pq), \\ \overline{a}_{\overline{p},\overline{q},\overline{r},\overline{s}} &= \overline{rs} - \overline{pq}, \\ b_{\overline{p},\overline{q},\overline{r},\overline{s}} &= 6\overline{pqrs}(\overline{r}+\overline{s}-\overline{p}-\overline{q}) + 4\alpha[\overline{rs}(\overline{r}+\overline{s}) - \overline{pq}(\overline{p}+\overline{q})] + 9\alpha(\overline{rs}-\overline{pq}), \end{aligned}$$

and $a_{p,q,r,s}\overline{b}_{\overline{p},\overline{q},\overline{r},\overline{s}} \neq \overline{a}_{\overline{p},\overline{q},\overline{r},\overline{s}} b_{p,q,r,s}$; $p, q, r, s, \overline{p}, \overline{q}, \overline{r}, \overline{s}$ are integers.

The $\frac{\alpha}{2}$-step shifted Grünwald–Letnikov formula

In [53], Tuan and Gorenflo introduced the following left fractional central difference operator; we also name it the $\frac{\alpha}{2}$-step shifted Grünwald–Letnikov formula:

$$_{C}\Delta^{\alpha}_{-\tau}u(x) = \sum_{k=0}^{\infty} \varpi^{(\alpha)}_{1,k} u\left(t - \left(k - \frac{\alpha}{2}\right)\tau\right). \tag{8}$$

It can be shown that

$$_{\mathrm{RL}}D^{\gamma}_{0,t}u(t) = \frac{1}{\tau^{\gamma}} \sum_{\ell=0}^{\infty} \varpi^{(\alpha)}_{1,\ell} u\left(t - \left(\ell - \frac{\gamma}{2}\right)\tau\right) + \mathcal{O}(\tau^2),$$

where $\varpi^{(\gamma)}_{1,\ell} = (-1)^{\ell}\binom{\gamma}{\ell}$.

Similarly, we define the following right fractional central difference operator:

$$_{C}\Delta^{\alpha}_{+\tau}u(t) = \sum_{k=0}^{\infty} \varpi^{(\alpha)}_{1,k} u\left(t + \left(k - \frac{\alpha}{2}\right)\tau\right). \tag{9}$$

Analogous to the integer-order finite difference formula, we define the following fractional average operator:

$$\mu^{\alpha}_{\pm\tau}u(t - s\tau) = \frac{u(t \pm (s - \frac{\alpha}{2})\tau) + u(t \pm (s + \frac{\alpha}{2})\tau)}{2}. \tag{10}$$

Then we can get the following fractional left and right average central difference operators, respectively, based on (8), (9), and (10):

$$\begin{aligned} {}_{AC}\Delta^{\alpha}_{-\tau}u(t) &= \mu^{\alpha}_{-\tau}({}_{C}\Delta^{\alpha}_{-\tau}u(t)) = \sum_{j=0}^{\infty}(-1)^j\binom{\alpha}{j}\mu^{\alpha}_{-\tau}\left(u\left(t - \left(j - \frac{\alpha}{2}\right)\tau\right)\right) \\ &= \frac{1}{2}\sum_{j=0}^{\infty}(-1)^j\binom{\alpha}{j}(u(t - j\tau) + u(t - (j-\alpha)\tau)) \end{aligned} \tag{11}$$

and

$$\begin{aligned} {}_{AC}\Delta^{\alpha}_{+\tau}u(t) &= \mu^{\alpha}_{+\tau}({}_{C}\Delta^{\alpha}_{+\tau}u(t)) = \sum_{j=0}^{\infty}(-1)^j\binom{\alpha}{j}\mu^{\alpha}_{+\tau}\left(u\left(t + \left(j - \frac{\alpha}{2}\right)\tau\right)\right) \\ &= \frac{1}{2}\sum_{j=0}^{\infty}(-1)^j\binom{\alpha}{j}(u(t + j\tau) + u(t + (j-\alpha)\tau)). \end{aligned} \tag{12}$$

Here, we always assume that $\mu^{\alpha}_{\pm\tau}$ can commute with the infinite summation. For the fractional left and right average central difference operators defined in (11) and (12), we have the following result.

Theorem 2. *Let $u(x)$ and the Fourier transforms of ${}_{\mathrm{RL}}D^{\alpha+2}_{-\infty,t}u(t)$ and ${}_{\mathrm{RL}}D^{\alpha+2}_{t,+\infty}u(t)$ be in $L_1(\mathbb{R})$. Then*

$$ {}_{\mathrm{RL}}D^{\alpha}_{-\infty,t}u(t) = \frac{{}_{\mathrm{AC}}\Delta^{\alpha}_{-\tau}u(t)}{\tau^{\alpha}} + \mathcal{O}(\tau^2) \tag{13} $$

and

$$ {}_{\mathrm{RL}}D^{\alpha}_{t,+\infty}u(t) = \frac{{}_{\mathrm{AC}}\Delta^{\alpha}_{+\tau}u(t)}{\tau^{\alpha}} + \mathcal{O}(\tau^2) $$

uniformly hold for $t \in \mathbb{R}$.

Next, we develop a second-order numerical scheme for the Riemann–Liouville derivative at non-grid points $t_{k+\frac{1}{2}}$ based on the above fractional central difference operator. Obviously, we obtain the following form at point $t_{k+\frac{1}{2}}$ in view of equation (13):

$$ {}_{\mathrm{RL}}D^{\gamma}_{0,t}u(t_{k+\frac{1}{2}}) = \frac{1}{\tau^{\gamma}}\sum_{\ell=0}^{\infty}\varpi^{(\gamma)}_{1,\ell}u\left(t_k - \left(\ell - \frac{\gamma+1}{2}\right)\tau\right) + \mathcal{O}(\tau^2). \tag{14} $$

Let $t_s = t_k - (\ell - \frac{\gamma+1}{2})\tau$, $t_p = t_k - (\ell - 1)\tau$, and $t_q = t_k - \ell\tau$. The formula

$$ u(t_s) = \frac{(t_s - t_q)u(t_p) + (t_p - t_s)u(t_q)}{t_p - t_q} + \mathcal{O}(|(t_p - t_s)(t_q - t_s)|) $$

gives the following second-order numerical formula, based on (14) and [12]:

$$ \begin{aligned} {}_{\mathrm{RL}}D^{\gamma}_{0,t}u(t_{k+\frac{1}{2}}) = \frac{1}{2\tau^{\gamma}}\sum_{\ell=0}^{\infty}\varpi^{(\gamma)}_{1,\ell}((1+\gamma)u(t_k - (\ell-1)\tau) \\ + (1-\gamma)u(t_k - \ell\tau)) + \mathcal{O}(\tau^2). \end{aligned} $$

Finally, we construct two classes of fourth-order difference schemes for the left and right Riemann–Liouville derivatives based on the above difference operators through the following theorem [16, 63].

Theorem 3. *Let $u(t)$ and the Fourier transforms of ${}_{\mathrm{RL}}D^{\alpha+4}_{-\infty,t}u(t)$ and ${}_{\mathrm{RL}}D^{\alpha+4}_{t,+\infty}u(t)$ be in $L_1(\mathbb{R})$. Then*

$$ {}_{\mathrm{RL}}D^{\alpha}_{-\infty,t}u(t) = \frac{1}{\tau^{\alpha}}\left(1 + \frac{\alpha}{24}\delta_t^2\right)^{-1} {}_{C}\Delta^{\alpha}_{-\tau}u(t) + \mathcal{O}(\tau^4), $$

$$ {}_{\mathrm{RL}}D^{\alpha}_{-\infty,t}u(t) = \frac{1}{\tau^{\alpha}}\left(1 + \frac{\alpha(3\alpha+1)}{24}\delta_t^2\right)^{-1} {}_{\mathrm{AC}}\Delta^{\alpha}_{-\tau}u(t) + O(\tau^4), $$

$$ {}_{\mathrm{RL}}D^{\alpha}_{t,+\infty}u(xt) = \frac{1}{\tau^{\alpha}}\left(1 + \frac{\alpha}{24}\delta_t^2\right)^{-1} {}_{C}\Delta^{\alpha}_{+\tau}u(t) + \mathcal{O}(\tau^4), $$

and

$$ {}_{\mathrm{RL}}D^{\alpha}_{t,+\infty}u(t) = \frac{1}{\tau^{\alpha}}\left(1 + \frac{\alpha(3\alpha+1)}{24}\delta_t^2\right)^{-1} {}_{\mathrm{AC}}\Delta^{\alpha}_{+\tau}u(t) + \mathcal{O}(\tau^4) $$

uniformly hold for $t \in \mathbb{R}$.

The second-order mid point approximation (MP$_2$) formula

For a certain type of time-fractional-order differential equations, such as the modified anomalous subdiffusion equation

$$ \frac{\partial u(x,t)}{\partial t} = (A_{0\mathrm{RL}}D^{1-\alpha_1}_{0,t} + B_{0\mathrm{RL}}D^{1-\alpha_2}_{0,t})\frac{\partial^2 u(x,t)}{\partial x^2}, \quad \alpha_1,\alpha_2 \in (0,1), $$

and the Cable equation

$$ \frac{\partial u(x,t)}{\partial t} = K_{1\mathrm{RL}}D^{1-\alpha_1}_{0,t}\frac{\partial^2 u(x,t)}{\partial x^2} - K_{2\mathrm{RL}}D^{1-\alpha_2}_{0,t}u(x,t), \quad \alpha_1,\alpha_2 \in (0,1), $$

with time derivatives of both integer and fractional order, we should consider the consistency of the convergence order of the two derivatives at the same time. If we use the first-order forward (or backward) difference formula for $\frac{\partial u(x,t)}{\partial t}$ and the Grünwald–Letnikov formula for the Riemann–Liouville derivative, the convergence order is only one in temporal direction [34, 41]. In order to improve the convergence order, we can choose the second-order backward difference formula for $\frac{\partial u(x,t)}{\partial t}$ and the WSGD formula for the Riemann–Liouville derivative. Then the convergence can reach the second order. However, in the obtained difference scheme, we need the additional starting values u_j^1 ($j = 1,2,\ldots,M-1$) besides the given initial values u_j^0 ($j = 1,2,\ldots,M-1$), so one must find other ways to calculate it, which is very inconvenient [33]. In order to overcome these defects, we consider establishing the finite difference schemes at points $(x, t_{k+1/2})$ ($k = 0,1,\ldots,N-1$). In this way, we can use the following central difference formula to approximate the first-order derivative:

$$ \left.\frac{\partial u(x,t)}{\partial t}\right|_{t=t_{k+\frac{1}{2}}} = \frac{u(x,t_{k+1}) - u(x,t_k)}{\tau} + \mathcal{O}(\tau^2), \quad k = 0,1,\ldots,N-1. $$

Therefore, the remaining task is to establish the second-order numerical approximation formula for the time Riemann–Liouville derivative at points $(x, t_{k+1/2})$ ($k = 0,1,\ldots,N-1$).

Now, we start to develop the second-order numerical approximation formula [18].

Theorem 4. *Denote*

$$ \mathscr{C}^{n+\alpha}(\mathbb{R}) = \left\{u \,\middle|\, u \in L_1(\mathbb{R}), \int_{\mathbb{R}} (1+|\omega|)^{n+\alpha}|\hat{u}(\omega)|\mathrm{d}\omega < \infty\right\}. $$

Define the operator

$$\delta^{\alpha}_{-\infty,t}u(t) = \tau^{-\alpha}\sum_{\ell=0}^{\infty}\varpi_{\ell}^{(\alpha)}u(t-\ell\tau),$$

where τ *is the temporal step size. If* $u \in \mathscr{C}^{2+\alpha}(\mathbb{R})$, *then one has*

$$_{\mathrm{RL}}D^{\alpha}_{-\infty,t}u\left(t+\frac{1}{2}\tau\right) = \delta^{\alpha}_{-\infty,t}u(t) + \mathcal{O}(\tau^2)$$

as $\tau \to 0$.

Here $\varpi_{\ell}^{(\alpha)}$ $(\ell = 0, 1, \ldots)$ *are the expansion coefficients of* $G(z)$, *that is,*

$$G(z) = \sum_{\ell=0}^{\infty}\varpi_{\ell}^{(\alpha)}z^{\ell}, \quad |z| < 1, \tag{15}$$

where

$$G(z) = \left(\frac{3\alpha+1}{2\alpha} - \frac{2\alpha+1}{\alpha}z + \frac{\alpha+1}{2\alpha}z^2\right)^{\alpha}.$$

The coefficients $\varpi_{\ell}^{(\alpha)}$ $(\ell = 0, 1, \ldots)$ in equation (15) can be expressed as follows:

$$\varpi_{\ell}^{(\alpha)} = \left(\frac{3\alpha+1}{2\alpha}\right)^{\alpha}\sum_{m=0}^{\ell}\left(\frac{\alpha+1}{3\alpha+1}\right)^{m}g_m^{(\alpha)}g_{\ell-m}^{(\alpha)}, \quad \ell = 0, 1, \ldots,$$

where

$$g_m^{(\alpha)} = (-1)^m\binom{\alpha}{m} = (-1)^m\frac{\Gamma(1+\alpha)}{\Gamma(m+1)\Gamma(1+\alpha-m)}, \quad m = 0, 1, \ldots, \ell.$$

For convenience, take the place of α in $\varpi_{\ell}^{(\alpha)}$ by $1-\alpha$, where $\alpha \in (0, 1)$. It is easy to get the following theorem.

Theorem 5. *The coefficients* $\varpi_{\ell}^{(1-\alpha)}$, $\ell = 0, 1, \ldots$, *can be computed recursively by the following formulas:*

$$\begin{cases} \varpi_0^{(1-\alpha)} = \left(\dfrac{4-3\alpha}{2(1-\alpha)}\right)^{1-\alpha}, \\ \varpi_1^{(1-\alpha)} = \dfrac{2(\alpha-1)(3-2\alpha)}{4-3\alpha}\varpi_0^{(1-\alpha)}, \\ \varpi_{\ell}^{(1-\alpha)} = \dfrac{1}{(4-3\alpha)\ell}[2(\ell+\alpha-2)(3-2\alpha)\varpi_{\ell-1}^{(1-\alpha)} \\ \qquad + (2-\alpha)(4-2\alpha-\ell)\varpi_{\ell-2}^{(1-\alpha)}], \quad \ell \geq 2. \end{cases}$$

The shifted pth-order fractional backward difference (SFBD$_p$) formula

Theorem 6 ([15]). *Suppose $u(x) \in C^{[\alpha]+3}(\mathbb{R})$ and all the derivatives of $u(x)$ up to order $[\alpha]+4$ belong to $L_1(\mathbb{R})$. Let*

$$^{L}\mathcal{B}_2^{\alpha}u(x) = \frac{1}{h^{\alpha}}\sum_{\ell=0}^{\infty}\kappa_{2,\ell}^{(\alpha)}u(x-(\ell-1)h),$$

where $\kappa_{2,\ell}^{(\alpha)}$ ($\ell = 0,1,\ldots$) are the coefficients of the novel generating function $\widetilde{W}_2(z) = (\frac{3\alpha-2}{2\alpha} - \frac{2(\alpha-1)}{\alpha}z + \frac{\alpha-2}{2\alpha}z^2)^{\alpha}$, that is,

$$\left(\frac{3\alpha-2}{2\alpha} - \frac{2(\alpha-1)}{\alpha}z + \frac{\alpha-2}{2\alpha}z^2\right)^{\alpha} = \sum_{\ell=0}^{\infty}\kappa_{2,\ell}^{(\alpha)}z^{\ell}, \quad |z|<1. \tag{16}$$

Then one has

$$_{\mathrm{RL}}D_{-\infty,x}^{\alpha}u(x) = {}^{L}\mathcal{B}_2^{\alpha}u(x) + \mathcal{O}(h^2)$$

as $h \to 0$.

As before, we also obtain

$$\kappa_{2,\ell}^{(\alpha)} = \left(\frac{3\alpha-2}{2\alpha}\right)^{\alpha}\sum_{m=0}^{\ell}\left(\frac{\alpha-2}{3\alpha-2}\right)^{m}\varpi_{1,m}^{(\alpha)}\varpi_{1,\ell-m}^{(\alpha)}, \quad \ell = 0,1,\ldots.$$

These coefficients have the following recursive relations:

$$\begin{cases} \kappa_{2,0}^{(\alpha)} = \left(\frac{3\alpha-2}{2\alpha}\right)^{\alpha}, \\ \kappa_{2,1}^{(\alpha)} = \frac{4\alpha(1-\alpha)}{3\alpha-2}\kappa_{2,0}^{(\alpha)}, \\ \kappa_{2,\ell}^{(\alpha)} = \frac{1}{\ell(3\alpha-2)}\left[4(1-\alpha)(\alpha-\ell+1)\kappa_{2,\ell-1}^{(\alpha)} + (\alpha-2)(2\alpha-\ell+2)\kappa_{2,\ell-2}^{(\alpha)}\right]. \end{cases}$$

In the following, we present the pth-order scheme ($p \ge 3$) [15].

Theorem 7. *Let $u(x) \in C^{[\alpha]+p+1}(\mathbb{R})$ and all the derivatives of $u(x)$ up to order $[\alpha]+p+2$ belong to $L_1(\mathbb{R})$. Set*

$$^{L}\mathcal{B}_p^{\alpha}u(x) = \frac{1}{h^{\alpha}}\sum_{\ell=0}^{\infty}\kappa_{p,\ell}^{(\alpha)}u(x-(\ell-1)h).$$

Then

$$_{\mathrm{RL}}D_{-\infty,x}^{\alpha}u(x) = {}^{L}\mathcal{B}_p^{\alpha}u(x) + \mathcal{O}(h^p), \quad p \ge 3.$$

Here the generating functions for coefficients $\kappa_{p,\ell}^{(\alpha)}$ $(\ell = 0, 1, \ldots)$ *with* $p \geq 3$ *are*

$$\widetilde{W}_p(z) = \left((1-z) + \frac{\alpha-2}{2\alpha}(1-z)^2 + \sum_{k=3}^{p} \frac{\lambda_{k-1,k-1}^{(\alpha)}}{\alpha}(1-z)^k\right)^{\alpha},$$

i. e.,

$$\widetilde{W}_p(z) = \sum_{\ell=0}^{\infty} \kappa_{p,\ell}^{(\alpha)} z^{\ell}, \quad |z| < 1,\ p \geq 3.$$

The parameters $\lambda_{k-1,k-1}^{(\alpha)}$ $(k = 3, 4, \ldots)$ *can be determined by the following equation:*

$$W_{k,s}(e^{-z})\frac{e^z}{z^{\alpha}} = 1 - \sum_{\ell=k}^{\infty} \lambda_{k,\ell}^{(\alpha)} z^{\ell}, \quad k = 2, 3, \ldots.$$

Here we only list the coefficients $\kappa_{p,\ell}^{(\alpha)}$ for $p = 3$ and $p = 4$.

(i) $p = 3$

The generating function for coefficients $\kappa_{3,\ell}^{(\alpha)}$ $(\ell = 0, 1, \ldots)$ reads as

$$\widetilde{W}_3(z) = (a_{31} + a_{32}z + a_{33}z^2 + a_{34}z^3)^{\alpha} = \sum_{\ell=0}^{\infty} \kappa_{3,\ell}^{(\alpha)} z^{\ell},$$

where

$$a_{31} = \frac{11\alpha^2 - 12\alpha + 3}{6\alpha^2}, \quad a_{32} = \frac{-6\alpha^2 + 10\alpha - 3}{2\alpha^2},$$
$$a_{33} = \frac{3\alpha^2 - 8\alpha + 3}{2\alpha^2}, \quad a_{34} = \frac{-2\alpha^2 + 6\alpha - 3}{6\alpha^2}.$$

Therefore,

$$\kappa_{3,\ell}^{(\alpha)} = a_{31}^{\alpha} \sum_{\ell_1=0}^{\ell} \sum_{\ell_2=0}^{[\frac{1}{2}\ell_1]} P(\alpha, \ell_1, \ell_2)\varpi_{1,\ell-\ell_1}^{(\alpha)}\varpi_{1,\ell_1-\ell_2}^{(\alpha)}, \quad \ell = 0, 1, \ldots,$$

where

$$P(\alpha, \ell_1, \ell_2) = \frac{(-1)^{\ell_1+\ell_2}(\ell_1 - \ell_2)!}{\ell_2!(\ell_1 - 2\ell_2)!} b_{31}^{\ell_1-2\ell_2} b_{32}^{\ell_2},$$
$$b_{31} = \frac{-7\alpha^2 + 18\alpha - 6}{11\alpha^2 - 12\alpha + 3}, \quad b_{32} = \frac{2\alpha^2 - 6\alpha + 3}{11\alpha^2 - 12\alpha + 3}.$$

In addition, we can get the following recursion relation by using the expressions of $\kappa_{3,\ell}^{(\alpha)}$ and automatic differentiation techniques:

$$\begin{cases} \kappa_{3,0}^{(\alpha)} = \left(\dfrac{11\alpha^2 - 12\alpha + 3}{6\alpha^2}\right)^{\alpha}, \\ \kappa_{3,1}^{(\alpha)} = -\dfrac{3\alpha(6\alpha^2 - 10\alpha + 3)}{11\alpha^2 - 12\alpha + 3}\kappa_{3,0}^{(\alpha)}, \\ \kappa_{3,2}^{(\alpha)} = \dfrac{3\alpha(108\alpha^5 - 402\alpha^4 + 520\alpha^3 - 312\alpha^2 + 87\alpha - 9)}{2(11\alpha^2 - 12\alpha + 3)^2}\kappa_{3,0}^{(\alpha)}, \\ \kappa_{3,\ell}^{(\alpha)} = \dfrac{1}{a_{31}\ell}\big[a_{32}(\alpha - \ell + 1)\kappa_{3,\ell-1}^{(\alpha)} + a_{33}(2\alpha - \ell + 2)\kappa_{3,\ell-2}^{(\alpha)} \\ \qquad + a_{34}(3\alpha - \ell + 3)\kappa_{3,\ell-3}^{(\alpha)}\big], \quad \ell \geq 3. \end{cases}$$

(ii) $p = 4$

The generating function for coefficients $\kappa_{4,\ell}^{(\alpha)}$ ($\ell = 0, 1, \ldots$) reads as follows:

$$\widetilde{W}_4(z) = (a_{41} + a_{42}z + a_{43}z^2 + a_{44}z^3 + a_{45}z^4)^{\alpha} = \sum_{\ell=0}^{\infty} \kappa_{4,\ell}^{(\alpha)} z^{\ell},$$

in which

$$a_{41} = \frac{25\alpha^3 - 35\alpha^2 + 15\alpha - 2}{12\alpha^3}, \quad a_{42} = \frac{-24\alpha^3 + 52\alpha^2 - 27\alpha + 4}{6\alpha^3},$$
$$a_{43} = \frac{6\alpha^3 - 19\alpha^2 + 12\alpha - 2}{2\alpha^3}, \quad a_{44} = \frac{-8\alpha^3 + 28\alpha^2 - 21\alpha + 4}{6\alpha^3},$$
$$a_{45} = \frac{3\alpha^3 - 11\alpha^2 + 9\alpha - 2}{12\alpha^3}.$$

Similarly, one can get

$$\kappa_{4,\ell}^{(\alpha)} = a_{41}^{\alpha} \sum_{\ell_1=0}^{\ell} \sum_{\ell_2=0}^{[\frac{2}{3}\ell_1]} \sum_{\ell_3=\max\{0, 2\ell_2-\ell_1\}}^{[\frac{1}{2}\ell_2]} P(\alpha, \ell_1, \ell_2, \ell_3)\varpi_{1,\ell-\ell_1}^{(\alpha)}\varpi_{1,\ell_1-\ell_2}^{(\alpha)},$$

where

$$P(\alpha, \ell_1, \ell_2, \ell_3) = \frac{(-1)^{\ell_1+\ell_2}(\ell_1 - \ell_2)!}{\ell_3!(\ell_2 - 2\ell_3)!(\ell_1 + \ell_3 - 2\ell_2)!} b_{41}^{\ell_1+\ell_3-2\ell_2} b_{42}^{\ell_2-2\ell_3} b_{43}^{\ell_3}$$

and

$$b_{41} = \frac{-23\alpha^3 + 69\alpha^2 - 39\alpha + 6}{25\alpha^3 - 35\alpha^2 + 15\alpha - 2}, \quad b_{42} = \frac{13\alpha^3 - 45\alpha^2 + 33\alpha - 6}{25\alpha^3 - 35\alpha^2 + 15\alpha - 2},$$
$$b_{43} = \frac{-3\alpha^3 + 11\alpha^2 - 9\alpha + 2}{25\alpha^3 - 35\alpha^2 + 15\alpha - 2}.$$

The cases for $p \geq 5$ can be similarly derived howbeit they are very complicated. We omit them here.

The fractional-compact forms of the shifted pth-order fractional backward difference (FC-SFBD$_p$) formula

The third-order fractional-compact formula I

Firstly, we denote the shifted second-order fractional backward difference (SFBD$_2$) operator as

$$ {}^{L}\mathcal{B}_2^{\alpha}u(x) = \frac{1}{h^{\alpha}} \sum_{\ell=0}^{\infty} \kappa_{2,\ell}^{(\alpha)} u(x-(\ell-1)h). $$

Next, we give the following asymptotic expansion formula for the difference operator ${}^{L}\mathcal{B}_2^{\alpha}$, which plays an important role in the establishment of high-order algorithms for Riemann–Liouville derivatives [13].

Theorem 8. *Let $u(x) \in C^{[\alpha]+n+1}(\mathbb{R})$ and all the derivatives of $u(x)$ up to order $[\alpha]+n+2$ belong to $L_1(\mathbb{R})$. Then*

$$ {}^{L}\mathcal{B}_2^{\alpha}u(x) = {}_{\mathrm{RL}}D_{-\infty,x}^{\alpha}u(x) + \sum_{\ell=1}^{n-1}(\sigma_{\ell}^{(\alpha)}\,{}_{\mathrm{RL}}D_{-\infty,x}^{\alpha+\ell}u(x))h^{\ell} + \mathcal{O}(h^{n}), \quad n \geq 2, $$

holds uniformly on $\mathbb{R}$. Here coefficients $\sigma_{\ell}^{(\alpha)}$ $(\ell = 1,2,\ldots)$ satisfy the equation $\frac{e^{z}}{z^{\alpha}}W_2(e^{-z}) = 1+\sum_{\ell=1}^{\infty}\sigma_{\ell}^{(\alpha)}z^{\ell}$. Especially, the first three coefficients are

$$ \sigma_1^{(\alpha)} = 0, \quad \sigma_2^{(\alpha)} = -\frac{2\alpha^2-6\alpha+3}{6\alpha}, \quad \sigma_3^{(\alpha)} = \frac{3\alpha^3-11\alpha^2+12\alpha-4}{12\alpha^2}. $$

Define the difference operator $\mathcal{L}$ as

$$ \mathcal{L}u(x) = (1+\sigma_2^{(\alpha)}h^2\delta_x^2)u(x), $$

where δ_x^2 is the second-order central difference operator defined by

$$ \delta_x^2 u(x) = \frac{u(x+h)-2u(x)+u(x-h)}{h^2}. $$

The difference operator $\mathcal{L}$ can be regarded as a fractional-compact operator by borrowing the nomenclature of the integer-order case. Accordingly, the main result is presented as follows.

Theorem 9. *Let $u(x) \in C^{[\alpha]+4}(\mathbb{R})$ and let all the derivatives of $u(x)$ up to order $[\alpha]+5$ belong to $L_1(\mathbb{R})$. Then we have*

$$ {}^{L}\mathcal{B}_2^{\alpha}u(x) = \mathcal{L}\,{}_{\mathrm{RL}}D_{-\infty,x}^{\alpha}u(x) + \mathcal{O}(h^3), $$

uniformly for $x \in \mathbb{R}$.

The third-order fractional-compact formula II

If we choose another new generating function,

$$\widetilde{W}_2(z) = \left(\frac{3\alpha+2}{2\alpha} - \frac{2(\alpha+1)}{\alpha}z + \frac{\alpha+2}{2\alpha}z^2\right)^{\alpha},$$

then we can define another shifted second-order fractional backward difference (SFBD_2) operator as follows:

$${}^{L}\widetilde{\mathcal{B}}_2^{\alpha}u(x) = \frac{1}{h^{\alpha}}\sum_{\ell=0}^{\infty}\widetilde{\kappa}_{2,\ell}^{(\alpha)}u(x-(\ell+1)h).$$

Here, the coefficients are

$$\widetilde{\kappa}_{2,\ell}^{(\alpha)} = \left(\frac{3\alpha+2}{2\alpha}\right)^{\alpha}\sum_{m=0}^{\ell}\left(\frac{\alpha+2}{3\alpha+2}\right)^{m}\varpi_{1,m}^{(\alpha)}\varpi_{1,\ell-m}^{(\alpha)}, \quad \ell = 0,1,\ldots,$$

which have the following recursion expressions:

$$\begin{cases}\widetilde{\kappa}_{2,0}^{(\alpha)} = (\dfrac{3\alpha+2}{2\alpha})^{\alpha},\\ \widetilde{\kappa}_{2,1}^{(\alpha)} = \dfrac{-4\alpha(1+\alpha)}{3\alpha+2}\widetilde{\kappa}_{2,0}^{(\alpha)},\\ \widetilde{\kappa}_{2,\ell}^{(\alpha)} = \dfrac{-4\alpha(1+\alpha)(\alpha-\ell+1)}{(3\alpha+2)\ell}\widetilde{\kappa}_{2,\ell-1}^{(\alpha)} + \dfrac{(\alpha+2)(2\alpha-\ell+2)}{(3\alpha+2)\ell}\widetilde{\kappa}_{2,\ell-2}^{(\alpha)}, \quad \ell \geq 2.\end{cases}$$

Similar to the previous discussion, we can obtain the following results.

Theorem 10. *Let $u(x) \in C^{[\alpha]+n+1}(\mathbb{R})$ and all the derivatives of $u(x)$ up to order $[\alpha]+n+2$ belong to $L_1(\mathbb{R})$. Then*

$${}^{L}\widetilde{\mathcal{B}}_2^{\alpha}u(x) = {}_{\mathrm{RL}}D_{-\infty,x}^{\alpha}u(x) + \sum_{\ell=1}^{n-1}(\widetilde{\sigma}_{\ell}^{(\alpha)}\,{}_{\mathrm{RL}}D_{-\infty,x}^{\alpha+\ell}u(x))h^{\ell} + \mathcal{O}(h^n), \quad n \geq 2,$$

holds uniformly on $\mathbb{R}$. Here coefficients $\widetilde{\sigma}_{\ell}^{(\alpha)}$ ($\ell = 1,2,\ldots$) satisfy the equation $\frac{\mathrm{e}^{-z}}{z^{\alpha}}\widetilde{W}_2(\mathrm{e}^{-z}) = 1 + \sum_{\ell=1}^{\infty}\widetilde{\sigma}_{\ell}^{(\alpha)}z^{\ell}$, $|z| < 1$. Especially, the first three coefficients are explicitly expressed as

$$\widetilde{\sigma}_1^{(\alpha)} = 0, \quad \widetilde{\sigma}_2^{(\alpha)} = -\frac{2\alpha^2+6\alpha+3}{6\alpha}, \quad \widetilde{\sigma}_3^{(\alpha)} = \frac{3\alpha^3+11\alpha^2+12\alpha+4}{12\alpha^2}.$$

Define another fractional-compact difference operator $\widetilde{\mathcal{L}}$ as

$$\widetilde{\mathcal{L}}u(x) = (1+\widetilde{\sigma}_2^{(\alpha)}h^2\delta_x^2)u(x).$$

Then the corresponding theorem is stated below.

Theorem 11. *Let $u(x) \in C^{[\alpha]+4}(\mathbb{R})$ and all the derivatives of $u(x)$ up to order $[\alpha]+5$ belong to $L_1(\mathbb{R})$. Then we have*

$$ {}^{L}\widetilde{\mathcal{B}}_2^{\alpha} u(x) = \widetilde{\mathcal{L}}_{\mathrm{RL}} D_{-\infty,x}^{\alpha} u(x) + \mathcal{O}(h^3), $$

uniformly for $x \in \mathbb{R}$.

The generalized numerical algorithm formulas

As far as we know, most of the numerical formulas for Riemann–Liouville derivatives deal with the grid points. If we want to know the approximate values on non-grid points, then these schemes are not applicable. In order to distinguish the schemes for approximating values on non-grid points from the known ones evaluating fractional derivatives at grid points, we call these approximations generalized numerical algorithm formulas. Here, we firstly establish the generalized numerical formula for Riemann–Liouville derivatives [13].

Theorem 12 (Generalized numerical approximation formula for Riemann–Liouville derivatives). *Let $u(x) \in C^{[\alpha]+p+1}(\mathbb{R})$ and let all the derivatives of $u(x)$ up to order $[\alpha]+p+2$ belong to $L_1(\mathbb{R})$. For any $s \in \mathbb{R}$, set*

$$ {}^{L}\mathcal{B}_{p,s}^{\alpha} u(x) = \frac{1}{h^{\alpha}} \sum_{\ell=0}^{\infty} \mu_{p,\ell}^{(\alpha,s)} u(x - (\ell + s)h). $$

Here coefficients $\mu_{p,\ell}^{(\alpha,s)}$ $(\ell = 0, 1, \ldots)$ can be determined by the following generating functions $G_{p,s}(z)$:

$$ G_{p,s}(z) = \left((1 - z) + \sum_{k=2}^{p} \frac{\vartheta_{k-1,k-1}^{(\alpha,s)}}{\alpha} (1 - z)^k \right)^{\alpha}, $$

that is,

$$ G_{p,s}(z) = \sum_{\ell=0}^{\infty} \mu_{p,\ell}^{(\alpha,s)} z^{\ell}, \quad |z| < 1, $$

where parameters $\vartheta_{k-1,k-1}^{(\alpha,s)}$ $(k = 2, 3, \ldots)$ can be obtained by the equation

$$ G_{k,s}(e^{-z}) \frac{e^{-sz}}{z^{\alpha}} = 1 - \sum_{\ell=1}^{\infty} \vartheta_{k,\ell}^{(\alpha,s)} z^{\ell}, \quad k = 1, 2, \ldots, $$

and satisfy

$$ 1 + \sum_{k=2}^{p} \frac{\vartheta_{k-1,k-1}^{(\alpha,s)}}{\alpha} \neq 0. $$

Then the left Riemann–Liouville derivative at any point $x = x_j + sh$ can be approximated by

$$ {}_{\mathrm{RL}}D_{-\infty,x}^{\alpha} u(x)\big|_{x=x_j+sh} = {}^{L}\mathcal{B}_{p,s}^{\alpha} u(x_j + sh) + \mathcal{O}(h^p), \quad j = 0, 1, \ldots, \ p \geq 1. $$

Remark 2. The above theorem is called the generalized numerical approximation formula for Riemann–Liouville derivatives since the values at any point $x = x_j + sh$, $(j = 0, 1, \ldots)$ on the real axis can be calculated. Here, the value x can be determined by selecting the appropriate parameter s, and x_j represent the grid point.

Due to the fact that the case $p = 1$ is the same as the Grünwald–Letnikov formula and the fact that the cases for $p \geq 5$ can be similarly obtained in view of the above theorem, we carefully consider cases with $p = 2, 3$, and 4 as follows.

(i) $p = 2$

According to Theorem 12, we easily know that the generating function for $p = 2$ is

$$G_{2,s}(z) = \left((1-z) + \frac{\alpha + 2s}{2\alpha}(1-z)^2 \right)^{\alpha},$$

and the coefficients $\mu_{2,\ell}^{(\alpha,s)}$ $(\ell = 0, 1, \ldots)$ are read as

$$\mu_{2,\ell}^{(\alpha,s)} = d_{21}^{\alpha} \sum_{m=0}^{\ell} d_{22}^{m} \varpi_{1,m}^{(\alpha)} \varpi_{1,\ell-m}^{(\alpha)}, \quad \ell = 0, 1 \ldots,$$

where

$$d_{21} = \frac{3\alpha + 2s}{2\alpha}, \quad d_{22} = \frac{\alpha + 2s}{3\alpha + 2s},$$

and the parameter s satisfies $3\alpha + 2s \neq 0$.

Furthermore, the coefficients $\mu_{2,\ell}^{(\alpha)}$ $(\ell = 0, 1, \ldots)$ can be obtained by the following recurrence relationships:

$$\begin{cases} \mu_{2,0}^{(\alpha,s)} = \left(\dfrac{3\alpha + 2s}{2\alpha} \right)^{\alpha}, \\ \mu_{2,1}^{(\alpha,s)} = \dfrac{-4\alpha(\alpha + s)}{3\alpha + 2s} \mu_{2,0}^{(\alpha,s)}, \\ \mu_{2,\ell}^{(\alpha,s)} = \dfrac{1}{(3\alpha + 2s)\ell} [-4(\alpha + s)(\alpha - \ell + 1)\mu_{2,\ell-1}^{(\alpha,s)} \\ \qquad + (\alpha + 2s)(2\alpha - \ell + 2)\mu_{2,\ell-2}^{(\alpha,s)}], \quad \ell \geq 2. \end{cases}$$

(ii) $p = 3$

For this case, the generating function is given as follows:

$$G_{3,s}(z) = \left((1-z) + \frac{\alpha + 2s}{2\alpha}(1-z)^2 + \frac{2\alpha^2 + 6\alpha s + 3s^2}{6\alpha^2}(1-z)^3 \right)^{\alpha}.$$

The coefficients $\mu_{3,\ell}^{(\alpha,s)}$ $(\ell = 0, 1, \ldots)$ can be acquired by simple calculations, i. e.,

$$\mu_{3,\ell}^{(\alpha,s)} = d_{31}^{\alpha} \sum_{\ell_1=0}^{\ell} \sum_{\ell_2=0}^{[\frac{1}{2}\ell_1]} \frac{(-1)^{\ell_1+\ell_2}(\ell_1 - \ell_2)!}{\ell_2!(\ell_1 - 2\ell_2)!} d_{32}^{\ell_1 - 2\ell_2} d_{33}^{\ell_2} \varpi_{1,\ell-\ell_1}^{(\alpha)} \varpi_{1,\ell_1-\ell_2}^{(\alpha)}, \quad \ell = 0, 1, \ldots.$$

Here,

$$d_{31} = \frac{11\alpha^2 + 12\alpha s + 3s^2}{6\alpha^2}, \quad d_{32} = -\frac{7\alpha^2 + 18\alpha s + 6s^2}{11\alpha^2 + 12\alpha s + 3s^2},$$
$$d_{33} = \frac{2\alpha^2 + 6\alpha s + 3s^2}{11\alpha^2 + 12\alpha s + 3s^2},$$

and the parameter s satisfies $11\alpha^2 + 12\alpha s + 3s^2 \neq 0$.

The recursion relations for coefficients $\mu_{3,\ell}^{(\alpha,s)}$ $(\ell = 0, 1, \dots)$ read as

$$\begin{cases} \mu_{3,0}^{(\alpha,s)} = \left(\dfrac{11\alpha^2 + 12\alpha s + 3s^2}{6\alpha^2}\right)^{\alpha}, \\ \mu_{3,1}^{(\alpha,s)} = -\dfrac{3\alpha(6\alpha^2 + 10\alpha s + 3s^2)}{11\alpha^2 + 12\alpha s + 3s^2}\mu_{3,0}^{(\alpha,s)}, \\ \mu_{3,2}^{(\alpha,s)} = \dfrac{3\alpha}{2(11\alpha^2 + 12\alpha s + 3s^2)^2}(108\alpha^5 + 360\alpha^4 s - 42\alpha^4 + 408\alpha^3 s^2 \\ \qquad - 112\alpha^3 s + 180\alpha^2 s^3 - 132\alpha^2 s^2 + 27\alpha s^4 - 60\alpha s^3 - 9s^4)\mu_{3,0}^{(\alpha,s)}, \\ \mu_{3,\ell}^{(\alpha,s)} = \dfrac{1}{(11\alpha^2 + 12\alpha s + 3s^2)\ell}[-3(6\alpha^2 + 10\alpha s + 3s^2)(\alpha - \ell + 1)\mu_{3,\ell-1}^{(\alpha,s)} \\ \qquad + 3(3\alpha^2 + 8\alpha s + 3s^2)(2\alpha - \ell + 2)\mu_{3,\ell-2}^{(\alpha,s)} \\ \qquad - (2\alpha^2 + 6\alpha s + 3s^2)(3\alpha - \ell + 3)\mu_{3,\ell-3}^{(\alpha,s)}], \quad \ell \geq 3. \end{cases}$$

(iii) $p = 4$

As before, we can easily get the following generating function for $p = 4$:

$$G_{4,s}(z) = \left((1 - z) + \frac{\alpha + 2s}{2\alpha}(1 - z)^2 + \frac{2\alpha^2 + 6\alpha s + 3s^2}{6\alpha^2}(1 - z)^3 \right.$$
$$\left. + \frac{3\alpha^3 + 11\alpha^2 s + 9\alpha s^2 + 2s^3}{12\alpha^3}(1 - z)^4\right)^{\alpha}.$$

By the back-of-the-envelope calculations, we can get the expressions of coefficients $\mu_{4,\ell}^{(\alpha,s)}$ $(\ell = 0, 1, \dots)$ as follows:

$$\mu_{4,\ell}^{(\alpha,s)} = d_{41}^{\alpha} \sum_{\ell_1=0}^{\ell} \sum_{\ell_2=0}^{[\frac{2}{3}\ell_1]} \sum_{\ell_3=\max\{0,2\ell_2-\ell_1\}}^{[\frac{1}{2}\ell_2]} P(\alpha, \ell_1, \ell_2, \ell_3)\varpi_{1,\ell-\ell_1}^{(\alpha)}\varpi_{1,\ell_1-\ell_2}^{(\alpha)},$$

where

$$P(\alpha, \ell_1, \ell_2, \ell_3) = \frac{(-1)^{\ell_1+\ell_2}(\ell_1 - \ell_2)!}{\ell_3!(\ell_2 - 2\ell_3)!(\ell_1 + \ell_3 - 2\ell_2)!} d_{42}^{\ell_1+\ell_3-2\ell_2} d_{43}^{\ell_2-2\ell_3} d_{44}^{\ell_3}$$

and

$$d_{41} = \frac{25\alpha^3 + 35\alpha^2 s + 15\alpha s^2 + 2s^3}{12\alpha^3}, \quad d_{42} = -\frac{23\alpha^3 + 69\alpha^2 s + 39\alpha s^2 + 6s^3}{25\alpha^3 + 35\alpha^2 s + 15\alpha s^2 + 2s^3},$$

$$d_{43} = \frac{13\alpha^3 + 45\alpha^2 s + 33\alpha s^2 + 6s^3}{25\alpha^3 + 35\alpha^2 s + 15\alpha s^2 + 2s^3}, \qquad d_{44} = -\frac{3\alpha^3 + 11\alpha^2 s + 9\alpha s^2 + 2s^3}{25\alpha^3 + 35\alpha^2 s + 15\alpha s^2 + 2s^3}.$$

The parameter s satisfies $25\alpha^3 + 35\alpha^2 s + 15\alpha s^2 + 2s^3 \neq 0$.

The fractional-compact forms of the generalized numerical algorithm formulas

Firstly, the asymptotic expansion formula of operator ${}^{L}\mathcal{B}^{\alpha}_{p,s}$ is listed. This is the foundation for the establishment of the fractional-compact form of the generalized numerical algorithm.

Theorem 13. *Let $u(x) \in C^{[\alpha]+n+1}(\mathbb{R})$ and all the derivatives of $u(x)$ up to order $[\alpha]+n+2$ belong to $L_1(\mathbb{R})$. Then for any $s \in \mathbb{R}$ and $p \in \mathbb{N}$, the equality*

$${}^{L}\mathcal{B}^{\alpha}_{p,s}u(x) = {}_{\mathrm{RL}}D^{\alpha}_{-\infty,x}u(x) + \sum_{\ell=p}^{n-1}\left(\varrho_{\ell}^{(\alpha,s)}\,{}_{\mathrm{RL}}D^{\alpha+\ell}_{-\infty,x}u(x)\right)h^{\ell} + \mathcal{O}(h^{n}), \quad n \geq p+1,$$

holds uniformly on $\mathbb{R}$. Here the coefficients $\varrho_{\ell}^{(\alpha,s)}$ $(\ell = 1, 2, \dots)$ can be determined by the following equation:

$$\frac{e^{-sz}}{z^{\alpha}}G_{p,s}(e^{-z}) = 1 + \sum_{\ell=p}^{\infty}\varrho_{\ell}^{(\alpha,s)}z^{\ell}, \quad |z| < 1.$$

Define a generalized fractional difference operator $\mathcal{J}_{p,s}$ as

$$\mathcal{J}_{p,s}u(x) = (1 + \varrho_{p}^{(\alpha,s)}h^{p}\delta_{x}^{p})u(x),$$

where the difference operator δ_x^p is defined by

$$\delta_x^p u(x) = \frac{1}{h^p}\sum_{m=0}^{p}(-1)^m\binom{p}{m}u\left(x + \left(\frac{p}{2} - m\right)h\right), \quad p \in \mathbb{N}.$$

Note that ${}_{\mathrm{RL}}D^{\alpha+p}_{-\infty,x}u(x) = \frac{\mathrm{d}^p}{\mathrm{d}x^p}({}_{\mathrm{RL}}D^{\alpha}_{-\infty,x}u(x))$ and $\frac{\mathrm{d}^p u(x)}{\mathrm{d}x^p} = \delta_x^p u(x) + \mathcal{O}(h^p)$. Then the generalized fractional-compact numerical algorithm formula for the Riemann–Liouville derivative is stated below.

Theorem 14. *Let $u(x) \in C^{[\alpha]+p+1}(\mathbb{R})$ and let all the derivatives of $u(x)$ up to order $[\alpha]+p+2$ belong to $L_1(\mathbb{R})$. Then we have*

$${}^{L}\mathcal{B}^{\alpha}_{p,s}u(x) = \mathcal{J}_{p,s}\,{}_{\mathrm{RL}}D^{\alpha}_{-\infty,x}u(x) + \mathcal{O}(h^{p+1}),$$

uniformly for $x, s \in \mathbb{R}$ and $p \in \mathbb{N}$.

Remark 3. The operators $\mathcal{L}$, $\widetilde{\mathcal{L}}$, and $\mathcal{J}_{p,s}$ are related by $\mathcal{J}_{2,-1} = \mathcal{L}$ and $\mathcal{J}_{2,1} = \widetilde{\mathcal{L}}$.

2.3 The Riesz derivative

2.3.1 The indirect method

Since the Riesz derivative can be regarded as the linear combination of left-sided and right-sided Riemann–Liouville derivatives, any approximation to Riemann–Liouville derivatives can be extended to the case of Riesz derivatives [7, 8, 16, 11, 13, 56, 61].

2.3.2 The direct method

The symmetrical fractional central (SFC) difference method

In [44], Ortigueira introduced a symmetrical fractional central (SFC) difference operator, i. e.,

$$\Delta_h^\alpha u(x) = \sum_{k=-\infty}^{\infty} \frac{(-1)^k \Gamma(\alpha+1)}{\Gamma(\frac{\alpha}{2}-k+1)\Gamma(\frac{\alpha}{2}+k+1)} u(x-kh),$$

and showed that

$$\frac{\partial^\alpha u(x)}{\partial |x|^\alpha} = \lim_{h\to 0}\left(-\frac{\Delta_h^\alpha u(x)}{h^\alpha}\right).$$

Later on, Çelik and Duman [3] proved that the above SFC difference operator and the Riesz derivative satisfy the following relation:

$$\frac{\partial^\alpha u(x)}{\partial |x|^\alpha} = -\frac{1}{h^\alpha}\Delta_h^\alpha u(x) + O(h^2).$$

The weighted and shifted SFC (WSSFC) difference method

Define

$$\mathcal{H}_\theta u(x) = \sum_{k=-\infty}^{\infty} g_k^{(\alpha)} u(x-(k+\theta)h), \quad \theta \in \chi = \{0, \pm 1, \pm 2, \pm 3, \ldots\},$$

where

$$g_k^{(\alpha)} = \frac{(-1)^k \Gamma(\alpha+1)}{\Gamma(\frac{\alpha}{2}-k+1)\Gamma(\frac{\alpha}{2}+k+1)}.$$

Let

$$\frac{\partial^\alpha u(x)}{\partial |x|^\alpha} = \frac{1}{h^\alpha}\sum_{\theta\in\chi} Z_{\theta,p}\mathcal{H}_\theta u(x) + \mathcal{O}(h^p), \quad p \geq 2,$$

where $Z_{\theta,p}$ are coefficients determined by the Fourier transform method. Below, we firstly give a fourth-order formula via the following theorem [17].

Theorem 15. *Let $u(x)$ lie in $C^7(\mathbb{R})$ with its derivatives up to order seven belonging to $L_1(\mathbb{R})$. Then we have*

$$\frac{\partial^\alpha u(x)}{\partial|x|^\alpha} = \frac{1}{h^\alpha}\left[\frac{\alpha}{24}\mathcal{H}_{-1}u(x) - \left(1+\frac{\alpha}{12}\right)\mathcal{H}_0 u(x) + \frac{\alpha}{24}\mathcal{H}_1 u(x)\right] + O(h^4).$$

In fact, we can use almost the same method to construct more even higher-order difference schemes for the Riesz derivative, such as sixth-order and eighth-order schemes. Here, we list several commonly used high-order formulas. We have

$$\frac{\partial^\alpha u(x)}{\partial|x|^\alpha} = \frac{D_s^\alpha u(x)}{h^\alpha} + \mathcal{O}(h^6), \quad \frac{\partial^\alpha u(x)}{\partial|x|^\alpha} = \frac{D_e^\alpha u(x)}{h^\alpha} + \mathcal{O}(h^8)$$

and

$$\frac{\partial^\alpha u(x)}{\partial|x|^\alpha} = \frac{D_{te}^\alpha u(x)}{h^\alpha} + \mathcal{O}(h^{10}), \quad \frac{\partial^\alpha u(x)}{\partial|x|^\alpha} = \frac{D_{tw}^\alpha u(x)}{h^\alpha} + \mathcal{O}(h^{12}),$$

where

$$\begin{aligned}
D_s^\alpha u(x) &= \mathcal{A}_1\mathcal{H}_{-2}u(x) + \mathcal{A}_2\mathcal{H}_{-1}u(x) + \mathcal{A}_3\mathcal{H}_0 u(x) \\
&\quad + \mathcal{A}_2\mathcal{H}_1 u(x) + \mathcal{A}_1\mathcal{H}_2 u(x), \\
D_e^\alpha u(x) &= \mathcal{B}_1\mathcal{H}_{-3}u(x) + \mathcal{B}_2\mathcal{H}_{-2}u(x) + \mathcal{B}_3\mathcal{H}_{-1}u(x) + \mathcal{B}_4\mathcal{H}_0 u(x) \\
&\quad + \mathcal{B}_3\mathcal{H}_1 u(x) + \mathcal{B}_2\mathcal{H}_2 u(x) + \mathcal{B}_1\mathcal{H}_3 u(x), \\
D_{te}^\alpha u(x) &= \mathcal{C}_1\mathcal{H}_{-4}u(x) + \mathcal{C}_2\mathcal{H}_{-3}u(x) + \mathcal{C}_3\mathcal{H}_{-2}u(x) + \mathcal{C}_4\mathcal{H}_{-1}u(x) \\
&\quad + \mathcal{C}_5\mathcal{H}_0 u(x) + \mathcal{C}_4\mathcal{H}_1 u(x) + \mathcal{C}_3\mathcal{H}_2 u(x) + \mathcal{C}_2\mathcal{H}_3 u(x) + \mathcal{C}_1\mathcal{H}_4 u(x), \\
D_{tw}^\alpha u(x) &= \mathcal{D}_1\mathcal{H}_{-5}u(x) + \mathcal{D}_2\mathcal{H}_{-4}u(x) + \mathcal{D}_3\mathcal{H}_{-3}u(x) \\
&\quad + \mathcal{D}_4\mathcal{H}_{-2}u(x) + \mathcal{D}_5\mathcal{H}_{-1}u(x) + \mathcal{D}_6\mathcal{H}_0 u(x) + \mathcal{D}_5\mathcal{H}_1 u(x) \\
&\quad + \mathcal{D}_4\mathcal{H}_2 u(x) + \mathcal{D}_3\mathcal{H}_3 u(x) + \mathcal{D}_2\mathcal{H}_4 u(x) + \mathcal{D}_1\mathcal{H}_5 u(x),
\end{aligned}$$

with

$$\mathcal{A}_1 = -\left(\frac{\alpha}{1152} + \frac{11}{2880}\right)\alpha, \quad \mathcal{A}_2 = \left(\frac{\alpha}{288} + \frac{41}{720}\right)\alpha, \quad \mathcal{A}_3 = -\left(\frac{\alpha^2}{192} + \frac{17\alpha}{160} + 1\right),$$

$$\mathcal{B}_1 = \left(\frac{\alpha^2}{82944} + \frac{11\alpha}{69120} + \frac{191}{362880}\right)\alpha, \quad \mathcal{B}_2 = -\left(\frac{\alpha^2}{13824} + \frac{7\alpha}{3840} + \frac{211}{30240}\right)\alpha,$$

$$\mathcal{B}_3 = \left(\frac{5\alpha^2}{27648} + \frac{3\alpha}{512} + \frac{7843}{120960}\right)\alpha, \quad \mathcal{B}_4 = -\left(\frac{5\alpha^3}{20736} + \frac{29\alpha^2}{3456} + \frac{5297\alpha}{45360} + 1\right),$$

$$\mathcal{C}_1 = -\left(\frac{2497}{29030400} + \frac{10181\alpha}{348364800} + \frac{11\alpha^2}{3317760} + \frac{\alpha^3}{7962624}\right)\alpha,$$

$$\mathcal{C}_2 = \left(\frac{1469}{1209600} + \frac{17111\alpha}{43545600} + \frac{\alpha^2}{25920} + \frac{\alpha^3}{995328}\right)\alpha,$$

$$\mathcal{C}_3 = -\left(\frac{68119}{7257600} + \frac{32861\alpha}{12441600} + \frac{137\alpha^2}{829440} + \frac{7\alpha^3}{1990656}\right)\alpha,$$

$$\mathcal{C}_4 = \left(\frac{252769}{3628800} + \frac{46631\alpha}{6220800} + \frac{19\alpha^2}{51840} + \frac{7\alpha^3}{995328}\right)\alpha,$$

$$\mathcal{C}_5 = -\left(1 + \frac{118829\alpha}{967680} + \frac{51941\alpha^2}{4976640} + \frac{157\alpha^3}{331776} + \frac{35\alpha^4}{3981312}\right),$$

$$\mathcal{D}_1 = \left(\frac{14797}{958003200} + \frac{11693\alpha}{2090188800} + \frac{6361\alpha^2}{8360755200} + \frac{11\alpha^3}{238878720} + \frac{\alpha^4}{955514880}\right)\alpha,$$

$$\mathcal{D}_2 = -\left(\frac{230371}{958003200} + \frac{5563\alpha}{65318400} + \frac{9133\alpha^2}{836075520} + \frac{7\alpha^3}{11943936} + \frac{\alpha^4}{95551488}\right)\alpha,$$

$$\mathcal{D}_3 = \left(\frac{203257}{106444800} + \frac{449171\alpha}{696729600} + \frac{13529\alpha^2}{185794560} + \frac{49\alpha^3}{15925248} + \frac{\alpha^4}{21233664}\right)\alpha,$$

$$\mathcal{D}_4 = -\left(\frac{299093}{26611200} + \frac{24041\alpha}{7257600} + \frac{17869\alpha^2}{69672960} + \frac{\alpha^3}{110592} + \frac{\alpha^4}{7962624}\right)\alpha,$$

$$\mathcal{D}_5 = \left(\frac{11639731}{159667200} + \frac{431513\alpha}{49766400} + \frac{20953\alpha^2}{39813120} + \frac{133\alpha^3}{7962624} + \frac{7\alpha^4}{31850496}\right)\alpha,$$

$$\mathcal{D}_6 = -\left(1 + \frac{6742753\alpha}{53222400} + \frac{2303\alpha^2}{194400} + \frac{22061\alpha^3}{33177600} + \frac{203\alpha^4}{9953280} + \frac{7\alpha^5}{26542080}\right).$$

The fractional-compact form of the SFC (FCSFC) difference method

Theorem 16 ([14]). *Suppose that $u(x) \in C^{[\alpha]+2n+1}(\mathbb{R})$ and all the derivatives of $u(x)$ up to order $[\alpha] + 2n + 2$ exist and belong to $L_1(\mathbb{R})$. Then the following estimate holds:*

$$\frac{\partial^\alpha u(x)}{\partial |x|^\alpha} = (\delta_x^0 - b_{n-1}\delta_x^{2n-2})^{-1}\left(\sum_{\ell=0}^{n-2} b_\ell\, \delta_x^{2\ell}\right)\left(-\frac{\Delta_h^\alpha u(x)}{h^\alpha}\right) + \mathcal{O}(h^{2n}), \quad n \in \mathrm{N}^+,$$

where

$$\Delta_h^\alpha u(x) = \frac{1}{h^\alpha} \sum_{k=-\infty}^{\infty} g_k^{(\alpha)} u(x - kh)$$

and

$$\delta_x^{2\ell}u(x_j)=\sum_{s=0}^{2\ell}(-1)^s\binom{2\ell}{s}u(x_{\ell+j-s}),\quad \ell\geq 0.$$

Specifically, δ_x^0 is the identity operator.

Here, the coefficients b_ℓ ($\ell=0,1,\dots,n-2$) satisfy the following equation:

$$\sum_{\ell=0}^{n-2}b_\ell\left(2\sum_{s=0}^{\ell-1}\sum_{q=0}^{n-1}\sum_{p=0}^{n-1-q}\frac{(-1)^{s+q}(\ell-s)^{2q}\binom{2\ell}{s}a_p}{(2q)!}|\omega h|^{2(p+q)}+(-1)^\ell\binom{2\ell}{\ell}\sum_{p=0}^{n-1}a_p|\omega h|^{2p}\right)$$
$$=1-b_{n-1}\left(\sum_{s=0}^{n-2}(-1)^s\binom{2n-2}{s}\frac{2(n-1-s)^{2n-2}}{(2n-2)!}\right)(-1)^{n-1}|\omega h|^{2n-2},$$

and coefficients a_p ($p=0,1,\dots$) satisfy

$$\sum_{p=0}^{\infty}a_p|\omega h|^{2p}=\left|\frac{2\sin(\frac{\omega h}{2})}{\omega h}\right|^\alpha=\left[1-\frac{\alpha}{24}|\omega h|^2+\left(\frac{1}{1920}+\frac{\alpha-1}{1152}\right)\alpha|\omega h|^4\right.$$
$$\left.-\left(\frac{1}{322560}+\frac{\alpha-1}{46080}+\frac{(\alpha-1)(\alpha-2)}{82944}\right)\alpha|\omega h|^6+\cdots\right].$$

Here, for future applications, we list a few even-order fractional-compact numerical differential formulas. We have

$$\frac{\partial^\alpha u(x)}{\partial|x|^\alpha}=(\delta_x^0-b_1\delta_x^2)^{-1}b_0\delta_x^0\left(-\frac{\Delta_h^\alpha u(x)}{h^\alpha}\right)+\mathcal{O}(h^4),\tag{17}$$
$$\frac{\partial^\alpha u(x)}{\partial|x|^\alpha}=(\delta_x^0-b_2\delta_x^4)^{-1}(b_0\delta_x^0+b_1\delta_x^2)\left(-\frac{\Delta_h^\alpha u(x)}{h^\alpha}\right)+\mathcal{O}(h^6),$$
$$\frac{\partial^\alpha u(x)}{\partial|x|^\alpha}=(\delta_x^0-b_3\delta_x^6)^{-1}(b_0\delta_x^0+b_1\delta_x^2+b_2\delta_x^4)\left(-\frac{\Delta_h^\alpha u(x)}{h^\alpha}\right)+\mathcal{O}(h^8),$$
$$\frac{\partial^\alpha u(x)}{\partial|x|^\alpha}=(\delta_x^0-b_4\delta_x^8)^{-1}(b_0\delta_x^0+b_1\delta_x^2+b_2\delta_x^4+b_3\delta_x^6)\left(-\frac{\Delta_h^\alpha u(x)}{h^\alpha}\right)+\mathcal{O}(h^{10}),$$
$$\frac{\partial^\alpha u(x)}{\partial|x|^\alpha}=(1-b_5\delta_x^{10})^{-1}(b_0+b_1\delta_x^2+b_2\delta_x^4+b_3\delta_x^6+b_4\delta_x^8)\left(-\frac{\Delta_h^\alpha u(x)}{h^\alpha}\right)+\mathcal{O}(h^{12}),$$

where

$$b_0=1,$$
$$b_1=-\frac{\alpha}{24},$$
$$b_2=\left(\frac{11}{2880}+\frac{\alpha}{1152}\right)\alpha,$$
$$b_3=-\left(\frac{191}{362880}+\frac{11\alpha}{69120}+\frac{\alpha^2}{82944}\right)\alpha,$$
$$b_4=\left(\frac{2497}{29030400}+\frac{10181\alpha}{348364800}+\frac{11\alpha^2}{3317760}+\frac{\alpha^3}{7962624}\right)\alpha,$$
$$b_5=-\left(\frac{14797}{958003200}+\frac{11693\alpha}{2090188800}+\frac{6361\alpha^2}{8360755200}+\frac{11\alpha^3}{238878720}+\frac{\alpha^4}{955514880}\right)\alpha.$$

3 High-order finite difference methods for fractional partial differential equations

For the finite difference methods, we know that the derived schemes share some similarities with the classical integer-order partial differential equations. Thus, the classical analysis techniques such as the Fourier method, energy method, and matrix method can be extended to those of numerical analysis for fractional partial differential equations. The difference is that the numerical analysis for the fractional case is much more complicated than the classical one. Some efficient numerical methods have been developed for solving fractional partial differential equations [4–6, 9, 10, 20, 24, 29–31, 34, 39, 48, 50, 54, 55, 58–60, 62, 65]. In particular, for the higher-order numerical algorithms, one can see the recent series of papers [2, 16, 17, 11, 13–15, 32]. Here, we only give an example to show an application of the higher-order numerical approximations in fractional partial differential equations [14]. Other applications are not going to be presented here due to the limited length of the chapter.

3.1 Introduction

In this section, we numerically study the Riesz spatial telegraph equation of the following form:

$$\frac{\partial^2 u(x,t)}{\partial t^2} + \nu \frac{\partial u(x,t)}{\partial t} = \kappa^2 \frac{\partial^\alpha u(x,t)}{\partial |x|^\alpha} + f(x,t), \quad l < x < L, \ 0 < t \le T, \tag{18}$$

with the following initial and boundary value conditions:

$$u(x,0) = \varphi(x), \quad \frac{\partial u(x,0)}{\partial t} = \phi(x), \quad l \le x \le L,$$
$$u(l,t) = 0, \quad u(L,t) = 0, \quad 0 \le t \le T,$$

where $1 < \alpha < 2$; $\nu > 0$ and κ^2 are two constants.

3.2 Construction of the numerical algorithm

Let $x_j = l + jh$, $j = 0, 1, \ldots, M$, and $t_s = s\tau$, $s = 0, 1, \ldots, N$, where $h = (L - l)/M$ and $\tau = T/N$, with M and N being two positive integers.

For convenience, denote

$$\mu_t u(x,t) = u\Big(x, t + \frac{\tau}{2}\Big) + u\Big(x, t - \frac{\tau}{2}\Big), \quad \delta_t u(x,t) = u\Big(x, t + \frac{\tau}{2}\Big) - u\Big(x, t - \frac{\tau}{2}\Big),$$
$$\mu_t \delta_t u(x,t) = u(x, t+\tau) - u(x, t-\tau), \qquad \delta_t^2 u(x,t) = u(x, t+\tau) - 2u(x,t) + u(x, t-\tau),$$

and

$$\delta_x^\alpha u(x_j,t)=\sum_{k=-(M-j)}^{j} g_k^{(\alpha)}u(x-kh,t).$$

For the equation $\frac{\partial^2 u(x,t)}{\partial t^2}+\nu\frac{\partial u(x,t)}{\partial t}=g(x,t)$, there exists a fourth-order compact formula

$$H_t^{-1}J_t u(x,t)=g(x,t)+\mathcal{O}(\tau^4),\tag{19}$$

where H_t and J_t are two operators defined by

$$H_t=1+\frac{1}{12}\left(\delta_t^2+\frac{\nu\tau}{2}\mu_t\delta_t\right),\quad J_t=\left(\frac{\nu^2}{12}+\frac{1}{\tau^2}\right)\delta_t^2+\frac{\nu}{2\tau}\mu_t\delta_t.$$

Replacing $g(x,t)$ by $\kappa^2\frac{\partial^\alpha u(x,t)}{\partial|x|^\alpha}+f(x,t)$ in equation (19) gives

$$H_t^{-1}J_t u(x,t)=\kappa^2\frac{\partial^\alpha u(x,t)}{\partial|x|^\alpha}+f(x,t)+\mathcal{O}(\tau^4).$$

Using the fourth-order fractional-compact numerical differential formula (17) to deal with the Riesz derivative $\frac{\partial^\alpha u(x,t)}{\partial|x|^\alpha}$ leads to

$$H_t^{-1}J_t u(x,t)=\kappa^2\left(\delta_x^0+\frac{\alpha}{24}\delta_x^2\right)^{-1}\left(-\frac{\Delta_h^\alpha u(x,t)}{h^\alpha}\right)+f(x,t)+R(x,t),$$

where $R(x,t)$ satisfies $|R(x,t)|\le C(\tau^4+h^4)$, with C being a positive constant. Neglecting the high-order term $R(x,t)$ and letting u_j^s be the approximate solution of function $u(x_j,t_s)$ gives

$$H_t^{-1}J_t u_j^s=-\frac{\kappa^2}{h^\alpha}\left(\delta_x^0+\frac{\alpha}{24}\delta_x^2\right)^{-1}\delta_x^\alpha u_j^s+f_j^s.\tag{20}$$

The finite difference scheme (20) is a three-level one which needs knowledge of the approximate value at $t=t_1$. Here, we provide a way to compute $u(x,t)$ at the first time level. For convenience, denote $\frac{\partial^m u(x,t)}{\partial t^m}|_{t=0}=\frac{\partial^m u(x,0)}{\partial t^m}$, $m=0,1,\ldots$.

Note that

$$\begin{aligned}\frac{\partial^2 u(x,0)}{\partial t^2}&=\kappa^2\frac{\partial^\alpha u(x,0)}{\partial|x|^\alpha}-\nu\frac{\partial u(x,0)}{\partial t}+f(x,0)\\&=\kappa^2\frac{\partial^\alpha\varphi(x)}{\partial|x|^\alpha}-\nu\phi(x)+f(x,0)\end{aligned}$$

and

$$\frac{\partial^3 u(x,0)}{\partial t^3}=\kappa^2\frac{\partial}{\partial t}\left(\frac{\partial^\alpha u(x,t)}{\partial|x|^\alpha}\right)\bigg|_{t=0}-\nu\frac{\partial^2 u(x,0)}{\partial t^2}+\frac{\partial f(x,0)}{\partial t}$$

$$= \frac{\kappa^2}{\tau}\left(\frac{\partial^\alpha u(x,t_1)}{\partial|x|^\alpha} - \frac{\partial^\alpha \varphi(x)}{\partial|x|^\alpha}\right) - \nu\left(\kappa^2\frac{\partial^\alpha \varphi(x)}{\partial|x|^\alpha} - \nu\phi(x) + f(x,0)\right)$$
$$+ \frac{\partial f(x,0)}{\partial t} + \mathcal{O}(\tau).$$

Then we can obtain the following equation by substituting the above two equations into the Taylor series expansion of function $u(x,t_1)$ at $t = 0$:

$$u(x,t_1) = u(x,0) + \tau\frac{\partial u(x,0)}{\partial t} + \frac{\tau^2}{2}\frac{\partial^2 u(x,0)}{\partial t^2} + \frac{\tau^3}{6}\frac{\partial^3 u(x,0)}{\partial t^3} + \mathcal{O}(\tau^4)$$
$$= \varphi(x) + \frac{\tau}{6}(6 - 3\nu\tau + \nu^2\tau^2)\phi(x) + \frac{\tau^2}{6}(3 - \nu\tau)f(x,0) + \frac{\tau^3}{6}\frac{\partial f(x,0)}{\partial t}$$
$$+ \frac{\kappa^2\tau^2}{6}(2-\nu^2)\left(\delta_x^0 + \frac{\alpha}{24}\delta_x^2\right)^{-1}\left(-\frac{\Delta_h^\alpha \varphi(x)}{h^\alpha}\right)$$
$$+ \frac{\kappa^2\tau^2}{6}\left(\delta_x^0 + \frac{\alpha}{24}\delta_x^2\right)^{-1}\left(-\frac{\Delta_h^\alpha u(x,t_1)}{h^\alpha}\right) + R(x,t_1),$$

where there exists a positive constant $\widetilde{C}$ such that $|R(x,t_1)| \le \widetilde{C}(\tau^4 + h^4)$.

Hence the first level value $u(x_j,t_1)$ can be numerically determined by the equation

$$\left(\delta_x^0 + \frac{\alpha}{24}\delta_x^2\right)u_j^1 + \frac{\kappa^2\tau^2}{6h^\alpha}\delta_x^\alpha u_j^1$$
$$= \left(\delta_x^0 + \frac{\alpha}{24}\delta_x^2\right)\left(\varphi(x_j) + \frac{\tau}{6}(6 - 3\nu\tau + \nu^2\tau^2)\phi(x_j) + \frac{\tau^2}{6}(3 - \nu\tau)f(x_j,0)\right.$$
$$\left. + \frac{\tau^3}{6}\frac{\partial f(x_j,0)}{\partial t}\right) - \frac{\kappa^2\tau^2}{6h^\alpha}(2-\nu^2)\sum_{k=-(M-j)}^{j} g_k^{(\alpha)}\varphi(x_{j-k}), \quad j = 1,\dots,M-1.$$

3.3 Theoretical analysis of numerical algorithm

3.3.1 Stability analysis

Denote

$$V_h = \{v \mid v = \{v_j\},\ j = 0,1,\dots,M\}, \quad \mathring{V}_h = \{v \mid v \in V_h,\ v_0 = v_M = 0\}.$$

For any $u, v \in \mathring{V}_h$, we define the inner products

$$(u,v) = h\sum_{j=1}^{M-1} u_j v_j, \quad (\delta_x u, \delta_x v) = h\sum_{j=1}^{M-1}(\delta_x u_{j-\frac{1}{2}})(\delta_x v_{j-\frac{1}{2}})$$

and the corresponding norms

$$\|u\| = \sqrt{(u,u)}, \quad \|\delta_x u\| = \sqrt{(\delta_x u, \delta_x v)}.$$

In this case, we readily obtain

$$(\delta_x^2 u, v) = -(\delta_x u, \delta_x v)$$

and

$$\|\delta_x u\|^2 \le 4\|u\|^2.$$

Next, we introduce several lemmas which will be used in the analysis of stability.

Lemma 1. *The operator* $(\delta_x^0 + \frac{\alpha}{24}\delta_x^2)$ *is self-adjoint, i. e., for any* $u, v \in \mathring{V}_h$, *one has*

$$\left(\left(\delta_x^0 + \frac{\alpha}{24}\delta_x^2\right)u, v\right) = \left(u, \left(\delta_x^0 + \frac{\alpha}{24}\delta_x^2\right)v\right).$$

Proof. Simple calculations give

$$\begin{aligned}\left(\left(\delta_x^0 + \frac{\alpha}{24}\delta_x^2\right)u, v\right) &= (u, v) + \frac{\alpha}{24}(\delta_x^2 u, v) = (u, v) - \frac{\alpha}{24}(\delta_x u, \delta_x v)\\ &= (u, v) + \frac{\alpha}{24}(u, \delta_x^2 v) = \left(u, \left(\delta_x^0 + \frac{\alpha}{24}\delta_x^2\right)v\right).\end{aligned}$$

The proof is completed. □

Lemma 2. *For any* $v \in \mathring{V}_h$, *one has the following estimate:*

$$\left(1 - \frac{\alpha}{6}\right)\|v\|^2 \le \left(\left(\delta_x^0 + \frac{\alpha}{24}\delta_x^2\right)v, v\right) \le \|v\|^2.$$

Proof. On the one hand,

$$\left(\left(\delta_x^0 + \frac{\alpha}{24}\delta_x^2\right)u, v\right) = \|v\|^2 - \frac{\alpha}{24}\|\delta_x v\|^2 \le \|v\|^2.$$

On the other hand, we have

$$\left(\left(\delta_x^0 + \frac{\alpha}{24}\delta_x^2\right)u, v\right) = \|v\|^2 - \frac{\alpha}{24}\|\delta_x v\|^2 \ge \|v\|^2 - \frac{\alpha}{6}\|v\|^2.$$

Thus the proof is finished. □

Lemma 3. *The fractional operator* δ_x^α *is self-adjoint, i. e., for any* $u, v \in \mathring{V}_h$, *one has*

$$(\delta_x^\alpha u, v) = (u, \delta_x^\alpha v).$$

Proof. Define the following matrix A_α as the corresponding matrix associated with the fractional operator δ_x^α:

$$A_\alpha = \begin{pmatrix} g_0^{(\alpha)} & g_{-1}^{(\alpha)} & g_{-2}^{(\alpha)} & \cdots & g_{4-M}^{(\alpha)} & g_{3-M}^{(\alpha)} & g_{2-M}^{(\alpha)} \\ g_1^{(\alpha)} & g_0^{(\alpha)} & g_{-1}^{(\alpha)} & g_{-2}^{(\alpha)} & \cdots & g_{4-M}^{(\alpha)} & g_{3-M}^{(\alpha)} \\ g_2^{(\alpha)} & g_1^{(\alpha)} & g_0^{(\alpha)} & g_{-1}^{(\alpha)} & g_{-2}^{(\alpha)} & \cdots & g_{4-M}^{(\alpha)} \\ \vdots & \ddots & \ddots & \ddots & \ddots & \ddots & \vdots \\ g_{M-4}^{(\alpha)} & \cdots & g_2^{(\alpha)} & g_1^{(\alpha)} & g_0^{(\alpha)} & g_{-1}^{(\alpha)} & g_{-2}^{(\alpha)} \\ g_{M-3}^{(\alpha)} & g_{M-4}^{(\alpha)} & \cdots & g_2^{(\alpha)} & g_1^{(\alpha)} & g_0^{(\alpha)} & g_{-1}^{(\alpha)} \\ g_{M-2}^{(\alpha)} & g_{M-3}^{(\alpha)} & g_{M-4}^{(\alpha)} & \cdots & g_2^{(\alpha)} & g_1^{(\alpha)} & g_0^{(\alpha)} \end{pmatrix}.$$

The matrix form of $\delta_x^\alpha v_j$ $(j = 1, 2, \ldots, M-1)$ can be denoted by $A_\alpha v$. A direct calculation leads to

$$(\delta_x^\alpha u, v) = v^T A_\alpha u = (v^T A_\alpha u)^T = u^T A_\alpha v = (u, \delta_x^\alpha v).$$

This ends the proof. □

Lemma 4. *The fractional operator δ_x^α is positive definite, i. e., for any $v \in \mathring{V}_h$, one has*

$$(\delta_x^\alpha v, v) > 0.$$

Furthermore, there exists a positive definite operator $\tilde{\delta}_x^\alpha$, such that

$$(\delta_x^\alpha v, v) = (\tilde{\delta}_x^\alpha v, \tilde{\delta}_x^\alpha v).$$

Proof. In view of the properties of the coefficients $g_k^{(\alpha)}$, one can see that the matrix A_α is symmetric.

Next, we prove the positivity of A_α. From the Gershgorin theorem, we know that the eigenvalues $\lambda_j(A_\alpha)$ $(j = 1, 2, \ldots, M-1)$ of the matrix A_α satisfy

$$|\lambda_j(A_\alpha) - g_0^{(\alpha)}| \le \sum_{k=-(M-2),k\neq 0}^{M-2} |g_k^{(\alpha)}|,$$

i. e.,

$$\begin{aligned} -\sum_{k=-(M-2),k\neq 0}^{M-2} |g_k^{(\alpha)}| \le \lambda_j(A_\alpha) - g_0^{(\alpha)} &\le \sum_{k=-(M-2),k\neq 0}^{M-2} |g_k^{(\alpha)}| \\ &< \sum_{k=-\infty,k\neq 0}^{\infty} |g_k^{(\alpha)}| \le \frac{2^{1+\alpha}}{\pi}, \end{aligned}$$

or

$$g_0^{(\alpha)} - \sum_{k=-(M-2),k\neq 0}^{M-2} |g_k^{(\alpha)}| \leq \lambda_j(A_\alpha) < g_0^{(\alpha)} + \frac{2^{1+\alpha}}{\pi}.$$

Noting that

$$g_0^{(\alpha)} - \sum_{k=-(M-2),k\neq 0}^{M-2} |g_k^{(\alpha)}| = \sum_{k=-(M-2)}^{M-2} g_k^{(\alpha)} > \sum_{k=-\infty}^{\infty} g_k^{(\alpha)} = 0,$$

one has

$$0 < \lambda_j(A_\alpha) < g_0^{(\alpha)} + \frac{2^{1+\alpha}}{\pi}.$$

It immediately follows that matrix A_α is positive definite, i. e.,

$$(\delta_x^\alpha v, v) = v^T A_\alpha v > 0.$$

Because matrix A_α is symmetric and positive definite, there exists a symmetric and positive definite matrix B_α such that

$$A_\alpha = B_\alpha^2.$$

Thus,

$$(\delta_x^\alpha v, v) = v^T A_\alpha v = (v^T B_\alpha)(B_\alpha v) = (B_\alpha v)^T (B_\alpha v) = (\tilde{\delta}_x^\alpha v, \tilde{\delta}_x^\alpha v),$$

where $\tilde{\delta}_x^\alpha$ is the associated fractional difference operator of matrix B_α. This completes the proof. □

Lemma 5. *For any $v \in \mathring{V}_h$, one has the following estimate:*

$$\lambda_{\min}(A_\alpha)\|v\|^2 \leq (\delta_x^\alpha v, v) \leq \lambda_{\max}(A_\alpha)\|v\|^2,$$

where $\lambda_{\min}(A_\alpha)$ and $\lambda_{\max}(A_\alpha)$ denote the smallest and the largest eigenvalues of matrix A_α, respectively.

Proof. For any symmetric matrix A_α and vector $v \neq 0$, in view of the property of the Rayleigh–Ritz ratio [23], one has

$$\lambda_{\min}(A_\alpha) \leq \frac{(A_\alpha v, v)}{(v, v)} \leq \lambda_{\max}(A_\alpha).$$

Because $A_\alpha v$ denotes the matrix form of $\delta_x^\alpha v_j$ ($j = 1, 2, \ldots, M-1$), one has

$$\lambda_{\min}(A_\alpha)\|v\|^2 \leq (\delta_x^\alpha v, v) \leq \lambda_{\max}(A_\alpha)\|v\|^2.$$

This finishes the proof. □

Lemma 6. *If*

$$\frac{\tau}{h^\alpha}\left(\frac{\tau\nu^2}{12}+\frac{1}{\tau}\right)^{-1} \le \frac{6-\alpha}{2\kappa^2 g_0^{(\alpha)}}, \tag{21}$$

then the operator $((\delta_x^0 + \frac{\alpha}{24}\delta_x^2) - \frac{\tau\kappa^2}{6h^\alpha}(\frac{\tau\nu^2}{12}+\frac{1}{\tau})^{-1}\delta_x^\alpha)$ *is positive definite, i. e., for any* $v \in \mathring{V}_h$, *we have*

$$\left(\left(\left(\delta_x^0 + \frac{\alpha}{24}\delta_x^2\right) - \frac{\tau\kappa^2}{6h^\alpha}\left(\frac{\tau\nu^2}{12}+\frac{1}{\tau}\right)^{-1}\delta_x^\alpha\right)v, v\right) > 0.$$

Proof. Denote the incidence matrix of operator $(\delta_x^0 + \frac{\alpha}{24}\delta_x^2)$ by

$$D_\alpha = \begin{pmatrix} 1-\frac{\alpha}{12} & \frac{\alpha}{24} & 0 & & \cdots & 0 \\ \frac{\alpha}{24} & 1-\frac{\alpha}{12} & \frac{\alpha}{24} & 0 & & \vdots \\ & \frac{\alpha}{24} & 1-\frac{\alpha}{12} & \frac{\alpha}{24} & \ddots & \\ \vdots & \ddots & \ddots & \ddots & \ddots & \vdots \\ 0 & \cdots & & & \frac{\alpha}{24} & 1-\frac{\alpha}{12} \end{pmatrix}.$$

Obviously, the incidence matrix of the operator

$$\left(\left(\delta_x^0 + \frac{\alpha}{24}\delta_x^2\right) - \frac{\tau\kappa^2}{6h^\alpha}\left(\frac{\tau\nu^2}{12}+\frac{1}{\tau}\right)^{-1}\delta_x^\alpha\right)$$

is

$$G_\alpha = D_\alpha - \frac{\tau\kappa^2}{6h^\alpha}\left(\frac{\tau\nu^2}{12}+\frac{1}{\tau}\right)^{-1} A_\alpha.$$

Its eigenvalues satisfy

$$\begin{aligned} &\left|\lambda_j(G_\alpha) - \left[\left(1-\frac{\alpha}{12}\right) - \frac{\tau\kappa^2}{6h^\alpha}\left(\frac{\tau\nu^2}{12}+\frac{1}{\tau}\right)^{-1} g_0^{(\alpha)}\right]\right| \\ &\quad\le \frac{\alpha}{12} + \frac{\tau\kappa^2}{6h^\alpha}\left(\frac{\tau\nu^2}{12}+\frac{1}{\tau}\right)^{-1} \sum_{k=-(M-2),k\ne 0}^{M-2} |g_k^{(\alpha)}| \\ &\quad< \frac{\alpha}{12} + \frac{\tau\kappa^2}{6h^\alpha}\left(\frac{\tau\nu^2}{12}+\frac{1}{\tau}\right)^{-1} \sum_{k=-\infty,k\ne 0}^{\infty} |g_k^{(\alpha)}|. \end{aligned}$$

Then we know that

$$\lambda_j(G_\alpha) > \left(1-\frac{\alpha}{6}\right) - \frac{\tau\kappa^2}{6h^\alpha}\left(\frac{\tau\nu^2}{12}+\frac{1}{\tau}\right)^{-1}\left(g_0^{(\alpha)} + \sum_{k=-\infty,k\ne 0}^{\infty} |g_k^{(\alpha)}|\right) \ge 0$$

under condition (21). Further, one has

$$\left(\left(\left(\delta_x^0 + \frac{\alpha}{24}\delta_x^2\right) - \frac{\tau\kappa^2}{6h^\alpha}\left(\frac{\tau v^2}{12} + \frac{1}{\tau}\right)^{-1}\delta_x^\alpha\right)v, v\right) = v^T G_\alpha v > 0.$$

The proof is ended. □

Lemma 7. *For any $v \in \mathring{V}_h$, the following inequality holds:*

$$\left\|\left(\delta_x^0 + \frac{\alpha}{24}\delta_x^2\right)v\right\| \le \|v\|.$$

Proof. According to the definition, we have

$$\begin{aligned}\left\|\left(\delta_x^0 + \frac{\alpha}{24}\delta_x^2\right)v\right\|^2 &= \left(\left(\delta_x^0 + \frac{\alpha}{24}\delta_x^2\right)v, \left(\delta_x^0 + \frac{\alpha}{24}\delta_x^2\right)v\right)\\ &= \|v\|^2 - \frac{\alpha}{12}\|\delta_x v\|^2 + \frac{\alpha^2}{576}\|\delta_x^2 v\|^2\\ &\le \|v\|^2 + \frac{\alpha^2 - 12\alpha}{144}\|\delta_x v\|^2\\ &\le \|v\|^2.\end{aligned}$$

Therefore the proof is completed. □

Now, we give the stability analysis in details.

Theorem 17. *The difference scheme* (20) *is stable with respect to the initial value under condition* (21).

Proof. Suppose that v_j^s is the solution of the following difference equation:

$$\begin{cases}\left(\delta_x^0 + \frac{\alpha}{24}\delta_x^2\right)J_t v_j^s = -\frac{\kappa^2}{h^\alpha}H_t\delta_x^\alpha v_j^s + \left(\delta_x^0 + \frac{\alpha}{24}\delta_x^2\right)H_t f_j^s, \quad j = 1,\dots,M-1,\\ \quad s = 2,\dots,N-1,\\ v_j^0 = \varphi(x_j) + \rho_j^0, \quad j = 1,\dots,M-1,\\ \left(\delta_x^0 + \frac{\alpha}{24}\delta_x^2\right)v_j^1 + \frac{\kappa^2\tau^2}{6h^\alpha}\delta_x^\alpha v_j^1 = \left(\delta_x^0 + \frac{\alpha}{24}\delta_x^2\right)\Big(\varphi(x_j) + \frac{\tau}{6}(6 - 3v\tau + v^2\tau^2)\phi(x_j)\\ \quad + \frac{\tau^2}{6}(3 - v\tau)f(x_j,0) + \frac{\tau^3}{6}\frac{\partial f(x_j,0)}{\partial t}\Big) - \frac{\kappa^2\tau^2}{6h^\alpha}(2 - v^2)\sum\limits_{k=-(M-j)}^{j} g_k^{(\alpha)}\varphi(x_{j-k}) + \rho_j^1,\\ \quad j = 1,2,\dots,M-1.\\ v_0^s = v_M^s = 0, \quad s = 0,1,\dots,N.\end{cases} \tag{22}$$

Let $\varepsilon_j^s = v_j^s - u_j^s$. Subtracting (20) from (22) yields the following perturbation system:

$$\begin{cases} \left(\delta_x^0 + \frac{\alpha}{24}\delta_x^2\right) J_t \varepsilon_j^s = -\frac{\kappa^2}{h^\alpha} H_t \delta_x^\alpha \varepsilon_j^s, & j = 1, \dots, M-1,\ s = 2, \dots, N-1, \\ \varepsilon_j^0 = \rho_j^0, & j = 1, \dots, M-1, \\ \left(\delta_x^0 + \frac{\alpha}{24}\delta_x^2\right) \varepsilon_j^1 + \frac{\kappa^2 \tau^2}{6h^\alpha} \delta_x^\alpha \varepsilon_j^1 = \rho_j^1, & j = 1, 2, \dots, M-1, \\ \varepsilon_0^s = \varepsilon_M^s = 0, & s = 0, 1, \dots, N. \end{cases} \tag{23}$$

Firstly, taking the inner product of the third equation of (23) leads to

$$\left(\left(\delta_x^0 + \frac{\alpha}{24}\delta_x^2\right)\varepsilon^1, \varepsilon^1\right) + \frac{\kappa^2\tau^2}{6h^\alpha}(\delta_x^\alpha \varepsilon^1, \varepsilon^1) = (\rho_j^1, \varepsilon^1).$$

In view of Lemmas 2 and 4, the above equation can be reduced to

$$\|\varepsilon^1\| \le \frac{6}{6-\alpha}\|\rho^1\|.$$

Next, taking the inner product of the first equation of (23) yields

$$\left(\left(\delta_x^0 + \frac{\alpha}{24}\delta_x^2\right) J_t \varepsilon^s, \mu_t \delta_t \varepsilon^s\right) + \frac{\kappa^2}{h^\alpha}(H_t \delta_x^\alpha \varepsilon^s, \mu_t \delta_t \varepsilon^s) = 0. \tag{24}$$

For the first term of equation (24), it follows from Lemma 1 that

$$\begin{aligned} \left(\left(\delta_x^0 + \frac{\alpha}{24}\delta_x^2\right) J_t \varepsilon^s, \mu_t \delta_t \varepsilon^s\right) &= \left(\frac{\nu^2}{12} + \frac{1}{\tau^2}\right)\left(\left(\delta_x^0 + \frac{\alpha}{24}\delta_x^2\right)\delta_t^2 \varepsilon^s, \mu_t \delta_t \varepsilon^s\right) \\ &\quad + \frac{\nu}{2\tau}\left(\left(\delta_x^0 + \frac{\alpha}{24}\delta_x^2\right)\mu_t \delta_t \varepsilon^s, \mu_t \delta_t \varepsilon^s\right) \\ &= \left(\frac{\nu^2}{12} + \frac{1}{\tau^2}\right)\left(\left(\delta_x^0 + \frac{\alpha}{24}\delta_x^2\right)\delta_t \varepsilon^{s+\frac{1}{2}}, \delta_t \varepsilon^{s+\frac{1}{2}}\right) \\ &\quad - \left(\frac{\nu^2}{12} + \frac{1}{\tau^2}\right)\left(\left(\delta_x^0 + \frac{\alpha}{24}\delta_x^2\right)\delta_t \varepsilon^{s-\frac{1}{2}}, \delta_t \varepsilon^{s-\frac{1}{2}}\right) \\ &\quad + \frac{\nu}{2\tau}\left(\left(\delta_x^0 + \frac{\alpha}{24}\delta_x^2\right)\mu_t \delta_t \varepsilon^s, \mu_t \delta_t \varepsilon^s\right). \end{aligned}$$

Similarly, for the second term of equation (24), Lemma 3 leads to

$$\begin{aligned} &\frac{\kappa^2}{h^\alpha}(H_t \delta_x^\alpha \varepsilon^s, \mu_t \delta_t \varepsilon^s) \\ &= \frac{\kappa^2}{h^\alpha}(\delta_x^\alpha \varepsilon^s, \mu_t \delta_t \varepsilon^s) + \frac{\kappa^2}{12h^\alpha}(\delta_x^\alpha \delta_t^2 \varepsilon^s, \mu_t \delta_t \varepsilon^s) + \frac{\nu\tau\kappa^2}{24h^\alpha}(\delta_x^\alpha \mu_t \delta_t \varepsilon^s, \mu_t \delta_t \varepsilon^s) \\ &= \frac{\kappa^2}{h^\alpha}(\delta_x^\alpha \varepsilon^s, \varepsilon^{s+1}) - \frac{\kappa^2}{h^\alpha}(\delta_x^\alpha \varepsilon^s, \varepsilon^{s-1}) + \frac{\kappa^2}{12h^\alpha}(\delta_x^\alpha \delta_t \varepsilon^{s+\frac{1}{2}}, \delta_t \varepsilon^{s+\frac{1}{2}}) \end{aligned}$$

$$- \frac{\kappa^2}{12h^\alpha}(\delta_x^\alpha \delta_t \varepsilon^{s-\frac{1}{2}}, \delta_t \varepsilon^{s-\frac{1}{2}}) + \frac{\nu\tau\kappa^2}{24h^\alpha}(\delta_x^\alpha \mu_t \delta_t \varepsilon^s, \mu_t \delta_t \varepsilon^s).$$

Denote

$$\begin{aligned} W^s &= \left(\frac{\tau\nu^2}{12} + \frac{1}{\tau}\right)\left(\left(\delta_x^0 + \frac{\alpha}{24}\delta_x^2\right)\delta_t \varepsilon^{s+\frac{1}{2}}, \delta_t \varepsilon^{s+\frac{1}{2}}\right) + \frac{\tau\kappa^2}{h^\alpha}(\delta_x^\alpha \varepsilon^s, \varepsilon^{s+1}) \\ &\quad + \frac{\tau\kappa^2}{12h^\alpha}(\delta_x^\alpha \delta_t \varepsilon^{s+\frac{1}{2}}, \delta_t \varepsilon^{s+\frac{1}{2}}). \end{aligned}$$

Then equation (24) can be rewritten as

$$W^s + \frac{\nu}{2}\left(\left(\delta_x^0 + \frac{\alpha}{24}\delta_x^2\right)\mu_t \delta_t \varepsilon^s, \mu_t \delta_t \varepsilon^s\right) + \frac{\nu\tau^2\kappa^2}{24h^\alpha}(\delta_x^\alpha \mu_t \delta_t \varepsilon^s, \mu_t \delta_t \varepsilon^s) = W^{s-1}.$$

By Lemmas 2 and 4, it follows from the above equation that

$$W^s \le W^{s-1} \le \cdots \le W^0.$$

In addition, recalling Lemmas 2 and 5, we have the following estimate:

$$\begin{aligned} W^s &= \left(\frac{\tau\nu^2}{12} + \frac{1}{\tau}\right)\left(\left(\delta_x^0 + \frac{\alpha}{24}\delta_x^2\right)\delta_t \varepsilon^{s+\frac{1}{2}}, \delta_t \varepsilon^{s+\frac{1}{2}}\right) + \frac{\tau\kappa^2}{h^\alpha}(\delta_x^\alpha \varepsilon^s, \varepsilon^{s+1}) \\ &\quad + \frac{\tau\kappa^2}{12h^\alpha}(\delta_x^\alpha \delta_t \varepsilon^{s+\frac{1}{2}}, \delta_t \varepsilon^{s+\frac{1}{2}}) \\ &\le 2\left(\frac{\tau\nu^2}{12} + \frac{1}{\tau}\right)(\|\varepsilon^{s+1}\|^2 + \|\varepsilon^s\|^2) + \frac{\tau\kappa^2}{2h^\alpha}(\|\delta_x^\alpha \varepsilon^s\|^2 + \|\varepsilon^{s+1}\|^2) \\ &\quad + \frac{\tau\kappa^2}{6h^\alpha}\lambda_{\max}(A_\alpha)(\|\varepsilon^{s+1}\|^2 + \|\varepsilon^s\|^2) \\ &\le 2\left(\frac{\tau\nu^2}{12} + \frac{1}{\tau}\right)(\|\varepsilon^{s+1}\|^2 + \|\varepsilon^s\|^2) + \frac{\tau\kappa^2}{2h^\alpha}(\|\delta_x^\alpha \varepsilon^s\|^2 + \|\varepsilon^{s+1}\|^2) \\ &\quad + \frac{2^{2+\alpha}\tau\kappa^2}{6\pi h^\alpha}(\|\varepsilon^{s+1}\|^2 + \|\varepsilon^s\|^2). \end{aligned}$$

Also,

$$\begin{aligned} W^s &= \left(\left(\left(\frac{\tau\nu^2}{12} + \frac{1}{\tau}\right)\left(\delta_x^0 + \frac{\alpha}{24}\delta_x^2\right) - \frac{\tau\kappa^2}{6h^\alpha}\delta_x^\alpha\right)\delta_t \varepsilon^{s+\frac{1}{2}}, \delta_t \varepsilon^{s+\frac{1}{2}}\right) \\ &\quad + \frac{\tau\kappa^2}{h^\alpha}(\delta_x^\alpha \varepsilon^s, \varepsilon^{s+1}) + \frac{\tau\kappa^2}{4h^\alpha}(\delta_x^\alpha \delta_t \varepsilon^{s+\frac{1}{2}}, \delta_t \varepsilon^{s+\frac{1}{2}}) \\ &= \left(\left(\left(\frac{\tau\nu^2}{12} + \frac{1}{\tau}\right)\left(\delta_x^0 + \frac{\alpha}{24}\delta_x^2\right) - \frac{\tau\kappa^2}{6h^\alpha}\delta_x^\alpha\right)\delta_t \varepsilon^{s+\frac{1}{2}}, \delta_t \varepsilon^{s+\frac{1}{2}}\right) \\ &\quad + \frac{\tau\kappa^2}{h^\alpha}(\tilde{\delta}_x^\alpha \varepsilon^s, \tilde{\delta}_x^\alpha \varepsilon^{s+1}) + \frac{\tau\kappa^2}{4h^\alpha}(\tilde{\delta}_x^\alpha \delta_t \varepsilon^{s+\frac{1}{2}}, \tilde{\delta}_x^\alpha \delta_t \varepsilon^{s+\frac{1}{2}}) \\ &= \left(\left(\left(\frac{\tau\nu^2}{12} + \frac{1}{\tau}\right)\left(\delta_x^0 + \frac{\alpha}{24}\delta_x^2\right) - \frac{\tau\kappa^2}{6h^\alpha}\delta_x^\alpha\right)\delta_t \varepsilon^{s+\frac{1}{2}}, \delta_t \varepsilon^{s+\frac{1}{2}}\right) \end{aligned}$$

$$
\begin{aligned}
&+\frac{\tau\kappa^2}{h^\alpha}(\widetilde{\delta}_x^\alpha\varepsilon^{s+\frac{1}{2}},\widetilde{\delta}_x^\alpha\varepsilon^{s+\frac{1}{2}})-\frac{\tau\kappa^2}{h^\alpha}[(\widetilde{\delta}_x^\alpha\varepsilon^{s+\frac{1}{2}},\widetilde{\delta}_x^\alpha\varepsilon^{s+\frac{1}{2}})-(\widetilde{\delta}_x^\alpha\varepsilon^{s},\widetilde{\delta}_x^\alpha\varepsilon^{s+1})]\\
&+\frac{\tau\kappa^2}{4h^\alpha}(\widetilde{\delta}_x^\alpha\delta_t\varepsilon^{s+\frac{1}{2}},\widetilde{\delta}_x^\alpha\delta_t\varepsilon^{s+\frac{1}{2}})\\
&=\left(\left(\left(\frac{\tau\nu^2}{12}+\frac{1}{\tau}\right)(\delta_x^0+\frac{\alpha}{24}\delta_x^2)-\frac{\tau\kappa^2}{6h^\alpha}\delta_x^\alpha\right)\delta_t\varepsilon^{s+\frac{1}{2}},\delta_t\varepsilon^{s+\frac{1}{2}}\right)\\
&+\frac{\tau\kappa^2}{h^\alpha}(\delta_x^\alpha\varepsilon^{s+\frac{1}{2}},\varepsilon^{s+\frac{1}{2}})\\
&\geq\left(\frac{\tau\nu^2}{12}+\frac{1}{\tau}\right)\left(\left(\left(\delta_x^0+\frac{\alpha}{24}\delta_x^2\right)-\frac{\tau\kappa^2}{6h^\alpha}\left(\frac{\tau\nu^2}{12}+\frac{1}{\tau}\right)^{-1}\delta_x^\alpha\right)\delta_t\varepsilon^{s+\frac{1}{2}},\delta_t\varepsilon^{s+\frac{1}{2}}\right)\\
&+\frac{\tau\kappa^2}{h^\alpha}\lambda_{\min}(A_\alpha)\|\varepsilon^{s+\frac{1}{2}}\|^2\\
&\geq\lambda_{\min}(G_\alpha)\left(\frac{\tau\nu^2}{12}+\frac{1}{\tau}\right)\|\delta_t\varepsilon^{s+\frac{1}{2}}\|^2+\frac{\tau\kappa^2}{h^\alpha}\lambda_{\min}(A_\alpha)\|\varepsilon^{s+\frac{1}{2}}\|^2\\
&\geq\lambda_{\min}(G_\alpha)\left(\frac{\tau\nu^2}{12}+\frac{1}{\tau}\right)\|\delta_t\varepsilon^{s+\frac{1}{2}}\|^2.
\end{aligned}
$$

Hence, one has

$$
\begin{aligned}
\|\delta_t\varepsilon^{s+\frac{1}{2}}\|^2\leq\frac{1}{\lambda_{\min}(G_\alpha)}\Bigg[&2(\|\varepsilon^1\|^2+\|\varepsilon^0\|^2)\\
&+\frac{6-\alpha}{4g_0^{(\alpha)}}(\|\varepsilon^1\|^2+\|\delta_x^\alpha\varepsilon^0\|^2)+\frac{2^\alpha(6-\alpha)}{3\pi g_0^{(\alpha)}}(\|\varepsilon^1\|^2+\|\varepsilon^0\|^2)\Bigg]
\end{aligned}
$$

due to condition (21). Note that $\|\varepsilon^0\|=\|\rho^0\|$ and $\|\varepsilon^1\|\leq\frac{6}{6-\alpha}\|\rho^1\|$. One has

$$
\begin{aligned}
\|\delta_t\varepsilon^{s+\frac{1}{2}}\|^2\leq\frac{1}{\lambda_{\min}(G_\alpha)}\Bigg[&\left(2+\frac{2^\alpha(6-\alpha)}{3\pi g_0^{(\alpha)}}\right)\|\rho^0\|^2+\frac{(6-\alpha)}{4g_0^{(\alpha)}}\|\delta_x^\alpha\rho^0\|^2\\
&+\frac{36}{(6-\alpha)^2}\left(2+\frac{(6-\alpha)}{4g_0^{(\alpha)}}+\frac{2^\alpha(6-\alpha)}{3\pi g_0^{(\alpha)}}\right)\|\rho^1\|^2\Bigg].
\end{aligned}
$$

This ends the proof. □

3.3.2 Convergence analysis

Theorem 18. *Under condition* (21)*, the finite difference scheme* (20) *is convergent with order* $\mathcal{O}(\tau^4+h^4)$.

Proof. Define the error grid functions $e_j^s=u(x_j,t_s)-u_j^s$, $j=1,2,\dots,M-1$, $s=0,1,\dots,N$. Then we can obtain the following error system:

$$\begin{cases} \left(\delta_x^0 + \frac{\alpha}{24}\delta_x^2\right)J_t e_j^s + \frac{\kappa^2}{h^\alpha} H_t \delta_x^\alpha e_j^s = \left(\delta_x^0 + \frac{\alpha}{24}\delta_x^2\right) H_t R_j^s, \\ \quad j = 1, \ldots, M-1,\ s = 2, \ldots, N-1, \\ e_j^0 = 0, \quad j = 1, \ldots, M-1, \\ \left(\delta_x^0 + \frac{\alpha}{24}\delta_x^2\right) e_j^1 + \frac{\kappa^2\tau^2}{6h^\alpha}\delta_x^\alpha e_j^1 = R_j^1, \quad j = 1, 2, \ldots, M-1, \\ e_0^s = e_M^s = 0, \quad s = 0, 1, \ldots, N. \end{cases} \tag{25}$$

First of all, taking the inner product of the third equation of (25) yields

$$\left(\left(\delta_x^0 + \frac{\alpha}{24}\delta_x^2\right)e^1, e^1\right) + \frac{\kappa^2\tau^2}{6h^\alpha}(\delta_x^\alpha e^1, e^1) = (R^1, e^1).$$

Applying Lemmas 2 and 4 to the above equation leads to

$$\left(1 - \frac{\alpha}{6}\right)\|e^1\| \le \|R^1\| \le c_1\sqrt{L-l}(\tau^4 + h^4),$$

i. e.,

$$\|e^1\| \le \frac{6c_1}{6-\alpha}\sqrt{L-l}(\tau^4 + h^4).$$

Secondly, taking the inner product of the first equation of (25) gives

$$\begin{aligned} &\left(\left(\delta_x^0 + \frac{\alpha}{24}\delta_x^2\right)J_t e^s, \mu_t\delta_t e^s\right) + \frac{\kappa^2}{h^\alpha}(H_t\delta_x^\alpha e^s, \mu_t\delta_t e^s) \\ &= \left(\left(\delta_x^0 + \frac{\alpha}{24}\delta_x^2\right)H_t R^s, \mu_t\delta_t e^s\right). \end{aligned} \tag{26}$$

Denote

$$\begin{aligned} E^s = &\left(\frac{\tau\nu^2}{12} + \frac{1}{\tau}\right)\left(\left(\delta_x^0 + \frac{\alpha}{24}\delta_x^2\right)\delta_t e^{s+\frac{1}{2}}, \delta_t e^{s+\frac{1}{2}}\right) + \frac{\tau\kappa^2}{h^\alpha}(\delta_x^\alpha e^s, e^{s+1}) \\ &+ \frac{\tau\kappa^2}{12h^\alpha}(\delta_x^\alpha\delta_t e^{s+\frac{1}{2}}, \delta_t e^{s+\frac{1}{2}}). \end{aligned}$$

Similar to the stability analysis, equation (26) can be rewritten as

$$\begin{aligned} E^s &+ \frac{\nu}{2}\left(\left(\delta_x^0 + \frac{\alpha}{24}\delta_x^2\right)\mu_t\delta_t e^s, \mu_t\delta_t e^s\right) + \frac{\nu\tau^2\kappa^2}{24h^\alpha}(\delta_x^\alpha\mu_t\delta_t e^s, \mu_t\delta_t e^s) \\ &= E^{s-1} + \tau\left(\left(\delta_x^0 + \frac{\alpha}{24}\delta_x^2\right)H_t R^s, \mu_t\delta_t e^s\right). \end{aligned} \tag{27}$$

As for the right-hand term of equation (27), by the Cauchy–Schwarz inequality and Lemma 7, we have

$$\tau\left(\left(\delta_x^0 + \frac{\alpha}{24}\delta_x^2\right)H_t R^s, \mu_t\delta_t e^s\right)$$

$$\begin{aligned}
&\leq \tau\Big(\|R^s\| + \frac{1}{12}\|\delta_t^2 R^s\| + \frac{\nu\tau}{2}\|\mu_t\delta_t R^s\|\Big)(\|\delta_t e^{s+\frac{1}{2}}\| + \|\delta_t e^{s-\frac{1}{2}}\|)\\
&\leq \frac{\tau}{\lambda_{\min}(G_\alpha)}\Big(\|R^s\|^2 + \frac{1}{12}\|\delta_t^2 R^s\|^2 + \frac{\nu\tau}{2}\|\mu_t\delta_t R^s\|^2\Big)\\
&\quad + \tau\lambda_{\min}(G_\alpha)\Big(\frac{13}{24} + \frac{\nu\tau}{4}\Big)(\|\delta_t e^{s+\frac{1}{2}}\|^2 + \|\delta_t e^{s-\frac{1}{2}}\|^2)\\
&\leq \frac{\tau}{\lambda_{\min}(G_\alpha)}\Big(2\|R^s\|^2 + \Big(\frac{1}{4} + \nu\tau\Big)(\|R^{s+1}\|^2 + \|R^{s-1}\|^2)\Big)\\
&\quad + \tau\lambda_{\min}(G_\alpha)\Big(\frac{13}{24} + \frac{\nu\tau}{4}\Big)(\|\delta_t e^{s+\frac{1}{2}}\|^2 + \|\delta_t e^{s-\frac{1}{2}}\|^2).
\end{aligned}$$

Then equation (27) can be rewritten as

$$\begin{aligned}
E^s &\leq E^{s-1} + \tau\lambda_{\min}(G_\alpha)\Big(\frac{13}{24} + \frac{\nu\tau}{4}\Big)(\|\delta_t e^{s+\frac{1}{2}}\|^2 + \|\delta_t e^{s-\frac{1}{2}}\|^2)\\
&\quad + \frac{\tau}{\lambda_{\min}(G_\alpha)}\Big(2\|R^s\|^2 + \Big(\frac{1}{4} + \nu\tau\Big)(\|R^{s+1}\|^2 + \|R^{s-1}\|^2)\Big).
\end{aligned} \tag{28}$$

Similarly, one also has

$$E^s \geq \lambda_{\min}(G_\alpha)\Big(\frac{\tau\nu^2}{12} + \frac{1}{\tau}\Big)\|\delta_t e^{s+\frac{1}{2}}\|^2 \geq \frac{\sqrt{3}}{3}\lambda_{\min}(G_\alpha)\|\delta_t e^{s+\frac{1}{2}}\|^2. \tag{29}$$

Combining (28) with (29) gives

$$\begin{aligned}
E^s &\leq E^{s-1} + \sqrt{3}\tau\Big(\frac{13}{24} + \frac{\nu\tau}{4}\Big)(E^s + E^{s-1})\\
&\quad + \frac{\tau}{\lambda_{\min}(G_\alpha)}\Big(2\|R^s\|^2 + \Big(\frac{1}{4} + \nu\tau\Big)(\|R^{s+1}\|^2 + \|R^{s-1}\|^2)\Big).
\end{aligned}$$

If $\tau(\frac{13}{24} + \frac{\nu\tau}{4}) \leq \frac{\sqrt{3}}{9}$, we further obtain

$$\begin{aligned}
E^s &\leq \Big(1 + 6\sqrt{3}\tau\Big(\frac{13}{24} + \frac{\nu\tau}{4}\Big)\Big)E^{s-1}\\
&\quad + \frac{3\tau}{2\lambda_{\min}(G_\alpha)}\Big(2\|R^s\|^2 + \Big(\frac{1}{4} + \nu\tau\Big)(\|R^{s+1}\|^2 + \|R^{s-1}\|^2)\Big).
\end{aligned}$$

Utilizing the Grönwall inequality, one has

$$\begin{aligned}
E^s &\leq e^{6\sqrt{3}(\frac{13}{24} + \frac{\nu\tau}{4})s\tau}\Big[E^0 + \frac{3\tau}{2\lambda_{\min}(G_\alpha)}\sum_{k=1}^{s}\Big(2\|R^k\|^2\\
&\quad + \Big(\frac{1}{4} + \nu\tau\Big)(\|R^{k+1}\|^2 + \|R^{k-1}\|^2)\Big)\Big]\\
&\leq e^{6\sqrt{3}(\frac{13}{24} + \frac{\nu}{4})T}\Big[E^0 + \frac{3T(L-l)}{2\lambda_{\min}(G_\alpha)}\Big(2 + \Big(\frac{1}{4} + \nu\Big)(c_1^2 + 2c_2^2)\Big)(\tau^4 + h^4)^2\Big].
\end{aligned}$$

At this point, we further have

$$\|\delta_t e^{s+\frac{1}{2}}\|^2 \le \frac{(L-l)e^{6\sqrt{3}(\frac{13}{24}+\frac{\nu}{4})T}}{\lambda_{\min}(G_\alpha)}\left[\frac{36c_1^2(L-l)}{(6-\alpha)^2}\left(2+\frac{(6-\alpha)}{4g_0^{(\alpha)}}+\frac{2^\alpha(6-\alpha)}{3\pi g_0^{(\alpha)}}\right)\right.$$
$$\left.+\frac{3T}{2\lambda_{\min}(G_\alpha)}\left(2+\left(\frac{1}{4}+\nu\right)(c_1^2+2c_2^2)\right)\right](\tau^4+h^4)^2,$$

i. e.,

$$\|\delta_t e^{s+\frac{1}{2}}\| \le C(\tau^4+h^4).$$

Here $C = C_1(C_2+C_3)$,

$$C_1 = \exp\left(3\sqrt{3}\left(\frac{13}{24}+\frac{\nu}{4}\right)T\right)\sqrt{\frac{(L-l)}{\lambda_{\min}(G_\alpha)}},$$
$$C_2 = \frac{6c_1}{6-\alpha}\sqrt{(L-l)\left(2+\frac{(6-\alpha)}{4g_0^{(\alpha)}}+\frac{2^\alpha(6-\alpha)}{3\pi g_0^{(\alpha)}}\right)},$$

and

$$C_3 = \sqrt{\frac{3T}{2\lambda_{\min}(G_\alpha)}\left(2+(\frac{1}{4}+\nu)(c_1^2+2c_2^2)\right)}.$$

The proof is thus completed. □

3.4 Numerical example

Example 1. Consider the following equation:

$$\frac{\partial^2 u(x,t)}{\partial t^2}+\frac{\partial u(x,t)}{\partial t} = \frac{\partial^\alpha u(x,t)}{\partial |x|^\alpha}+f(x,t), \quad 1<\alpha<2,\ 0<x<1,\ 0<t\le 1,$$

where

$$f(x,t) = 2(2t^2+t+1)\exp(t^2)x^6(1-x)^6+\frac{1}{2}\exp(t^2)\sec\left(\frac{\pi}{2}\alpha\right)$$
$$\times\sum_{\ell=0}^{6}(-1)^\ell\frac{(6)!(6+\ell)![x^{6+\ell-\alpha}+(1-x)^{6+\ell-\alpha}]}{\ell!(6-\ell)!\Gamma(6+\ell+1-\alpha)}.$$

The analytical solution is $u(x,t) = x^6(1-x)^6\exp(t)$, which satisfies the necessary initial and boundary value conditions.

Table 1: The absolute error $E(\tau, h)$ and temporal and spatial experimental convergence orders (TECO and SECO) of Example 1 by difference scheme (20).

α		$E(\tau, h)$	TECO (SECO)
1.2	$h=\frac{1}{4}, \tau=\frac{1}{4}$	6.629148e−005	–
	$h=\frac{1}{8}, \tau=\frac{1}{8}$	3.374742e−006	4.2960
	$h=\frac{1}{16}, \tau=\frac{1}{16}$	1.987545e−007	4.0857
	$h=\frac{1}{32}, \tau=\frac{1}{32}$	1.216316e−008	4.0304
	$h=\frac{1}{64}, \tau=\frac{1}{64}$	7.535224e−010	4.0127
	$h=\frac{1}{128}, \tau=\frac{1}{128}$	4.689844e−011	4.0060
1.4	$h=\frac{1}{4}, \tau=\frac{1}{4}$	7.814928e−005	–
	$h=\frac{1}{8}, \tau=\frac{1}{8}$	3.293314e−006	4.5686
	$h=\frac{1}{16}, \tau=\frac{1}{16}$	1.906135e−007	4.1108
	$h=\frac{1}{32}, \tau=\frac{1}{32}$	1.160778e−008	4.0375
	$h=\frac{1}{64}, \tau=\frac{1}{64}$	7.178332e−010	4.0153
	$h=\frac{1}{128}, \tau=\frac{1}{128}$	4.465339e−011	4.0068
1.6	$h=\frac{1}{4}, \tau=\frac{1}{4}$	9.026191e−005	–
	$h=\frac{1}{8}, \tau=\frac{1}{8}$	3.701763e−006	4.6078
	$h=\frac{1}{16}, \tau=\frac{1}{16}$	2.350544e−007	3.9771
	$h=\frac{1}{32}, \tau=\frac{1}{32}$	1.451435e−008	4.0174
	$h=\frac{1}{64}, \tau=\frac{1}{64}$	9.043217e−010	4.0045
	$h=\frac{1}{128}, \tau=\frac{1}{128}$	5.651652e−011	4.0001
1.8	$h=\frac{1}{4}, \tau=\frac{1}{4}$	1.027098e−004	–
	$h=\frac{1}{8}, \tau=\frac{1}{8}$	4.552933e−006	4.4956
	$h=\frac{1}{16}, \tau=\frac{1}{16}$	2.329253e−007	4.2889
	$h=\frac{1}{32}, \tau=\frac{1}{32}$	1.358572e−008	4.0997
	$h=\frac{1}{64}, \tau=\frac{1}{64}$	8.403641e−010	4.0149
	$h=\frac{1}{128}, \tau=\frac{1}{128}$	5.231801e−011	4.0056

Here, the absolute error is calculated by

$$E(\tau, h) = \max_{0\le j\le M, 0\le s\le N} |u(x_j, t_s) - u_j^s|.$$

The spatial experimental convergence order (SECO) and temporal experimental convergence order (TECO) are both computed by the following formula:

$$\text{SECO} = \text{TECO} = \log_2\left(\frac{E(2\tau, 2h)}{E(\tau, h)}\right).$$

The absolute error and temporal and spatial experimental convergence orders are listed in Table 1. It can be concluded from the table that the convergence order of the finite difference scheme (20) is $\mathcal{O}(\tau^4 + h^4)$, which is in line with the theoretical analysis.

Bibliography

[1] A. A. Alikhanov, A new difference scheme for the time fractional diffusion equation, *J. Comput. Phys.*, **280** (2015), 424–438.

[2] J. X. Cao, C. P. Li, and Y. Chen, High-order approximation to Caputo derivatives and Caputo-type advection-diffusion equations (II), *Fract. Calc. Appl. Anal.*, **18** (2015), 735–761.

[3] C. Çelik and M. Duman, Crank–Nicolson method for the fractional diffusion equation with the Riesz fractional derivative, *J. Comput. Phys.*, **231** (2012), 1743–1750.

[4] A. Chen and C. P. Li, A novel compact ADI scheme for the time-fractional subdiffusion equation in two space dimensions, *Int. J. Comput. Math.*, **93** (2016), 889–914.

[5] A. Chen and C. P. Li, Numerical solution of fractional diffusion-wave equation, *Numer. Funct. Anal. Optim.*, **37** (2016), 19–39.

[6] C. M. Chen, F. Liu, I. Turner, and V. Anh, A Fourier method for the fractional diffusion equation describing sub-diffusion, *J. Comput. Phys.*, **227** (2007), 886–897.

[7] M. H. Chen and W. H. Deng, Fourth order accurate scheme for the space fractional diffusion equations, *SIAM J. Numer. Anal.*, **52** (2014), 1418–1438.

[8] M. H. Chen, W. H. Deng, and Y. J. Wu, Superlinearly convergent algorithms for the two-dimensional space-time Caputo–Riesz fractional diffusion equation, *Appl. Numer. Math.*, **70** (2013), 22–41.

[9] M. R. Cui, Compact finite difference method for the fractional diffusion equation, *J. Comput. Phys.*, **228** (2009), 7792–7804.

[10] H. F. Ding and C. P. Li, Mixed spline function method for reaction-subdiffusion equations, *J. Comput. Phys.*, **242** (2013), 103–123.

[11] H. F. Ding and C. P. Li, High-order algorithms for Riesz derivative and their applications (III), *Fract. Calc. Appl. Anal.*, **19** (2016), 19–55.

[12] H. F. Ding and C. P. Li, High-order compact difference schemes for the modified anomalous subdiffusion equation, *Numer. Methods Partial Differ. Equ.*, **32**(1) (2016), 213–242.

[13] H. F. Ding and C. P. Li, Fractional-compact numerical algorithms for Riesz spatial fractional reaction-dispersion equations, *Fract. Calc. Appl. Anal.*, **20**(3) (2017), 722–764.

[14] H. F. Ding and C. P. Li, High-order algorithms for Riesz derivative and their applications (V), *Numer. Methods Partial Differ. Equ.*, **33**(5) (2017), 1754–1794.

[15] H. F. Ding and C. P. Li, High-order numerical algorithms for Riesz derivatives via constructing new generating functions, *J. Sci. Comput.*, **71**(2) (2017), 759–784.

[16] H. F. Ding, C. P. Li, and Y. Q. Chen, High-order algorithms for Riesz derivative and their applications (I), *Abstr. Appl. Anal.*, (2014).

[17] H. F. Ding, C. P. Li, and Y. Q. Chen, High-order algorithms for Riesz derivative and their applications (II), *J. Comput. Phys.*, **293** (2015), 218–237.

[18] H. F. Ding, C. P. Li, and Q. Yi, A new second-order midpoint approximation formula for Riemann–Liouville derivative: algorithm and its application, *IMA J. Appl. Math.*, **82**(5) (2017), 909–944.

[19] G. H. Gao, H. W. Sun, and Z. Z. Sun, Stability and convergence of finite difference schemes for a class of time-fractional sub-diffusion equations based on certain superconvergence, *J. Comput. Phys.*, **280** (2015), 510–528.

[20] G. H. Gao and Z. Z. Sun, A compact finite difference scheme for the fractional sub-diffusion equations, *J. Comput. Phys.*, **230** (2011), 586–595.

[21] G. H. Gao, Z. Z. Sun, and H. W. Zhang, A new fractional numerical differentiation formula to approximate the Caputo fractional derivative and its applications, *J. Comput. Phys.*, **259** (2014), 33–50.

[22] Z. P. Hao, Z. Z. Sun, and W. R. Cao, A fourth-order approximation of fractional derivatives with its applications, *J. Comput. Phys.*, **281** (2015), 787–805.

[23] R. A. Horn and C. R. Johnson, *Matrix Analysis*, Cambridge University Press, Cambridge, 1986.
[24] X. L. Hu and L. M. Zhang, An analysis of a second order difference scheme for the fractional subdiffusion system, *Appl. Math. Model.*, **40** (2016), 1634–1649.
[25] T. Langlands and B. Henry, The accuracy and stability of an implicit solution method for the fractional diffusion equation, *J. Comput. Phys.*, **205** (2005), 719–736.
[26] C. P. Li and M. Cai, High-order approximation to Caputo derivatives and Caputo-type advection-diffusion equations: Revisited, *Numer. Funct. Anal. Optim.*, **38** (2017), 861–890.
[27] C. P. Li and H. F. Ding, Higher order finite difference method for the reaction and anomalous-diffusion equation, *Appl. Math. Model.*, **38** (2014), 3802–3821.
[28] C. P. Li, R. F. Wu, and H. F. Ding, High-order approximation to Caputo derivatives and Caputo-type advection–diffusion equations, *Commun. Appl. Ind. Math.*, **6** (2014), e-536, pp. 1–32, 10.1685/journal.caim.536.
[29] C. P. Li, Y. J. Wu, and R. S. Ye (eds.), *Recent Advances in Applied Nonlinear Dynamics with Numerical Analysis*, World Scientific, Singapore, 2013.
[30] C. P. Li and F. H. Zeng, Finite difference methods for fractional differential equations, *Int. J. Bifurc. Chaos*, **22** (2012), 427–432.
[31] C. P. Li and F. H. Zeng, *Numerical Methods for Fractional Calculus*, Chapman and Hall/CRC Press, Boca Raton, USA, 2015.
[32] H. F. Li, J. X. Cao, and C. P. Li, High-order approximation to Caputo derivatives and Caputo-type advection-diffusion equations (III), *J. Comput. Appl. Math.*, **299** (2016), 159–175.
[33] Y. Li and D. Wang, Improved efficient difference method for the modified anomalous sub-diffusion equation with a nonlinear source term, *Int. J. Comput. Math.*, **94**(4) (2017), 821–840.
[34] F. Liu, C. Yang, and K. Burrage, Numerical method and analytical technique of the modified anomalous subdiffusion equation with a nonlinear source term, *J. Comput. Appl. Math.*, **231** (2009), 160–176.
[35] C. Lubich, Discretized fractional calculus, *SIAM J. Math. Anal.*, **17** (1986), 704–719.
[36] W. H. Luo, C. P. Li, T. Z. Huang, X. M. Gu, and G. C. Wu, A high-order accurate numerical scheme for the Caputo derivative with an application to fractional diffusion problems, *Numer. Funct. Anal. Optim.*, **39** (2018), 600–622.
[37] V. Lynch, B. Carreras, D. del Castillo-Negrete, K. Ferreira-Mejias, and H. Hicks, Numerical methods for the solution of partial differential equations of fractional order, *J. Comput. Phys.*, **192** (2003), 406–421.
[38] M. M. Meerschaert and C. Tadjeran, Finite difference approximations for fractional advection-dispersion flow equations, *J. Comput. Appl. Math.*, **172** (2004), 65–77.
[39] M. M. Meerschaert and C. Tadjeran, Finite difference approximations for two-sided space-fractional partial differential equations, *Appl. Numer. Math.*, **56** (2006), 80–90.
[40] R. Metzler and J. Klafter, The random walk's guide to anomalous diffusion: a fractional dynamics approach, *Phys. Rep.*, **339** (2000), 1–77.
[41] A. Mohebbi, M. Abbaszadeh, and M. Dehghan, A high-order and unconditionally stable scheme for the modified anomalous fractional sub-diffusion equation with a nonlinear source term, *J. Comput. Phys.*, **240** (2013), 36–48.
[42] H. M. Nasir, B. L. K. Gunawardana, and H. M. N. P. Abeyrathna, A second order finite difference approximation for the fractional diffusion equation, *Int. J. Appl. Phys. Math.*, **3** (2013), 237–242.
[43] K. B. Oldham and J. Spanier, *The Fractional Calculus*, Academic Press, New York, 1974.
[44] M. D. Ortigueira, Riesz potential operators and inverses via fractional centred derivatives, *Int. J. Math. Math. Sci.*, **2006** (2006), 48391, 12 pages.

[45] I. Podlubny, *Fractional Differential Equations*, Academic Press, New York, 1998.

[46] L. B. Rall, *Perspectives on Automatic Differentiation: Past, Present, and Future? Automatic Differentiation: Applications, Theory, and Implementations*, Springer, Berlin, 2006.

[47] S. G. Samko, A. A. Kilbas, and O. I. Marichev, *Fractional Integrals and Derivatives: Theory and Applications*, Gordon and Breach, Amsterdam, 1993.

[48] E. Sousa, Finite difference approximations for a fractional advection diffusion problem, *J. Comput. Phys.*, **228** (2009), 4038–4054.

[49] E. Sousa and C. Li, A weighted finite difference method for the fractional diffusion equation based on the Riemann–Liouville derivative, *Appl. Numer. Math.*, **90** (2015), 22–37.

[50] H. Sun, Z. Z. Sun, and G. H. Gao, Some high order difference schemes for the space and time fractional Bloch–Torrey equations, *Appl. Math. Comput.*, **281** (2016), 356–380.

[51] Z. Z. Sun and X. N. Wu, A fully discrete difference scheme for a diffusion-wave system, *Appl. Numer. Math.*, **56** (2006), 193–209.

[52] W. Y. Tian, H. Zhou, and W. H. Deng, A class of second order difference approximations for solving space fractional diffusion equations, *Math. Comput.*, **84** (2015), 1703–1727.

[53] V. K. Tuan and R. Gorenflo, Extrapolation to the limit for numerical fractional differentiation, *ZAMM J. Appl. Math. Mech.*, **75** (1995), 646–648.

[54] Y. M. Wang, A compact finite difference method for solving a class of time fractional convection-subdiffusion equations, *BIT Numer. Math.*, **55** (2015), 1187–1217.

[55] Z. B. Wang and S. Vong, Compact difference schemes for the modified anomalous fractional sub-diffusion equation and the fractional diffusion-wave equation, *J. Comput. Phys.*, **277** (2014), 1–15.

[56] Q. Yang, F. Liu, and I. Turner, Numerical methods for fractional partial differential equations with Riesz space fractional derivatives, *Appl. Math. Model.*, **34** (2010), 200–218.

[57] Y. Y. Yu, W. H. Deng, and Y. J. Wu, Fourth order quasi-compact difference schemes for (tempered) space fractional diffusion equations, *Commun. Math. Sci.*, **15**(5) (2017), 1183–1209.

[58] S. B. Yuste, Weighted average finite difference methods for fractional diffusion equations, *J. Comput. Phys.*, **216** (2006), 264–274.

[59] F. H. Zeng, Second-order stable finite difference schemes for the time-fractional diffusion-wave equation, *J. Sci. Comput.*, **65** (2015), 411–430.

[60] F. H. Zeng, C. P. Li, F. Liu, and I. Turner, Numerical algorithms for time-fractional subdiffusion equation with second-order accuracy, *SIAM J. Sci. Comput.*, **37** (2015), A55–A78.

[61] F. H. Zeng, F. Liu, C. P. Li, K. Burrage, I. Turner, and V. Anh, A Crank–Nicolson ADI spectral method for a two-dimensional Riesz space fractional nonlinear reaction-diffusion equation, *SIAM J. Numer. Anal.*, **52** (2014), 2599–2622.

[62] Y. N. Zhang and Z. Z. Sun, Error analysis of a compact ADI scheme for the 2D fractional subdiffusion equation, *J. Sci. Comput.*, **59** (2014), 104–128.

[63] Y. X. Zhang, H. F. Ding, and J. C. Luo, Fourth-order compact difference schemes for the Riemann–Liouville and Riesz derivatives, *Abstr. Appl. Anal.*, (2014), 540692.

[64] L. J. Zhao and W. H. Deng, A series of high-order quasi-compact schemes for space fractional diffusion equations based on the superconvergent approximations for fractional derivatives, *Numer. Methods Partial Differ. Equ.*, **31** (2015), 1345–1381.

[65] X. Zhao, Z. Z. Sun, and Z. P. Hao, A fourth-order compact ADI scheme for two-dimensional nonlinear space fractional Schrödinger equation, *SIAM J. Sci. Comput.*, **36** (2014), A2865–A2886.

[66] H. Zhou, W. Y. Tian, and W. H. Deng, Quasi-compact finite difference schemes for space fractional diffusion equations, *J. Sci. Comput.*, **56** (2013), 45–66.

Chuanju Xu

Spectral methods for some kinds of fractional differential equations

Traditional and Müntz spectral methods

Abstract: In this chapter, we design and analyze two types of spectral methods for fractional differential equations. The first spectral method makes use of the traditional polynomials and follows the standard Galerkin framework, while the second method is based on the Müntz polynomials and weighted Galerkin approach. For the first method we will first introduce suitable functional spaces and develop a theoretical framework for the weak solution of the STFDE. This allows to apply the existing theory for elliptic problems to prove the existence and uniqueness of the weak solution. Then we construct a Galerkin spectral method for efficiently solving the space-time fractional diffusion problem. Optimal error estimates are derived under certain regularity conditions on the exact solution. For the second method, we first introduce a new class of generalized fractional Jacobi polynomials (GFJPs), also known as Müntz polynomials, and investigate their approximation properties. Then we propose an efficient scheme using GFJPs for the time-fractional diffusion equation. Our theoretical or numerical investigation shows that the proposed scheme is exponentially convergent for general right hand side functions, even though the exact solution has very limited regularity. Implementation details are also provided, along with a series of numerical examples to show the efficiency of the proposed methods.

Keywords: Time-space fractional differential equations, weak solution, well-posedness, spectral methods, Müntz polynomials

MSC 2010: 35S10, 35A05, 65M70, 65M12, 65N35, 45K05, 41A05, 41A10, 41A25

1 Introduction

As is well known, any discretization including low-order approaches of a fractional derivative has to take into account its non-local structure, which results in a full linear system and a high storage requirement. Therefore it is very natural to consider a global method, such as the spectral method, since the high accuracy of spectral methods may significantly reduce the storage requirements.

Chuanju Xu, School of Mathematical Sciences and Fujian Provincial Key Laboratory of Mathematical Modeling and High Performance Scientific Computing, Xiamen University, 361005 Xiamen, China, e-mail: cjxu@xmu.edu.cn

https://doi.org/10.1515/9783110571684-004

In this chapter, we consider two space-time fractional diffusion equations (STFDEs) and construct efficient space-time spectral methods for their numerical solutions. Generally speaking, compared to the spectral method in space for classical integer-order differential equations, the spectral method in time is relatively rare. This is mainly due to the fact that for classical parabolic problems the time derivative is of the first order, which makes the construction of efficient spectral methods quite troublesome. However, as we will see below, the fractional differential operator of order α, with $0 < \alpha < 1$, possesses some features of elliptic operators, as long as suitable spaces and norms are chosen. This makes it possible to use the standard Galerkin formulation. It is known that a suitable variational formulation is essential for a spectral method to be efficient, especially if we want a good numerical analysis. Moreover, formulation of weak problems also strongly relies on the choice of suitable spaces and norms. The main ingredient of this chapter includes two parts: (1) supposing the exact solution is regular enough we propose and analyze a standard Galerkin spectral method using the traditional polynomials for the STFDE; (2) for the solution having weak singularity on the boundary, we construct a fractional spectral method based on the Müntz polynomials and weighted Galerkin approach for the time-fractional diffusion equation (TFDE).

- For the traditional spectral STFDE method, we first introduce suitable fractional Sobolev spaces and investigate basic properties of these spaces. A weak problem and its well-posedness for STFDE for both the Riemann–Liouville derivative and the Caputo derivative are established. Then we construct and analyze an efficient spectral method for numerical approximations to the weak solution. Based on the weak formulation and the polynomial approximation results in the related Sobolev spaces, optimal error estimates are derived.
- For the fractional Jacobi spectral method for TFDE, we will first develop a kind of fractional Jacobi polynomials as the approximating space and derive basic approximation results for some weighted projection operators defined in suitable weighted Sobolev spaces. Then a detailed convergence analysis is carried out, and several error estimates are established. The main novelty of this method is that the exponential convergence can be attained for non-smooth solutions, which is the typical case of many fractional differential equations with smooth data.

2 Functional spaces

The first part of this chapter is devoted to providing basic tool for the investigation of the well-posedness of weak solutions. We first introduce some notations that will be used throughout this chapter. Let $\Omega = (-1,1)^d$, $I = (0,T)$, $Q = \Omega \times I$, where $d \geq 1$ is the space dimension. We use the symbol $\mathcal{O}$ to denote a domain which may stand for Ω, I, Q, or $\mathbb{R}$. Let $L^2(\mathcal{O})$ be the space of measurable functions whose square is Lebesgue integrable in $\mathcal{O}$. The inner product and norm of $L^2(\mathcal{O})$ are defined by

$$(u,v)_{\mathcal{O}} = \int_{\mathcal{O}} uv\,d\mathcal{O}, \quad \|u\|_{0,\mathcal{O}} = (u,u)_{\mathcal{O}}^{1/2}, \quad \forall u,v \in L^2(\mathcal{O}).$$

For a non-negative real number s, we use $H^s(\mathcal{O})$ and $H_0^s(\mathcal{O})$ to denote the usual Sobolev spaces, whose norm is denoted by $\|\cdot\|_{s,\mathcal{O}}$. Let $C_0^\infty(\mathcal{O})$ stand for the space of all functions having continuous derivatives of all orders and compactly supported in $\mathcal{O}$. For the Sobolev space X with norm $\|\cdot\|_X$, let

$$H^s(I;X) := \{v;\ \|v(\cdot,t)\|_X \in H^s(I)\},$$

endowed with the norm

$$\|v\|_{H^s(I;X)} := \big\|\|v(\cdot,t)\|_X\big\|_{s,I}.$$

Particularly, when X is $H^\sigma(\Omega)$ or $H_0^\sigma(\Omega)$, $\sigma \geqslant 0$, the norm of the space $H^s(I;X)$ will be denoted by $\|\cdot\|_{\sigma,s,Q}$. Hereafter, in cases where no confusion would arise, the domain symbols Ω, I, or Q may be dropped from the notations.

We then recall some definitions of fractional derivatives and fractional integrals; see [11, 12]. Let $\Gamma(\cdot)$ denote the Gamma function. For any positive integer n and $n-1 \le s < n$, $-\infty \le a < b \le \infty$, v is defined in the interval $[a,b]$, and the Caputo derivative, Riemann–Liouville derivative, and fractional integral of order s of v are respectively defined as

$$\text{left Caputo derivative:}\quad {}_a^C D_t^s v(t) = \frac{1}{\Gamma(n-s)} \int_a^t \frac{v^{(n)}(\tau)d\tau}{(t-\tau)^{s-n+1}} \quad \forall t \in [a,b], \tag{D1}$$

$$\text{right Caputo derivative:}\quad {}_t^C D_b^s v(t) = \frac{(-1)^n}{\Gamma(n-s)} \int_t^b \frac{v^{(n)}(\tau)\mathrm{d}\tau}{(\tau-t)^{s-n+1}} \quad \forall t \in [a,b], \tag{D2}$$

$$\text{left R–L derivative:}\quad {}_a^R D_t^s v(t) = \frac{1}{\Gamma(n-s)} \frac{\mathrm{d}^n}{\mathrm{d}t^n} \int_a^t \frac{v(\tau)\mathrm{d}\tau}{(t-\tau)^{s-n+1}} \quad \forall t \in [a,b], \tag{D3}$$

$$\text{right R–L derivative:}\quad {}_t^R D_t^s v(t) = \frac{(-1)^n}{\Gamma(n-s)} \frac{\mathrm{d}^n}{\mathrm{d}t^n} \int_t^b \frac{v(\tau)\mathrm{d}\tau}{(\tau-t)^{s-n+1}} \quad \forall t \in [a,b], \tag{D4}$$

$$\text{fractional integral:}\quad {}_a I_t^s v(t) = \frac{1}{\Gamma(s)} \int_0^t \frac{v(\tau)\mathrm{d}\tau}{(t-\tau)^{1-s}} \quad \forall t \in [a,b]. \tag{I1}$$

The same definitions apply for the derivatives with respect to the spatial variable x.

Let c stand for a generic positive constant independent of any functions and of any discretization parameters. In what follows, the expression $A \lesssim B$ (respectively,

$A \gtrsim B$) means that $A \leqslant cB$ (respectively, $A \geqslant cB$), and the expression $A \cong B$ means that $A \lesssim B \lesssim A$.

Let $\Lambda = (a, b)$, which may stand for I or Ω. For any real $s \geq 0$, we define the spaces

$$ {}^{l}H^{s}(\Lambda) := \{v; \|v\|_{{}^{l}H^{s}(\Lambda)} < \infty\}, \tag{2.1}$$

with

$$\|v\|_{{}^{l}H^{s}(\Lambda)} := \left(\|v\|_{0,\Lambda}^{2} + |v|_{{}^{l}H^{s}(\Lambda)}^{2}\right)^{\frac{1}{2}}, \quad |v|_{{}^{l}H^{s}(\Lambda)} := \|{}^{R}D_{z}^{s}v\|_{0,\Lambda}, \tag{2.2}$$

and

$$ {}^{r}H^{s}(\Lambda) := \{v; \|v\|_{{}^{r}H^{s}(\Lambda)} < \infty\}, \tag{2.3}$$

with

$$\|v\|_{{}^{r}H^{s}(\Lambda)} := \left(\|v\|_{0,\Lambda}^{2} + |v|_{{}^{r}H^{s}(\Lambda)}^{2}\right)^{\frac{1}{2}}, \quad |v|_{{}^{r}H^{s}(\Lambda)} := \|{}_{z}^{R}D^{s}v\|_{0,\Lambda}. \tag{2.4}$$

Let ${}^{l}H_{0}^{s}(\Lambda)$ and ${}^{r}H_{0}^{s}(\Lambda)$ be the closures of $C_{0}^{\infty}(\Lambda)$ with respect to the norms $\|v\|_{{}^{l}H^{s}(\Lambda)}$ and $\|v\|_{{}^{r}H^{s}(\Lambda)}$, respectively. In the above notations the exponents "l" and "r" have been used to indicate the left and right fractional derivatives in the norm definitions, respectively. The definitions of these spaces differ from the ones of the usual Sobolev spaces. However, as we are going to see, these spaces are indeed equivalent for $s \neq n - 1/2$ in a sense to be specified later. To this end, in the usual Sobolev space $H_{0}^{s}(\Lambda)$, we also define

$$|v|_{H_{0}^{s}(\Lambda)}^{*} := \left(\frac{({}^{R}D_{z}^{s}v, {}_{z}^{R}D^{s}v)_{\Lambda}}{\cos(\pi s)}\right)^{\frac{1}{2}}, \quad \forall v \in H_{0}^{s}(\Lambda).$$

We will prove that this functional is well-defined and equivalent to the usual semi-norm $|\cdot|_{H_{0}^{s}(\Lambda)}$ in the sense that

$$|v|_{H_{0}^{s}(\Lambda)}^{*} \lesssim |v|_{H_{0}^{s}(\Lambda)} \lesssim |v|_{H_{0}^{s}(\Lambda)}^{*}, \quad \forall v \in H_{0}^{s}(\Lambda).$$

In fact, a more general result will be proven in Lemma 6. Nevertheless, we emphasize that $|\cdot|_{H_{0}^{s}(\Lambda)}^{*}$ is not a semi-norm as it does not satisfy the triangular inequality [9]. We first recall the following result.

Lemma 1 ([8]). *Let $s > 0$, $s \neq n - \frac{1}{2}$; the semi-norms $|\cdot|_{{}^{l}H^{s}(\Lambda)}$, $|\cdot|_{{}^{r}H^{s}(\Lambda)}$ and $|\cdot|_{H_{0}^{s}(\Lambda)}$ are all equivalent to $|\cdot|_{H_{0}^{s}(\Lambda)}^{*}$ in space $C_{0}^{\infty}(\Lambda)$.*

We list below a number of useful properties related to the fractional derivatives and integrals. The detailed proofs can be found in Li and Xu [9].

Lemma 2. *If $0 < p < 1/2$, $v \in L^{2}(\Lambda)$, or if $1/2 \leq p < 1$, $v \in H^{s}(\Lambda)$, $p - 1/2 < s < 1/2$, then we have*

$$|I_{z}^{1-p}v(z)|_{z=a^{+}} = 0. \tag{2.5}$$

Lemma 3. *For real s,* $0 < s < 1$, *if* $w \in {}^{l}H^{s}(\Lambda) \cap H^{s}(\Lambda)$, $v \in C^{\infty}(\Lambda)$, *then*

$$({}^{R}D_{z}^{s}w(z), v(z))_{\Lambda} = (w(z), {}_{z}^{R}D^{s}v(z))_{\Lambda}. \tag{2.6}$$

For general positive real s, we have the following result.

Lemma 4. *For all positive real s, if* $w \in {}^{l}H^{s}(\Lambda)$, $v \in C_{0}^{\infty}(\Lambda)$, *then*

$$({}^{R}D_{z}^{s}w(z), v(z))_{\Lambda} = (w(z), {}_{z}^{R}D^{s}v(z))_{\Lambda}. \tag{2.7}$$

For a given positive real s, the fractional derivative ${}^{R}D_{z}^{s}v$ can be generalized for all $v \in L^{2}(\Lambda)$ in the following way: for $v \in L^{2}(\Lambda)$, we define the linear functional, denoted still by ${}^{R}D_{z}^{s}v\colon C_{0}^{\infty}(\Lambda) \to \mathbb{R}$, through

$${}^{R}D_{z}^{s}v(\phi) := \int_{\Lambda} v\, {}_{z}^{R}D^{s}\phi \,\mathrm{d}z \quad \forall \phi \in C_{0}^{\infty}(\Lambda). \tag{2.8}$$

It was proven in [9] that ${}^{R}D_{z}^{s}v(\phi)$ is a continuous functional in $C_{0}^{\infty}(\Lambda)$. Thus ${}^{R}D_{z}^{s}v$ is a distribution. This means that for $v \in L^{2}(\Lambda)$, ${}^{R}D_{z}^{s}v$ can be defined as a distribution, which, by virtue of Lemma 4, coincides with the standard definition (D3) if $v \in {}^{l}H^{s}(\Lambda)$.

Thanks to these properties, we are able to prove the following fundamental results; see [9] for the detailed proofs.

Lemma 5. *For all positive real s, spaces* ${}^{l}H^{s}(\Lambda)$ *and* ${}^{r}H^{s}(\Lambda)$ *are complete.*

Lemma 6. *For* $s > 0$, $s \neq n - 1/2$, *the spaces* ${}^{l}H_{0}^{s}(\Lambda)$, ${}^{r}H_{0}^{s}(\Lambda)$, *and* $H_{0}^{s}(\Lambda)$ *are equal and their semi-norms are all equivalent to* $|\cdot|_{H_{0}^{s}(\Lambda)}^{*}$.

Lemma 7. *For* $0 < s < 2$, $s \neq 1$, $w \in H_{0}^{\frac{s}{2}}(\Lambda)$, ${}^{R}D_{z}^{s}w = {}^{R}D_{z}^{\frac{s}{2}}\,{}^{R}D_{z}^{\frac{s}{2}}w$ *holds in the distribution sense, i. e.,*

$$\langle {}^{R}D_{z}^{s}w(z), \phi(z)\rangle_{\Lambda} = \langle {}^{R}D_{z}^{\frac{s}{2}}\,{}^{R}D_{z}^{\frac{s}{2}}w(z), \phi(z)\rangle_{\Lambda}, \quad \forall \phi \in C_{0}^{\infty}(\Lambda).$$

Furthermore, ${}^{R}D_{z}^{s}w \in H^{-\frac{s}{2}}(\Lambda)$.

Lemma 8. *If* $0 < s < 2$, $s \neq 1$, $w, v \in H_{0}^{\frac{s}{2}}(\Lambda)$, *then*

$$\langle {}^{R}D_{z}^{s}w(z), v(z)\rangle_{\Lambda} = ({}^{R}D_{z}^{\frac{s}{2}}w(z), {}_{z}^{R}D^{\frac{s}{2}}v(z))_{\Lambda}, \tag{2.9}$$

$$\langle {}_{z}^{R}D^{s}w(z), v(z)\rangle_{\Lambda} = ({}_{z}^{R}D^{\frac{s}{2}}w(z), {}^{R}D_{z}^{\frac{s}{2}}v(z))_{\Lambda}. \tag{2.10}$$

Remark 1. Property (2.9) has first been proven in Lemma 2.6 in [8] under a stronger assumption, i. e., assuming w is a function in $H^{1}(\Lambda)$ with $w(a) = 0$. The present result, which was proven in [9], is an improvement by removing this restriction. This is an important point which allows establishing suitable weak formulations for the problems considered in this section.

3 Time-space fractional diffusion equation

We consider the following two space-time fractional diffusion problems: $0 < \alpha < 1$, $1 < \beta < 2$,

$$\begin{cases} {}^R D_t^{\alpha} u(x,t) - p_1 {}^R D_x^{\beta} u(x,t) - p_2 {}^R_x D^{\beta} u(x,t) = f(x,t) & \forall (x,t) \in Q, \\ u(x,t)|_{\partial\Omega} = 0 & \forall t \in I, \\ I_t^{1-\alpha} u(x,0) = 0 & \forall x \in \Omega \end{cases} \tag{3.1}$$

and

$$\begin{cases} {}^C D_t^{\alpha} u(x,t) - p_1 {}^R D_x^{\beta} u(x,t) - p_2 {}^R_x D^{\beta} u(x,t) = f(x,t) & \forall (x,t) \in Q, \\ u(x,t)|_{\partial\Omega} = 0 & \forall t \in I, \\ u(x,0) = u_0(x) & \forall x \in \Omega, \end{cases} \tag{3.2}$$

where p_1, p_2 are two constants satisfying

$$p_1 + p_2 = 1, \quad 0 < p_1, p_2 < 1.$$

Note that different fractional time derivatives, Riemann–Liouville and Caputo, correspond to different initial conditions, the reason of which will become clear after the well-posedness analysis to be carried out later. It is first notable that, according to Lemma 2, the initial condition in (3.1) will be automatically satisfied (thus becomes unnecessary) if $u(x,\cdot) \in L^2(I)$ for $0 < \alpha < 1/2$ or $u(x,\cdot) \in H^s(I)$ for $1/2 \leq \alpha < 1$, $\alpha - 1/2 < s < 1/2$.

We define the space

$$B^{s,\sigma}(Q) := H^s(I; L^2(\Omega)) \cap L^2(I; H_0^{\sigma}(\Omega))$$

equipped with the norm

$$\|v\|_{B^{s,\sigma}} := \left(\|v\|^2_{H^s(I;L^2(\Omega))} + \|v\|^2_{L^2(I;H^{\sigma}(\Omega))}\right)^{1/2}.$$

3.1 Weak formulation with Riemann–Liouville derivative

We consider the weak formulation of problem (3.1) as follows: for $f \in B^{\frac{\alpha}{2},\frac{\beta}{2}}(Q)'$, the dual space of $B^{\frac{\alpha}{2},\frac{\beta}{2}}(Q)$, find $u \in B^{\frac{\alpha}{2},\frac{\beta}{2}}(Q)$ such that

$$\mathcal{A}(u,v) = \mathcal{F}_1(v), \quad \forall v \in B^{\frac{\alpha}{2},\frac{\beta}{2}}(Q), \tag{3.3}$$

where the bilinear form $\mathcal{A}(\cdot,\cdot)$ is defined by

$$\mathcal{A}(u,v) := \left({}^R D_t^{\frac{\alpha}{2}} u, {}^R_t D^{\frac{\alpha}{2}} v\right)_Q - p_1\left({}^R D_x^{\frac{\beta}{2}} u, {}^R_x D^{\frac{\beta}{2}} v\right)_Q - p_2\left({}^R_x D^{\frac{\beta}{2}} u, {}^R D_x^{\frac{\beta}{2}} v\right)_Q \tag{3.4}$$

and the functional $\mathcal{F}_1(\cdot)$ is given by

$$\mathcal{F}_1(v) := \langle f, v\rangle_Q.$$

The well-posedness of the weak problem (3.3) is provided in the following theorem [9] by using the functional tool given in the Section 2.

Theorem 1. *For all* $0 < \alpha < 1, 1 < \beta < 2$, *and* $f \in B^{\frac{\alpha}{2},\frac{\beta}{2}}(Q)'$, *problem* (3.3) *admits a unique solution. Furthermore, the solution u satisfies*

$$\|u\|_{B^{\frac{\alpha}{2},\frac{\beta}{2}}(Q)} \lesssim \|f\|_{B^{\frac{\alpha}{2},\frac{\beta}{2}}(Q)'}. \tag{3.5}$$

The link between variational problem (3.3) and problem (3.1) is established below.

Theorem 2. *For all* $0 < \alpha < 1$, $1 < \beta < 2, f \in B^{\frac{\alpha}{2},\frac{\beta}{2}}(Q)'$, *if u is a solution of problem* (3.1), *then u is also a solution of weak form* (3.3). *Reciprocally, if u is the solution of weak form* (3.3), *then it is also a solution of problem* (3.1) *in the distribution sense.*

Proof. First, if u is a solution of problem (3.1), then obviously $u \in B^{\frac{\alpha}{2},\frac{\beta}{2}}(Q)$. By multiplying the first equation of (3.1) by any $v \in B^{\frac{\alpha}{2},\frac{\beta}{2}}(Q)$, integrating the resulting equation over Q, and then using Lemma 8 respectively with $z = t$ for the first term and with $z = x$ for the second and third terms, we obtain (3.3).

Inversely, if u is the solution of weak form (3.3), then by using Lemma 8 we get the first equation of (3.1) in the distribution sense. The boundary condition is guaranteed by the fact that $u(\cdot, t) \in H_0^{\beta/2}(\Omega)$ for almost every $t \in I$. The initial condition is derived from Lemma 2. □

3.2 Weak formulation with Caputo derivative

The construction of weak formulations for the problem with Caputo derivative is more delicate. We may think about defining a similar space as ${}^lH^s(I)$ and ${}^rH^s(I)$ in (2.1) and (2.3) with Caputo derivatives. For example, we may try to define the weak solution space for Caputo problem (3.2) as follows:

$${}^l\tilde{H}^s(I) := \{v; \|v\|_{{}^l\tilde{H}^s(I)} < \infty\},$$

with

$$\|v\|_{{}^l\tilde{H}^s(I)} := \left(\|v\|_{0,I}^2 + |v|_{{}^l\tilde{H}^s(I)}^2\right)^{\frac{1}{2}}, \quad |v|_{{}^l\tilde{H}^s(I)} := \|{}^CD_t^s v\|_{0,I},$$

where we have used the Caputo derivative ${}^CD_t^s$ instead of ${}^RD_t^s$ (see (2.2)) to define the norm. This idea seems natural, but would not work for the fact that for $0 < s < 1/2$ the space ${}^l\tilde{H}^s(I)$ is not complete; see [9] for a detailed explanation. This indicates that the space ${}^{l}\tilde{H}^{s}(I)$ is not a suitable solution space for the Caputo problem, and therefore we are led to consider the following weak formulation for problem (3.2): for $f \in B^{\frac{\alpha}{2},\frac{\beta}{2}}(Q)'$, find $u \in B^{\frac{\alpha}{2},\frac{\beta}{2}}(Q)$ such that

$$\mathcal{A}(u,v) = \mathcal{F}_2(v), \quad \forall v \in B^{\frac{\alpha}{2},\frac{\beta}{2}}(Q), \tag{3.6}$$

where $\mathcal{A}(\cdot,\cdot)$ is defined in (3.4) and the functional $\mathcal{F}_2(\cdot)$ is given by

$$\mathcal{F}_2(v) := \langle f, v\rangle_Q + \left(\frac{u_0(x)t^{-\alpha}}{\Gamma(1-\alpha)}, v\right)_Q.$$

In order to establish the link between variational formulation (3.6) and problem (3.2), we need the following two lemmas.

Lemma 9 ([9]). *If $0 < \alpha < 1$, $w \in H^1(I)$, $v \in H_0^{\frac{\alpha}{2}}(I)$, then*

$$({}^{C}D_t^{\alpha} w(t), v(t))_I = ({}^{R}D_t^{\frac{\alpha}{2}} w(t), {}_t^{R}D^{\frac{\alpha}{2}} v(t))_I - \left(\frac{w(0)t^{-\alpha}}{\Gamma(1-\alpha)}, v(t)\right)_I. \tag{3.7}$$

Lemma 10 (Hardy–Littlewood lemma, [11]). *For $0 < s < 1$, $1 < p < 1/s$, the fractional integration operator $I_t^s v(t)$ is bounded from L^p into L^q with $q = p/(1-sp)$.*

The investigation of the weak problem (3.6) is justified in the theorem below.

Theorem 3. *Suppose $0 < \alpha < 1, 1 < \beta < 2$. If u is a classical solution of problem (3.2), then u is a weak solution of (3.6). Reciprocally, if u is a weak solution of (3.6), and for almost every x, $u(x,\cdot) \in H^1(I)$, ${}^{R}D_x^{\beta}u(x,\cdot)$, ${}_x^{R}D^{\beta}u(x,\cdot)$, $f(x,\cdot) \in L^{\frac{1}{\alpha}}(I)$, then u is also a solution of (3.2).*

Proof. If u is a classical solution of problem (3.2), we have

$$({}^{C}D_t^{\alpha}u, v)_Q - p_1({}^{R}D_x^{\beta}u, v)_Q - p_2({}_x^{R}D^{\beta}u, v)_Q = \langle f, v\rangle_Q \quad \forall v \in B^{\frac{\alpha}{2},\frac{\beta}{2}}(Q).$$

Then (3.6) can be derived by employing Lemma 9 with respect to t for the first term and Lemma 8 with respect to x for the second and third terms.

Inversely, if u is a solution of weak form (3.6), i. e., u satisfies

$$\begin{aligned}&({}^{R}D_t^{\frac{\alpha}{2}}u, {}_t^{R}D^{\frac{\alpha}{2}}v)_Q - p_1({}^{R}D_x^{\frac{\beta}{2}}u, {}_x^{R}D^{\frac{\beta}{2}}v)_Q - p_2({}_x^{R}D^{\frac{\beta}{2}}u, {}^{R}D_x^{\frac{\beta}{2}}v)_Q \\ &= \langle f, v\rangle_Q + \left(\frac{u_0(x)t^{-\alpha}}{\Gamma(1-\alpha)}, v\right)_Q, \quad \forall v \in B^{\frac{\alpha}{2},\frac{\beta}{2}}(Q),\end{aligned}$$

then employing Lemmas 9 and 8 yields

$$\begin{aligned}&({}^{C}D_t^{\alpha}u, v)_Q - p_1\langle {}^{R}D_x^{\beta}u, v\rangle_Q - p_2\langle {}_x^{R}D^{\beta}u, v\rangle_Q \\ &= \langle f, v\rangle_Q - \left(\frac{(u(x,0) - u_0(x))t^{-\alpha}}{\Gamma(1-\alpha)}, v\right)_Q, \quad \forall v \in C_0^{\infty}(Q).\end{aligned}$$

Thus we have in the distribution sense

$$ {}^C D_t^\alpha u - p_1 {}^R D_x^\beta u - p_2 {}_x^R D^\beta u = f - \frac{(u(x,0) - u_0(x)) t^{-\alpha}}{\Gamma(1-\alpha)}. $$

Furthermore, if $u(x,\cdot) \in H^1(I)$, then $\partial_t u(x,\cdot) \in L^2(I)$, and by Hardy–Littlewood's lemma, Lemma 10, with $s = 1 - \alpha$ and $1 < p < \min\{1/1-\alpha, 2\}$, we have ${}^C D_t^\alpha u = I_t^{1-\alpha}\partial_t u(x,t) \in L^{\frac{p}{1-(1-\alpha)p}} \subset L^{\frac{1}{\alpha}}(I)$ for almost every $x \in \Omega$. This result, together with the assumption on ${}^R D_x^\beta u(x,\cdot)$, ${}_x^R D^\beta u(x,\cdot)$, and $f(x,\cdot)$, leads to

$$ \frac{(u(x,0) - u_0(x)) t^{-\alpha}}{\Gamma(1-\alpha)} = f - {}^C D_t^\alpha u + p_1 {}^R D_x^\beta u + p_2 {}_x^R D^\beta u \in L^{\frac{1}{\alpha}}(I), \qquad \text{a.e. } x \in \Omega. $$

On the other side, it is readily seen that $t^{-\alpha}$ does not belong to $L^{\frac{1}{\alpha}}(I)$. Thus, we necessarily have

$$ u(x,0) = u_0(x), \qquad \text{a.e. } x \in \Omega. $$

The other direct consequence of the above is

$$ {}^C D_t^\alpha u - p_1 {}^R D_x^\beta u - p_2 {}_x^R D^\beta u = f. $$

We finally conclude that u satisfies (3.2) since the boundary condition $u(x,t)|_{\partial\Omega} = 0$, $\forall t \in I$, is evident. □

Remark 2. In the weak solution space for both problems (3.3) and (3.6) the solutions are not required to satisfy any initial conditions. In fact there is no sense to define the trace at time $t = 0$ for functions in $B^{\frac{\alpha}{2},\frac{\beta}{2}}(Q)$ with $\alpha < 1$. The initial conditions imposed in (3.1) and (3.2) are obtained only if the weak solutions are regular enough. In the Riemann case, this is guaranteed by Lemma 2 as long as $u(x,\cdot) \in H^{\alpha/2}(I)$ for fixed $x \in \Omega$. In the Caputo case, a regularity requirement sufficient to guarantee the initial condition is given in Theorem 3.

Theorem 4. *For all $0 < \alpha < 1$ and $f \in B^{\frac{\alpha}{2},\frac{\beta}{2}}(Q)'$, problem (3.6) admits a unique solution. Furthermore, the solution u satisfies*

$$ \|u\|_{B^{\frac{\alpha}{2},\frac{\beta}{2}}(Q)} \lesssim \|f\|_{B^{\frac{\alpha}{2},\frac{\beta}{2}}(Q)'} + \|u_0\|_{L^2(\Omega)} \|t^{-\alpha}\|_{L^q(I)}, \quad q = \frac{2}{1+\alpha}. \tag{3.8} $$

Proof. The proof of the existence and uniqueness of the solution is similar to that in Theorem 1. We only derive the stability (3.8). By taking $v = u$ in (3.6) and then using the coercivity of $\mathcal{A}$ and the Hölder inequality, we get

$$ \|u\|^2_{B^{\frac{\alpha}{2},\frac{\beta}{2}}(Q)} \lesssim \langle f, u\rangle_Q + \frac{1}{\Gamma(1-\alpha)} (u_0(x) t^{-\alpha}, u)_Q $$

$$
\begin{aligned}
&\lesssim \|f\|_{B^{\frac{\alpha}{2},\frac{\beta}{2}}(Q)'} \|u\|_{B^{\frac{\alpha}{2},\frac{\beta}{2}}(Q)} + \int_I t^{-\alpha} \int_\Omega u_0(x)u(x,t)\mathrm{d}x\mathrm{d}t \\
&\lesssim \|f\|_{B^{\frac{\alpha}{2},\frac{\beta}{2}}(Q)'} \|u\|_{B^{\frac{\alpha}{2},\frac{\beta}{2}}(Q)} + \int_I t^{-\alpha} \|u(\cdot,t)\|_{L^2(\Omega)} \|u_0\|_{L^2(\Omega)} \mathrm{d}t \\
&\lesssim \|f\|_{B^{\frac{\alpha}{2},\frac{\beta}{2}}(Q)'} \|u\|_{B^{\frac{\alpha}{2},\frac{\beta}{2}}(Q)} + \|u_0\|_{L^2(\Omega)} \|t^{-\alpha}\|_{L^q(I)} \|u\|_{L^{q'}(I;L^2(\Omega))},
\end{aligned} \tag{3.9}
$$

where $q = \frac{2}{1+\alpha}$, $q' = \frac{2}{1-\alpha}$. Furthermore, by the embedding theorem [1], we know that

$$
H^{\frac{\alpha}{2}}(I) \hookrightarrow L^{q'}(I).
$$

Thus

$$
\|u\|_{L^{q'}(I;L^2(\Omega))} \lesssim \|u\|_{H^{\frac{\alpha}{2}}(I;L^2(\Omega))} \lesssim \|u\|_{B^{\frac{\alpha}{2},\frac{\beta}{2}}(Q)}. \tag{3.10}
$$

Finally, combining (3.9) and (3.10) yields (3.8). □

4 Galerkin spectral method for the time-space fractional diffusion equation

In this section we propose and analyze a spectral Galerkin method to solve the initial boundary value problems of STFDE expressed in the weak forms. For the sake of simplification, we only consider the problem with the Riemann derivative (3.3).

4.1 Galerkin spectral method

We define $P_M(\Omega)$ (respectively $P_N(I)$) as the polynomials spaces of degree less than or equal to M (respectively N) with respect to x (respectively t). Let

$$
P_M^0(\Omega) := P_M(\Omega) \cap H_0^{\frac{\beta}{2}}(\Omega), \quad S_{M,N} := P_M^0(\Omega) \otimes P_N(I).
$$

We now consider the following space-time spectral method to problem (3.3): find $u_{M,N} \in S_{M,N}$, such that

$$
\mathcal{A}(u_{M,N}, v_{M,N}) = \mathcal{F}(v_{M,N}), \quad \forall v_{M,N} \in S_{M,N}.
$$

Since $S_{M,N}$ is a subspace of $B^{\frac{\alpha}{2},\frac{\beta}{2}}(Q)$, the well-posedness of problem (4.1) is immediate.

In order to derive the error estimate for the numerical solution of (4.1), we introduce some approximation operators and present their approximation properties.

Theorem 5 ([10]). *Let p and s be two real numbers such that $p \neq n + 1/2$, $0 \leq s \leq p$; there exists an operator $\Pi^{s,0}_{p,N}$, from $H^p(\Lambda) \cap H^s_0(\Lambda)$ onto $P^s_N(\Lambda)$ such that, for any $\varphi \in H^\sigma(\Lambda) \cap H^s_0(\Lambda)$ with $\sigma \geq p$, we have*

$$\|\varphi - \Pi^{s,0}_{p,N}\varphi\|_\nu \lesssim N^{\nu-\sigma}\|\varphi\|_\sigma, \forall 0 \leq \nu \leq p. \tag{4.1}$$

In the next lemma, we give the property of the composite approximation operator $\Pi^{0,0}_{\frac{\alpha}{2},N}\Pi^{\frac{\beta}{2},0}_{\frac{\beta}{2},M}$, where the operation $\Pi^{\frac{\beta}{2},0}_{\frac{\beta}{2},M}v(x,t)$ acts on the space variable, while $\Pi^{0,0}_{\frac{\alpha}{2},N}$ acts on the time variable.

Lemma 11 ([9]). *For $0 < \alpha < 1, 1 < \beta < 2, \gamma > 1, \sigma \geqslant 1$. If $v \in H^{\frac{\alpha}{2}}(I; H^\sigma(\Omega)) \cap H^\gamma(I; H^{\frac{\beta}{2}}_0(\Omega))$, then we have*

$$\left\|{}^RD_x^{\frac{\beta}{2}}\left(v - \Pi^{0,0}_{\frac{\alpha}{2},N}\Pi^{\frac{\beta}{2},0}_{\frac{\beta}{2},M}v\right)\right\|_{0,0} \lesssim M^{\frac{\beta}{2}-\sigma}\|v\|_{\sigma,0} + N^{-\gamma}\|v\|_{\frac{\beta}{2},\gamma},$$

$$\left\|{}^RD_t^{\frac{\alpha}{2}}\left(v - \Pi^{0,0}_{\frac{\alpha}{2},N}\Pi^{\frac{\beta}{2},0}_{\frac{\beta}{2},M}v\right)\right\|_{0,0} \lesssim N^{\frac{\alpha}{2}-\gamma}\|v\|_{0,\gamma} + M^{-\sigma}\|v\|_{\sigma,\frac{\alpha}{2}}.$$

We are now in a position to derive the error estimate for the space-time spectral discrete solution.

Theorem 6. *Let $0 < \alpha < 1, 1 < \beta < 2, \gamma > 1, \sigma \geqslant 1$ and let u, $u_{M,N}$ be the solutions of (3.1) and (4.1), respectively. If $u \in H^{\frac{\alpha}{2}}(I; H^\sigma(\Omega)) \cap H^\gamma(I; H^{\frac{\beta}{2}}_0(\Omega))$, then we have*

$$\|u - u_{M,N}\|_{B^{\frac{\alpha}{2},\frac{\beta}{2}}(Q)} \lesssim N^{\frac{\alpha}{2}-\gamma}\|u\|_{0,\gamma} + M^{-\sigma}\|u\|_{\sigma,\frac{\alpha}{2}} + M^{\frac{\beta}{2}-\sigma}\|u\|_{\sigma,0} + N^{-\gamma}\|u\|_{\frac{\beta}{2},\gamma}.$$

Proof. First, from the standard procedure of error estimation for Galerkin methods follows that

$$\|u - u_{M,N}\|_{B^{\frac{\alpha}{2},\frac{\beta}{2}}(Q)} \leqslant \inf_{v_{M,N} \in S_{M,N}} \|u - v_{M,N}\|_{B^{\frac{\alpha}{2},\frac{\beta}{2}}(Q)}.$$

Then by taking $v_{M,N} = \Pi^{0,0}_{\frac{\alpha}{2},N}\Pi^{\frac{\beta}{2},0}_{\frac{\beta}{2},M}u$ in the right-hand side and employing Lemma 11, we obtain immediately the estimate given in Theorem 6. □

4.2 Numerical results

In this subsection, we present some numerical results to demonstrate the efficiency of the proposed space-time spectral method. We refer the readers to [9] for some implementation details. In order to confirm the theoretical result predicted by the error estimate in Theorem 6, we consider the following fabricated solutions having limited regularity:

$$u(x, y, t) = t^3(1 - x^2)x^\gamma \sin(\pi y),$$

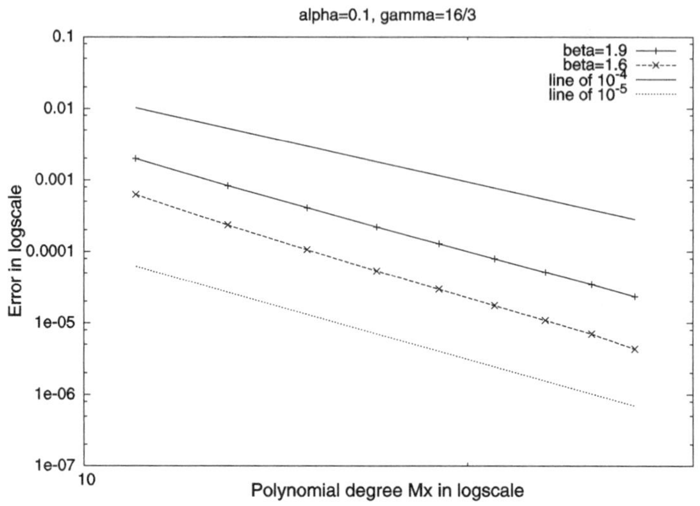

Figure 1: B-errors versus M_x with $N = 5$, $M_y = 20$ for varying β.

where γ is a constant. It can be verified that the solution belongs to $H^{\gamma+1/2}$ with respect to the space variable. We plot in Figure 1 the error decay rates in the $B^{\frac{\alpha}{2},\frac{\beta}{2}}$-norm versus the polynomial degrees M_x with $\alpha = 0.1$, $\gamma = 16/3$ for the two different values of $\beta = 1.6, 1.9$. The M_x^{-4} and M_x^{-5} decay rates are also shown for comparison. It is observed that all the error curves are straight lines in this log-log representation, which indicates the algebraic convergence for this solution of limited regularity. Moreover it is seen that the errors decrease with rates in accordance with the estimate in Theorem 6, which predicts, in Figure 1, an $M_x^{-5.0}$ decay rate for $\beta = 1.6$ and $M_x^{-4.8}$ decay rate for $\beta = 1.9$.

The investigation of the convergence behavior for less regular solutions is done by decreasing γ. The results are presented in Figure 2, where the errors versus M_x are plotted for the three different values of $\gamma = 16/3, 10/3, 4/3$. As expected, the convergence rate slows down as γ, i. e., the regularity of the solution, decreases.

5 Müntz spectral methods for the time-fractional diffusion equation

An efficient spectral method, along with a functional and variational framework, has been developed in the Section 4. It was proven that the exponential convergence of the proposed method is attainable for smooth solutions. Several other works focused on spectral methods for some related fractional equations for non-smooth solutions. Among these methods, Zheng et al. [16] extended the above spectral method to the multi-term TFDE. Zayernouri et al. [14] considered a Petrov–Galerkin spectral method

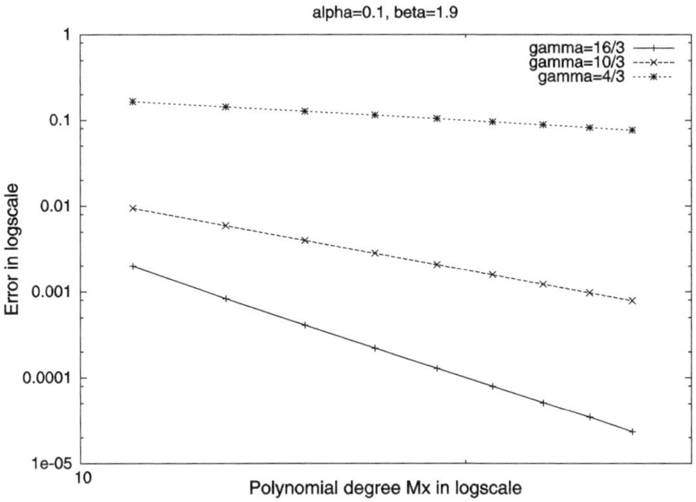

Figure 2: B-errors versus M_x with $N = 5$, $M_y = 20$ for varying γ.

using the so-called polyfractonomials, introduced in Zayernouri and Karniadakis [15]. Chen et al. [4] used generalized Jacobi functions to construct Petrov–Galerkin methods for a class of fractional initial/boundary value problems. Numerical experiments and theoretical analysis presented therein have shown exponential convergence for the exact solution to be non-smooth but having specific singularity. This section aims at providing a fractional Jacobi spectral method that is capable of handling a family of the solutions in a more efficient way.

5.1 Generalized fractional Jacobi polynomials

As our basic approximation functions, we present below the generalized fractional Jacobi polynomials and a number of useful properties related to these polynomials, some of which can be found in [6].

For $\alpha, \beta \geq -1$, $0 < \lambda \leq 1$, the generalized fractional Jacobi polynomials in I of degree $n + l$ are defined as

$$J_{n+l}^{\alpha,\beta,\lambda}(t) = \begin{cases} J_n^{\alpha,\beta}(2t^\lambda - 1), & \alpha, \beta > -1, \\ \frac{n+\alpha+1}{n+1} t^\lambda J_n^{\alpha,1}(2t^\lambda - 1), & \alpha > -1,\ \beta = -1, \\ \frac{n+\beta+1}{n+1}(1 - t^\lambda) J_n^{1,\beta}(2t^\lambda - 1), & \alpha = -1,\ \beta > -1, \\ -(1 - t^\lambda) t^\lambda J_n^{1,1}(2t^\lambda - 1), & \alpha = \beta = -1, \end{cases}$$

where $J_n^{\alpha,\beta}(t)$ denotes the classical Jacobi polynomial of index $\{\alpha, \beta\}$ of degree $n = 0, 1, 2, \ldots$ and

$$l = \begin{cases} 0, & \alpha, \beta > -1, \\ 1, & \alpha = -1,\ \beta > -1 \text{ or } \alpha > -1,\ \beta = -1, \\ 2, & \alpha = \beta = -1. \end{cases} \tag{5.1}$$

It is readily seen that when $\lambda = 1$, $\{J_{n+l}^{\alpha,\beta,1}(t)\}_{n=0}^{\infty}$ are the shifted generalized Jacobi polynomials, which are orthogonal polynomials in I with respect to the weight $(1-t)^{\alpha}t^{\beta}$. The following lemmas can be found in or can be proven by using the materials given in [6].

Lemma 12. *The generalized fractional Jacobi polynomials $J_{n+l}^{\alpha,\beta,\lambda}(t)$ in I are mutually orthogonal with respect to the weight function $\omega^{\alpha,\beta,\lambda}(t) = \lambda(1-t^{\lambda})^{\alpha}t^{(\beta+1)\lambda-1}$ for $\alpha, \beta \geq -1$, $0 < \lambda \leq 1$, i. e.,*

$$\int_0^1 \omega^{\alpha,\beta,\lambda}(t) J_{n+l}^{\alpha,\beta,\lambda}(t) J_{m+l}^{\alpha,\beta,\lambda}(t)\,dt = \gamma_{n+l}^{\alpha,\beta}\delta_{m+l,n+l}, \tag{5.2}$$

where

$$\gamma_{n+l}^{\alpha,\beta} = \frac{\Gamma(n+l+\alpha+1)\Gamma(n+l+\beta+1)}{(2(n+l)+\alpha+\beta+1)(n+l)!\,\Gamma(n+l+\alpha+\beta+1)} \tag{5.3}$$

with n, l being given in (5.1).

Lemma 13. *The first derivatives of the generalized fractional Jacobi polynomials are orthogonal to each other with respect to the weight function $\hat{\omega}^{\alpha,\beta,\lambda}(t) = \lambda^{-1}(1-t^{\lambda})^{\alpha+1}t^{\beta\lambda+1}$, $\alpha, \beta \geq -1$, $0 < \lambda \leq 1$, i. e.,*

$$\int_{-1}^1 \hat{\omega}^{\alpha,\beta,\lambda}(t) \frac{d}{dt}J_{n+l}^{\alpha,\beta,\lambda}(t) \frac{d}{dt}J_{m+l}^{\alpha,\beta,\lambda}(t)\,dt = \sigma_{n+l}^{\alpha,\beta}\gamma_{n+l}^{\alpha,\beta}\delta_{m+l,n+l}, \tag{5.4}$$

where $m, n = 0, 1, 2, \ldots, l$ is given in (5.1), $\gamma_{n+l}^{\alpha,\beta}$ is defined in (5.3), and $\sigma_{n+l}^{\alpha,\beta}$ is given by

$$\sigma_{n+l}^{\alpha,\beta} = (n+l)(n+l+\alpha+\beta+1). \tag{5.5}$$

Moreover, we have

$$\frac{d}{dt}J_{n+l}^{\alpha,\beta,\lambda}(t) = (n+l+\alpha+\beta+1)\lambda t^{\lambda-1}J_{n+l-1}^{\alpha+1,\beta+1,\lambda}(t). \tag{5.6}$$

Lemma 14. *The generalized fractional Jacobi polynomials satisfy the following Sturm–Liouville problem:*

$$(\omega^{\alpha,\beta,\lambda}(t))^{-1}\frac{d}{dt}\left\{\lambda^{-1}(1-t^{\lambda})^{\alpha+1}t^{\beta\lambda+1}\frac{d}{dt}J_{n+l}^{\alpha,\beta,\lambda}(t)\right\} = -\sigma_{n+l}^{\alpha,\beta}J_{n+l}^{\alpha,\beta,\lambda}(t).$$

5.2 Fractional spectral methods and analysis

We introduce the following differential operators:

$$D_\lambda^0 := I_d, \quad D_\lambda := \frac{d}{dx^\lambda} := \frac{d}{\lambda x^{\lambda-1}dx}, \quad D_\lambda^2 := D_\lambda D_\lambda, \quad \dots, \quad D_\lambda^k := \overbrace{D_\lambda \cdots D_\lambda}^{k}, \quad k \in \mathbb{N}.$$

Define the non-uniform fractional Jacobi-weighted Sobolev spaces

$$B^m_{\omega^{\alpha,\beta,\lambda}}(I) := \{v : D_\lambda^k v \in L^2_{\omega^{\alpha+k,\beta+k,\lambda}}(I), 0 \le k \le m\}, \quad m \in \mathbb{N},$$

equipped with the inner product, norm, and semi-norm, respectively, as follows:

$$(u,v)_{B^m_{\omega^{\alpha,\beta,\lambda}}} = \sum_{k=0}^{m} (D_\lambda^k u, D_\lambda^k v)_{\omega^{\alpha+k,\beta+k,\lambda}}, \quad \|v\|_{m,\omega^{\alpha,\beta,\lambda}} = (v,v)^{1/2}_{B^m_{\omega^{\alpha,\beta,\lambda}}},$$

$$|v|_{m,\omega^{\alpha,\beta,\lambda}} = \|D_\lambda^m v\|_{0,\omega^{\alpha+m,\beta+m,\lambda}}.$$

The special case $\lambda = 1$ gives the classical non-uniform Jacobi-weighted Sobolev spaces, i. e.,

$$B^m_{\omega^{\alpha,\beta,1}}(I) := \{v : \partial_x^k v \in L^2_{\omega^{\alpha+k,\beta+k,1}}(I), 0 \le k \le m\}, \quad m \in \mathbb{N}.$$

One can check that for any $0 < s < \frac{1}{2}$, $\alpha > -1$, we have ${}_0H^s(I) \subset L^2_{\omega^{\alpha,-1,\lambda}}(I)$.

5.2.1 Galerkin fractional spectral methods

Let $I = (0,1)$, $\Lambda = (-1,1)$, $\Omega = \Lambda \times I$, $0 < \mu < 1$. We consider the TFDE

$${}_0D_t^\mu u(x,t) - \partial_x^2 u(x,t) = f(x,t), \quad \forall (x,t) \in \Omega, \tag{5.7}$$

subject to the homogeneous initial and boundary conditions. Here we consider a right-hand side function $f \in L^p(I, L^2(\Lambda))$ with $p > 1/\mu$.

The first fractional spectral method to be proposed is based on the Galerkin formulation of the TFDE (5.7). To this end, we need some more notations. We define the space ${}_0H^s(I)$ as the closure of ${}_0C^\infty(I)$ with respect to the norm $\|\cdot\|_{s,I}$, where

$${}_0C^\infty(I) = \{v \mid v \in C^\infty(I) \text{ having compact support in } (0,1]\}.$$

For a Sobolev space X on Λ equipped with the norm $\|\cdot\|_X$, we denote

$$\begin{aligned} H^s(I;X) &:= \{v \mid \|v(\cdot,t)\|_X \in H^s(I)\}, \quad s \ge 0,\\ {}_0H^s(I;X) &:= \{v \mid \|v(\cdot,t)\|_X \in {}_0H^s(I)\}, \quad s \ge 0,\\ B^m_{\omega^{\alpha,\beta,\lambda}}(I;X) &:= \{v \mid \|v(\cdot,t)\|_X \in B^m_{\omega^{\alpha,\beta,\lambda}}(I)\}, \quad m \in \mathbb{N}, \end{aligned}$$

endowed respectively with the norms

$$\|v\|_{H^s(I;X)} := \big\|\|v(\cdot,t)\|_X\big\|_{s,I},$$
$$\|v\|_{B^m_{\omega^{\alpha,\beta,\lambda}}(I;X)} := \big\|\|v(\cdot,t)\|_X\big\|_{B^m_{\omega^{\alpha,\beta,\lambda}}}.$$

When X is $H^\sigma(\Lambda)$ or $H^\sigma_0(\Lambda)$, $\sigma \geq 0$, the norm of the space $H^s(I;X)$ will be denoted by $\|\cdot\|_{\sigma,s,\Omega}$. Then we define the spaces

$$\mathcal{H}^s(\Omega) := {}_0H^s(I,L^2(\Lambda)) \cap L^2(I,H^1_0(\Lambda)),$$
$$\mathcal{B}^m_{\omega^{\alpha,\beta,\lambda}}(\Omega) := B^m_{\omega^{\alpha,\beta,\lambda}}(I,L^2(\Lambda)) \cap L^2_{\omega^{\alpha,\beta,\lambda}}(I,H^1_0(\Lambda)),$$

equipped respectively with the norms

$$\begin{aligned}\|v\|_{\mathcal{H}^s(\Omega)} &:= \left(\|v\|^2_{0,s} + \|v\|^2_{1,0}\right)^{1/2} := \left(\|v\|^2_{H^s(I,L^2(\Lambda))} + \|v\|^2_{L^2(I,H^1_0(\Lambda))}\right)^{1/2},\\ \|v\|_{\mathcal{B}^m_{\omega^{\alpha,\beta,\lambda}}(\Omega)} &:= \left(\|v\|^2_{B^m_{\omega^{\alpha,\beta,\lambda}}(I,L^2(\Lambda))} + \|v\|^2_{L^2_{\omega^{\alpha,\beta,\lambda}}(I,H^1_0(\Lambda))}\right)^{1/2}.\end{aligned} \tag{5.8}$$

It can be proven that $\mathcal{H}^s(\Omega)$ and $\mathcal{B}^m_{\omega^{\alpha,\beta,1}}(\Omega)$ are both Banach spaces.

Given f satisfying ${}_0I^{\mu/2}_t f(x,t) \in L^2(\Omega)$, we consider the weak form of (5.7) as follows: find $u \in \mathcal{H}^{\mu/2}(\Omega)$ such that

$$\mathcal{A}(u,v) = \mathcal{F}(v), \quad \forall v \in \mathcal{H}^{\mu/2}(\Omega), \tag{5.9}$$

where the bilinear form $\mathcal{A}(\cdot,\cdot)$ is defined by

$$\mathcal{A}(u,v) := \left({}_0D^{\mu/2}_t u, {}_tD^{\mu/2}_1 v\right)_\Omega + (\partial_x u, \partial_x v)_\Omega$$

and the functional $\mathcal{F}(\cdot)$ is given by

$$\mathcal{F}(v) := (f,v)_\Omega.$$

Theorem 7 ([5]). *For any $0 < \mu < 1$ and ${}_0I^{\mu/2}_t f \in L^2(\Omega)$, the problem (5.9) is well-posed. Furthermore, the solution u satisfies*

$$\|u\|_{\mathcal{H}^{\mu/2}(\Omega)} \lesssim \|{}_0I^{\mu/2}_t f\|_{0,\Omega}. \tag{5.10}$$

Now we consider numerical approximations to the weak problem (5.9). Let $P_M(\Lambda)$ be the space of polynomials of degree less than or equal to M defined in Λ and let $P^\lambda_N(I)$ be the fractional polynomial space, defined by

$$P^\lambda_N(I) := \text{span}\{1, t^\lambda, t^{2\lambda}, \dots, t^{N\lambda}\}.$$

Let

$$P^0_M(\Lambda) := P_M(\Lambda) \cap H^1_0(\Lambda),$$
$$S^{\alpha,-1}_{N,\lambda}(I) := \text{span}\{v \in P^\lambda_{N+1}(I) | v(0) = 0\} = \text{span}\{J^{\alpha,-1,\lambda}_{i+1}(x),\ i = 0,1,\dots,N\},\ \alpha > -1.$$

Then we denote $L := (M, N)$, and

$$S_L^\alpha(\Omega) := P_M^0(\Lambda) \otimes S_{N,\lambda}^{\alpha,-1}(I). \tag{5.11}$$

We propose the fractional spectral Galerkin method for (5.9) as follows: find $u_L \in S_L^\alpha(\Omega)$, such that

$$\mathcal{A}(u_L, v_L) = \mathcal{F}(v_L), \quad \forall v_L \in S_L^\alpha(\Omega). \tag{5.12}$$

It order to analyze the error of the fractional spectral Galerkin method, we will need to introduce some approximation operators. The orthogonal projector $\pi_M^{1,0} : H_0^1(\Lambda) \to P_M^0(\Lambda)$ is defined by $\forall v \in H_0^1(\Lambda)$, $\pi_M^{1,0} v \in P_M^0(\Lambda)$, such that

$$((\pi_M^{1,0} v - v)', \varphi)_\Lambda = 0, \quad \forall \varphi \in P_M^0(\Lambda).$$

Then, for all $v \in H^m(\Lambda) \cap H_0^1(\Lambda)$, $m \geq 1$, the following optimal error estimates have been well known; see, e. g., Theorem 1.7 in [3]:

$$\|\pi_M^{1,0} v - v\|_{0,\Lambda} \leq cM^{-m} \|v\|_{m,\Lambda}, \tag{5.13}$$

$$|\pi_M^{1,0} v - v|_{1,\Lambda} \leq cM^{1-m} \|v\|_{m,\Lambda}. \tag{5.14}$$

For $\alpha > -1$, we define the $L^2_{\omega^{\alpha,-1,\lambda}}(I) \to S_{N,\lambda}^{\alpha,-1}(I)$ orthogonal projector $\pi_{N,\lambda}^{\alpha,-1}$ as follows: for all $v \in L^2_{\omega^{\alpha,-1,\lambda}}(I)$, $\pi_{N,\lambda}^{\alpha,-1} v \in S_{N,\lambda}^{\alpha,-1}(I)$, such that

$$(v - \pi_{N,\lambda}^{\alpha,-1} v, v_N)_{\omega^{\alpha,-1,\lambda}} = 0, \quad \forall v_N \in S_{N,\lambda}^{\alpha,-1}(I). \tag{5.15}$$

It was proven in [6] that for a function $v(t)$, if there exists λ, $0 < \lambda \leq 1$, such that $v(t^{\frac{1}{\lambda}}) \in B^m_{\omega^{\alpha,-1,1}}(I)$, $m \geq 1$, then the operator $\pi_{N,\lambda}^{\alpha,-1} v$ admits the following error estimates:

$$\|v - \pi_{N,\lambda}^{\alpha,-1} v\|_{0,\omega^{\alpha,-1,\lambda}} \lesssim N^{-m} \|\partial_t^m v(t^{\frac{1}{\lambda}})\|_{0,\omega^{m+\alpha,m-1,1}}, \tag{5.16}$$

$$\|\partial_t (v - \pi_{N,\lambda}^{\alpha,-1} v)\|_{0,\omega^{\alpha+1,2/\lambda-2,\lambda}} \lesssim N^{1-m} \|\partial_t^m v(t^{\frac{1}{\lambda}})\|_{0,\omega^{m+\alpha,m-1,1}}. \tag{5.17}$$

Next we will derive a weighted H^s-error estimate for $\pi_{N,\lambda}^{\alpha,-1}$. This is done with the help of two lemmas, which we present below.

Lemma 15 ([2] or [13]). *Let $\omega(t)$, $\sigma(t)$, and $v(t)$ be functions defined on $(0, \infty)$, $1 \leq p \leq q \leq \infty$, $\frac{1}{p} + \frac{1}{p'} = 1$, $0 < s < 1$. Suppose there exists γ, $0 \leq \gamma \leq 1$, such that*

$$\sup_{t>0} A_{s,p,q,\gamma}(t) < \infty, \tag{5.18}$$

where

$$A_{s,p,q,\gamma}(t) = \left(\int_t^\infty (\tau - t)^{(s-1)q(1-\gamma)} |\omega(\tau)|^q d\tau \right)^{1/q} \left(\int_0^t (t - \tau)^{(s-1)p'\gamma} |\sigma(\tau)|^{-p'} d\tau \right)^{1/p'}. \tag{5.19}$$

Then we have the following weighted inequality of the Hardy type for the fractional Riemann–Liouville integral of order s:

$$\left(\int_0^\infty |\omega(t)_0 I_t^s v|^q dt\right)^{1/q} \lesssim \left(\int_0^\infty |\sigma(t)v(t)|^p dt\right)^{1/p}. \tag{5.20}$$

The above lemma can be applied to functions defined on the bounded domain I, resulting in the following lemma [5].

Lemma 16. *Let $v(t)$ be any function defined in I, $0 < s < 1$, $0 < \lambda \le 1$, $-1 < \alpha \le \frac{s-1}{2}$. Then we have*

$$\int_0^1 (1-t^\lambda)^{-\alpha} t |_0 I_t^s v(t)|^2 dt \lesssim \int_0^1 (1-t^\lambda)^{\alpha+1} t^{1-\lambda} v(t)^2 dt. \tag{5.21}$$

Theorem 8. *For any $v(t) \in {}_0H^s(I)$, $0 < s < \frac{1}{2}$, and $-1 < \alpha \le -s$, if there exists a real λ, $0 < \lambda \le 1$, such that $v(t^{\frac{1}{\lambda}}) \in B^m_{\omega^{\alpha,-1,1}}(I)$, $m \ge 1$, then we have*

$$\|v - \pi_{N,\lambda}^{\alpha,-1} v\|_{s,I} \lesssim N^{1/2-m} \|\partial_t^m v(t^{1/\lambda})\|_{0,\omega^{\alpha+m,m-1,1}}. \tag{5.22}$$

In the next theorem, we present properties of the composite approximation operator $\pi_{N,\lambda}^{\alpha,-1}\pi_M^{1,0}$.

Theorem 9 ([5]). *Let $0 < s < \frac{1}{2}$, $-1 < \alpha \le -s$, $v(x,t) \in \mathcal{H}^s(\Omega)$. Suppose there exists $0 < \lambda \le 1$, such that $v(x, t^{1/\lambda}) \in B^m_{\omega^{\alpha,-1,1}}(I, H^\sigma(\Lambda)) \cap B^m_{\omega^{\alpha,-1,1}}(I, H_0^1(\Lambda))$, $m \ge 1$, $\sigma \ge 1$. Then we have*

$$\begin{aligned}
\|v - \pi_{N,\lambda}^{\alpha,-1}\pi_M^{1,0} v\|_{1,0} &\lesssim M^{1-\sigma}\|v\|_{\sigma,0} + N^{-m}\big\|\|\partial_t^m v(\cdot, t^{1/\lambda})\|_{1,\Lambda}\big\|_{0,\omega^{\alpha+m,m-1,1}}, \\
\|v - \pi_{N,\lambda}^{\alpha,-1}\pi_M^{1,0} v\|_{0,s} &\lesssim M^{-\sigma}\|v\|_{\sigma,s} + N^{\frac{1}{2}-m} M^{-\sigma}\big\|\|\partial_t^m v(\cdot, t^{1/\lambda})\|_{\sigma,\Lambda}\big\|_{0,\omega^{\alpha+m,m-1,1}} \\
&\quad + N^{\frac{1}{2}-m}\big\|\|\partial_t^m v(\cdot, t^{1/\lambda})\|_{0,\Lambda}\big\|_{0,\omega^{\alpha+m,m-1,1}}.
\end{aligned} \tag{5.23}$$

The error estimate for the solution of the time-fractional spectral/space spectral approximation is given in the following theorem.

Theorem 10. *Let $0 < \mu < 1$, $-1 < \alpha \le -\mu/2$; u is the solution of (5.9). Suppose there exists $0 < \lambda \le 1$, such that $u(x, t^{1/\lambda}) \in B^m_{\omega^{\alpha,-1,1}}(I, H^\sigma(\Lambda)) \cap B^m_{\omega^{\alpha,-1,1}}(I, H_0^1(\Lambda))$, $m \ge 1$, $\sigma \ge 1$. Then the solution u_L of the fractional spectral approximation (5.12) admits the following error estimate:*

$$\begin{aligned}
\|u - u_L\|_{\mathcal{H}^{\mu/2}(\Omega)} &\lesssim N^{\frac{1}{2}-m}\Big[\big\|\|\partial_t^m u(\cdot, t^{1/\lambda})\|_{0,\Lambda}\big\|_{0,\omega^{\alpha+m,m-1,1}} + M^{-\sigma}\big\|\|\partial_t^m u(\cdot, t^{1/\lambda})\|_{\sigma,\Lambda}\big\|_{0,\omega^{\alpha+m,m\ 1,1}}\Big] \\
&\quad + M^{-\sigma}\|u\|_{\sigma,s} + M^{1-\sigma}\|u\|_{\sigma,0} + N^{-m}\big\|\|\partial_t^m u(\cdot, t^{1/\lambda})\|_{1,\Lambda}\big\|_{0,\omega^{\alpha+m,m-1,1}}.
\end{aligned}$$

Proof. The error estimate is a direct consequence of the standard estimate for the Galerkin method and the approximation result given in Theorem 9. In fact, from (5.9) and (5.12) follows

$$\|u - u_L\|_{\mathcal{H}^{\mu/2}(\Omega)} \le \inf_{v_L \in S_L^\alpha(\Omega)} \|u - v_L\|_{\mathcal{H}^{\mu/2}(\Omega)} \le \|u - \pi_{N,\lambda}^{\alpha,-1}\pi_M^{1,0} u\|_{\mathcal{H}^{\mu/2}(\Omega)}.$$

Then, according to the norm definition in (5.8), we have

$$\|u - u_L\|_{\mathcal{H}^{\mu/2}(\Omega)} \le \|u - \pi_{N,\lambda}^{\alpha,-1}\pi_M^{1,0} u\|_{0,\mu/2} + \|u - \pi_{N,\lambda}^{\alpha,-1}\pi_M^{1,0} u\|_{1,0}. \tag{5.24}$$

Finally, the desired result follows from (5.23) and (5.24). □

5.2.2 Petrov–Galerkin fractional spectral methods

In this subsection we propose a fractional spectral method based on the Petrov–Galerkin formulation. First we notice that the TFDE (5.7) can be reformulated into [7]

$$\partial_t u(x,t) - {}_0D_t^{1-\mu}\partial_x^2 u(x,t) = {}_0D_t^{1-\mu} f(x,t), \quad \forall (x,t) \in \Omega. \tag{5.25}$$

We define the dual fractional polynomial space of $S_{N,\lambda}^{\alpha,-1}(I)$ as follows:

$$V_{N,\lambda}^{-\alpha-1,0}(I) := \operatorname{span}\{(1-t^\lambda)^{\alpha+1} J_j^{\alpha+1,0,\lambda}(t),\ j = 0,1,2,\dots,N\}.$$

For the $L^2_{\omega^{\alpha,-1,\lambda}}$-orthogonal projection $\pi_{N,\lambda}^{\alpha,-1}$ defined in (5.15), we have the following property.

Lemma 17 ([6], Lemma 3.2). *The projection operator $\pi_{N,\lambda}^{\alpha,-1}$ satisfies, for all $v \in B^1_{\omega^{\alpha,-1,\lambda}}(I)$,*

$$(\partial_t(v - \pi_{N,\lambda}^{\alpha,-1} v), v_N) = 0, \quad \forall v_N \in V_{N,\lambda}^{-\alpha-1,0}(I).$$

Now we consider the following Petrov–Galerkin-based fractional spectral method: Find $u_L \in S_L^\alpha(\Omega)$, such that

$$(\partial_t u_L, v_L) + ({}_0D_t^{1-\mu}\partial_x u_L, \partial_x v_L) = ({}_0D_t^{1-\mu} f, v_L), \quad \forall v_L \in V_L^\alpha(\Omega), \tag{5.26}$$

where $S_L^\alpha(\Omega)$ was defined in (5.11), $V_L^\alpha(\Omega) := P_M^0(\Lambda) \otimes V_{N,\lambda}^{-\alpha-1,0}(I)$. From Lemma 17, we see that without the second-order derivative term in x the solution of (5.26) is nothing more than the projection $\pi_{N,\lambda}^{\alpha,-1}$ of the exact solution, i. e., $\pi_{N,\lambda}^{\alpha,-1} u$. Therefore the error estimate of the numerical solution u_L follows directly from the result for the $\pi_{N,\lambda}^{\alpha,-1}$ operator. However with the presence of the second-order derivative in space, the stability of the bilinear defined by $(\partial_t u_L, v_L) + ({}_0D_t^{1-\mu}\partial_x u_L, \partial_x v_L)$ remains unknown. Nevertheless, the performance of this fractional spectral method can be investigated numerically through a series of numerical experiments. We will show that the accuracy of the Petrov–Galerkin-based method is comparable to the Galerkin-based method (5.12). In particular, the Petrov–Galerkin-type method (5.26) allows a fast solver.

5.3 Numerical results

We will present in this subsection some numerical examples to demonstrate the accuracy of the proposed fractional spectral schemes (5.12) and (5.26). The most important feature of our schemes is, as indicated by Theorem 10, that the convergence of the numerical solution is exponential with respect to the degree of the fractional polynomial if $u(x, t^{1/\lambda})$ is smooth, where $u(x, t)$ is the exact solution of the problem (5.7). The numerical tests that follow will aim at confirming this feature. We first give some implementation details of the Galerkin-based fractional spectral method. We denote by $L_i(x)$ the Legendre polynomial of degree i, and let

$$\phi_i(x) = \frac{1}{\sqrt{4i+6}}(L_i(x) - L_{i+2}(x)).$$

Then we have

$$S_L^\alpha(\Omega) = \operatorname{span}\{\phi_i(x)J_{j+1}^{\alpha,-1,\lambda}(t);\ i = 0, 1, \dots, M-2;\ j = 0, 1, \dots, N\}. \tag{5.27}$$

Thus the numerical solution u_L can be expressed by

$$u_L(x,t) = \sum_{i=0}^{M-2}\sum_{j=0}^{N} u_{ij}\phi_i(x)J_{j+1}^{\alpha,-1,\lambda}(t).$$

Let the test function v_L in (5.12) go through all the basis functions in (5.27). Then we obtain the following linear system:

$$A_1UB_1^T + A_2UB_2 = F,$$

where

$$(A_1)_{ij} = (\phi_j, \phi_i) = \begin{cases} \frac{-1}{\sqrt{(2j+3)(2j+7)(2j+5)}}, & i = j+2,\ j = 0, 1, \dots, M-4, \\ \frac{2}{(2j+1)(2j+5)}, & i = j,\ j = 0, 1, \dots, M-2, \\ \frac{-1}{\sqrt{(2j-1)(2j+3)(2j+1)}}, & i = j-2,\ j = 2, 3, \dots, M-2, \\ 0, & \text{others}, \end{cases}$$

$$(A_2)_{ij} = (\partial_x\phi_j, \partial_x\phi_i) = \delta_{i,j}, \quad i, j = 0, 1, \dots, M-2,$$

$$(B_1)_{mn} = ({}_0D_t^\mu J_{n+1}^{\alpha,-1,\lambda}(t), J_{m+1}^{\alpha,-1,\lambda}(t)), \quad m, n = 0, 1, \dots, N,$$

$$(B_2)_{mn} = (J_{n+1}^{\alpha,-1,\lambda}(t), J_{m+1}^{\alpha,-1,\lambda}(t)), \quad m, n = 0, 1, \dots, N,$$

$$(U)_{kl} = u_{kl}, \quad (F)_{kl} = (f, \phi_k(x)J_{l+1}^{\alpha,-1,\lambda})_\Omega, \quad k = 0, 1, \dots, M-2;\ l = 0, 1, \dots, N.$$

The entries of the matrices B_1 and B_2 can be calculated by using (5.6) and variable change.

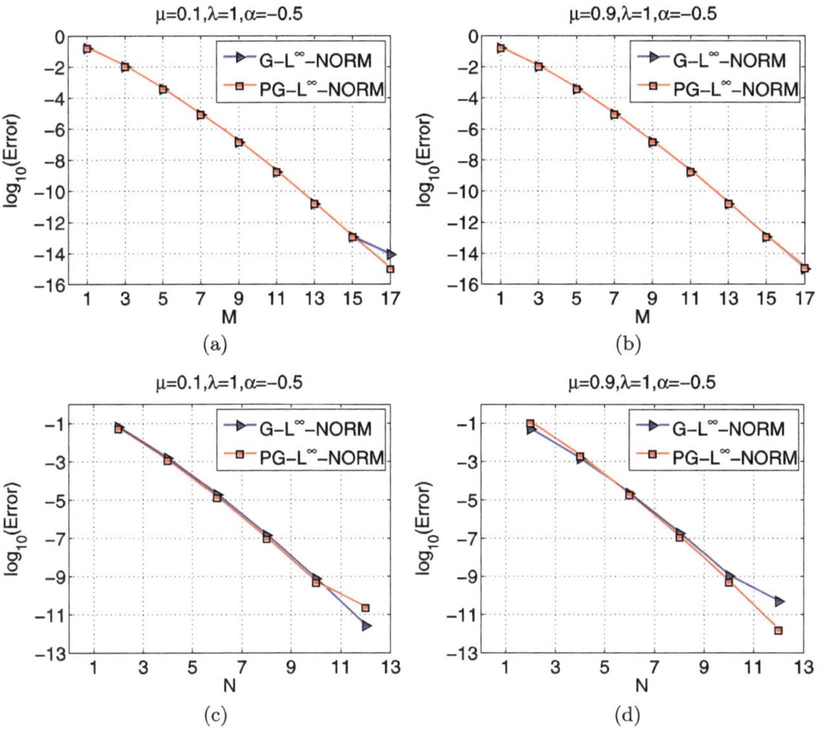

Figure 3: Error decays of the numerical solutions with respect to the polynomial degrees for the smooth exact solution. (a) $\mu = 0.1$ with $N = 20$. (b) $\mu = 0.9$ with $N = 20$. (c) $\mu = 0.1$ with $M = 20$. (d) $\mu = 0.9$ with $M = 20$.

Example 1. We consider the problem (5.7) for two values of μ, i. e., 0.1 and 0.9. The fabricated exact solution is:

$$u(x,t) = \sin \pi t \sin \pi x. \tag{5.28}$$

Since the exact solution is smooth in both x and t, it is reasonable to take $\lambda = 1$ in the schemes (5.12) and (5.26), which correspond to the standard spectral methods using classical polynomials. In Figure 3(a)–(d) we plot the L^∞-errors in semi-log scale with respect to the polynomial degrees M and N for $\alpha = -0.5$. Here "G-" in the figures means Galerkin-based method (5.12), while "PG-" stands for Petrov–Galerkin-based method (5.26). The first two figures, i. e., Figure 1(a) and (b), show the error behavior as a function of the polynomial degree M with respect to the spatial variable with fixed $N = 20$ for both schemes. The fact that all error curves are straight lines in these semi-log plots indicates that the numerical solutions are exponentially convergent when the polynomial degree M increases.

The error behavior for the approximation in time is presented in Figure 1(c) and (d), where the L^∞-error in log scale versus the degree of polynomial N with

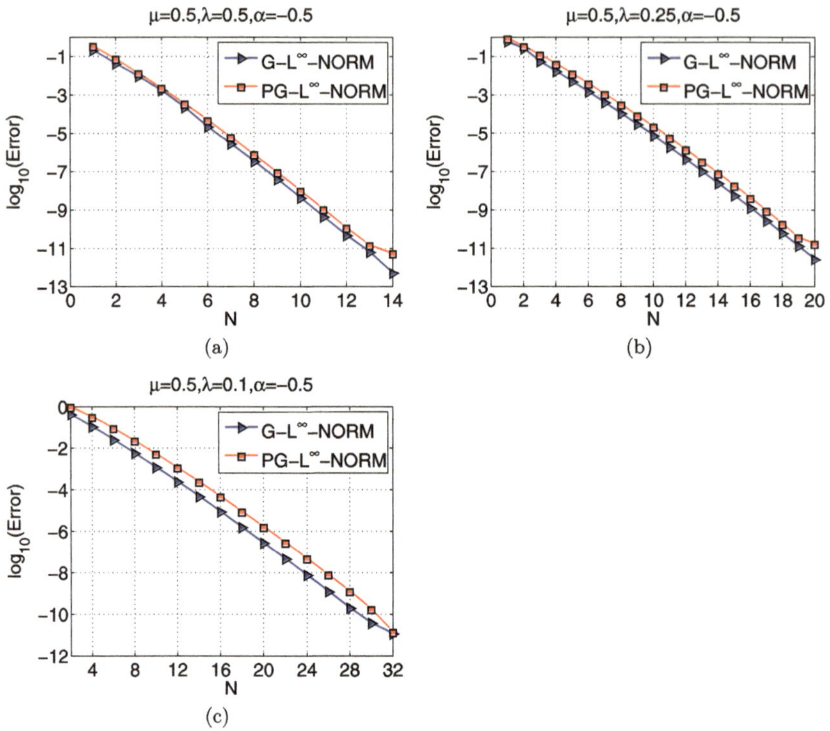

Figure 4: Error decays versus N for the non-smooth solution $u(x,t) = \sin(\pi x)(E_{\mu,1}(t^\mu) - 1)$. (a) $\lambda = 0.5$. (b) $\lambda = 0.25$. (c) $\lambda = 0.1$.

fixed $M = 20$ is plotted. It is observed that the convergence rate of both methods is exponential with respect to the polynomial degree in the time variable.

Example 2. Consider the fabricated source term

$$f(x,t) = \sin(\pi x)E_{\mu,1}(t^\mu) + \pi^2 \sin(\pi x)(E_{\mu,1}(t^\mu) - 1),$$

such that the exact solution is $u(x,t) = \sin(\pi x)(E_{\mu,1}(t^\mu) - 1)$.

It is readily seen that the exact solution has limited regularity in time direction at $t = 0$. The accuracy in time of the proposed methods is investigated in this test. The results computed by the schemes (5.12) and (5.26) with $\mu = 0.5$, $\alpha = -0.5$, and $\lambda = 1/2, 1/4, 1/10$ are shown in Figure 4. Once again, the error curves are all straight lines with respect to the polynomial degree N in semi-log plots. This clearly indicates the spectral accuracy of the obtained numerical solutions by both methods for all tested values of λ. This observation is in perfect agreement with the theoretical prediction given in Theorem 10 stating that the convergence of the numerical solution is exponential if $u(x, t^{1/\lambda})$ is smooth.

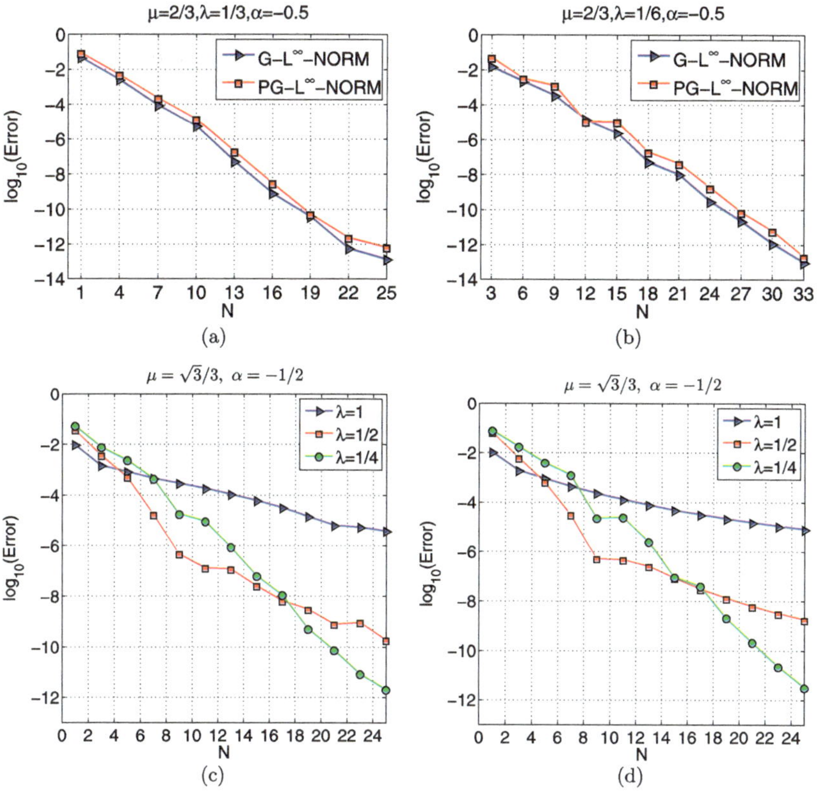

Figure 5: Errors versus N for different μ and λ. (a) $\mu = 2/3$, $\lambda = 1/3$. (b) $\mu = 2/3$, $\lambda = 1/6$. (c) Galerkin L^∞-norm with $\mu = \sqrt{3}/3$, $\lambda = 1, 1/2, 1/4$. (d) Petrov–Galerkin L^∞-norm with $\mu = \sqrt{3}/3$, $\lambda = 1, 1/2, 1/4$.

Example 3. Take an arbitrary smooth force function $f(x,t) = \sin(\pi x)\sin(\pi t)$, for which the exact solution to the problem (5.7) is unknown.

In order to investigate the convergence behavior of the numerical solutions, we will serve the numerical solution calculated with some large degrees, i. e., $M = 40$, $N = 100$, as the “exact” solution. In Figure 5(a) and (b), we plot the errors as functions of the polynomial degree N with fixed $M = 20$ for $\mu = 2/3$. According to the well-known solution expression, for a rational number $\mu = p/q$ with p and q being positive integer numbers, we can take $\lambda = 1/kq$, k is an integer, such that $u(x, t^{1/\lambda})$ is smooth. The numerical results presented in Figure 5(a) and (b) for $\lambda = 1/3, 1/6$ demonstrate, as expected, an exponential decay of the errors when the fractional polynomial degree N increases. For an irrational number μ, although there is no possibility to make $u(x, t^{1/\lambda})$ smooth by choosing λ, we can try to take λ small enough such that $u(x, t^{1/\lambda})$ is smooth enough. By doing so it is still possible to catch the solution with high accuracy. This is confirmed in Figure 5(c) and (d), where we present the error history versus N for the irrational value $\mu = \sqrt{3}/3$ and $\lambda = 1, 1/2, 1/4$. It is observed that for small enough λ,

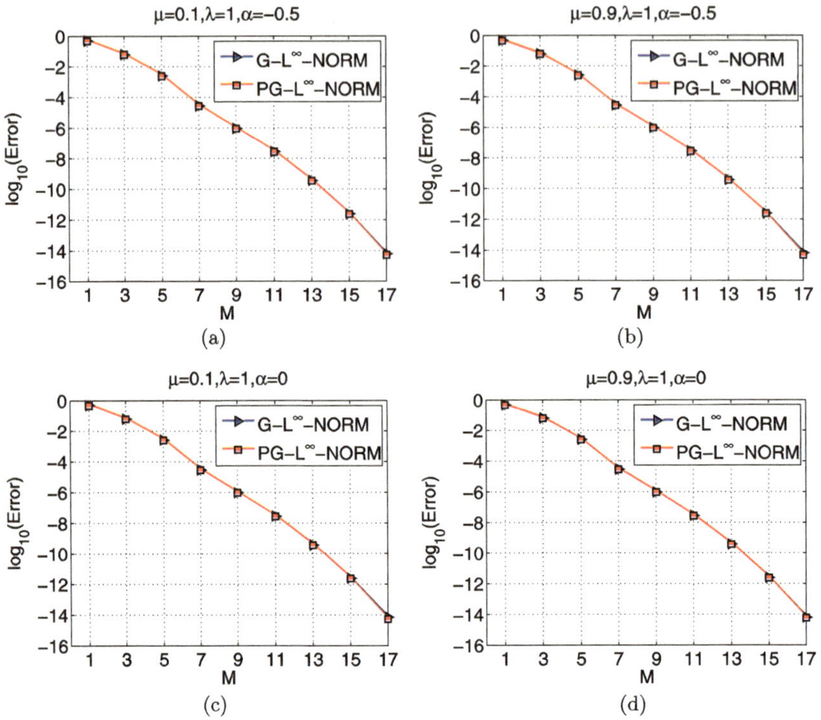

Figure 6: Error decays with respect to the polynomial degree M with fixed $N = 20$ for the exact solution $u(x,t) = e^x \sin \pi x \sin \pi t$.

say $\lambda = 1/4$, the convergence rate for both Galerkin-based and Petrov–Galerkin-based methods becomes almost exponential with respect to the fractional polynomial degree N.

Example 4. We consider the problem (5.7) for $\mu = 0.1, 0.9$ and the exact solution

$$u(x,t) = e^x \sin \pi x \sin \pi t. \tag{5.29}$$

In this example, the solution with respect to the spatial variable is no longer an eigenfunction of the one-dimensional Dirichlet Laplacian, thus representing essentially a true two-dimensional problem. We repeat the convergence behavior investigation as for Example 3. The computed results for $\lambda = 1$, $\alpha = 0, -0.5$ are shown in Figure 6. We clearly observe the spectral convergence of the numerical solution with respect to the polynomial degree in space.

Remark 3. In the error analysis, i. e., Theorem 10, we have assumed $-1 < \alpha < -\mu/2$. However, our numerical tests showed that the proposed method works for all $\alpha > -1$. The choice of λ is also of importance for the efficiency of the method, which can be made according to the following strategy.

Case I: If the solution is smooth, the optimal value is $\lambda = 1$.

Case II: If the source term $f(x,t)$ is smooth, then (a) when μ is a rational number p/q, the best choice is $\lambda = 1/q$. Theoretically $\lambda = 1/nq$ $(n = 1,2,\dots)$ works too, but larger n leads to larger calculations; (b) when μ is an irrational number, there is no suitable value of λ to make $u(x,t^\lambda)$ smooth. In this case, we can take $\lambda = 1/q$ with a reasonably large q such that $u(x,t^{1/\lambda})$ is smooth enough.

6 Concluding remarks

We have established a variational framework for numerical solutions of the STFDE. Suitable fractional Sobolev spaces are introduced and investigated. We proved the well-posedness of the weak formulation of the STFDE. Based on this weak formulation, we constructed and analyzed a space-time spectral method using traditional polynomials for the numerical solution. The optimal error estimates were derived under necessary smoothness assumption on the exact solution.

For the TFDE having weakly singular solutions, we proposed and analyzed a fractional spectral method based on the fractional Jacobi polynomials, also known as Müntz polynomials, to approximate the solution. The significance of this approach is that the fractional polynomials can much better capture typical solution structures of the TFDE than the traditional polynomials. Our analysis and numerical experiments have demonstrated that the fractional spectral method is capable of achieving spectral convergence for low regular solutions.

Bibliography

[1] R. A. Adams, *Sobolev Spaces*, Academic Press, New York, 1975.

[2] K. F. Andersen and H. P. Heinig, Weighted norm inequalities for certain integral operators, *SIAM J. Math. Anal.*, **14**(14) (1983), 834–844.

[3] C. Bernardi and Y. Maday, Spectral methods, *Handbook of Numerical Analysis*, **5** (1997), 209–485.

[4] S. Chen, J. Shen, and L. L. Wang, Generalized Jacobi functions and their applications to fractional differential equations, *Math. Comput.*, **85**(300) (2016), 1603–1638.

[5] D. M. Hou, M. T. Hasan, and C. J. Xu, Müntz spectral methods for the time-fractional diffusion equation, *Comput. Methods Appl. Math.*, **18**(1) (2018), 43–62.

[6] D. M. Hou and C. J. Xu, A fractional spectral method with applications to some singular problems, *Adv. Comput. Math.*, (2017), 1–34.

[7] T. A. M. Langlands and B. I. Henry, The accuracy and stability of an implicit solution method for the fractional diffusion equation, *J. Comput. Phys.*, **205**(2) (2005), 719–736.

[8] X. Li and C. Xu, A space-time spectral method for the time fractional diffusion equation, *SIAM J. Numer. Anal.*, **47**(3) (2009), 2108–2131.

[9] X. Li and C. Xu, Existence and uniqueness of the weak solution of the space-time fractional diffusion equation and a spectral method approximation, *Commun. Comput. Phys.*, **8**(5) (2010), 1016–1051.
[10] Y. Maday, Analysis of spectral projections in one dimensional domains, *Math. Comput.*, **55**(192) (1990), 537–562.
[11] K. B. Oldham and J. Spanier, *The Fractional Calculus*, Academic Press, New York, London, 1974.
[12] I. Podlubny, *Fractional Differential Equations*, Academic Press, 1999.
[13] V. D. Stepanov, Weight inequalities of the Hardy type for fractional Riemann–Liouville integrals, *Sib. Math. J.*, **31**(3) (1990), 513–522.
[14] M. Zayernouri, M. Ainsworth, and G. E. Karniadakis, A unified Petrov–Galerkin spectral method for fractional PDEs, *Comput. Methods Appl. Math.*, **283**(1) (2015), 1545–1569.
[15] M. Zayernouri and G. E. Karniadakis, Fractional Sturm–Liouville eigen-problems: theory and numerical approximation, *J. Comput. Phys.*, **252** (2014), 495–517.
[16] M. Zheng, F. Liu, V. Anh, and I. Turner, A high-order spectral method for the multi-term time-fractional diffusion equations, *Appl. Math. Model.*, **40**(7–8) (2016), 4970–4985.

Jie Shen and Changtao Sheng

Spectral methods for fractional differential equations using generalized Jacobi functions

Abstract: We present essential properties of the generalized Jacobi functions (GJFs) and their application to construct efficient and accurate spectral methods for a class of fractional differential equations. In particular, it is shown that GJFs allow us to effortlessly compute the stiffness matrices and resolve the leading singular term for a general class of fractional differential equations.

Keywords: Fractional differential equations, generalized Jacobi functions, spectral method, Petrov–Galerkin method

MSC 2010: 65N35, 65E05, 65M70, 41A05, 41A10

1 Introduction

Fractional differential equations (FDEs) have attracted considerable attention in recent years due to their ability to model certain processes which cannot be adequately described by usual partial differential equations. Two main difficulties for dealing with FDEs are (i) fractional derivatives are non-local operators; (ii) fractional derivatives involve singular kernel/weight functions, and the solutions of FDEs are usually weakly singular near the boundaries and at initial time. Hence, a straightforward extension of usual polynomial-based numerical methods for FDEs is not effective as it usually involves dense matrices even for the simplest FDEs and suffers from low convergence rate due to the weak singularity. Thus, one needs to develop non-standard and non-polynomial-based numerical methods to effectively deal with the difficulties associated with FDEs.

In this chapter, we shall focus on some recently developed spectral methods using the generalized Jacobi functions (GJFs) for solving FDEs. Some pioneer work on using spectral methods for solving FDEs has been carried out [17, 18]. However, the methods in these papers use the usual polynomial approximations which do not particularly address the two difficulties mentioned above. A breakthrough is made in [32], in which

Acknowledgement: This work is supported in part by AFOSR FA9550-16-1-0102 and NSF DMS-1620262, DMS-1720442.

Jie Shen, Department of Mathematics, Purdue University, West Lafayette, IN, USA, e-mail: shen7@purdue.edu
Changtao Sheng, School of Mathematical Sciences and Fujian Provincial Key Laboratory on Mathematical Modeling & High Performance Scientific Computing, Xiamen University, Xiamen, China, e-mail: ctsheng8@gmail.com

https://doi.org/10.1515/9783110571684-005

the authors introduced the so-called poly-fractonomials, which are eigenfunctions of some fractional Sturm–Liouville operator. So using them as basis functions greatly simplifies the computation of fractional derivatives and leads to sparse matrices for some simple model equations. Furthermore, one can choose suitable parameters in poly-fractonomials so that its leading term is in fact the leading singular term of the corresponding FDE. Thus, the spectral method using poly-fractonomials can also resolve the leading singularity in an FDE. It turns out that the poly-fractonomials introduced in [32] coincide, within certain parameter range, with the GJFs introduced in [10]. In [6], the authors re-examined the GJFs in the context of FDEs and derived optimal approximation results with norms suitable for fractional derivatives. In particular, it is shown in [6] that well-constructed spectral methods using GJFs for some typical FDEs can lead to exponential convergence despite the fact that their solutions are weakly singular.

Thanks to the aforementioned remarkable properties of the poly-fractonomials/GJFs with regard to fractional derivatives and fractional FDEs, there is now a significant number of recent work on using the poly-fractonomials/GJFs for different kinds of FDEs and singular integral equations, including, for example, spectral Galerkin or spectral Petrov–Galerkin methods [31, 11, 36, 23, 21, 28, 15]; spectral collocation methods [35, 36, 15, 12, 14, 34, 13]; and discontinuous Galerkin spectral element methods [33, 16].

The aim of this chapter is not to provide an exhaustive review of all developments with regard to spectral methods for FDEs; rather, it aims to present essential properties of the GJFs, including relations to the fractional derivatives and their approximation results in properly weighted Sobolev spaces, and how one can use them to construct efficient and accurate spectral methods for FDEs.

In order to illustrate the idea and advantage of spectral methods using GJFs, we shall consider the following FDE:

$$\begin{cases} \gamma\, {}^{C}D_t^{\alpha}u(x,t) - p\,{}^{R}D_x^{\beta}u(x,t) - (1-p)\,{}_x^{R}D^{\beta}u(x,t) = f(x,t), \\ u(x,t)|_{\partial\Lambda} = 0, \quad \forall t \in I := (0,T), \\ u(x,0) = u_0(x), \quad \forall x \in \Lambda = (a,b), \end{cases} \tag{1.1}$$

where $0 < \alpha < 1$, $1 < \beta < 2$, $p \in [0,1]$; ${}^{C}D_t^{\alpha}$, ${}^{R}D_x^{\beta}$, and ${}_x^{R}D^{\beta}$ are the Caputo and left and right Riemann–Liouville fractional derivative operator, respectively. More precisely, we shall consider, successively, the following four cases: (i) $\gamma = 0$, $p = 1$: an initial value problem and a boundary value problem with one-sided fractional derivative; (ii) $\gamma = 0$, $p = 1/2$: Riesz FDE; (iii) $\gamma = 0$, $p \neq 0, 1/2, 1$: a boundary value problem with general two-sided fractional derivatives; and (iv) $\gamma \neq 0$ and $p = 1/2$: a space-time fractional diffusion equation.

The rest of the chapter is organized as follows. In the next section, we introduce the GJFs corresponding to the one-sided fractional derivative, Riesz derivative, and two-sided fractional derivatives with different coefficients, summarize their special properties, particularly with regard to fractional derivatives, and present approximation results using these GJFs. In Section 3, we construct successively efficient spectral methods using GJFs for the four cases above and derive their error estimates. We provide some concluding remarks in Section 4.

Due to the space constraint, we shall not present numerical results in this chapter. However, theoretical results presented in this chapter have been validated by ample numerical experiments in the correspondingly cited papers.

2 Generalized Jacobi functions and their approximation properties

In this section, we collect some basic relations and approximation properties of GJFs with respect to fractional derivatives from [21, 23, 22]. These results will play essential roles in developing efficient algorithms for FDEs and in deriving the corresponding error estimates in the next section.

2.1 Fractional integrals/derivatives

We introduce the definitions of fractional integrals/derivatives. Let $a < b$ and denote $\Lambda = (a, b)$.

Definition 1 (One-sided fractional integrals and derivatives [24, 7]). For $\rho \in \mathbb{R}^+$, the left and right fractional integrals are respectively defined as

$$
\begin{aligned}
I_x^\rho v(x) &= \frac{1}{\Gamma(\rho)} \int_a^x \frac{v(y)}{(x-y)^{1-\rho}} dy, \quad x \in \Lambda, \\
{}_xI^\rho v(x) &= \frac{1}{\Gamma(\rho)} \int_x^b \frac{v(y)}{(y-x)^{1-\rho}} dy, \quad x \in \Lambda,
\end{aligned} \tag{2.1}
$$

where $\Gamma(\cdot)$ is the usual Gamma function.

For $s \in [k-1, k)$ with $k \in \mathbb{N}$, the left-sided Riemann–Liouville fractional derivative of order s is defined by

$$
{}^R D_x^s v(x) = \frac{1}{\Gamma(k-s)} \frac{d^k}{dx^k} \int_a^x \frac{v(y)}{(x-y)^{s-k+1}} dy, \quad x \in \Lambda, \tag{2.2}
$$

and the right-sided Riemann–Liouville fractional derivative of order s is defined by

$$ {}_{x}^{R}D^{s}v(x) = \frac{(-1)^k}{\Gamma(k-s)} \frac{d^k}{dx^k} \int_x^b \frac{v(y)}{(y-x)^{s-k+1}} dy, \quad x \in \Lambda. \tag{2.3} $$

For $s \in [k-1, k)$ with $k \in \mathbb{N}$, the left-sided Caputo fractional derivative of order s is defined by

$$ {}^{C}D_x^s v(x) := \frac{1}{\Gamma(k-s)} \int_a^x \frac{v^{(k)}(y)}{(x-y)^{s-k+1}} dy, \quad x \in \Lambda, \tag{2.4} $$

and the right-sided Caputo fractional derivatives of order s is defined by

$$ {}_{x}^{C}D^{s}v(x) := \frac{(-1)^k}{\Gamma(k-s)} \int_x^b \frac{v^{(k)}(y)}{(y-x)^{s-k+1}} dy, \quad x \in \Lambda. \tag{2.5} $$

According to [7, Theorem 2.14], for any absolutely integrable function v and real $s \geq 0$ we have

$$ {}^{R}D_x^s I_x^s v(x) = v(x), \quad {}_{x}^{R}D^{s}\,{}_{x}I^{s} v(x) = v(x), \quad \text{a. e.} \quad \text{in } \Lambda. \tag{2.6} $$

The following lemma shows the relationship between the Riemann–Liouville and Caputo fractional derivatives (see, e. g., [7, 24]).

Lemma 1. *For $s \in [k-1, k)$ with $k \in \mathbb{N}$, we have*

$$ {}^{R}D_t^s u(t) = {}^{C}D_t^s u(t) + \sum_{j=0}^{k-1} \frac{u^{(j)}(a)}{\Gamma(1+j-s)} (t-a)^{j-s}. \tag{2.7} $$

Definition 2 (Riesz fractional integrals and derivatives [25]). For $\rho \in [0, 1)$, the Riesz fractional integral of order ρ is defined as

$$ I^{\rho} v(x) := \frac{1}{2\cos(\pi\rho/2)} (I_x^{\rho} + {}_{x}I^{\rho}) v(x), \quad x \in \Lambda. \tag{2.8} $$

For $\rho \in [2k-1, 2k)$ with $k \in \mathbb{N}$, the Riesz fractional derivative of order ρ is defined by

$$ D^{\rho} v(x) := D^{2k} I^{2k-\rho} v(x). \tag{2.9} $$

The following results [9, 17] play fundamental roles in the analysis of FDEs.

Lemma 2. *For all $0 < \alpha < 2$ and $\alpha \neq 1$, we have*

$$ ({}^{R}D_t^{\alpha} v, w)_{\Lambda} = ({}^{R}D_t^{\frac{\alpha}{2}} v, {}_{t}^{R}D^{\frac{\alpha}{2}} w)_{\Lambda}, \quad \forall v, w \in H_0^{\alpha/2}(\Lambda). \tag{2.10} $$

Lemma 3. *For all $\alpha > 0$ and if $\alpha - 1/2$ is not an integer, we have*

$$ ({}^{R}D_t^{\frac{\alpha}{2}} v, {}_{t}^{R}D^{\frac{\alpha}{2}} v)_{\Lambda} \cong \|v\|^2_{H_0^{\alpha/2}(\Lambda)}, \quad \forall v \in H_0^{\alpha/2}(\Lambda). \tag{2.11} $$

In the sequel, we use c to denote a generic constant.

2.2 GJFs for one-sided fractional derivatives

We start by considering the one-sided fractional derivatives. Unless otherwise specified, we set $\Lambda = (-1, 1)$.

2.2.1 Jacobi polynomials and Bateman's fractional integral formula

According to [30, (4.21.2)], the classical Jacobi polynomials with parameters $\alpha, \beta \in \mathbb{R}$ can be defined by

$$P_n^{(\alpha,\beta)}(x) = \frac{(\alpha+1)_n}{n!}\, {}_2F_1\Big(-n, n+\alpha+\beta+1; \alpha+1; \frac{1-x}{2}\Big), \tag{2.12}$$

where ${}_2F_1(a, b; c; x)$ is the hypergeometric function, and the rising factorial in the Pochhammer symbol, for $a \in \mathbb{R}$ and $j \in \mathbb{N}_0$, is defined by

$$(a)_0 = 1, \quad (a)_j := a(a+1)\cdots(a+j-1) = \frac{\Gamma(a+j)}{\Gamma(a)}, \quad \text{for } j \geq 1. \tag{2.13}$$

For $\alpha, \beta > -1$, the classical Jacobi polynomials are orthogonal with respect to the Jacobi weight function $\omega^{(\alpha,\beta)}(x) = (1-x)^\alpha(1+x)^\beta$, namely,

$$\int_{-1}^{1} P_n^{(\alpha,\beta)}(x) P_{n'}^{(\alpha,\beta)}(x)\omega^{(\alpha,\beta)}(x)dx = \gamma_n^{(\alpha,\beta)}\delta_{nn'}, \tag{2.14}$$

where $\delta_{nn'}$ is the Dirac delta symbol, and the normalization constant is given by

$$\gamma_n^{(\alpha,\beta)} = \frac{2^{\alpha+\beta+1}\Gamma(n+\alpha+1)\Gamma(n+\beta+1)}{(2n+\alpha+\beta+1)n!\Gamma(n+\alpha+\beta+1)}. \tag{2.15}$$

The following fractional integral formula of hypergeometric functions due to Bateman [3] (also see [2, p. 313]) plays an important role in the computation of fractional integrals/derivatives: for real $c, \rho \geq 0$,

$${}_2F_1(a, b; c+\rho; x) = \frac{\Gamma(c+\rho)}{\Gamma(c)\Gamma(\rho)} x^{1-(c+\rho)} \int_0^x t^{c-1}(x-t)^{\rho-1}{}_2F_1(a, b; c; t)\, dt, \quad |x| < 1. \tag{2.16}$$

One derives easily from (2.12) and (2.16) the following results (cf. [30, p. 96]).

Lemma 4. *Let $\rho \in \mathbb{R}^+$, $n \in \mathbb{N}_0$, and $x \in \Lambda$.*
(i) *For $\alpha > -1$ and $\beta \in \mathbb{R}$,*

$$(1-x)^{\alpha+\rho}\frac{P_n^{(\alpha+\rho,\beta-\rho)}(x)}{P_n^{(\alpha+\rho,\beta-\rho)}(1)} = \frac{\Gamma(\alpha+\rho+1)}{\Gamma(\alpha+1)\Gamma(\rho)} \int_x^1 \frac{(1-y)^\alpha}{(y-x)^{1-\rho}} \frac{P_n^{(\alpha,\beta)}(y)}{P_n^{(\alpha,\beta)}(1)}\, dy. \tag{2.17}$$

(ii) *For $\alpha \in \mathbb{R}$ and $\beta > -1$,*

$$(1+x)^{\beta+\rho}\frac{P_n^{(\alpha-\rho,\beta+\rho)}(x)}{P_n^{(\beta+\rho,\alpha-\rho)}(1)} = \frac{\Gamma(\beta+\rho+1)}{\Gamma(\beta+1)\Gamma(\rho)}\int_{-1}^{x}\frac{(1+y)^{\beta}}{(x-y)^{1-\rho}}\frac{P_n^{(\alpha,\beta)}(y)}{P_n^{(\beta,\alpha)}(1)}\,dy. \tag{2.18}$$

Using the notation in definition (2.1), we can rewrite the formulas in Lemma 4 as follows.

Lemma 5. *Let $\rho \in \mathbb{R}^+$, $n \in \mathbb{N}_0$, and $x \in \Lambda$.*
- *For $\alpha > -1$ and $\beta \in \mathbb{R}$,*

$$ {}_xI^{\rho}\{(1-x)^{\alpha}P_n^{(\alpha,\beta)}(x)\} = \frac{\Gamma(n+\alpha+1)}{\Gamma(n+\alpha+\rho+1)}(1-x)^{\alpha+\rho}P_n^{(\alpha+\rho,\beta-\rho)}(x). \tag{2.19}$$

- *For $\alpha \in \mathbb{R}$ and $\beta > -1$,*

$$ I_x^{\rho}\{(1+x)^{\beta}P_n^{(\alpha,\beta)}(x)\} = \frac{\Gamma(n+\beta+1)}{\Gamma(n+\beta+\rho+1)}(1+x)^{\beta+\rho}P_n^{(\alpha-\rho,\beta+\rho)}(x). \tag{2.20}$$

Thanks to (2.6), we obtain from Lemma 5 the following useful "inverse" rules.

Lemma 6. *Let $s \in \mathbb{R}^+$, $n \in \mathbb{N}_0$, and $x \in \Lambda$.*
- *For $\alpha > -1$ and $\beta \in \mathbb{R}$,*

$$ {}_x^RD^{s}\{(1-x)^{\alpha+s}P_n^{(\alpha+s,\beta-s)}(x)\} = \frac{\Gamma(n+\alpha+s+1)}{\Gamma(n+\alpha+1)}(1-x)^{\alpha}P_n^{(\alpha,\beta)}(x). \tag{2.21}$$

- *For $\alpha \in \mathbb{R}$ and $\beta > -1$,*

$$ {}^RD_x^{s}\{(1+x)^{\beta+s}P_n^{(\alpha-s,\beta+s)}(x)\} = \frac{\Gamma(n+\beta+s+1)}{\Gamma(n+\beta+1)}(1+x)^{\beta}P_n^{(\alpha,\beta)}(x). \tag{2.22}$$

The above lemmas are remarkable in the sense that fractional integrals/derivatives of functions in the form of $(1\pm x)^{\alpha}P_n^{(\alpha,\beta)}$ can be expressed in the same form with a set of different parameters (α,β). In other words, the **global** fractional integral/derivative operators become **local** operators in suitable "spectral" spaces. In particular, if $\alpha = 0$ in (2.21), the fractional derivative operator ${}_x^RD^s$ takes $(1-x)^sP_n^{(s,\beta-s)}(x)$ to the polynomial $P_n^{(0,\beta)}(x)$; conversely, if $\alpha+s = k \in \mathbb{N}_0$, ${}_x^RD^s$ takes the polynomial $(1-x)^kP_n^{(k,\beta-s)}(x)$ to $(1-x)^{k-s}P_n^{(k-s,\beta)}(x)$. Such remarkable properties are essential for constructing efficient spectral algorithms for FDEs.

Definition 3 (One-sided generalized Jacobi functions [6]). Define

$$ {}^{+}J_n^{(-\alpha,\beta)}(x) := (1-x)^{\alpha}P_n^{(\alpha,\beta)}(x), \quad \text{for } \alpha > -1,\ \beta \in \mathbb{R}, \tag{2.23}$$

and

$$ {}^{-}J_n^{(\alpha,-\beta)}(x) := (1+x)^{\beta}P_n^{(\alpha,\beta)}(x), \quad \text{for } \alpha \in \mathbb{R},\ \beta > -1, \tag{2.24}$$

for all $x \in \Lambda$ and $n \in \mathbb{N}_0$.

2.2.2 Some important properties of GJFs

It follows straightforwardly from (2.14) and Definition 3 that for $\alpha, \beta > -1$,

$$\int_{-1}^{1} {}^{+}J_n^{(-\alpha,\beta)}(x)\,{}^{+}J_{n'}^{(-\alpha,\beta)}(x)\,\omega^{(-\alpha,\beta)}(x)\,dx = \int_{-1}^{1} {}^{-}J_n^{(\alpha,-\beta)}(x)\,{}^{-}J_{n'}^{(\alpha,-\beta)}(x)\,\omega^{(\alpha,-\beta)}(x)\,dx = \gamma_n^{(\alpha,\beta)}\delta_{nn'}. \tag{2.25}$$

Similarly, for $\alpha > -1$, $k \in \mathbb{N}$, and $n, n' \geq k$, we have

$$\int_{-1}^{1} {}^{+}J_n^{(-\alpha,-k)}(x)\,{}^{+}J_{n'}^{(-\alpha,-k)}(x)\,\omega^{(-\alpha,-k)}(x)\,dx = \int_{-1}^{1} {}^{-}J_n^{(-k,-\alpha)}(x)\,{}^{-}J_{n'}^{(-k,-\alpha)}(x)\,\omega^{(-k,-\alpha)}(x)\,dx = \gamma_n^{(\alpha,-k)}\delta_{nn'}, \tag{2.26}$$

where $\gamma_n^{(\alpha,\beta)}$ and $\gamma_n^{(\alpha,-k)}$ are some constants which can be found in [6].

With the above definitions, we can rewrite Lemma 6 as follows.

Theorem 1. *Let $s \in \mathbb{R}^+$, $n \in \mathbb{N}_0$, and $x \in \Lambda$.*

- *For $\alpha > s-1$ and $\beta \in \mathbb{R}$,*

$$ {}_{x}^{R}D^{s}\{{}^{+}J_n^{(-\alpha,\beta)}(x)\} = \frac{\Gamma(n+\alpha+1)}{\Gamma(n+\alpha-s+1)}\,{}^{+}J_n^{(-\alpha+s,\beta+s)}(x). \tag{2.27}$$

- *For $\alpha \in \mathbb{R}$ and $\beta > s-1$,*

$$ {}^{R}D_{x}^{s}\{{}^{-}J_n^{(\alpha,-\beta)}(x)\} = \frac{\Gamma(n+\beta+1)}{\Gamma(n+\beta-s+1)}\,{}^{-}J_n^{(\alpha+s,-\beta+s)}(x). \tag{2.28}$$

The analysis of GJFs essentially relies on the orthogonality of fractional derivatives of GJFs. Recall the derivative formula of the classical Jacobi polynomials (see, e. g., [29, p. 72]): for $\alpha, \beta > -1$ and $n \geq l$,

$$D^{l}P_n^{(\alpha,\beta)}(x) = \kappa_{n,l}^{(\alpha,\beta)}P_{n-l}^{(\alpha+l,\beta+l)}(x), \tag{2.29}$$

where $\kappa_{n,l}^{(\alpha,\beta)}$ is some constant. Noting that ${}_{x}^{R}D^{s+l} = (-1)^{l}D^{l}\,{}_{x}^{R}D^{s}$ and ${}^{R}D_{x}^{s+l} = D^{l}\,{}^{R}D_{x}^{s}$ for $s \in \mathbb{R}^+$ and $l \in \mathbb{N}$, we derive from (2.14) and (2.27)–(2.29) the following orthogonality relations.

- For $\alpha > 0$, $\alpha+\beta > -1$, and $n, n' \geq l \geq 0$,

$$\int_{-1}^{1} {}_{x}^{R}D^{\alpha+l}\,{}^{+}J_n^{(-\alpha,\beta)}(x)\,{}_{x}^{R}D^{\alpha+l}\,{}^{+}J_{n'}^{(-\alpha,\beta)}(x)\,\omega^{(l,\alpha+\beta+l)}(x)\,dx = h_{n,l}^{(\alpha,\beta)}\delta_{nn'}, \tag{2.30}$$

where

$$h_{n,l}^{\alpha,\beta} := \frac{2^{\alpha+\beta+1}\Gamma^2(n+\alpha+1)\Gamma(n+\alpha+\beta+l+1)}{(2n+\alpha+\beta+1)n!(n-l)!\Gamma(n+\alpha+\beta+1)}. \tag{2.31}$$

– For $\alpha+\beta > -1$, $\beta > 0$, and $n, n' \geq l \geq 0$,

$$\int_{-1}^{1} {}^R D_x^{\beta+l}\, {}^{-}J_n^{(\alpha,-\beta)}(x)\, {}^R D_x^{\beta+l}\, {}^{-}J_{n'}^{(\alpha,-\beta)}(x)\, \omega^{(\alpha+\beta+l,l)}(x)\, dx = h_{n,l}^{(\beta,\alpha)} \delta_{nn'}. \tag{2.32}$$

2.2.3 Approximation by the one-sided GJFs

We show below that approximation by GJFs can lead to truly spectral convergence for functions in properly weighted Sobolev spaces involving fractional derivatives.

For simplicity of presentation, we only provide the results for $\{{}^{-}J_n^{(\alpha,-\beta)}\}$. Similar results can be established for $\{{}^{+}J_n^{(-\alpha,\beta)}\}$ [6].

Let $\mathcal{P}_N$ be the set of all algebraic (real-valued) polynomials of degree at most N. Let $\varpi(x) > 0$, $x \in \Lambda$, be a generic weight function. The weighted space $L^2_\varpi(\Lambda)$ is defined as in Adams [1] with the inner product and norm

$$(u, v)_\varpi = \int_\Lambda u(x)v(x)\varpi(x)dx, \quad \|u\|_\varpi = (u, u)_\varpi^{1/2}.$$

If $\varpi \equiv 1$, we omit the weight function in the notation.

We define the finite-dimensional fractional-polynomial space

$${}^{-}\mathcal{F}_N^{(\alpha,-\beta)}(\Lambda) = \{\phi = (1+x)^\beta \psi : \psi \in \mathcal{P}_N\} = \operatorname{span}\{{}^{-}J_n^{(\alpha,-\beta)} : 0 \leq n \leq N\}.$$

By the orthogonality (2.25), we can expand any $u \in L^2_{\omega^{(\alpha,-\beta)}}$ as

$$u(x) = \sum_{n=0}^{\infty} \hat{u}_n^{(\alpha,-\beta)}\, {}^{-}J_n^{(\alpha,-\beta)}(x), \tag{2.33}$$

where

$$\hat{u}_n^{(\alpha,-\beta)} = \frac{1}{\gamma_n^{(\alpha,\beta)}} \int_{-1}^{1} u\, {}^{-}J_n^{(\alpha,-\beta)} \omega^{(\alpha,-\beta)} dx,$$

and the Parseval identity holds, i. e.,

$$\|u\|^2_{\omega^{(\alpha,-\beta)}} = \sum_{n=0}^{\infty} \gamma_n^{(\alpha,\beta)} |\hat{u}_n^{(\alpha,-\beta)}|^2. \tag{2.34}$$

Consider the following $L^2_{\omega^{(\alpha,-\beta)}}$-orthogonal projection onto ${}^{-}\mathcal{F}_N^{(\alpha,-\beta)}(\Lambda)$:

$$({}^{-}\pi_N^{(\alpha,-\beta)}u - u, v_N)_{\omega^{(\alpha,-\beta)}} = 0, \quad \forall v_N \in {}^{-}\mathcal{F}_N^{(\alpha,-\beta)}(\Lambda). \tag{2.35}$$

Then, it is easy to derive from (2.28) that for any $l \in \mathbb{N}_0$, we have

$$({}^R D_x^{\beta+l}({}^{-}\pi_N^{(\alpha,-\beta)}u - u), D^l w_N)_{\omega^{(\alpha+\beta+l,l)}} = 0, \quad \forall w_N \in \mathcal{P}_N(\Lambda). \tag{2.36}$$

To describe the projection error, we define

$${}^{-}\mathcal{B}^m_{\alpha,\beta}(\Lambda) := \{u \in L^2_{\omega^{(\alpha,-\beta)}}(\Lambda) \; : \; {}^R D_x^{\beta+l}u \in L^2_{\omega^{(\alpha+\beta+l,l)}}(\Lambda) \text{ for } 0 \le l \le m\}. \tag{2.37}$$

By (2.32) and (2.33), for $(\alpha,\beta) \in {}^{-}\Sigma^{\alpha,\beta} := \{(\alpha,\beta) \; : \; \beta > 0, \; \alpha > -1\}$ and $l \in \mathbb{N}_0$ we have

$$\|{}^R D_x^{\beta+l}u\|^2_{\omega^{(\alpha+\beta+l,l)}} = \sum_{n=l}^{\infty} h_{n,l}^{(\beta,\alpha)} |\hat{u}_n^{(\alpha,-\beta)}|^2. \tag{2.38}$$

Theorem 2. *Let $(\alpha,\beta) \in {}^{-}\Sigma^{\alpha,\beta}$ and $u \in {}^{-}\mathcal{B}^m_{\alpha,\beta}(\Lambda)$.*
- *For $0 \le l \le m$, we have*

$$\|{}^R D_x^{\beta+l}({}^{-}\pi_N^{(\alpha,-\beta)}u - u)\|_{\omega^{(\alpha+\beta+l,l)}} \le cN^{l-m}\|{}^R D_x^{\beta+m}u\|_{\omega^{(\alpha+\beta+m,m)}}. \tag{2.39}$$

- *For $0 \le m$, we also have the $L^2_{\omega^{(\alpha,-\beta)}}$-estimate*

$$\|{}^{-}\pi_N^{(\alpha,-\beta)}u - u\|_{\omega^{(\alpha,-\beta)}} \le cN^{-(\beta+m)}\|{}^R D_x^{\beta+m}u\|_{\omega^{(\alpha+\beta+m,m)}}. \tag{2.40}$$

Proof. By (2.33), (2.35), and (2.38), we have

$$\begin{aligned}\|{}^R D_x^{\beta+l}({}^{-}\pi_N^{(\alpha,-\beta)}u - u)\|^2_{\omega^{(\alpha+\beta+l,l)}} &= \sum_{n=N+1}^{\infty} h_{n,l}^{(\beta,\alpha)} |\hat{u}_n^{(\alpha,-\beta)}|^2 \\ &= \sum_{n=N+1}^{\infty} \frac{h_{n,l}^{(\beta,\alpha)}}{h_{n,m}^{(\beta,\alpha)}} h_{n,m}^{(\beta,\alpha)} |\hat{u}_n^{(\alpha,-\beta)}|^2 \le \frac{h_{N+1,l}^{(\beta,\alpha)}}{h_{N+1,m}^{(\beta,\alpha)}} \|{}^R D_x^{\beta+m}u\|^2_{\omega^{(\alpha+\beta+m,m)}}.\end{aligned} \tag{2.41}$$

We now estimate the constant factor. By (2.13), (2.31), and a direct calculation, for $0 \le l \le m \le N$ we find

$$\begin{aligned}\frac{h_{N+1,l}^{(\beta,\alpha)}}{h_{N+1,m}^{(\beta,\alpha)}} &= \frac{\Gamma(N+\alpha+\beta+l+2)(N-m+1)!}{\Gamma(N+\alpha+\beta+m+2)(N-l+1)!} \\ &= \frac{1}{(N+\alpha+\beta+2+l)\cdots(N+\alpha+\beta+1+m)} \frac{(N-m+1)!}{(N-l+1)!} \\ &\le N^{l-m}\frac{(N-m+1)!}{(N-l+1)!},\end{aligned} \tag{2.42}$$

where we used the fact that $\alpha + \beta > -1$. Thus, we obtain (2.39) from (2.41), (2.42), and the property of the Gamma function.

The $L^2_{\omega^{(\alpha,-\beta)}}$-estimates can be obtained by using the same argument. We sketch the derivation below. By (2.34) and (2.38),

$$\begin{aligned}\|{}^{-}\pi_N^{(\alpha,-\beta)}u - u\|^2_{\omega^{(\alpha,-\beta)}} &= \sum_{n=N+1}^{\infty} \gamma_n^{(\alpha,\beta)} |\hat{u}_n^{(\alpha,-\beta)}|^2 \\ &= \sum_{n=N+1}^{\infty} \frac{\gamma_n^{(\alpha,\beta)}}{h_{n,m}^{(\beta,\alpha)}} h_{n,m}^{(\beta,\alpha)} |\hat{u}_n^{(\alpha,-\beta)}|^2 \le \frac{\gamma_{N+1}^{(\alpha,\beta)}}{h_{N+1,m}^{(\beta,\alpha)}} \|{}^R D_x^{\beta+m} u\|^2_{\omega^{(\alpha+\beta+m,m)}}.\end{aligned} \tag{2.43}$$

Working out the constants by (2.15) and (2.31), we use the property of the Gamma function again to obtain

$$\frac{\gamma_{N+1}^{(\alpha,\beta)}}{h_{N+1,m}^{(\beta,\alpha)}} = \frac{\Gamma(N+\alpha+2)\Gamma(N+m+2)(N-m+1)!}{\Gamma(N+\beta+2)\Gamma(N+\alpha+\beta+m+2)(N+m+1)!} \le cN^{-2(\beta+m)}. \tag{2.44}$$

□

Remark 1. Note that the error estimates, in the above theorem and in subsequent theorems, depend on the smoothness of fractional derivatives of the function, instead of the usual smoothness. To better understand the above results, we consider a typical solution of a one-sided FDE

$$u(x) = (1-x)^{\beta} g(x), \quad \beta \in \mathbb{R}^+, \; x \in \Lambda, \tag{2.45}$$

where g is a smooth function, and compare the GJF approximation with the Legendre approximation. Recall the best L^2-approximation of u by its orthogonal projection $\pi_N^L u$ (see, e. g., [29, Ch. 3]):

$$\|\pi_N^L u - u\| \le cN^{1-m} \|D^m u\|_{\omega^{(m,m)}}.$$

If β is not an integer, a direct calculation shows that the right-hand side is only bounded for $m < 1 + 2\beta - \epsilon$. On the other hand, using the explicit formulas for fractional integral/derivative of $(1+x)^{\beta}$ and the Leibniz formula [7, Ch. 2], we find that ${}^R D_x^{\beta+m} u$ is integrable for any $m \in \mathbb{N}_0$, so the convergence by GJF approximation is faster than any algebraic rate.

Remark 2. The results in Theorem 2 can be extended to some other $(\alpha, \beta) \notin {}^{-}\Sigma^{\alpha,\beta}$; we refer to [6] for more detail.

2.3 GJFs for Riesz derivatives

The Riesz derivatives include both the left- and right-sided fractional derivatives, so the one-sided GJFs defined above are not suitable. Instead, we define a new class of

GJFs. We have

$$\mathcal{J}_n^{-\mu,-\nu}(x) = (1-x)^{\mu}(1+x)^{\nu}P_n^{(\mu,\nu)}(x), \quad \mu,\nu > -1. \tag{2.46}$$

It can be derived from (2.14) that the general Jacobi functions $\mathcal{J}_n^{-\mu,-\nu}(x)$ are mutually orthogonal, i. e.,

$$\int_{-1}^{1} \mathcal{J}_n^{-\mu,-\nu}(x)\mathcal{J}_m^{-\mu,-\nu}(x)\omega^{(-\mu,-\nu)}(x) = \gamma_n^{(\mu,\nu)}\delta_{mn}, \tag{2.47}$$

and

$$\mathbb{F}_N^{-\mu,-\nu}(\Lambda) := \{\mathcal{J}_n^{-\mu,-\nu}(x) : n = 0,1,\ldots,N\}. \tag{2.48}$$

For Riesz derivatives, we shall use $\mathcal{J}_n^{-\alpha,-\alpha}(x)$ which satisfies the following.

Theorem 3. *If $s \in (2k-1,2k)$ with $k \in \mathbb{N}$, then*

$$I^{2k-s}\mathcal{J}_m^{-\frac{s}{2},-\frac{s}{2}}(x) = (-1)^k \frac{\Gamma(m+s+1-2k)}{2^{-2k}m!}P_{m+2k}^{(\frac{s}{2}-2k,\frac{s}{2}-2k)}(x), \tag{2.49}$$

and for $j = 0,1,\ldots,2k-1$,

$$D^{s-j}\mathcal{J}_m^{-\frac{s}{2},-\frac{s}{2}}(x) = 2^j(-1)^k \frac{\Gamma(m-j+1+s)}{m!}P_{m+j}^{(\frac{s}{2}-j,\frac{s}{2}-j)}(x). \tag{2.50}$$

We point out in particular that with $j = 0$ in (2.50), we have the following.

Corollary 1. *If $s \in (2k-1,2k)$ with $k \in \mathbb{N}$, then*

$$D^{s}\mathcal{J}_m^{-\frac{s}{2},-\frac{s}{2}}(x) = (-1)^k \frac{\Gamma(m+1+s)}{m!}P_{m}^{(\frac{s}{2},\frac{s}{2})}(x). \tag{2.51}$$

Let $s \in (2k-1,2k)$ with $k \in \mathbb{N}$ and $l \in \mathbb{N}_0$. We denote for any $m \in \mathbb{N}_0$

$$\mathcal{B}_s^m(\Lambda) := \{u \in L^2_{\omega^{(-\frac{s}{2},-\frac{s}{2})}}(\Lambda) : D^{s-k+l}u \in L^2_{\omega^{(\frac{s}{2}-k+l,\frac{s}{2}-k+l)}}(\Lambda), \text{ for } 0 \le l \le m\}. \tag{2.52}$$

By the orthogonality (2.47), we can expand any $u \in L^2_{\omega^{(-\alpha,-\alpha)}}(\Lambda)$ as

$$u(x) = \sum_{n=0}^{\infty} \hat{u}_n^{(-\alpha,-\alpha)}\mathcal{J}_n^{-\alpha,-\alpha}(x), \tag{2.53}$$

and the Parseval identity holds, i. e.,

$$\|u\|^2_{\omega^{(-\alpha,-\alpha)}} = \sum_{n=0}^{\infty} \gamma_n^{(\alpha,\alpha)}|\hat{u}_n^{(-\alpha,-\alpha)}|^2. \tag{2.54}$$

Moreover, for any $2\alpha \in (2k-1,2k)$ with $k \in \mathbb{N}$, we have

$$\|D^{2\alpha-k+l}u\|^2_{\omega^{(\alpha-k+l,\alpha-k+l)}} = \sum_{n=l}^{\infty} h_{n,l}^{(\alpha,\alpha)}|\hat{u}_n^{(-\alpha,-\alpha)}|^2. \tag{2.55}$$

Theorem 4. *Assume $2\alpha \in (2k-1, 2k)$ and $k \in \mathbb{N}$. Let $u \in \mathcal{B}^m_{2\alpha}(\Lambda)$. We have*

$$\|D^{2\alpha-k+l}(\pi_N^{(-\alpha,-\alpha)}u - u)\|_{\omega^{(\alpha-k+l,\alpha-k+l)}} \le cN^{l-m}\|D^{2\alpha-k+m}u\|_{\omega^{(\alpha-k+m,\alpha-k+m)}} \tag{2.56}$$

and

$$\|\pi_N^{(-\alpha,-\alpha)}u - u\|_{\omega^{(-\alpha,-\alpha)}} \le cN^{k-(2\alpha+m)}\|D^{2\alpha-k+m}u\|_{\omega^{(\alpha-k+m,\alpha-k+m)}}. \tag{2.57}$$

Proof. By (2.68) (with $\mu = \nu = \alpha$) and (2.55), we have

$$\begin{aligned}\|D^{2\alpha-k+l}(\pi_N^{(-\alpha,-\alpha)}u - u)\|^2_{\omega^{(\alpha-k+l,\alpha-k+l)}} &= \sum_{n=N+1}^{\infty} h_{n,l}^{(\alpha,\alpha)}|\hat{u}_n^{(-\alpha,-\alpha)}|^2 \\ = \sum_{n=N+1}^{\infty}\frac{h_{n,l}^{(\alpha,\alpha)}}{h_{n,m}^{(\alpha,\alpha)}}h_{n,m}^{(\alpha,\alpha)}|\hat{u}_n^{(-\alpha,-\alpha)}|^2 &\le \frac{h_{N+1,l}^{(\alpha,\alpha)}}{h_{N+1,m}^{(\alpha,\alpha)}}\|D^{2\alpha-k+m}u\|_{\omega^{(\alpha-k+m,\alpha-k+m)}}.\end{aligned} \tag{2.58}$$

We now estimate the constant factor. Similar to the proof in Theorem 2, we find that for $0 \le l \le m \le N$,

$$\frac{h_{N+1,l}^{(\alpha,\alpha)}}{h_{N+1,m}^{(\alpha,\alpha)}} \le N^{l-m}\frac{(N+k-m+1)!}{(N+k-l+1)!}. \tag{2.59}$$

Thus, we obtain (2.56) from (2.58), (2.59), and the property of the Gamma function.

The $L^2_{\omega^{(-\alpha,-\alpha)}}$-estimates can be obtained by using the same argument. We sketch the derivation below. By (2.54) and (2.55),

$$\begin{aligned}\|\pi_N^{(-\alpha,-\alpha)}u - u\|^2_{\omega^{(-\alpha,-\alpha)}} &= \sum_{n=N+1}^{\infty}\gamma_n^{(\alpha,\alpha)}|\hat{u}_n^{(-\alpha,-\alpha)}|^2 \\ &= \sum_{n=N+1}^{\infty}\frac{\gamma_n^{(\alpha,\alpha)}}{h_{n,m}^{(\alpha,\alpha)}}h_{n,m}^{(\alpha,\alpha)}|\hat{u}_n^{(\alpha,\alpha)}|^2 \le \frac{\gamma_{N+1}^{(\alpha,\alpha)}}{h_{N+1,m}^{(\alpha,\alpha)}}\|D^{2\alpha-k+m}u\|_{\omega^{(\alpha-k+m,\alpha-k+m)}}.\end{aligned}$$

Similarly, by (2.15) and (2.31), we obtain

$$\begin{aligned}\frac{\gamma_{N+1}^{(\alpha,\alpha)}}{h_{N+1,m}^{(\alpha,\alpha)}} &= \frac{(N+1)!(N+k+m+1)!(N+k-m+1)!}{\Gamma(N+2\alpha+2)\Gamma(N+2\alpha-k+m+2)(N+k+m+1)!} \\ &\le N^{2k-(4\alpha+2m)}.\end{aligned}$$

This completes the proof. □

2.4 GJFs for two-sided fractional derivatives with different coefficients

For $1 < \beta < 2$, $0 < \mu, \nu < \beta$, $\mu + \nu = \beta$, $0 \le p \le 1$, and $p \ne 1/2$, we define the two-sided fractional integral operator

$$\mathcal{I}_p^{\mu,\nu,\rho} := C_{\beta,p}(pI_x^\rho + (1-p)_xI^\rho), \tag{2.60}$$

where

$$C_{\beta,p} := C(\beta,\mu,\nu) = \frac{\sin(\pi\mu) + \sin(\pi\nu)}{\sin(\pi\beta)}.$$

Then for $s \in (k-1,k)$, we define the two-sided fractional derivative operator

$$\mathcal{D}_p^{\mu,\nu,s} := \frac{d^k}{dx^k}\mathcal{I}_p^{\mu,\nu,k-s}. \tag{2.61}$$

It turns out that suitable basis functions for dealing with two-sided FDEs with different coefficients are of the form $\mathcal{J}_n^{-\mu,-\nu}(x)$, where (μ,ν) are determined as follows [8, 22].

Theorem 5. *Given (p,β) such that $0 \le p \le 1$ and $1 < \beta < 2$, let (μ,ν) be determined from*

$$\mu + \nu = \beta, \quad p\sin(\pi\mu) = (1-p)\sin(\pi\nu). \tag{2.62}$$

Then for $n = 0,1,2,\ldots$, we have

$$\mathcal{I}_p^{\mu,\nu,2-\beta}\mathcal{J}_n^{-\mu,-\nu}(x) = 4\frac{\Gamma(n+\beta-1)}{n!}P_{n+2}^{(\nu-2,\mu-2)}(x), \tag{2.63}$$

and for $k = 1,2,\ldots,n+2$, we have

$$\mathcal{D}_p^{\mu,\nu,k+\beta-2}\mathcal{J}_n^{-\mu,-\nu}(x) = \frac{\Gamma(n+k+\beta-1)}{2^{k-2}n!}P_{n+2-k}^{(\nu-2+k,\mu-2+k)}(x). \tag{2.64}$$

In particular, for $k = 2$,

$$\mathcal{D}_p^{\mu,\nu,\beta}\mathcal{J}_n^{-\mu,-\nu}(x) = \frac{\Gamma(n+\beta+1)}{n!}P_n^{(\nu,\mu)}(x). \tag{2.65}$$

For the sake of simplicity, we shall denote $\mathcal{I}_p^{\mu,\nu,\rho}$ and $\mathcal{D}_p^{\mu,\nu,\rho}$ by $\mathcal{I}_p^{\rho}$ and $\mathcal{D}_p^{\rho}$, respectively.

By virtue of (2.14), a consequent result of equation (2.64) is the orthogonality of $\mathcal{D}_p^{\beta+l}$ for $l = -1,0,1,\ldots,\min\{m,n\}$. If (μ,ν) and (p,β) satisfy the conditions of Theorem 5, then

$$\int_{-1}^{1} \mathcal{D}_p^{\beta+l}\mathcal{J}_m^{-\mu,-\nu}(x)\mathcal{D}_p^{\beta+l}\mathcal{J}_n^{-\mu,-\nu}(x)\omega^{(\nu+l,\mu+l)}(x)dx = 0, \quad \forall n \ne m, \tag{2.66}$$

$$\int_{-1}^{1} \mathcal{D}_{1-p}^{\beta+l}\mathcal{J}_m^{-\mu,-\nu}(x)\mathcal{D}_{1-p}^{\beta+l}\mathcal{J}_n^{-\mu,-\nu}(x)\omega^{(\nu+l,\mu+l)}(x)dx = 0, \quad \forall n \ne m. \tag{2.67}$$

We define

$$(\pi_N^{(-\mu,-\nu)}u - u, v_N)_{\omega^{(-\mu,-\nu)}} = 0, \quad \forall v_N \in \mathbb{F}_N^{-\mu,-\nu}(\Lambda), \tag{2.68}$$

and for μ, ν satisfying (2.62), we denote

$$\tilde{\mathcal{B}}^m_{\beta,p}(\Lambda) := \{u \in L^2_{\omega^{(-\mu,-\nu)}}(\Lambda) : \mathcal{D}^{\beta+l}_p u \in L^2_{\omega^{(\nu+l,\mu+l)}}(\Lambda), \text{ for } -1 \le l \le m\}. \tag{2.69}$$

Then, we have the approximation results for the projection errors.

Theorem 6. *Assume* $1 < \beta < 2$ *and let* $u \in \tilde{\mathcal{B}}^m_{\beta,p}(\Lambda)$ *with* $m \in \mathbb{N}$. *Then for a given* p, $0 \le p \le 1$, *if* $0 < \mu, \nu < \beta$, *and* μ, ν *satisfying* (2.62), *we have, for* $-1 \le l \le m \le N$,

$$\|\mathcal{D}^{\beta+l}_p(\pi^{(-\mu,-\nu)}_N u - u)\|_{\omega^{(\nu+l,\mu+l)}} \le cN^{l-m}\|\mathcal{D}^{\beta+m}_p u\|_{\omega^{(\nu+m,\mu+m)}} \tag{2.70}$$

and

$$\|\pi^{(-\mu,-\nu)}_N u - u\|_{\omega^{(-\mu,-\nu)}} \le cN^{-(\beta+m)}\|\mathcal{D}^{\beta+m}_p u\|_{\omega^{(\nu+m,\mu+m)}}. \tag{2.71}$$

Proof. The proof is similar to that of Theorem 4. □

3 Spectral methods for FDEs based on generalized Jacobi functions

In this section, we present spectral methods using the GJFs defined in Section 2 to solve several typical FDEs.

3.1 Fractional differential equations with one-sided fractional derivative

We consider first a fractional initial value problem (FIVP), followed by a fractional boundary value problem with one-sided fractional derivative.

3.1.1 A fractional initial value problem (FIVP)

As the first example, we consider the following FIVP of order $s \in (0, 1)$:

$${}^C D^s_t u(t) = f(t), \quad t \in I := (0, T), \quad u(0) = u_0. \tag{3.1}$$

For the non-homogeneous initial conditions $u(0) = u_0$, we first decompose the solution $u(t)$ into two parts as

$$u(t) = u^h(t) + u_0, \tag{3.2}$$

with $u^h(0) = 0$. By definition, ${}^C D_t^s u_0 = 0$. We then derive from (2.7) that equation (3.1) is equivalent to the following equation with Riemann–Liouville fractional derivative:

$$ {}^R D_t^s u^h(t) = f(t), \quad t \in I := (0,T), \quad u^h(0) = 0. \tag{3.3} $$

A Petrov–Galerkin scheme for (3.3) is: find $u_N^h \in {}^-\mathcal{F}_N^{(-s,-s)}(I)$ (defined in (2.2.3)) such that

$$ ({}^R D_t^s u_N^h, v_N) = (f, v_N), \quad \forall v_N \in \mathcal{P}_N(I). \tag{3.4} $$

We expand $f(t)$ as

$$ f(t) = \sum_{n=0}^{\infty} \tilde{f}_n \tilde{P}_n(t), \tag{3.5} $$

where $\tilde{P}_n(t) := P_n^{(0,0)}(x(t))$, $x(t) = (2t - T)/T$ is the shifted Legendre polynomial of degree n on I, and write

$$ u_N^h(t) = \sum_{n=0}^{N} \tilde{u}_n^{(s)} \, {}^-\tilde{J}_n^{(-s,-s)}(t) \in {}^-\mathcal{F}_N^{(-s,-s)}(I). \tag{3.6} $$

Taking $v_N = \tilde{P}_l(t)$ in (3.4), we obtain from (2.28) and the orthogonality of Legendre polynomials

$$ \tilde{u}_n^{(s)} = \frac{n!}{\Gamma(n+s+1)} \tilde{f}_n, \quad 0 \le n \le N. \tag{3.7} $$

Therefore, we obtain the numerical solution u_N^h without solving any algebraic equation. Hence, the method is very efficient. As for the error estimate, we have the following result [6].

Theorem 7. *Let u^h and u_N^h be the solution of* (3.3) *and* (3.4), *respectively. Then*

$$ \|{}^R D_t^s (u^h - u_N^h)\| \le cN^{-m} \|f^{(m)}\|_{\omega^{(m-1,m-1)}}. \tag{3.8} $$

Proof. Let ${}^-\pi_N^{(-s,-s)} u^h$ be as defined in (2.35) for $0 < s < 1$. By (2.28), we have

$$ ({}^R D_t^s ({}^-\pi_N^{(-s,-s)} u^h - u^h), \psi) = 0, \quad \forall \psi \in \mathcal{P}_N. $$

Then by (3.3),

$$ (f - {}^R D_t^s {}^-\pi_N^{(-s,-s)} u^h, \psi) = ({}^R D_t^s u^h - {}^R D_t^s {}^-\pi_N^{(-s,-s)} u^h, \psi) = 0, \quad \forall \psi \in \mathcal{P}_N. $$

Let $\pi_N f$ be the L^2-orthogonal projection of f upon $\mathcal{P}_N$. We infer from the above that ${}^R D_t^s {}^-\pi_N^{(-s,-s)} u^h = \pi_N f = {}^R D_t^s u_N^h$. Therefore,

$$ \begin{aligned} \|{}^R D_t^s (u^h - u_N^h)\| &= \|{}^R D_t^s (u^h - {}^-\pi_N^{(-s,-s)} u^h)\| \\ &\le \|{}^R D_t^s (u^h - {}^-\pi_N^{(-s,-s)} u^h)\| + \|\pi_N f - f\|. \end{aligned} \tag{3.9} $$

It follows from Theorem 2 (with $\alpha = -\beta = s$ and $0 < s < 1$) and the Legendre approximation results (see, e. g., [29, Chapter 3]) that

$$\|{}^R D_t^s (u^h - u_N^h)\| \le cN^{-m} \big(\|{}^R D_t^{s+m} u^h\|_{\omega^{(m,m)}} + \|f^{(m)}\|_{\omega^{(m-1,m-1)}} \big). \tag{3.10}$$

We deduce from (3.1) that

$$\|{}^R D_t^{s+m} u^h\|_{\omega^{(m,m)}} \le c\|f^{(m)}\|_{\omega^{(m-1,m-1)}}. \qquad \square$$

3.1.2 One-sided fractional boundary value problems

Now we consider a one-sided fractional boundary value problem,

$${}_x^R D^\nu u(x) = f(x), \quad x \in \Lambda = (-1,1), \quad u(\pm 1) = 0, \tag{3.11}$$

where $\nu \in (1,2)$.

Let $s = \nu - 1$ and introduce the solution and test function spaces

$$\begin{aligned} U &:= \{u \in L^2_{\omega^{(-s,-1)}}(\Lambda) : {}_x^R D^s u \in L^2_{\omega^{(0,s-1)}}(\Lambda)\}, \\ V &:= \{v \in L^2_{\omega^{(-1,-s)}}(\Lambda) : Dv \in L^2_{\omega^{(0,1-s)}}(\Lambda)\}, \end{aligned} \tag{3.12}$$

equipped with the norms

$$\begin{aligned} \|u\|_U &= \big(\|u\|^2_{\omega^{(-s,-1)}} + \|{}_x^R D^s u\|^2_{\omega^{(0,s-1)}}\big)^{1/2}, \\ \|v\|_V &= \big(\|v\|^2_{\omega^{(-1,-s)}} + \|Dv\|^2_{\omega^{(0,1-s)}}\big)^{1/2}. \end{aligned} \tag{3.13}$$

For $u \in U$ and $v \in V$, we write

$$\begin{aligned} u(x) &= \sum_{n=1}^{\infty} \hat u_n {}^+J_n^{(-s,-1)}(x) = (1-x)^s(1+x) \sum_{n=1}^{\infty} \tilde u_n P_{n-1}^{(s,1)}(x), \\ v(x) &= \sum_{n=1}^{\infty} \hat v_n {}^-J_n^{(-1,-s)}(x) = (1-x)(1+x)^s \sum_{n=1}^{\infty} \tilde v_n P_{n-1}^{(1,s)}(x). \end{aligned} \tag{3.14}$$

With the above set-up, we can build in the homogeneous boundary conditions and also perform fractional integration by parts. Hence, a weak form of (3.11) is to find $u \in U$ such that

$$a(u,v) := ({}_x^R D^s u, Dv) = (f,v), \quad \forall v \in V. \tag{3.15}$$

Let ${}^+\mathcal{F}_N^{(-s,-1)}(\Lambda)$ and ${}^-\mathcal{F}_N^{(-1,-s)}(\Lambda)$ be the finite-dimensional spaces as defined in Section 2. Then the GJF-Petrov–Galerkin scheme for (3.15) is to find $u_N \in {}^+\mathcal{F}_N^{(-s,-1)}(\Lambda)$ such that

$$a(u_N,v_N) = ({}_x^R D^s u_N, Dv_N) = (f,v_N), \quad \forall v_N \in {}^-\mathcal{F}_N^{(-1,-s)}(\Lambda). \tag{3.16}$$

We derive from (2.27) and ${}^R D_x^1 = \frac{d}{dx}$ that

$$a({}^+J_n^{(-s,-1)}(x), {}^-J_m^{(-1,-s)}(x)) = C_{n,m}^s({}^+J_n^{(0,s-1)}(x), {}^-J_m^{(0,1-s)}(x)) = 0 \quad \forall n \neq m. \tag{3.17}$$

Hence, we can obtain u_N directly without solving any algebraic equation.

To characterize the regularity of u, we define for any $m \in \mathbb{N}_0$,

$${}^+\mathcal{B}_{\alpha,\beta}^m(\Lambda) := \{u \in L^2_{\omega^{(-\alpha,\beta)}}(\Lambda) : {}_x^R D^{\alpha+l} u \in L^2_{\omega^{(l,\alpha+\beta+l)}}(\Lambda) \text{ for } 0 \le l \le m\}. \tag{3.18}$$

Theorem 8. *Let $s \in (0,1)$ and let u and u_N be the solutions of (3.15) and (3.16), respectively. If $u \in U \cap {}^+\mathcal{B}_{s,-1}^m(\Lambda)$ with $0 \le m \le N$, then we have the error estimates*

$$\|u - u_N\|_U \le cN^{-m} \|{}_x^R D^{s+m} u\|_{\omega^{(m,s-1+m)}}. \tag{3.19}$$

In particular, if $f^{(m-1)} \in L^2_{\omega^{(m,s-1+m)}}(\Lambda)$ for $m \ge 1$, we have

$$\|u - u_N\|_U \le cN^{-m} \|f^{(m-1)}\|_{\omega^{(m,s-1+m)}}. \tag{3.20}$$

Here, c is a positive constant independent of u, N, and m.

Proof. We derive from (3.15) and (3.16) that

$$a(u - u_N, v_N) = 0 \quad \forall v_N \in {}^-\mathcal{F}_N^{(-1,-s)}(\Lambda),$$

which, along with (3.17), implies immediately that $u_N = {}^+\pi_N^{(-s,-1)} u$. Hence, the desired results follow from Theorem 4.1 in [6] and the fact that ${}_x^R D^\nu u = {}_x^R D^{s+1} u = f$. □

3.2 Fractional boundary value problems with Riesz derivatives

We consider first the so-called Riesz FDEs, followed by a more general case with a zeroth-order term.

3.2.1 Riesz fractional differential equations

We consider the following Riesz fractional equation of order $2\alpha \in (2k-1, 2k)$ with $k \in \mathbb{N}$:

$$\begin{aligned} (-1)^k D^{2\alpha} u(x) &= f(x), \quad x \in \Lambda, \\ u^{(l)}(\pm 1) &= 0, \quad l = 0, 1, \dots, k-1. \end{aligned} \tag{3.21}$$

A Petrov–Galerkin spectral method for (3.21) is: find $u_N \in \mathbb{F}_N^{-\alpha,-\alpha}$ such that

$$(-1)^k (D^{2\alpha} u_N, v_N)_{\omega^{(\alpha,\alpha)}} = (f, v_N)_{\omega^{(\alpha,\alpha)}}, \quad \forall v_N \in P_N. \tag{3.22}$$

The solution to this discrete problem can be found directly as follows. Given

$$f(x) = \sum_{m=0}^{\infty} f_m P_m^{(\alpha,\alpha)}(x), \tag{3.23}$$

write

$$u_N(x) = \sum_{n=0}^{N} \hat{u}_n \mathcal{J}_n^{-\alpha,-\alpha}(x). \tag{3.24}$$

Plugging the above in (3.22), using (2.51) and the orthogonality of $\{P_m^{(\alpha,\alpha)}\}$ in $L^2_{\omega^{(\alpha,\alpha)}}(\Lambda)$, we find

$$\hat{u}_n = f_n \Big/ \left(2\cos(\pi\alpha) \frac{\Gamma(n+1+2\alpha)}{n!} \right), \quad \forall 0 \le n \le N. \tag{3.25}$$

As for the error estimate, we have the following.

Theorem 9. *Assuming $f^{(j)} \in L^2_{\omega^{\alpha+j,\alpha+j}}(\Lambda)$ for $0 \le j \le m$, we have*

$$\|u - u_N\|_{\omega^{(-\alpha,-\alpha)}} \le cN^{-2\alpha-m} \|f^{(m)}\|_{\omega^{(\alpha+m,\alpha+m)}}, \tag{3.26}$$

$$\|D^{2\alpha}(u - u_N)\|_{\omega^{(\alpha,\alpha)}} \le cN^{-m} \|f^{(m)}\|_{\omega^{(\alpha+m,\alpha+m)}}. \tag{3.27}$$

Proof. It is clear from (3.25) that $u_N = \pi_N^{(-\alpha,-\alpha)} u$. Hence, by (2.57), we have

$$\|u - u_N\|_{\omega^{(-\alpha,-\alpha)}} = \|u - \pi_N^{(-\alpha,-\alpha)} u\|_{\omega^{(-\alpha,-\alpha)}} \le cN^{-2\alpha-m} \|D^{2\alpha+m} u\|_{\omega^{(\alpha+m,\alpha+m)}}.$$

On the other hand, we derive from (2.56) with $k = l = 0$ that

$$\|D^{2\alpha} u - u_N\|_{\omega^{(\alpha,\alpha)}} = \|D^{2\alpha}(u - \pi_N^{(-\alpha,-\alpha)} u)\|_{\omega^{(\alpha,\alpha)}} \le cN^{-m} \|D^{2\alpha+m} u\|_{\omega^{(\alpha+m,\alpha+m)}}.$$

Since $D^{2\alpha+m} u = f^{(m)}$, we obtain the desired results from the above two inequalities. □

3.2.2 A more general case

In the previous examples, we have developed optimal spectral methods using GJFs in the sense that (i) the numerical solution can be determined directly without solving any algebraic equation and (ii) the error converges faster than any algebraic rate as long as the right-hand side function f is smooth, despite the fact that the solution is weakly singular at the end point(s).

However, it is not possible to construct such optimal spectral methods for more general FDEs. Nevertheless, using proper GJFs still allows us to (i) deal with fractional derivatives efficiently and (ii) resolve the leading singular term in the solution. Consider for example

$$\begin{aligned} &\rho u(x) - D^{2\alpha} u(x) = f, \quad x \in \Lambda = (-1,1), \\ &u(\pm 1) = 0, \end{aligned} \tag{3.28}$$

where $\rho > 0$ and $2\alpha \in (1,2)$. A GJF spectral Galerkin approximation to (3.28) is: find $u_N \in \mathbb{F}_N^{-\alpha,-\alpha}(\Lambda)$ such that

$$a(u_N, v_N) := \rho(u_N, v_N) - (D^{2\alpha} u_N, v_N) = (f, v_N), \quad \forall v_N \in \mathbb{F}_N^{-\alpha,-\alpha}(\Lambda). \tag{3.29}$$

Set

$$u_N(x) = \sum_{n=0}^{N} \tilde{u}_n \mathcal{J}_n^{-\alpha,-\alpha}(x). \tag{3.30}$$

Set

$$\begin{aligned} & u_N(x) = \sum_{n=0}^{N} \tilde{u}_n \mathcal{J}_n^{-\alpha,-\alpha}(x), \quad U = (\tilde{u}_0, \tilde{u}_1, \ldots, \tilde{u}_N), \\ & m_{jk} = (\mathcal{J}_k^{-\alpha,-\alpha}(x), \mathcal{J}_j^{-\alpha,-\alpha}(x)), \quad s_{jk} = -(D^{2\alpha} \mathcal{J}_k^{-\alpha,-\alpha}(x), \mathcal{J}_j^{-\alpha,-\alpha}(x)), \\ & M = (m_{jk}), \quad S = (s_{jk}), \quad \tilde{f}_j = (f, \mathcal{J}_j^{-\alpha,-\alpha}(x)), \quad F = (\tilde{f}_0, \tilde{f}_1, \ldots, \tilde{f}_N)^T. \end{aligned} \tag{3.31}$$

Then, (3.29) reduces to the following matrix system:

$$(\rho M + S)U = F.$$

We recall from (2.50), (2.14) that S is a diagonal matrix. We have

$$s_{jk} = \frac{2^{2\alpha+1}\Gamma(k + \alpha + 1)^2}{(k!)^2(2k + 2\alpha + 1)} \delta_{jk}. \tag{3.32}$$

The mass matrix M is full but its entries can be evaluated exactly (cf. [4]), i. e.,

$$\begin{aligned} m_{jk} &= \int_{-1}^{1} (1 - x^2)^\alpha P_j^{(\alpha,\alpha)}(x)(1 - x^2)^\alpha P_k^{(\alpha,\alpha)}(x) dx \\ &= \frac{(-1)^{\frac{j-k}{2}}(j + k)!}{2^{j+k} j! k!} \frac{(-j - \alpha)_{\frac{j+k}{2}}(-k - \alpha)_{\frac{j+k}{2}}}{(2\alpha + \frac{3}{2})_{\frac{j+k}{2}} \frac{j+k}{2}!} \frac{\sqrt{\pi}\Gamma(2\alpha + 1)}{\Gamma(2\alpha + \frac{3}{2})}, \end{aligned} \tag{3.33}$$

where the Pochhammer symbol $(a)_\nu = \frac{\Gamma(a+\nu)}{\Gamma(a)}$.

Lemma 7. *For all* $u \in \{u \in L^2_{\omega^{-\alpha,-\alpha}}(\Lambda) : D^{2\alpha} u \in L^2(\Lambda)\}$, *we have*

$$\|u\|^2_{\omega^{(-\alpha,-\alpha)}} \le -(D^{2\alpha} u, u). \tag{3.34}$$

Proof. We write $u = \sum_{i=0}^{\infty} \tilde{u}_i \mathcal{J}_i^{-\alpha,-\alpha}(x)$, so that

$$\|u\|^2_{\omega^{(-\alpha,-\alpha)}} = \sum_{i=0}^{\infty} \tilde{u}_i^2 \gamma_i^{(\alpha,\alpha)}.$$

By Theorem 3 and (2.14), we have

$$
\begin{aligned}
-(D^{2\alpha}u,u) &= \Big(-\sum_{i=0}^{\infty}\tilde{u}_i D^{2\alpha}\mathcal{J}_i^{-\alpha,-\alpha}, \sum_{j=0}^{\infty}\tilde{u}_j\mathcal{J}_j^{-\alpha,-\alpha}\Big)\\
&= \Big(\sum_{i=0}^{\infty}\tilde{u}_i\frac{\Gamma(i+2\alpha+1)}{i!}P_i^{(\alpha,\alpha)}, \sum_{j=0}^{\infty}\tilde{u}_jP_j^{(\alpha,\alpha)}\Big)_{\omega^{(\alpha,\alpha)}}\\
&= \sum_{i=0}^{\infty}\frac{\Gamma(i+2\alpha+1)}{i!}\tilde{u}_i^2\gamma_i^{(\alpha,\alpha)}.
\end{aligned}
$$

We can easily get the desired result (3.34) by comparing the above results. □

We define the energy norm associated with (3.28) by

$$\|u\|_{B^\alpha}^2 = \rho(u,u) - (D^{2\alpha}u,u). \tag{3.35}$$

Theorem 10. *Assume* $2\alpha \in (1,2)$. *Let* u *and* u_N *be the solution of* (3.28) *and* (3.29). *Then we have*

$$\|u-u_N\|_{B^\alpha} \le N^{1-m}\|D^{2\alpha-1+m}u\|_{\omega^{(\alpha-1+m,\alpha-1+m)}}. \tag{3.36}$$

Proof. Let us denote $\tilde{u}_N := \pi_N^{(-\alpha,-\alpha)}u$ and $e_N := \tilde{u}_N - u_N$. We derive from (3.28) and (3.29) that

$$
\begin{aligned}
a(e_N,v_N) &= \rho(e_N,v_N) - (D^{2\alpha}e_N,v_N)\\
&= \rho(\tilde{u}_N-u,v_N) - (D^{2\alpha}(\tilde{u}_N-u),v_N), \quad \forall v\in \mathbb{F}_N^{-\alpha,-\alpha}(\Lambda).
\end{aligned}
$$

Taking $v_N = e_N$ in the above, we derive from (3.35) and the Cauchy–Schwarz inequality that

$$
\begin{aligned}
\|e_N\|_{B^\alpha}^2 := \rho(e_N,e_N) - (D^{2\alpha}e_N,e_N) &\le \rho\|\pi_N^{(-\alpha,-\alpha)}u-u\|_{\omega^{(\alpha,\alpha)}}\|e_N\|_{\omega^{(-\alpha,-\alpha)}}\\
&+ \|D^{2\alpha}(\pi_N^{(-\alpha,-\alpha)}u-u)\|_{\omega^{(\alpha,\alpha)}}\|e_N\|_{\omega^{(-\alpha,-\alpha)}}.
\end{aligned}
$$

The above and Lemma 7 lead to

$$\|e_N\|_{B^\alpha} \le \rho\|\pi_N^{(-\alpha,-\alpha)}u-u\|_{\omega^{(-\alpha,-\alpha)}} + \|D^{2\alpha}(\pi_N^{(-\alpha,-\alpha)}u-u)\|_{\omega^{(\alpha,\alpha)}}.$$

Then, by Theorem 4, the right-hand side terms of the above equation can be estimated as

$$\|\pi_N^{(-\alpha,-\alpha)}u-u\|_{\omega^{(-\alpha,-\alpha)}} \le cN^{1-(2\alpha+m)}\|D^{2\alpha-1+m}u\|_{\omega^{(\alpha-1+m,\alpha-1+m)}} \tag{3.37}$$

and

$$\|D^{2\alpha}(\pi_N^{(-\alpha,-\alpha)}u-u)\|_{\omega^{(\alpha,\alpha)}} \le cN^{1-m}\|D^{2\alpha-1+m}u\|_{\omega^{(\alpha-1+m,\alpha-1+m)}}, \tag{3.38}$$

which implies that

$$\|e_N\|_{B^\alpha} \le cN^{1-m}\|D^{2\alpha-1+m}u\|_{\omega^{(\alpha-1+m,\alpha-1+m)}}. \tag{3.39}$$

On the other hand,

$$\begin{aligned}&-(D^{2\alpha}(u-\pi_N^{(-\alpha,-\alpha)}u), u-\pi_N^{(-\alpha,-\alpha)}u)\\ &\le \|D^{2\alpha}(\pi_N^{(-\alpha,-\alpha)}u-u)\|_{\omega^{(\alpha,\alpha)}}\|u-\pi_N^{(-\alpha,-\alpha)}u\|_{\omega^{(-\alpha,-\alpha)}}.\end{aligned}$$

Since

$$\|u-\pi_N^{(-\alpha,-\alpha)}u\| \le \|u-\pi_N^{(-\alpha,-\alpha)}u\|_{\omega^{(-\alpha,-\alpha)}},$$

we find from the above that

$$\|u-\pi^{(-\alpha,-\alpha)}u\|_{B^\alpha} \le \rho\|u-\pi_N^{(-\alpha,-\alpha)}u\|_{\omega^{(-\alpha,-\alpha)}} + \|D^{2\alpha}(\pi_N^{(-\alpha,-\alpha)}u-u)\|_{\omega^{(\alpha,\alpha)}}.$$

Finally, since $u - u_N = u - \tilde{u}_N + e_N$, combining the above with (3.39), (3.37), and (3.38), we obtain the desired result. □

Remark 3. We emphasize that, unlike in previous examples, here we cannot easily bound the errors in terms of f. In particular, the smoothness of f does not imply that the right-hand side of (3.36) is bounded for any m. Hence, only an algebraic convergence rate can be achieved in this case.

3.3 Two-sided fractional differential equations with different coefficients

We consider the two-sided FDE

$$\begin{aligned}\mathcal{D}_p^\beta u(x) := C_{\beta,p}(p\,{}^R D_x^\beta u(x) + (1-p){}_x^R D^\beta u(x)) &= f(x), \quad x \in \Lambda,\\ u(\pm 1) &= 0,\end{aligned} \tag{3.40}$$

where $1 < \beta < 2$, $0 \le p \le 1$, $C_{\beta,p}$ defined in (2.4), and $f(x)$ is a given function.

Let μ, ν satisfy (2.62) and $0 < \mu, \nu < \beta$. A Petrov–Galerkin spectral method for (3.40) is: find $u_N \in \mathbb{F}_N^{-\mu,-\nu}$ such that

$$(\mathcal{D}_p^\beta u_N, v_N)_{\omega^{(\nu,\mu)}} = (f, v_N)_{\omega^{(\nu,\mu)}}, \quad \forall v_N \in \mathcal{P}_N. \tag{3.41}$$

The solution to this discrete problem can be found directly as follows. We expand $f(x)$ as

$$f(x) = \sum_{m=0}^{\infty} f_m P_m^{(\nu,\mu)}(x), \tag{3.42}$$

and we write

$$u_N(x) = \sum_{n=0}^{N} \hat{u}_n \mathcal{J}_n^{-\mu,-\nu}(x). \tag{3.43}$$

Plugging (3.43) and (3.42) in (3.41), using (2.65) and the orthogonality of $\{P_m^{(\nu,\mu)}\}$ in $L^2_{\omega^{(\nu,\mu)}}(\Lambda)$, we find

$$\hat{u}_n = f_n \Big/ \Big(\frac{\Gamma(n+1+\beta)}{n!} \Big), \quad \forall\, 0 \le n \le N. \tag{3.44}$$

As for the error estimate, we have the following.

Theorem 11. *Assuming $f^{(j)} \in L^2_{\omega^{(\nu+j,\mu+j)}}(\Lambda)$ for $0 \le j \le m$, let u and u_N be the solution of* (3.40) *and* (3.41). *Then we have*

$$\|u - u_N\|_{\omega^{(-\mu,-\nu)}} \le cN^{-\beta-m} \|f^{(m)}\|_{\omega^{(\nu+m,\mu+m)}}, \tag{3.45}$$

$$\|\mathcal{D}_p^{\beta}(u - u_N)\|_{\omega^{(\nu,\mu)}} \le cN^{-m} \|f^{(m)}\|_{\omega^{(\nu+m,\mu+m)}}. \tag{3.46}$$

Proof. It is clear from (3.44) that $u_N = \pi_N^{(-\mu,-\nu)} u$. Hence, by (2.71), we have

$$\|u - u_N\|_{\omega^{(-\mu,-\nu)}} = \|u - \pi_N^{(-\mu,-\nu)} u\|_{\omega^{(-\mu,-\nu)}} \le cN^{-\beta-m} \|\mathcal{D}_p^{\beta+m} u\|_{\omega^{(\nu+m,\mu+m)}}.$$

On the other hand, we derive from (2.70) with $l = 0$ that

$$\|\mathcal{D}_p^{\beta}(u - u_N)\|_{\omega^{(\nu,\mu)}} = \|\mathcal{D}_p^{\beta}(u - \pi_N^{(-\mu,-\nu)} u)\|_{\omega^{(\nu,\mu)}} \le cN^{-m} \|\mathcal{D}_p^{\beta+m} u\|_{\omega^{(\nu+m,\mu+m)}}.$$

Since $\mathcal{D}_p^{\beta+m} u = f^{(m)}$, we obtain the desired results from the above. □

3.4 Space-time fractional differential equations

As the last example, we consider

$$\begin{aligned} &{}^{C}D_t^{\alpha} u(x,t) - D^{2\beta} u(x,t) = f(x,t), \quad \forall (x,t) \in Q = \Lambda \times I, \\ &u(x,t)|_{\partial\Lambda} = 0, \quad \forall t \in I := (0 < T), \\ &u(x,0) = u_0(x) \quad \forall x \in \Lambda, \end{aligned} \tag{3.47}$$

where $\alpha \in (0,1)$, $2\beta \in (1,2)$. To deal with the non-homogeneous initial condition, we first decompose the solution $u(x,t)$ into two parts, i. e.,

$$u(x,t) = u^h(x,t) + u_0(x), \tag{3.48}$$

with $u^h(x,0) = 0$. Hence, equation (3.47) is equivalent to the following with Riemann–Liouville fractional derivative:

$$\begin{aligned} {}^R D_t^{\alpha} u^h(x,t) - D^{2\beta} u^h(x,t) &= g(x,t), \quad \forall (x,t) \in Q, \\ u^h(x,0) &= 0, \quad \forall x \in \Lambda, \\ u^h(x,t)|_{\partial\Lambda} &= 0, \quad \forall t \in I, \end{aligned} \tag{3.49}$$

where

$$g(x,t) = f(x,t) + D^{2\beta} u_0(x).$$

We consider the following weak formulation of (3.49): find $u \in X$, such that

$$\mathcal{A}(u^h, v) := ({}^R D_t^{\alpha} u^h, v)_Q - (D^{2\beta} u^h, v)_Q = (g, v)_Q, \quad \forall v \in Y, \tag{3.50}$$

and describe a space-time spectral method for solving the above equation.

For the time variable, we shall use the shifted fractional polynomial space. Let $x(t) = 2t/T - 1$. We denote $\tilde{P}_n^{(\alpha,\beta)}(t) = P_n^{(\alpha,\beta)}(x(t))$, ${}^{-}\tilde{J}_n^{(-\alpha,-\alpha)}(t) := t^{\alpha} \tilde{P}_n^{(-\alpha,\alpha)}(t)$, and

$${}^{-}\mathcal{F}_N^{(-\alpha,-\alpha)}(I) = \mathrm{span}\{ {}^{-}\tilde{J}_n^{(-\alpha,-\alpha)}(t) : 0 \le n \le N \}. \tag{3.51}$$

For the space variable, we shall use $\mathbb{F}_M^{-\beta,-\beta}$ defined in (2.48). Then, the Petrov–Galerkin method for (3.49) is: find $u_L^h(x,t) := u_{MN}^h \in \mathbb{F}_M^{-\beta,-\beta} \otimes {}^{-}\mathcal{F}_N^{(-\alpha,-\alpha)}$ such that

$$\mathcal{A}(u_L^h, v) := ({}^R D_t^{\alpha} u_L^h, v)_Q - (D^{2\beta} u_L^h, v)_Q = (g, v)_Q, \quad \forall v \in \mathbb{F}_M^{-\beta,-\beta} \otimes \mathcal{P}_N. \tag{3.52}$$

We first describe an efficient algorithm for solving (3.52) similar to the one used in [23]. We write

$$u_L^h(x,t) = \sum_{m=0}^{M} \sum_{n=0}^{N} \tilde{u}_{mn}^h \mathcal{J}_m^{-\beta,-\beta}(x)\, {}^{-}\tilde{J}_n^{(-\alpha,-\alpha)}(t). \tag{3.53}$$

For the test function, we take $v_L = \mathcal{J}_p^{-\beta,-\beta}(x) L_q^{(\alpha)}(t)$, where $L_q^{(\alpha)}(t) := \kappa_{q,\alpha} \tilde{P}_q^{(0,0)}(t)$ with $\kappa_{q,\alpha} = \frac{q!(2q+1)}{T \cdot \Gamma(q+\alpha+1)}$. Substituting the above into (3.52), we obtain

$$\begin{aligned} &\sum_{m=0}^{M} \sum_{n=0}^{N} \tilde{u}_{mn}^h \{ (\mathcal{J}_m^{-\beta,-\beta}, \mathcal{J}_p^{-\beta,-\beta}) ({}^R D_t^{\alpha}\, {}^{-}\tilde{J}_n^{(-\alpha,-\alpha)}, L_q^{(\alpha)}) - (D^{2\beta} \mathcal{J}_m^{-\beta,-\beta}, \mathcal{J}_p^{-\beta,-\beta}) ({}^{-}\tilde{J}_n^{(-\alpha,-\alpha)}, L_q^{(\alpha)}) \} \\ &= (g, \mathcal{J}_p^{-\beta,-\beta} L_q^{(\alpha)})_Q. \end{aligned} \tag{3.54}$$

Denote

$$\begin{aligned} &g_{pq} = (g, \mathcal{J}_p^{-\beta,-\beta}(x) L_q^{(\alpha)}(t))_Q, \quad G = (g_{pq})_{0 \le p \le M, 0 \le q \le N}, \\ &s_{pq}^t = \int_I {}^R D_t^{\alpha}\, {}^{-}\tilde{J}_q^{(-\alpha,-\alpha)}(t) L_p^{(\alpha)}(t)\,dt, \quad m_{pq}^t = \int_I {}^{-}\tilde{J}_q^{(-\alpha,-\alpha)}(t) L_p^{(\alpha)}(t)\,dt, \\ &U = (\tilde{u}_{mn}^h)_{0 \le m \le M, 0 \le n \le N}, \quad \boldsymbol{S}^t = (s_{pq}^t)_{0 \le p,q \le N}, \quad \boldsymbol{M}^t = (m_{pq}^t)_{0 \le p,q \le N}. \end{aligned}$$

Note that $\boldsymbol{S}^t$ is a diagonal matrix, $\boldsymbol{M}^t$ is not sparse but its entries can be accurately computed by Jacobi–Gauss quadrature with index $(0, \alpha)$.

Then, from (3.4), we find that (3.52) is equivalent to the following linear system:

$$\boldsymbol{M}^x U(\boldsymbol{S}^t)^T + \boldsymbol{S}^x U(\boldsymbol{M}^t)^T = G, \tag{3.55}$$

where $\boldsymbol{M}^x$ and $\boldsymbol{S}^x$ are the mass and stiffness matrix in the x-direction defined in (3.31); $\boldsymbol{S}^x$ is a diagonal matrix and $\boldsymbol{M}^x$ is full but symmetric. The linear system (3.55) can be solved efficiently by using the matrix diagonalization method [26]. Indeed, let $\boldsymbol{E} := (\bar{e}_0, \dots, \bar{e}_N) = (e_{pq})_{p,q=0,\dots,N}$ be the matrix formed by the orthonormal eigenvectors of the generalized eigenvalue problem $\boldsymbol{M}^x \bar{e}_j = \lambda_j \boldsymbol{S}^x \bar{e}_j$ and $\Lambda = \mathrm{diag}(\lambda_0, \dots, \lambda_N)$, i. e.,

$$\boldsymbol{M}^x \boldsymbol{E} = \boldsymbol{S}^x \boldsymbol{E}\Lambda. \tag{3.56}$$

Setting $U = \boldsymbol{E}V$ and multiplying both sides of (3.55) by $(\boldsymbol{S}^x\boldsymbol{E})^{-1} = \boldsymbol{E}^T\boldsymbol{S}^x$, we arrive at

$$\Lambda V(\boldsymbol{S}^t)^T + V(\boldsymbol{M}^t)^T = H := \boldsymbol{E}^T\boldsymbol{S}^x G. \tag{3.57}$$

Hence, let $\mathbf{v}_m$ and $\mathbf{h}_m$ be the mth row of V and H, respectively. Then the above matrix equation becomes

$$(\lambda_m \boldsymbol{S}^t + \boldsymbol{M}^t)\mathbf{v}_m = \mathbf{h}_m, \quad 0 \le m \le M, \tag{3.58}$$

which we solve directly with LU decomposition. Once we obtain V, we set $U = \boldsymbol{E}V$. Finally, we obtain the numerical solutions of (3.47) by $u_L = u_L^h + u_0$.

We now turn to the error estimate. Thanks to (3.34), we can define the following norm:

$$\|v\|_{X^{\alpha,\beta}(Q)} := \left(\|{}^R D_t^\alpha v\|_Q^2 - (D^{2\beta} v, v)_Q\right)^{1/2}.$$

Theorem 12. *Let u^h and u_L^h be the solutions of* (3.50) *and* (3.52), *respectively. Then we have the following error estimates:*

$$\begin{aligned}
\|u - u_L\|_{X^{\alpha,\beta}(Q)} \lesssim\ & M^{1-(2\beta+m)} \|D^{2\beta-1+m}({}^R D_t^\alpha u)\|_{L^2_{\omega^{(\beta-1+m,\beta-1+m)}}(\Lambda; L^2(I))} \\
& + N^{-(\alpha+n)} \|{}^R D_t^{\alpha+n}(D^{2\beta} u)\|_{L^2(\Lambda; L^2_{\omega^{(n,n)}}(I))} \\
& + M^{1-m} \|D^{2\beta-1+m} u\|_{L^2_{\omega^{(\beta-1+m,\beta-1+m)}}(\Lambda; L^2_{\omega^{(-\alpha,-\alpha)}}(I))} \\
& + N^{-n} \|{}^R D_t^{\alpha+n} u\|_{L^2_{\omega^{(-\beta,-\beta)}}(\Lambda; L^2_{\omega^{(n,n)}}(I))}.
\end{aligned}$$

Proof. Let us denote $\tilde{u}_L^h := {}^-\pi_N^{(-\alpha,-\alpha)} \pi_M^{(-\beta,-\beta)} u^h = \pi_M^{(-\beta,-\beta)}\, {}^-\pi_N^{(-\alpha,-\alpha)} u^h$ and $e_L := \tilde{u}_L^h - u_L^h$. We derive from (3.49) and (3.52) that for all $\forall v_L \in \mathbb{F}_M^{-\beta,-\beta} \otimes \mathcal{P}_N$, we have

$$\begin{aligned}
b(e_L, v_L) :&= ({}^R D_t^\alpha e_L, v)_Q - (D^{2\beta} e_L, v_L)_Q \\
&= ({}^R D_t^\alpha(\tilde{u}_L^h - u^h), v_L)_Q - (D^{2\beta}(\tilde{u}_L^h - u^h), v_L)_Q \\
&= ({}^R D_t^\alpha(\pi_M^{(-\beta,-\beta)} u^h - u^h), v_L)_Q - (D^{2\beta}({}^-\pi_N^{(-\alpha,-\alpha)} u^h - u^h), v_L)_Q.
\end{aligned}$$

Taking $v_L = {}^R D_t^\alpha e_L (\in \mathbb{F}_M^{-\beta,-\beta} \otimes \mathcal{P}_N)$ in the above equation, we obtain

$$\begin{aligned}&({}^R D_t^\alpha e_L, {}^R D_t^\alpha e_L)_Q - (D^{2\beta} e_L, {}^R D_t^\alpha e_L)_Q \\ &\quad = ({}^R D_t^\alpha (\pi_M^{(-\beta,-\beta)} u^h - u^h), {}^R D_t^\alpha e_L)_Q - (D^{2\beta}({}^-\pi_N^{(-\alpha,-\alpha)} u^h - u^h), {}^R D_t^\alpha e_L)_Q. \end{aligned} \tag{3.59}$$

Thanks to the generalized Poincaré inequality [9],

$$\|u\|_{L^2(I)} \le c \|{}^R D_t^\alpha u\|_{L^2(I)},$$

we derive from Lemmas 2–3 and (3.59) that

$$\begin{aligned}(D^{2\beta} e_L, e_L)_Q &\lesssim (D^{2\beta}({}^R D_t^{\frac{\alpha}{2}} e_L), {}^R D_t^{\frac{\alpha}{2}} e_L)_Q \\ &\cong (D^{2\beta}({}_t^R D^{\frac{\alpha}{2}} e_L), {}^R D_t^{\frac{\alpha}{2}} e_L)_Q = (D^{2\beta} e_L, {}^R D_t^\alpha e_L)_Q. \end{aligned} \tag{3.60}$$

This, along with equation (3.59), yields

$$\begin{aligned}\|{}^R D_t^\alpha e_L\|_Q^2 + (D^{2\beta} e_L, e_L)_Q &\le \|{}^R D_t^\alpha (\pi_M^{(-\beta,-\beta)} u^h - u^h)\|_Q \|{}^R D_t^\alpha e_L\|_Q \\ &\quad + \|D^{2\beta}({}^-\pi_N^{(-\alpha,-\alpha)} u^h - u^h)\|_Q \|{}^R D_t^\alpha e_L\|_Q,\end{aligned}$$

which implies

$$\begin{aligned}\|e_L\|_{X^{\alpha,\beta}(Q)}^2 &:= \|{}^R D_t^\alpha e_L\|_Q^2 + (D^{2\beta} e_L, e_L)_Q \\ &\lesssim \|{}^R D_t^\alpha (\pi_M^{(-\beta,-\beta)} u^h - u^h)\|_Q^2 + \|D^{2\beta}({}^-\pi_N^{(-\alpha,-\alpha)} u^h - u^h)\|_Q^2.\end{aligned}$$

The two terms at the right-hand side can be bounded by using Lemma 2 and Lemma 4 as follows:

$$\begin{aligned}\|{}^R D_t^\alpha (\pi_M^{(-\beta,-\beta)} u^h - u^h)\|_Q &\lesssim \|{}^R D_t^\alpha (\pi_M^{(-\beta,-\beta)} u^h - u^h)\|_{L^2_{\omega^{(-\beta,-\beta)}}(\Lambda; L^2(I))} \\ &\lesssim M^{1-(2\beta+m)} \|D^{2\beta-1+m}({}^R D_t^\alpha u)\|_{L^2_{\omega^{(\beta-1+m,\beta-1+m)}}(\Lambda; L^2(I))}\end{aligned}$$

and

$$\begin{aligned}\|D^{2\beta}({}^-\pi_N^{(-\alpha,-\alpha)} u^h - u^h)\|_Q &\lesssim \|D^{2\beta}({}^-\pi_N^{(-\alpha,-\alpha)} u^h - u^h)\|_{L^2(\Lambda; L^2_{\omega^{(-\alpha,-\alpha)}}(I))} \\ &\lesssim N^{-(\alpha+n)} \|{}^R D_t^{\alpha+n}(D^{2\beta} u)\|_{L^2(\Lambda; L^2_{\omega^{(n,n)}}(I))}.\end{aligned}$$

Combining the above estimates, we arrive at

$$\begin{aligned}\|e_L\|_{X^{\alpha,\beta}(Q)} &\lesssim M^{1-(2\beta+m)} \|D^{2\beta-1+m}({}^R D_t^\alpha u)\|_{L^2_{\omega^{(\beta-1+m,\beta-1+m)}}(\Lambda; L^2(I))} \\ &\quad + N^{-(\alpha+n)} \|{}^R D_t^{\alpha+n}(D^{2\beta} u)\|_{L^2(\Lambda; L^2_{\omega^{(n,n)}}(I))}.\end{aligned} \tag{3.61}$$

On the other hand, we have $u^h - u_L^h = u - \tilde{u}_L^h + e_L$. Then, using Lemma 2 and Lemma 4 again yields

$$
\begin{aligned}
(D^{2\beta}(\tilde{u}^h - u^h), \tilde{u}^h - u^h)_Q &\le \|D^{2\beta}(\tilde{u}^h - u^h)\|_{L^2_{\omega^{(\beta,\beta)}}(\Lambda;L^2_{\omega^{(\alpha,\alpha)}}(I))} \|\tilde{u}^h - u^h\|_{L^2_{\omega^{(-\beta,-\beta)}}(\Lambda;L^2_{\omega^{(-\alpha,-\alpha)}}(I))} \\
&\le \|D^{2\beta}(\tilde{u}^h - u^h)\|^2_{L^2_{\omega^{(\beta,\beta)}}(\Lambda;L^2_{\omega^{(\alpha,\alpha)}}(I))} + \|\tilde{u}^h - u^h\|^2_{L^2_{\omega^{(-\beta,-\beta)}}(\Lambda;L^2_{\omega^{(-\alpha,-\alpha)}}(I))} \\
&:= I_1 + I_2.
\end{aligned}
$$

Similarly, these two terms can be estimated as follows:

$$
\begin{aligned}
I_1 &\le \|D^{2\beta}\pi_M^{(-\beta,-\beta)}({}^{-}\pi_N^{(-\alpha,-\alpha)}u^h - u^h)\|^2_{L^2_{\omega^{(\beta,\beta)}}(\Lambda;L^2_{\omega^{(\alpha,\alpha)}}(I))} \\
&\quad + \|D^{2\beta}(\pi_M^{(-\beta,-\beta)}u^h - u^h)\|^2_{L^2_{\omega^{(\beta,\beta)}}(\Lambda;L^2_{\omega^{(\alpha,\alpha)}}(I))} \\
&\le \|D^{2\beta}({}^{-}\pi_N^{(-\alpha,-\alpha)}u^h - u^h)\|^2_{L^2_{\omega^{(\beta,\beta)}}(\Lambda;L^2_{\omega^{(-\alpha,-\alpha)}}(I))} \\
&\quad + \|D^{2\beta}(\pi_M^{(-\beta,-\beta)}u^h - u^h)\|^2_{L^2_{\omega^{(\beta,\beta)}}(\Lambda;L^2_{\omega^{(\alpha,\alpha)}}(I))} \\
&\le cN^{-2(\alpha+n)}\|{}^R D_t^{\alpha+n}(D^{2\beta}u)\|^2_{L^2(\Lambda;L^2_{\omega^{(n,n)}}(I))} \\
&\quad + M^{2-2m}\|D^{2\beta-1+m}u\|^2_{L^2_{\omega^{(\beta-1+m,\beta-1+m)}}(\Lambda;L^2(I))}
\end{aligned}
$$

and

$$
\begin{aligned}
I_2 &\le \|{}^{-}\pi_N^{(-\alpha,-\alpha)}(\pi_M^{(-\beta,-\beta)}u^h - u^h)\|^2_{L^2_{\omega^{(-\beta,-\beta)}}(\Lambda;L^2_{\omega^{(-\alpha,-\alpha)}}(I))} \\
&\quad + \|{}^{-}\pi_N^{(-\alpha,-\alpha)}u^h - u^h\|^2_{L^2_{\omega^{(-\beta,-\beta)}}(\Lambda;L^2_{\omega^{(-\alpha,-\alpha)}}(I))} \\
&\le \|\pi_M^{(-\beta,-\beta)}u^h - u^h\|^2_{L^2_{\omega^{(-\beta,-\beta)}}(\Lambda;L^2_{\omega^{(-\alpha,-\alpha)}}(I))} \\
&\quad + \|{}^{-}\pi_N^{(-\alpha,-\alpha)}u^h - u^h\|^2_{L^2_{\omega^{(-\beta,-\beta)}}(\Lambda;L^2_{\omega^{(-\alpha,-\alpha)}}(I))} \\
&\le cM^{2-2(2\beta+m)}\|D^{2\beta-1+m}u\|^2_{L^2_{\omega^{(\beta-1+m,\beta-1+m)}}(\Lambda;L^2_{\omega^{(-\alpha,-\alpha)}}(I))} \\
&\quad + cN^{-2(\alpha+n)}\|{}^R D_t^{\alpha+n}u\|^2_{L^2_{\omega^{(-\beta,-\beta)}}(\Lambda;L^2_{\omega^{(n,n)}}(I))}.
\end{aligned}
$$

Moreover, we have

$$
\begin{aligned}
&\|{}^R D_t^{\alpha}(\pi_M^{(-\beta,-\beta)}{}^{-}\pi_N^{(-\alpha,-\alpha)}u^h - u^h)\|_Q \\
&\quad \le \|{}^R D_t^{\alpha}{}^{-}\pi_N^{(-\alpha,-\alpha)}(\pi_M^{(-\beta,-\beta)}u^h - u^h)\|_Q + \|{}^R D_t^{\alpha}({}^{-}\pi_N^{(-\alpha,-\alpha)}u^h - u^h)\|_Q \\
&\quad \le \|{}^R D_t^{\alpha}(\pi_M^{(-\beta,-\beta)}u^h - u^h)\|_Q + \|{}^R D_t^{\alpha}({}^{-}\pi_N^{(-\alpha,-\alpha)}u^h - u^h)\|_Q \\
&\quad \lesssim cM^{1-(2\beta+m)}\|D^{2\beta-1+m}({}^R D_t^{\alpha}u)\|_{L^2_{\omega^{(\beta-1+m,\beta-1+m)}}(\Lambda;L^2(I))} \\
&\qquad + cN^{-n}\|{}^R D_t^{\alpha+n}u\|_{L^2(\Lambda;L^2_{\omega^{(n,n)}}(I))}.
\end{aligned}
$$

Consequently, the desired result follows from the above estimates and the triangle inequality. □

Remark 4. Note that the error estimate in the above theorem cannot be easily expressed in terms of the data (f, u_0). In particular, the space-time Petrov–Galerkin method (3.52) will not lead to high-order convergence, even if f and u_0 are sufficiently smooth, due to the singularities of the solution at $t = 0$ and $x = \pm 1$. However, the leading singular term in time and in space is included in our approximation space so our method will lead to a better convergence rate than those based on the polynomial approximations.

4 Concluding remarks

We presented in this chapter essential properties of the GJFs and their application to a class of FDEs. In particular, we showed that (i) by using suitable GJFs, the non-local fractional operators become local operators in the space spanned by GJFs; (ii) for simple FDEs, the spectral methods using GJFs can lead to an exponential convergence rate despite the non-smoothness of the solution in usual Sobolev spaces; and (iii) for more general FDEs, a suitable spectral method using GJFs is still very efficient as the non-local fractional stiffness matrices can be easily computed, and furthermore, it is also more accurate than using a usual polynomial-based method as the GJFs include the leading singular term of the underlying FDEs.

We only consider one-dimensional FDEs in this chapter. For multi-dimensional fractional partial differential equations with only fractional derivative in time, one can couple the GJF spectral method in time with a usual spatial approximation to construct a space-time Petrov–Galerkin method. It can still be efficiently solved by using the matrix diagonalization method as in Subsection 3.4; we refer to [27] for more details. As for FDEs with multi-dimensional fractional operators in space, one has to construct appropriate numerical methods with respect to the specific definitions of the fractional operator. In particular, the GJFs for the one-dimensional Riesz equation can be extended to deal with fractional Laplacian by the integral definition on the multi-dimensional balls [20]. On the other hand, for fractional Laplacians defined through the spectral decomposition of the Laplacian operator, one can use the Caffarelli–Silvestre extension to cast the fractional Laplacian equation in d-dimension into an extended problem in $d + 1$-dimension with regular derivatives and a weakly singular weight in the extended direction. Then, one can construct an efficient and accurate spectral method in the extended direction to couple with any consistent approximation in space; for more detail, we refer to [5]. For other types of multi-dimensional fractional partial differential equations, we refer to [19] for a nice presentation on different definitions of fractional Laplacians and their numerical treatments.

Bibliography

[1] R. A. Adams, *Sobolev Spaces*, Academic Press, New York, 1975.

[2] G. E. Andrews, R. Askey, and R. Roy et al., *Special Functions*, Encyclopedia of Mathematics and Its Applications, 1999.

[3] H. Bateman, The solution of linear differential equations by means of definite integrals, *Trans. Camb. Philos. Soc.*, **21** (1909), 171–196.

[4] L. Chen, Z. Mao, and H. Li, Jacobi–Galerkin spectral method for eigenvalue problems of Riesz fractional differential equations, arXiv:1803.03556.

[5] S. Chen and J. Shen, An efficient and accurate method for the fractional Laplacian equation using the Caffarelli–Silvestre extension, Preprint.

[6] S. Chen, J. Shen, and L. L. Wang, Generalized Jacobi functions and their applications to fractional differential equations, *Math. Comput.*, **85**(300) (2016), 1603–1638.

[7] K. Diethelm, *The Analysis of Fractional Differential Equations*, Springer-Verlag, Berlin, 2010.

[8] V. J. Ervin, N. Heuer, and J. P. Roop, Regularity of the solution to 1-D fractional order diffusion equations, Preprint, 2016.

[9] V. J. Ervin and J. P. Roop, Variational formulation for the stationary fractional advection dispersion equation, *Numer. Methods Partial Differ. Equ.*, **22**(3) (2006), 558–576.

[10] B.-Y. Guo, J. Shen, and L.-L. Wang, Generalized Jacobi polynomials/functions and their applications, *Appl. Numer. Math.*, **59**(5) (2009), 1011–1028.

[11] D. Hou and C. Xu, A fractional spectral method with applications to some singular problems, *Adv. Comput. Math.*, **43**(5) (2017), 911–944.

[12] C. Huang, Y. Jiao, L.-L. Wang, and Z. Zhang, Optimal fractional integration preconditioning and error analysis of fractional collocation method using nodal generalized Jacobi functions, *SIAM J. Numer. Anal.*, **54**(6) (2016), 3357–3387.

[13] C. Huang and M. Stynes, A spectral collocation method for a weakly singular Volterra integral equation of the second kind, *Adv. Comput. Math.*, **42**(5) (2016), 1015–1030.

[14] Y. Jiao, L.-L. Wang, and C. Huang, Well-conditioned fractional collocation methods using fractional Birkhoff interpolation basis, *J. Comput. Phys.*, **305** (2016), 1–28.

[15] E. Kharazmi, M. Zayernouri, and G. E. Karniadakis, Petrov–Galerkin and spectral collocation methods for distributed order differential equations, *SIAM J. Sci. Comput.*, **39**(3) (2017), A1003–A1037.

[16] E. Kharazmi, M. Zayernouri, and G. E. Karniadakis, A Petrov–Galerkin spectral element method for fractional elliptic problems, *Comput. Methods Appl. Mech. Eng.*, **324** (2017), 512–536.

[17] X. Li and C. Xu, A space-time spectral method for the time fractional diffusion equation, *SIAM J. Numer. Anal.*, **47**(3) (2009), 2108–2131.

[18] X. Li and C. Xu, Existence and uniqueness of the weak solution of the space-time fractional diffusion equation and a spectral method approximation, *Commun. Comput. Phys.*, **8**(5) (2010), 1016.

[19] A. Lischke, G. Pang, M. Gulian, F. Song, C. Glusa, X. Zheng, Z. Mao, M. M. Meerschaert, W. Cai, M. Ainsworth, and G. E. Karniadakis, What is the fractional Laplacian? arXiv:1801.09767.

[20] S. Ma, H. Li, and Z. Zhang, Efficient spectral methods for the fractional Laplacian, Preprint.

[21] Z. Mao, S. Chen, and J. Shen, Efficient and accurate spectral method using generalized Jacobi functions for solving Riesz fractional differential equations, *Appl. Numer. Math.*, **106** (2016), 165–181.

[22] Z. Mao and G. E. Karniadakis, A spectral method (of exponential convergence) for singular solutions of the diffusion equation with general two-sided fractional derivative, *SIAM J. Numer. Anal.*, **56**(1) (2018), 24–49.

[23] Z. Mao and J. Shen, Efficient spectral-Galerkin methods for fractional partial differential equations with variable coefficients, *J. Comput. Phys.*, **307** (2016), 243–261.
[24] I. Podlubny, *Fractional Differential Equations*, Academic Press, New York, 1999.
[25] S. G. Samko, A. A. Kilbas, and O. I. Marichev, *Fractional Integrals and Derivatives: Theory and Applications*, Gordon and Breach Science, Yverdon, Switzerland, 1993.
[26] J. Shen, Efficient spectral-Galerkin method I. Direct solvers of second-and fourth-order equations using Legendre polynomials, *SIAM J. Sci. Comput.*, **15**(6) (1994), 1489–1505.
[27] J. Shen and C. Sheng, An efficient space-time method for time fractional diffusion equation, Preprint.
[28] J. Shen, C. Sheng, and Z. Wang, Generalized Jacobi spectral-Galerkin method for nonlinear Volterra integral equations with weakly singular kernels, *J. Math. Study*, **48**(4) (2015), 315–329.
[29] J. Shen, T. Tang, and L.-L. Wang, *Spectral Methods: Algorithms, Analysis and Applications*, vol. 41, Springer Science & Business Media, 2011.
[30] G. Szegö, *Orthogonal Polynomials*, 4th edn., AMS Coll. Publ., vol. 23, 1975.
[31] M. Zayernouri, M. Ainsworth, and G. E. Karniadakis, A unified Petrov–Galerkin spectral method for fractional PDEs, *Comput. Methods Appl. Mech. Eng.*, **283** (2015), 1545–1569.
[32] M. Zayernouri and G. E. Karniadakis, Fractional Sturm–Liouville eigen-problems: theory and numerical approximation, *J. Comput. Phys.*, **252** (2013), 495–517.
[33] M. Zayernouri and G. E. Karniadakis, Exponentially accurate spectral and spectral element methods for fractional ODEs, *J. Comput. Phys.*, **257**(part A) (2014), 460–480.
[34] M. Zayernouri and G. E. Karniadakis, Fractional spectral collocation methods for linear and nonlinear variable order FPDEs, *J. Comput. Phys.*, **293** (2015), 312–338.
[35] F. Zeng, Z. Mao, and G. E. Karniadakis, A generalized spectral collocation method with tunable accuracy for fractional differential equations with end-point singularities, *SIAM J. Sci. Comput.*, **39**(1) (2017), A360–A383.
[36] Z. Zhang, F. Zeng, and G. E. Karniadakis, Optimal error estimates of spectral Petrov–Galerkin and collocation methods for initial value problems of fractional differential equations, *SIAM J. Numer. Anal.*, **53**(4) (2015), 2074–2096.

Anna Lischke, Mohsen Zayernouri, and Zhongqiang Zhang

Spectral and spectral element methods for fractional advection–diffusion–reaction equations

Abstract: We review recent advances in spectral and spectral element methods for a class of fractional partial differential equations. We focus on linear advection–diffusion–reaction equations in one and two dimensions. In particular, we discuss how to deal with boundary and interior singularity of solutions on finite intervals, on the half line, and on two-dimensional domains. Regularity theory for equations with fractional Laplacian is also presented. Finally, we present two approaches to discretize equations with space-time fractional derivatives. Numerical results are presented for both one- and two-dimensional problems.

Keywords: Spectral Galerkin methods, multi-domain spectral methods, fractional Laplacian, time-fractional derivatives, space-fractional derivatives, multi-term fractional equations

MSC 2010: 65N30 , 65N35 , 65M20 , 65M60 , 65M70 , 65L60 , 41A30, 26A33

1 Introduction

Traditional spectral element methods (SEMs) combine the high-order accuracy of spectral methods with the flexible domain partitions of finite element methods. As fractional operators are well-suited for modeling anomalous transport in heterogeneous materials (see [64] and the references therein), methods which are not cumbersome for problems with non-smooth surfaces are needed. Therefore, efficient SEMs for solving fractional differential equations (FDEs) are particularly valuable. Development of such methods is still in its early stages, and we report existing state-of-the-art approaches in this chapter.

Acknowledgement: AL was supported by the ARO/MURI grant W911NF-15-1-0562. MZ was supported by the AFOSR Young Investigator Program (YIP) award (FA9550- 17-1-0150). MZ and ZZ were also partially supported by MURI/ARO (W911NF- 15-1-0562). The code for the example of Section 4.3.1 was kindly provided by F. Song, an author of [78].

Anna Lischke, Division of Applied Mathematics, Brown University, Providence, RI 02912, USA, e-mail: anna_lischke@brown.edu
Mohsen Zayernouri, Department of Computational Mathematics, Science, and Engineering, Michigan State University, 428 S. Shaw Lane, East Lansing, MI 48824, USA, e-mail: zayern@msu.edu
Zhongqiang Zhang, Department of Mathematical Sciences, Worcester Polytechnic Institute, Worcester, MA 01609, USA, e-mail: zzhang7@wpi.edu

https://doi.org/10.1515/9783110571684-006

The main bottlenecks in solving FDEs computationally are their *global* nature, memory-dependent characteristics, and singularities, which may occur both on the boundaries and in the interior domain. These hurdles result in excessive computer memory storage and insufficient accuracy for long-time integration of non-linear models, which can originate, for example, from disordered heterogeneous media and viscoelasticity. Compared to finite difference and finite element methods, which are inherently local discretization techniques, spectral methods are *global*, and single- and multi-domain spectral methods naturally fit the non-local processes described by FDEs. To address the aforementioned computational challenges, we discuss in this chapter recent advancements in both spectral methods and SEMs for fractional equations using both standard and singular basis functions.

In this chapter, we consider space-time fractional partial differential equations (FPDEs) of the form

$$\partial_t^\beta u = \mathcal{L}^\alpha u + \mu_1 D u + \mu_2 u + f, \quad x \in \Omega \subset \mathbb{R}^d, \ \alpha \in (1,2), \tag{1.1}$$

with Dirichlet boundary conditions. Here, $\partial_t^\beta u$ is a fractional derivative (either the Riemann–Liouville or the Caputo definition) of order β, $\mathcal{L}^\alpha$ is an appropriate fractional diffusion operator, and D is a first-order derivative operator in x. The coefficients μ_1 and μ_2 are constants, and $\Omega \subset \mathbb{R}^d$ is a bounded or unbounded domain with dimension $d = 1$ or 2.

The time-fractional derivative appears in applications in which memory is an important aspect, such as viscoelastic materials [63] and acoustic waves in complex media [44], and in some cases, a time-fractional equation arises through duality with a space-fractional equation [45]. The space-fractional operators have recently become important tools for modeling anomalous transport in many applications, from fractional ultrasound models and transport in porous media to fractional quantum mechanics [83, 23, 7, 6, 49, 81, 22, 69].

We start with FDEs with time-fractional derivatives. Since a time-fractional derivative is one-dimensional, we know explicitly the structure of solutions to time-fractional FDEs. Indeed, solutions usually have a singularity at time $t = 0$. Accordingly, spectral methods employ some non-polynomial bases (e. g., poly-fractonomials) to accommodate the singularity, which we discuss in Section 2, where we present three classes of spectral methods with three families of non-standard bases. Both spectral Galerkin and collocation methods are discussed. Also, a spectral Galerkin method for FDEs with multi-term time-fractional derivatives is discussed.

In Section 3, we discuss a boundary value problem with space-fractional derivatives in one dimension ($d = 1$). The structure of solutions to such equations is similar to those of time-fractional equations, while in this case, solutions have weak singularities at both boundary points. The weak singularity is actually characterized by pseudo-eigenfunctions of the leading operator. Some regularity estimates of the solutions to these equations are presented in this section. Spectral methods with non-standard bases for a steady advection–diffusion–reaction equation are presented, accompanied by some numerical examples. SEMs for this equation are also presented for a steady equation, where solutions also have a weak singularity in the interior of the bounded interval.

In dimension $d > 1$, we consider the operator $\mathcal{L}^\alpha = (-\Delta)^\alpha$ in (1.1), which is known as the *fractional Laplacian*. The fractional Laplacian in $\mathbb{R}^d$ can be defined in many equivalent ways [48], but for computational experiments, we often need to formulate these definitions in bounded domains. The resulting formulations are not necessarily equivalent, and different numerical approaches are required for each fractional Laplacian. In Section 4, we focus on spectral element approaches that have been used to approximate the fractional Laplacian.

In Section 5, we discuss single-domain spectral methods for FDEs with both space- and time-fractional derivatives. For time-fractional derivatives, most researchers prefer to apply fast-convolution approaches [75] or Laplace transforms [58], as these approaches often exhibit lower computational complexity and are easier to implement than spectral and spectral element methods. However, it can be advantageous to use spectral and spectral element discretizations in time as in [90, 86, 88] to obtain high accuracy. Most studies on spectral discretizations for FPDEs have been devoted to single-domain spectral methods; see e. g., [52, 51, 46, 20, 82, 25, 36, 59, 55].

The remainder of this chapter is organized as follows. In Section 2, we briefly review a simple FDE and spectral Galerkin and collocation methods for this model using standard and non-standard basis functions. We also discuss spectral methods for FDEs with multi-term fractional derivatives in Section 2.1. In Section 3, we present a spectral Galerkin method and an SEM for a one-dimensional linear advection–diffusion equation with a two-sided fractional derivative. In Section 4, we discuss SEMs for two definitions of the fractional Laplacian in two and higher dimensions, regularity properties, and formulations of different types of boundary conditions. In Section 5, we present two approaches to solving space-time FPDEs: the method of lines using finite difference in time and spectral or spectral elements in space, and spectral methods in both time and space. Some numerical results are presented.

2 From polynomials to generalized polynomials

Standard spectral methods for FDEs employ Jacobi polynomials (e. g., Legendre, Chebyshev) as a basis for approximation. The Jacobi polynomials have the following orthogonal property:

$$\int_{-1}^{1} P_l^{\gamma,\beta}(x) P_n^{\gamma,\beta}(x) \omega^{\gamma,\beta}(x)\, dx = h_n^{\gamma,\beta} \delta_{nl}, \quad \gamma, \beta > -1. \tag{2.1}$$

Here n and l are integers, and $\delta_{nl} = 1$ if $n = l$ and is zero otherwise,

$$\omega^{\gamma,\beta} = (1-x)^{\gamma}(1+x)^{\beta}, \quad \text{and} \quad h_n^{\gamma,\beta} = \frac{2^{\gamma+\beta+1}}{2n+\gamma+\beta+1}\frac{\Gamma(n+\gamma+1)\Gamma(n+\beta+1)}{\Gamma(n+\gamma+\beta+1)n!}.$$

In spectral methods, we usually approximate the solution to an FPDE by a truncation of a so-called Fourier–Jacobi expansion, $u_N = \sum_{k=0}^{N} a_k P_k^{\gamma,\beta}(x)$. Here, γ and β are typically chosen as $\gamma = \beta = 0$ (Legendre polynomials) or $\gamma = \beta = -\frac{1}{2}$ (Chebyshev polynomials). There is extensive literature along these lines; see, e. g., [19, 50, 52, 61, 92]. For standard spectral methods, one important issue is to compute the fractional integral (or derivative) of orthogonal polynomials. In [50], these integrals (or derivatives) are evaluated using three-term recurrence relations inherited from the three-term relations for orthogonal polynomials. For special cases of Jacobi polynomials, their fractional integrals can be found in [9].

One disadvantage of the standard spectral methods is their low accuracy when solutions have intrinsic weak singularities at the boundary points. Let us consider the following simplified version of (1.1) with a zero initial value:

$$ {}_{-1}^{\mathrm{RL}}\mathcal{D}_x^{\mu} u + u = f. \tag{2.2}$$

Here, ${}_{-1}^{\mathrm{RL}}\mathcal{D}_x^{\mu} u$ is the Riemann–Liouville derivative of u (see, e. g., [68, 74]), i. e.,

$$ {}_{-1}^{\mathrm{RL}}\mathcal{D}_x^{\mu} u(x) = \frac{1}{\Gamma(n-\mu)}\frac{d^n}{dx^n}\int_{-1}^{x}\frac{u(\xi)}{(x-\xi)^{\mu-n+1}}d\xi, \quad x > -1, \tag{2.3}$$

with $n-1 < \mu < n$, where n is a natural number. Suppose u is a solution to the steady equation (2.2), where f is some smooth function. It is well known that the Green's function of this equation is exactly the Mittag-Leffler function [68, 74]. Then the solution can be written as a linear combination of terms $(1+x)^{j\mu}$, $j = 1, 2, \ldots$, plus a smooth function. The solution to (2.2) clearly has a weak singularity at $x = -1$; indeed, its derivative does not exist at $x = -1$.

To capture the singularities at the boundary points, some non-standard bases have been used in spectral methods. The first type is called *Jacobi poly-fractonomials* (see, e. g., [89, 85]), $\mathcal{P}^{(1)}(x) = (1+x)^{\mu} P_n^{-\mu,\mu}(x)$, where $n \geq 0$ is an integer and μ is the order of the leading fractional operator in the underlying equation. This family of bases satisfies the following family of *fractional Sturm–Liouville problems* on $(-1, 1)$:

$$ {}_{x}^{\mathrm{RL}}\mathcal{D}_1^{\mu}\left[(1-x)^{\alpha+1}(1+x)^{\beta+1}\,{}_{-1}^{C}\mathcal{D}_x^{\mu}\mathcal{P}^{(1)}(x)\right] + \lambda(1-x)^{\alpha+1-\mu}(1+x)^{\beta+1-\mu}\,\mathcal{P}^{(1)}(x) = 0,$$

and enjoys several useful properties such as orthogonality, recurrence formulas, exact fractional derivatives, and spectral approximability. These singularity capturing bases were successfully employed in high-order and efficient Petrov–Galerkin spectral and spectral element methods for linear and non-linear fractional ordinary differential equations/partial differential equations [87, 86, 88, 91, 55]. Convergence analysis of spectral Galerkin methods using Jacobi poly-fractonomials is presented in [20, 42, 96].

Collocation methods with Jacobi poly-fractonomials, also known as fractional spectral collocation methods, are used for variable-order FDEs [94], where the factor $(1+x)^{\beta}$ is chosen with $\beta < 0$ as a tuning parameter such that better accuracy can be obtained. A similar approach has been applied to variable-order FDEs with solutions exhibiting two end point singularities [93]. A modification of this idea using negative weights is investigated in [31]. Computing cost and pre-conditioning in solving FDEs with collocation methods are discussed in [41]. Convergence analysis of collocation methods can be found in [41, 96]. Superconvergence points of fractional collocation methods are discussed in [98].

The second non-standard basis has the form (see, e. g., [32]) $(1+x)^{j\mu}P_n(x)$, where $j, n \geq 0$ are integers, $P_n(x)$ can be any Jacobi polynomials, and $0 \leq n + j\alpha \leq r$ such that a convergence order r is expected. This basis is motivated by the structure of the Mittag-Leffler functions. Solutions are approximated by $\sum_{j,n\geq 0, n+j\alpha\geq r} b_{n,j}(1+x)^{j\mu}P_n(x)$. This basis has been also employed in [16] for Volterra integral equations.

The third non-standard basis is called *Müntz polynomials*; see, e. g., [30, 40]. The use of this basis is also motivated by the structure of solutions to FDEs. It is observed that $u((1+x)^{\frac{1}{\mu}})$ can be smooth enough, where $0 < \mu < 1$ and u is a solution to (2.2), if f is smooth enough. Then the basis is chosen to be $P_n^{\gamma,\beta}((1+x)^{\frac{1}{\mu}})$.

Remark 2.1. Suppose that we replace the left-sided Riemann–Liouville derivative with the right-sided Riemann–Liouville derivative

$$ {}_{x}^{\mathrm{RL}}\mathcal{D}_{1}^{\mu}u = \frac{(-1)^n}{\Gamma(n-\mu)}\frac{d^n}{dx^n}\int_x^1 \frac{u(\xi)}{(\xi - x)^{\mu-n+1}}d\xi, \quad x < 1,\ n-1 < \mu < n. \tag{2.4}$$

The solution to $u + {}_{x}^{\mathrm{RL}}\mathcal{D}_{1}^{\mu}u = f$ will have singularity at $x = 1$ with no derivatives at $x = 1$. We only need to change the variable to $y = -x$. Then we have $v + {}_{-1}^{\mathrm{RL}}\mathcal{D}_{y}^{\mu}v = f(-y)$, where $v = u(-x)$, and the approaches discussed above for the left-sided derivative may be applied.

2.1 Spectral methods for multi-term FDEs

The study of multi-term fractional equations can be motivated as an approximation of distributed-order fractional operators, in which the order of the fractional derivative is determined by a probability distribution [10, 66, 17, 18, 77]. In bounded domains, a spectral method for fractional multi-term equations has been developed in [99], and *tempered* multi-term equations were solved using a generalized Jacobi spectral method in [21]. On the half line, a spectral method for tempered fractional equations was developed using *generalized Laguerre* polynomials in [97]. However, these spectral methods, when applied to multi-term equations, have cubic complexity, i. e., $\mathcal{O}(N^3)$ operations are required to compute the solution, where N represents the degrees of freedom of the numerical method. In the Laguerre Petrov–Galerkin approach due to Lischke et al. [55], a factorization of the discrete linear system reduced this complexity to $\mathcal{O}(N\log(N))$, as described below.

Consider the following multi-term FDEs:

$$\sum_{i=1}^{K} b_i \, {}_0\mathcal{D}_t^{\nu_i}(u-u_0)(t) = f(t), \quad u(0) = u_0. \tag{2.5}$$

Here, the coefficients $\{b_i\}_{i=1}^K$ are positive and ${}_0\mathcal{D}_t^{\nu_i}$ represents the Riemann–Liouville fractional derivative of order ν_i, where $\nu_1 > \nu_2 > \cdots > \nu_K \in (0,1)$ and the domain for t can be either bounded intervals or the half line $(0,+\infty)$.

To simplify the presentation, we consider a *Laguerre Petrov–Galerkin spectral method* for the following two-term FDE with a vanishing initial value:

$${}_0\mathcal{D}_t^{\nu_1}u(t) + {}_0\mathcal{D}_t^{\nu_2}u(t) = f(t), \quad t \in (0,+\infty), \tag{2.6}$$

where $\nu_1,\nu_2 \in (0,1)$. We use the generalized Laguerre functions to approximate the solution. We have $u(t) \approx u_N(t) = \sum_{n=1}^N a_n \phi_n^{\alpha_1,1}(t)$, with $\{a_n\}_{n=1}^N$ the unknown coefficients. The trial and test functions are defined, respectively, by

$$\phi_n^{\alpha_1,1}(t) = t^{\alpha_1} L_{n-1}^{(\alpha_1)}(t), \quad \phi_k^{\alpha_2,2}(t) = e^{-t} L_{k-1}^{(\alpha_2)}(t). \tag{2.7}$$

Here, $\alpha_1 + \alpha_2 = \nu_1$, $\alpha_1 > 0$, and $\alpha_2 > -1$, and $L_k^{(\alpha_i)}(t)$s are the generalized Laguerre polynomials with the property

$$\int_0^\infty t^\alpha e^{-t} L_n^{(\alpha)}(t) L_m^{(\alpha)}(t)\,dt = \frac{\Gamma(n+\alpha+1)}{n!}\delta_{n,m}.$$

The variational form for the Petrov–Galerkin spectral method is

$$\sum_{n=1}^N a_n \int_0^\infty \phi_k^{\alpha_2,2}(t) {}_0\mathcal{D}_t^{\nu_1}\phi_n^{\alpha_1,1}(t)\,dt + \sum_{n=1}^N a_n \int_0^\infty \phi_k^{\alpha_2,2}(t) {}_0\mathcal{D}_t^{\nu_2}\phi_n^{\alpha_1,1}(t)\,dt = \int_0^\infty f(t)\phi_k^{\alpha_2,2}(t)\,dt.$$

By the calculation of the entries of the resulting linear system, it is shown in [55] that the linear system is a Toeplitz-like system, which can be solved in $O(N\log(N))$ operations. FDEs with several fractional differential operators are also discussed in [55] while the resulting linear systems keep a Toeplitz-like structure and can be solved in $O(N\log(N))$ operations.

Remark 2.2. If, instead, a bounded domain for t is used, we can change the time domain to $[-1,1]$ by a linear transform and use Jacobi poly-fractonomials.

While the non-standard bases can capture the weak singularity at the boundary points $x = 1$ or $x = -1$, they may be unable to accommodate singularity at the interior of the interval $(-1, 1)$, which leads to low accuracy and slow convergence of the spectral methods. A standard way to resolve this issue is to use *spectral element methods*, also called *multi-domain spectral methods*, where we apply the finite element methods and use high-degree piece-wise orthogonal polynomials as basis functions. The advantage of SEMs is that they can be adapted to exponential convergence orders, especially with the *h*- and *p*-refinements.

In the next section, we take a careful look at a linear fractional advection–diffusion–reaction equation in which the leading operator is a linear combination of left-sided and right-sided fractional derivatives, resulting in a solution exhibiting weak singularities at both boundary points, which will be incorporated into the numerical method directly.

3 A linear advection–diffusion–reaction equation in 1D

Consider the following fractional advection–diffusion–reaction equation:

$$\mathcal{L}_\theta^\alpha u + \mu_1 Du + \mu_2 u = f(x), \quad x \in I = (-1,1),\ \alpha \in (1,2), \tag{3.1}$$

with homogeneous boundary conditions $u(-1) = u(1) = 0$, where $\mu_1 \in \mathbb{R}$ and $\mu_2 \geq 0$. Here $\mathcal{L}_\theta^\alpha := -[(\theta\, {}_{-1}^{\mathrm{RL}}\mathcal{D}_x^\alpha + 1 - \theta)\, {}_x^{\mathrm{RL}}\mathcal{D}_1^\alpha]$, where the left-sided Riemann–Liouville derivative ${}_{-1}^{\mathrm{RL}}\mathcal{D}_x^\alpha$ is defined in (2.3) and the right-sided Riemann–Liouville derivative ${}_x^{\mathrm{RL}}\mathcal{D}_1^\alpha$ is defined in (2.4).

The following proposition illustrates that the solution to (3.1) may have a weak singularity at both boundary points $x = \pm 1$.

Proposition 3.1 ([29]). *For the nth-order Jacobi polynomial $P_n^{\sigma,\sigma^*}(x)$, we have*

$$\mathcal{L}_\theta^\alpha[\omega^{\sigma,\sigma^*} P_n^{\sigma,\sigma^*}(x)] = \lambda_{\theta,n}^\alpha P_n^{\sigma^*,\sigma}(x), \quad \mathcal{L}_\theta^\alpha[\omega^{\sigma-1,\sigma^*-1}(x)] = 0, \tag{3.2}$$

where

$$\lambda_{\theta,n}^\alpha = -\frac{\sin(\pi\alpha)}{\sin(\pi\sigma) + \sin(\pi\sigma^*)}\frac{\Gamma(\alpha+n+1)}{\Gamma(n+1)} \quad \text{and} \quad \sigma^* = \alpha - \sigma.$$

Here σ is determined by $\theta = \frac{\sin(\pi\sigma^*)}{\sin(\pi\sigma^*)+\sin(\pi\sigma)}$.

When $\theta = 1/2$, the relation (3.2) was derived independently in [59].

If we have $\mu_1 = \mu_2 = 0$ in (3.1) and f is smooth, Proposition 3.1 suggests that the solution has weak singularity at $x = \pm 1$, which is of the type $(1+x)^\sigma(1-x)^{\sigma^*}$. In [29], convergence analysis of spectral methods for equation (3.1) with $\mu_1 = \mu_2 = 0$ is presented when a weighted basis (non-standard spectral basis) of the form $(1+x)^{\sigma}(1-x)^{\sigma^*}\mathbb{P}_{\mathbb{N}}$ is employed, where $\mathbb{P}_{\mathbb{N}}$ is the collection of algebraic polynomials of order less than $N+1$; see also [60]. When $\mu_1 = 0$ but $\mu_2 \neq 0$, regularity analysis and convergence analysis of the non-standard basis is detailed in [38].

We will discuss a spectral Galerkin method in Section 3.2. Let us first discuss some results about the regularity of the solution to (3.1).

3.1 Regularity of solutions to equation (3.1)

To incorporate singularities at the end points, we introduce a non-uniformly Jacobi-weighted Sobolev space (see, e. g., [11, 37]). Let $L^2_{\omega^{\gamma,\beta}}(I)$ be the space with the inner product and norm defined by $(u,v)_{\omega^{\gamma,\beta}} = \int_I uv\omega^{\gamma,\beta}dx$, $\|u\|_{\omega^{\gamma,\beta}} = (u,u)^{\frac{1}{2}}_{\omega^{\gamma,\beta}}$. When $\gamma = \beta = 0$, we write $(u,v) = (u,v)_{\omega^{0,0}}$ and $\|u\| = (u,u)^{1/2}_{\omega^{0,0}}$. The weighted space is defined by

$$B^m_{\omega^{\gamma,\beta}}(I) = \{u \mid \partial^k_x u \in L^2_{\omega^{\gamma+k,\beta+k}}(I),\ k = 0,1,\dots,m\},\quad m \text{ is a non-negative integer,}$$

which is equipped with the following norm: $\|u\|_{B^m_{\omega^{\gamma,\beta}}} = (\sum_{k=0}^m |u|^2_{B^k_{\omega^{\gamma,\beta}}})^{1/2}$ and $|u|_{B^k_{\omega^{\gamma,\beta}}} = \|\partial^k_x u\|_{\omega^{\gamma+k,\beta+k}}$. When s is not an integer, the space can be defined via the classical space interpolation method.

The following theorem states smoothness of the solution to the linear advection–diffusion–reaction equation (3.1) with the Riesz derivative ($\theta = 1/2$).

Theorem 3.2 ([39]). *Assume that $f \in B^r_{\omega^{\alpha/2-1}}(I)$ with $r \geq 0$. Then*
- *(diffusion only) $\omega^{-\alpha/2}u \in B^{r+\alpha}_{\omega^{\alpha/2}}(I)$ and $u \in H^{\frac{\alpha+1}{2}-\epsilon}(I)$ when $\mu_1 = \mu_2 = 0$;*
- *(diffusion–reaction) $\omega^{-\alpha/2}u \in B^{(3\alpha/2+1)\wedge r+\alpha}_{\omega^{\alpha/2}}(I)$ when $\mu_1 = 0$ and $\mu_2 > 0$ (here $\alpha \wedge r = \min\{\alpha, r\}$);*
- *(advection–diffusion–reaction) $\omega^{-\alpha/2}u \in B^{(3\alpha/2-1-\epsilon)\wedge r+\alpha}_{\omega^{\alpha/2}}(I)$, where $\alpha > 4/3$ and $\epsilon > 0$. If $\alpha \leq 4/3$, we have $\omega^{\alpha/2}u \in B^{(\alpha-1)\wedge r+\alpha}_{\omega^{\alpha/2}}(I)$. In this case, $\mu_1 \neq 0$ and $\mu_2 \geq 0$.*

For the problem (3.1) with $\mu_1 \neq 0$ and $\mu_2 > 0$, if $f \in L^2(I) \cap B^r_{\omega^{\alpha/2}}(I)$, then $\tilde{u} = \omega^{-\alpha/2}u \in B^{(\alpha-1)\wedge r+\alpha}_{\omega^{\alpha/2}}(I)$. For $\alpha > 4/3$ we have $\tilde{u} \in B^{(3\alpha/2-1-\epsilon)\wedge r+\alpha}_{\omega^{\alpha/2}}(I)$ with ϵ arbitrarily small.

Since the solution to (3.1) belongs to $H^{\frac{\alpha+1}{2}-\epsilon}(I)$ when $\mu_1 = \mu_2 = 0$, we cannot expect higher regularity when $\mu_1^2 + \mu_2^2 \neq 0$.

3.2 A spectral Galerkin method and numerical examples

When $\theta = 1/2$, the solution to (3.1) has singularities at both left and right end points, which can be represented by $u = (1-x^2)^{\alpha/2}\tilde{u}$. In numerical methods, we can use this form to find an accurate approximation of u. Specifically, we approximate $\tilde{u}$ by using a basis of Jacobi polynomials, $P_n^{\alpha/2,\alpha/2}(x)$, $n \le N$.

The spectral Galerkin method is to find

$$u_N = (1-x^2)^{\alpha/2}\tilde{u}_N, \quad \text{where } \tilde{u}_N = \sum_{j=0}^{N} a_j P_j^{\alpha/2,\alpha/2}(x) \in \mathbb{P}_N$$

such that for all $v \in \mathbb{P}_N$,

$$(\mathcal{L}_{1/2}^{\alpha} u_N + \mu_1 D u_N + \mu_2 u_N, v\omega^{\alpha/2}) = (f, v\omega^{\alpha/2}). \tag{3.3}$$

From Proposition 3.1 with $\theta = \frac{1}{2}$, we have

$$\mathcal{L}_{1/2}^{\alpha}[(1-x^2)^{\alpha/2} P_n^{\alpha/2,\alpha/2}(x)] = \lambda_{\frac{1}{2},n}^{\alpha} P_n^{\alpha/2,\alpha/2}(x), \quad \lambda_{\frac{1}{2},n}^{\alpha} = -\cos(\pi\alpha/2)\frac{\Gamma(\alpha+n+1)}{n!}.$$

Then taking $v = P_k^{\alpha/2,\alpha/2}(x)$, where $0 \le k \le N$, we obtain from the orthogonality of Jacobi polynomials that for $k = 0, 1, 2, \ldots, N$,

$$a_k \lambda_{\frac{1}{2},n}^{\alpha} h_k^{\alpha/2} + \mu_1 \sum_{n=0}^{N} a_n A_{n,k} + \mu_2 \sum_{n=0}^{N} a_n M_{n,k} = (f, P_k^{\alpha/2,\alpha/2}(1-x^2)^{\alpha/2}), \tag{3.4}$$

where $M_{n,k} = (P_n^{\alpha/2,\alpha/2}, P_k^{\alpha/2,\alpha/2}(1-x^2)^{\alpha/2})$ and $A_{n,k} = (\partial_x P_n^{\alpha/2,\alpha/2}, P_k^{\alpha/2,\alpha/2}(1-x^2)^{\alpha/2})$. All these integrals can be evaluated using appropriate Gauss–Jacobi quadrature rules.

Let us look at two *numerical examples*. In these examples, we take a smooth function f in the first example and a non-smooth function f. In simulations, we use reference solutions $u_{\text{ref}} =: u_{512}$, which are computed with (3.3). We take $\theta = 1/2$ and measure the error by $E(N) = \|u_{\text{ref}} - u_N\|_{\omega^{-\alpha/2}}$, $\omega^{-\alpha/2}(x) = (1-x^2)^{-\alpha/2}$.

Example 3.3. Consider $f = \sin x$. Here f is analytical and belongs to $B^{\infty}_{\omega^{\alpha/2}}$.

From Theorem 3.2, we expect convergence orders of the spectral Galerkin method (3.3) to be $5\alpha/2 - 1 - \epsilon$. In Table 1 we observe that the convergence orders of the spectral Galerkin method are about $5\alpha/2 - 1$ for the fractional advection–diffusion problem when the order $\alpha = 1.4, 1.6, 1.8$.

Example 3.4. Take $f = |\sin x|$. The function f has a weak singularity at $x = 0$ and $f \in B^{1.5-\epsilon}_{\omega^{\alpha/2-1}}$ for any $\epsilon > 0$.

By Theorem 3.2, $\omega^{\alpha/2} u \in B^{\alpha+(3\alpha/2-1)\wedge 1.5-\epsilon}_{\omega^{\alpha/2}}$ for (3.1). The convergence orders for the spectral Galerkin method (3.3) are expected to be $\alpha + (3\alpha/2 - 1) \wedge 1.5 - \epsilon$. Indeed, we observe in Table 2 that the convergence order for the spectral Galerkin method (3.3) is $\alpha + (3\alpha/2 - 1) \wedge 1.5 - \epsilon$ when the order $\alpha = 1.4, 1.6, 1.8$. When $\alpha = 1.2$, the observed convergence order is higher than the regularity index from Theorem 3.2 but is less than $\alpha + (3\alpha/2 - 1) \wedge 1.5 - \epsilon$. The convergence orders for $\alpha = 1.2$ suggest possible improvement in regularity when $\alpha < 4/3$.

Table 1: Convergence orders and errors of the spectral Galerkin method (3.3) for Example 3.3 with $\mu_1 = \mu_2 = 1$ and $f = \sin x$.

	$\alpha = 1.2$		$\alpha = 1.4$		$\alpha = 1.6$		$\alpha = 1.8$	
N	$E(N)$	rate	$E(N)$	rate	$E(N)$	rate	$E(N)$	rate
16	7.98e−03		3.51e−04		2.58e−05		2.25e−06	
32	2.95e−03	1.44	6.83e−05	2.36	3.78e−06	2.77	2.47e−07	3.18
64	9.95e−04	1.57	1.27e−05	2.43	5.17e−07	2.87	2.46e−08	3.33
128	3.10e−04	1.68	2.30e−06	2.46	6.78e−08	2.93	2.32e−09	3.41
Order	(theory)	1.40		2.50		3.00		3.50

Table 2: Convergence orders and errors of the spectral Galerkin method (3.3) for Example 3.4 with $\mu_1 = \mu_2 = 1$ and $f = |\sin x|$.

	$\alpha = 1.2$		$\alpha = 1.4$		$\alpha = 1.6$		$\alpha = 1.8$	
N	$E(N)$	rate	$E(N)$	rate	$E(N)$	rate	$E(N)$	rate
16	8.55e−02		2.13e−03		1.86e−04		6.32e−05	
32	3.12e−02	1.45	4.05e−04	2.39	2.61e−05	2.84	7.95e−06	2.99
64	1.05e−02	1.58	7.39e−05	2.46	3.38e−06	2.95	9.03e−07	3.14
128	3.25e−03	1.69	1.32e−05	2.48	4.21e−07	3.00	9.71e−08	3.22
Order	(theory)	1.60		2.50		3.00		3.30

In conclusion, the convergence orders of the spectral methods in these two examples are at most $\alpha + (3\alpha/2 - 1) \wedge r$ no matter how large r is. From the second example, we observe the convergence of spectral methods is even more restricted due to the limited regularity in the interior interval. A standard treatment is applying multi-element spectral methods while capturing the region of limited regularity adaptively.

3.3 A spectral element method

For equation (3.1) with $\theta = 1$, a hybrid SEM for fractional boundary value problems has been developed in [76]. Here we present a C^0-continuous Petrov–Galerkin SEM for elliptic problems of the form from [47] when $\theta = 1$ in (3.1). We have

$$ {}_{-1}^{\mathrm{RL}}\mathcal{D}_x^{\alpha} u(x) - \lambda u(x) = f(x), \quad \alpha = 1 + \mu \in (1, 2]. \tag{3.5}$$

Integration by parts results in the weak form $a(u, v) = l(v)$, where

$$ a(u, v) = \left(\frac{du}{dx}, {}_{x}^{\mathrm{RL}}\mathcal{D}_1^{\mu} v \right)_{\Omega} - \lambda (u, v)_{\Omega}, \tag{3.6}$$

and $l(v) = (f, v)_{\Omega}$, where $(\cdot, \cdot)_{\Omega}$ denotes the usual L^2 inner product.

We partition the domain into N_{el} non-overlapping elements $\Omega_e = [x_{e-1}, x_e]$ and $\Omega = \bigcup_{e=1}^{N_{el}} \Omega_e$. The bilinear form (3.6) can be approximated as

$$a(u,v) \approx a(u^\delta, v^\delta) = \sum_{e=1}^{N_{el}} \left(\frac{du_N^{(e)}}{dx}, {}^{\mathrm{RL}}_{\ x}\mathcal{D}_1^{\mu} v^\delta \right)_{\Omega_e} - \lambda \sum_{e=1}^{N_{el}} (u_N^{(e)}, v^\delta)_{\Omega_e}, \tag{3.7}$$

where $u_N^{(e)}(x) = \sum_{p=0}^{P} \hat{u}_p^{(e)} \psi_p(x)$, $x \in \Omega_e$, using the basis functions $\psi_p(x)$ in the element. Thus, the approximated solution is $u \approx u^\delta(x) = \sum_{e=1}^{N_{el}} \sum_{p=0}^{P} \hat{u}_p^{(e)} \psi_p(x)$. The modal bases $\psi_p(x)$ are defined in the standard (reference) domain $[-1, 1]$ as

$$\psi_p(x(\zeta)) = \begin{cases} \frac{1-\zeta}{2}, & p = 0, \\ (\frac{1-\zeta}{2})(\frac{1+\zeta}{2}) P_{p-1}^{1,1}(\zeta), & p = 1, 2, \dots, P-1, \\ \frac{1+\zeta}{2}, & p = P. \end{cases} \tag{3.8}$$

The choice of basis functions is the same as in standard SEMs for integer-order partial differential equations (see, e. g., [43]). Moreover, we choose the test functions v^δ defined *locally* for $\varepsilon = 1, 2, \dots, N_{el}$ as follows:

$$v_k^{\mathrm{local}}(x) = v_k^{\varepsilon}(x) = \begin{cases} {}^{(2)}\mathcal{P}_{k+1}^{\mu}(x^\varepsilon), & \forall x \in \Omega_\varepsilon, \\ 0, & \text{otherwise}, \end{cases} \qquad k = 0, 1, \dots, P,$$

in which ${}^{(2)}\mathcal{P}_{k+1}^{\mu}(x^\varepsilon)$ represents the Jacobi *poly-fractonomial* of second kind, defined in the corresponding intervals $\Omega_\varepsilon = [x_{\varepsilon-1}, x_\varepsilon]$. Next, the bilinear form (3.7) can be written as

$$\sum_{e=1}^{\varepsilon-1} \sum_{p=0}^{P} \hat{u}_p^{(e)} \hat{\mathbf{S}}_{kp}^{(e,\varepsilon)} + \sum_{p=0}^{P} \hat{u}_p^{(\varepsilon)} [\mathbf{S}_{kp}^{(\varepsilon)} - \lambda \mathbf{M}_{kp}^{(\varepsilon)}] = \mathbf{f}_k^{(\varepsilon)}, \qquad \begin{cases} \varepsilon = 1, 2, \dots, N_{el}, \\ k = 0, 1, \dots, P, \end{cases}$$

in which

$$\hat{\mathbf{S}}_{kp}^{(e,\varepsilon)} = \left(\frac{d\psi_p}{dx}, H_k^{(\varepsilon)}(x) \right)_{\Omega_e}, \qquad e = 1, 2, \dots, \varepsilon - 1,$$

$$\mathbf{S}_{kp}^{(\varepsilon)} = \left(\frac{d\psi_p}{dx}, {}^{\mathrm{RL}}_{\ x}\mathcal{D}_{x_\varepsilon}^{\mu} [{}^{(2)}\mathcal{P}_{k+1}^{\mu}(x)] \right)_{\Omega_\varepsilon},$$

$$\mathbf{M}_{kp}^{(\varepsilon)} = (\psi_p(x), {}^{(2)}\mathcal{P}_{k+1}^{\mu}(x))_{\Omega_\varepsilon}, \qquad \mathbf{f}_k^{(\varepsilon)} = (f, {}^{(2)}\mathcal{P}_{k+1}^{\mu}(x))_{\Omega_\varepsilon}$$

are the history, local stiffness, local mass matrices, and local force vector, respectively. The implementation of this scheme is similar to that in finite/spectral element methods; see [47] for details, where the authors address the main challenge in developing *hp*-methods for FPDEs, the computation of non-local contributions and global assembly of linear systems.

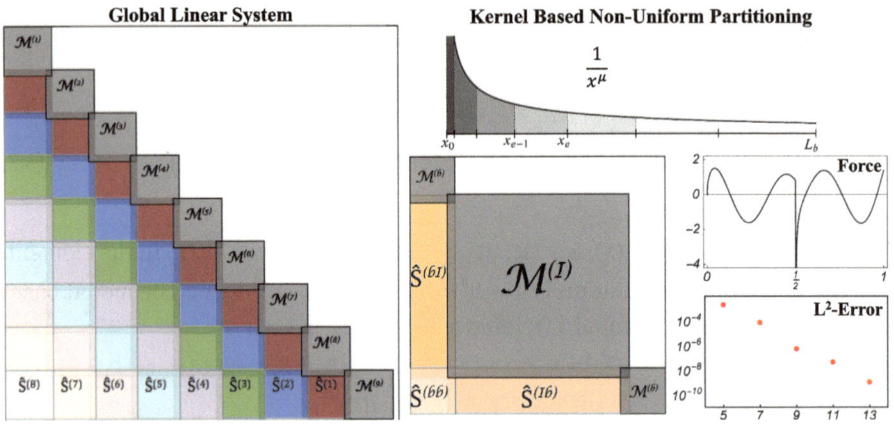

Figure 1: Spectral element method: (Left) Non-local assembly of local $\mathcal{M}^{(i)}$ and $\mathcal{O}(N_{el})$ history matrices; similarly colored blocks represent the same history matrix $\hat{S}^{(i)}$. (Top-middle) Kernel-based grid generation near the boundary. (Right) Capturing singularities in the interior domain.

In Figure 1, the structure of the resulting linear system is illustrated. The size of elements is governed by the power-law distribution, mimicking the underlying kernel for singularity treatment (Figure 1 (top-middle)). A fast decay in accuracy is presented.

The boundary singularity can also be dealt with using a geometric mesh. In [62], implementations of SEMs on a geometric mesh are discussed for (3.1) with $\mu_1 = 0$ and $\mu_2 > 0$. Error estimates and numerical examples are also provided in [62].

4 Fractional Laplacian in high dimensions: two definitions

As mentioned at the beginning of this chapter, the different definitions of the fractional Laplacian in bounded domains are not necessarily equivalent, and different numerical approaches are required for each fractional Laplacian. In this section, we focus on spectral element approaches that have been developed for the *spectral* and *integral* formulations of the fractional Laplacian. We will also present regularity properties of these operators to aid practitioners developing new (or modifying existing) SEMs that discretize the fractional Laplacian. We will also describe the implementation and present the solution of a numerical example for the spectral fractional diffusion equation in two dimensions.

4.1 Spectral fractional Laplacian

The *spectral* definition of the fractional Laplacian, denoted $(-\Delta)^\alpha$ with $0 < \alpha < 1$, can be understood as the fractional spectral power of the integer-order Laplacian. In order to write this definition, we must use the spectrum of the integer Laplacian $-\Delta$, denoted $\sigma(-\Delta)$, which in $\mathbb{R}^d$ is $|\xi|^2$ with $\xi \in \mathbb{R}^d$. The corresponding eigenfunctions are $e^{-i\xi\cdot x}$. Next, we define the projection-valued measure

$$dE = \frac{1}{(2\pi)^d}(\cdot, e^{-i\xi\cdot x})_{L^2} e^{i\xi\cdot x} d\xi, \tag{4.1}$$

where $(u, v)_{L^2} = \int_{\mathbb{R}^d} uv dx$. Finally, the spectral fractional Laplacian in $\mathbb{R}^d$ is written

$$(-\Delta)^\alpha u(x) = \int |\xi|^\alpha u(dE) = \frac{1}{(2\pi)^d} \int |\xi|^\alpha (u, e^{-i\xi\cdot x})_{L^2} e^{i\xi\cdot x} d\xi, \tag{4.2}$$

which is equal to $\mathcal{F}^{-1}\{|\xi|^\alpha \mathcal{F}\{u\}(\xi)\}(x)$, where $\mathcal{F}$ and $\mathcal{F}^{-1}$ denote the Fourier and inverse Fourier transforms, respectively. This representation and the necessity of the projection-valued measure are due to the spectral theorem (see [70], p. 263 or [71], p. 368) and are explained in detail in [54].

In a bounded domain Ω, the spectrum $\sigma(-\Delta)$ is discrete and depends on the domain Ω and the boundary condition of the associated eigenproblem. For the operator $-\Delta$, we represent the eigenvalues as $\{\lambda_k\}_{k=1}^\infty$ and the eigenfunctions as $\{\phi_k\}_{k=1}^\infty$. Specifically, the spectrum of the Laplacian is defined by the problem

$$\begin{aligned} -\Delta\phi_k &= \lambda_k \phi_k, \quad x \in \Omega, \\ \mathcal{B}\phi_k &= g(x), \quad x \in \partial\Omega. \end{aligned} \tag{4.3}$$

Here, the boundary operator $\mathcal{B}$ represents the Dirichlet boundary operator $\mathcal{B}\phi_k = \phi_k$ (although, in principle, any appropriate boundary condition may be used, which will lead to another version of the spectral fractional Laplacian). Then, the series representation for the Dirichlet spectral fractional Laplacian is

$$(-\Delta_{\Omega,g})^\alpha u = \sum_{k=1}^\infty \left(\lambda_k^\alpha \int_\Omega u\phi_k + \lambda_k^{\alpha-1} \int_{\partial\Omega} g \frac{\partial \phi_k}{\partial n} \right) \phi_k. \tag{4.4}$$

We adopt the notation $(-\Delta_{\Omega,g})^\alpha$ of [8] to emphasize the dependence of the definition on the domain Ω and the boundary condition g. It is important to note here that under suitable regularity conditions, we can find functions w and v so that $u = w + v$, where v is chosen to satisfy $-\Delta v = 0$ in the weak sense and $v|_{\partial\Omega} = g$, and w is some function that admits zero Dirichlet boundary conditions. Then the inhomogeneous spectral fractional Laplacian can be written as

$$(-\Delta_{\Omega,g})^\alpha u = (-\Delta_{\Omega,0})^\alpha [u - v]. \tag{4.5}$$

We refer to the equivalence (4.5) as *lifting* to the homogeneous spectral fractional Laplacian. Non-harmonic lifting functions may also be used (in place of the harmonic function v), as discussed in [54].

Another way to represent the spectral fractional Laplacian is through the *heat semi-group* formulation [79, 24]. In [24], the symbol $\mathcal{L}^{\alpha}_{\Omega,g}u$ was used to represent the spectral fractional Laplacian as

$$\mathcal{L}^{\alpha}_{\Omega,g}u = -\frac{1}{\Gamma(-\alpha)}\int_0^{\infty}(e^{t\Delta_{\Omega,g}}u(x) - u(x))\frac{dt}{t^{1+2\alpha}}, \tag{4.6}$$

where $e^{t\Delta}$ is the propagator of the heat equation, i. e., the function $w(x,t) = e^{t\Delta}u(x)$ is the solution of the integer-order heat equation

$$\begin{aligned}\partial_t w(x,t) - \Delta w(x,t) &= 0, \quad x \in \Omega \times [0,+\infty),\\ w(x,t=0) &= u(x), \quad x \in \Omega,\\ w(x,t) &= g(x), \quad x \in \partial\Omega \times [0,+\infty),\end{aligned} \tag{4.7}$$

where $g(x)$ defines the Dirichlet boundary values of $u(x)$. We refer to $e^{t\Delta}$ as the heat semi-group. Notice that this operator is defined for Dirichlet boundary conditions. In order to modify this definition for other types of boundary conditions, one simply needs to modify the boundary condition of equation (4.7). All numerical methods developed for the homogeneous spectral fractional Laplacian (i. e., where zero Dirichlet boundary values are prescribed) may be applied directly to the inhomogeneous case, because

$$\mathcal{L}^{\alpha}_{\Omega,g}u = (-\Delta_{\Omega,0})^{\alpha}[u - v](x), \tag{4.8}$$

where $v|_{\partial\Omega} = g$, as shown in [54, 24].

4.2 Sobolev regularity for the spectral fractional Poisson equation

As we develop spectral elements for these operators, it is important to know what regularity we can expect, i. e., in which Sobolev spaces will the solution $u(x)$ reside. In this section, we focus on the Sobolev regularity for the fractional Poisson equation

$$\begin{aligned}(-\Delta)^{\alpha}u(x) &= f(x), \quad x \in \Omega,\\ u(x) &= 0, \quad x \in \partial\Omega \text{ or } \mathbb{R}^d \setminus \Omega,\end{aligned} \tag{4.9}$$

where the boundary condition is assigned for spectral fractional Laplacian and the exterior boundary condition is for the so-called integral fractional Laplacian, introduced in Section 4.5. We summarize below the regularity of the solution to (4.9). See Table 3.

Table 3: Sobolev regularity results for the homogeneous Dirichlet spectral fractional Poisson equation (4.9).

Regularity: Dirichlet spectral fractional Poisson equation [35]	
$s < 1/2$	$f \in H^s(\Omega)$ implies $u \in \mathbb{H}^{s+2\alpha}(\Omega)$
$s = 1/2$	$f \in H^s_{00}(\Omega)$ implies $u \in \mathbb{H}^{s+2\alpha}(\Omega)$
$1/2 < s < 3/2$	$f \in H^s(\Omega)$ implies $u \in \mathbb{H}^{1/2-\varepsilon+2\alpha}(\Omega)$
$1/2 < s < 3/2$	$f \in H^s(\Omega)$ and $f \equiv 0$ on $\partial\Omega$ implies $u \in \mathbb{H}^{s+2\alpha}(\Omega)$

Recall the Lions–Magenes Sobolev space $H^{1/2}_{00}(\Omega)$ [53], defined by

$$H^{1/2}_{00}(\Omega) := \left\{ u \in H^{1/2}(\Omega) : \int_\Omega \frac{u^2(x)}{\operatorname{dist}(x, \partial\Omega)} dx < \infty \right\}. \tag{4.10}$$

As noted in [8], the fractional Sobolev space $\mathbb{H}^s(\Omega)$ can be understood in terms of the usual fractional Sobolev spaces $H^s(\Omega)$ and the Lions–Magenes space $H^s_{00}(\Omega)$. We have

$$\mathbb{H}^s(\Omega) = \begin{cases} H^s(\Omega) = H^s_0(\Omega) & \text{if } 0 < s < 1/2, \\ H^{1/2}_{00}(\Omega) & \text{if } s = 1/2, \\ H^s_0(\Omega) & \text{if } 1/2 < s. \end{cases} \tag{4.11}$$

In the integer-order Poisson equation, we expect that for $f \in H^s(\Omega)$, the solution u has regularity $u \in \mathbb{H}^{2+s}(\Omega)$ when the boundary of Ω is smooth. We can see that the analogous fractional property holds in cases 1 and 2, where $s \le 1/2$. When $s > 1/2$, this lifting only holds if $f \equiv 0$ on $\partial\Omega$. If f does not have zero boundary values, the solution $u \in \mathbb{H}^{\alpha+1/2-\varepsilon}(\Omega)$ (for any $\varepsilon > 0$), similar to the one-dimensional case in Section 3. The result implies that the solution does not gain any regularity as s is increased. A similar property holds for the integral fractional Laplacian, as explained in Section 4.5.

4.3 A spectral element method for the spectral fractional diffusion equation

Next, we discuss an SEM of Song et al. [78] that has been applied to definition (4.4) with zero Dirichlet boundary values. We refer to this method as "direct" to contrast this approach with the elliptic extension [67, 33, 4] and the heat semi-group [24] approaches.

In [78], the authors consider the fractional diffusion equation

$$\begin{aligned} \partial_t u &= -\mu(-\Delta)^\alpha u, \quad x \in \Omega \subseteq \mathbb{R}^2, \\ u(x,t)|_{\partial\Omega} &= 0, \\ u(x,0) &= u_0(x), \quad x \in \Omega, \end{aligned} \tag{4.12}$$

where Ω is bounded and $\mu > 0$. The SEM proceeds as in the following steps.

1. The domain Ω is mapped to a rectangular reference domain, which is portioned into rectangular elements $\{\Omega_m\}$, and a grid of Gauss–Lobatto–Legendre points are generated for each Ω_m.
2. The eigenfunctions $\{\phi_i\}$ and eigenvalues $\{\lambda_i\}$ of the Laplacian with zero Dirichlet boundary conditions are approximated using a Legendre SEM applied to the eigenproblem (4.3) within each element Ω_m. This spectral method is described in detail in [78].
3. The computed eigenfunctions are orthonormalized according to a weighted Gram–Schmidt procedure, i. e., the weighted QR algorithm, with weights w defined as the diagonal entries of the mass matrix M from the SEM mentioned in item 2, and we write the normalized eigenfunctions as $\{\tilde{\phi}_i\}$. This step is crucial for the spectral convergence observed in the numerical examples of [78]. Without the weighted Gram–Schmidt procedure, this method may converge slowly or fail to converge in some cases.
4. Using analytical techniques, we obtain the solution of (4.12) as

$$u(x,t) = \sum_{j=0}^{\infty} e^{-\mu\lambda_j^\alpha t} c_j(0)\tilde{\phi}_j, \quad \text{where } c_j(0) = \int_\Omega u_0(x)\phi_j \, dx, \tag{4.13}$$

   and $\{\lambda_j, \phi_j\}$ are the analytical eigenpairs of the Laplacian on Ω.
5. Using the numerical eigenpairs $\{\lambda_i, \tilde{\phi}_i\}$, we get the approximate solution of (4.12) as

$$u_{\mathcal{N}}(x,t) = \sum_{i=1}^{\mathcal{N}} e^{-\mu\lambda_i^\alpha t} c_i(0)\tilde{\phi}_i, \tag{4.14}$$

   where $c_i(0) = (u_0, \tilde{\phi}_i)_{\mathcal{N}}$ and $\mathcal{N}$ is the number of degrees of freedom of the spectral element approximation.

The SEM is flexible for different geometries of the domain Ω. The main contribution to the computational burden of this approach is the approximation of the eigenpairs in each element Ω_m. However, these need to be computed only once and can be re-used at each step of the online phase of the method. At the time of this writing, error and convergence analysis of this method has not been reported.

4.3.1 Numerical example

Consider the following fractional diffusion equation with zero Dirichlet boundary values:

$$\partial_t u(x,y,t) = -(-\Delta)^\alpha u(x,y,t), \quad (x,y,t) \in [-1,1]^2 \times [0,T], \tag{4.15}$$

with the initial condition $u(x,y,0) = (1-x^2)(1-y^2)$ and the fractional order $\alpha = 0.5$. First, we generate a mesh of rectangular elements, plotted in Figure 2. Then, we consider the eigenproblem

$$\begin{aligned} -\Delta u - \lambda u &= 0, \quad x \in \Omega = [-1,1]^2, \\ u|_{\partial\Omega} &= 0. \end{aligned} \tag{4.16}$$

We discretize the domain Ω into $El = 16$ elements $\{\Omega_m\}$, and each element is overlaid with an $N \times N$ grid of Gauss–Lobatto–Legendre (GLL) points $\{\boldsymbol{\xi}_i^m\}_{i=1}^N$, with $N = 60$ for this example. The trial functions $\eta_i(x,y)$ and the test functions $\psi_i(x,y)$, with $i = 1,\dots,N$, are Legendre polynomials of order i, which are used in the Galerkin SEM used to solve the eigenproblem, which is formulated as follows: Let $\mathcal{S}_\mathcal{N}$ be the space of piece-wise polynomials on Ω with order $\mathcal{N}$. Find $u_\mathcal{N} \in \mathcal{S}_\mathcal{N}$ such that

$$(\nabla u_\mathcal{N}, \nabla v)_\mathcal{N} - \lambda(u_\mathcal{N}, v)_\mathcal{N} = 0, \quad \forall v \in \mathcal{S}_\mathcal{N}. \tag{4.17}$$

We choose $v = \psi$ and

$$u_\mathcal{N}(x,y) = \sum_{i=1}^{\mathcal{N}} \hat{\boldsymbol{u}}_i \eta(x,y). \tag{4.18}$$

The entries of the mass matrix $M_\mathcal{N}$ are computed using GLL quadrature to evaluate the inner products (η_j, ψ_k), resulting in a diagonal matrix due to the orthogonality of $\{\eta_i\}$ and $\{\psi_i\}$. Similarly, the stiffness matrix $A_\mathcal{N}$ has entries $(\nabla\eta_j, \nabla\psi_k)$, also computed using GLL quadrature. Then, we have the linear system $A_\mathcal{N} u_\mathcal{N} - \lambda M_\mathcal{N} u_\mathcal{N} = 0$. As $M_\mathcal{N}$ is diagonal, we can write this system as

$$(M_\mathcal{N})^{-1} A_\mathcal{N} u_\mathcal{N} - \lambda u_\mathcal{N} = 0. \tag{4.19}$$

The QR algorithm is then applied to $(M_\mathcal{N})^{-1} A_\mathcal{N}$, which is referred to in [78] as the "weighted" Gram–Schmidt method, resulting in the eigenvalues $\{\lambda_i\}_{i=1}^{\mathcal{N}}$ and the normalized eigenfunctions $\{\tilde{\phi}_i\}$. Finally, the solution $u_\mathcal{N}(x,y,t_k)$ can be realized according to equation (4.14).

After the mesh has been generated, we solve the eigenproblem on each subdomain and assemble the solution according to equation (4.14). The initial condition $u(x,y,0)$ is plotted in Figure 2. In Figure 3, we plot the solution $u(x,y,t)$ for times $t = [0.1, 0.3, 0.5]$. For all details of this approach, see [78].

While convergence analysis of this SEM has not yet been reported, the authors of [78] demonstrated the spectral accuracy of this approach numerically. This spectral accuracy occurs in spite of the fact that the approximation of the higher-frequency eigenpairs decreases in accuracy. For an overview of some approaches that avoid the explicit computation of the eigenvalues and eigenfunctions, see the following discussion on the elliptic extension as well as the previous discussion of the heat semi-group approach.

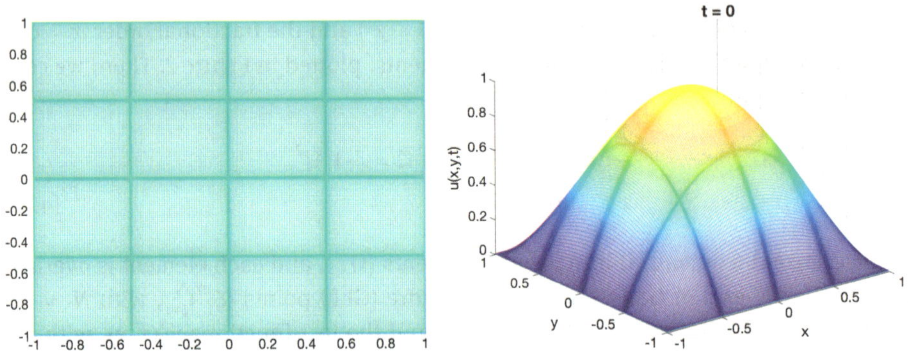

Figure 2: (*Left*) Grid of subdomains and Guass–Lobatto grids generated to solve equation (4.15). (*Right*) Initial condition $u(x, y, 0) = (1 - x^2)(1 - y^2)$.

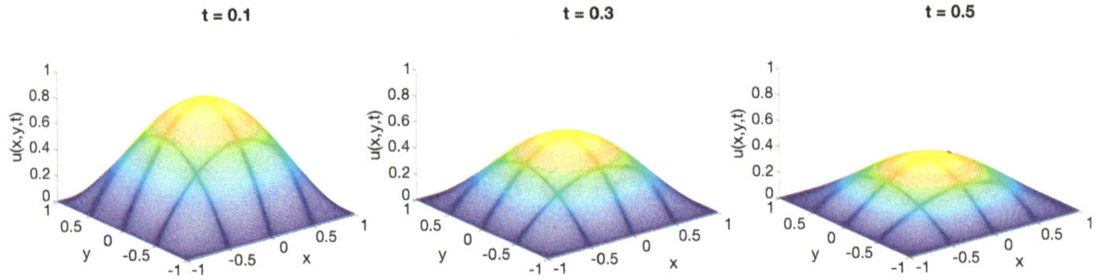

Figure 3: Solutions of fractional diffusion equation with $\alpha = 0.5$ and initial condition $u(x, y, 0) = (1 - x^2)(1 - y^2)$, at times $t = [0.1, 0.3, 0.5]$.

4.4 Elliptic extension for spectral fractional Laplacian

The *elliptic extension method*, also known as the *Dirichlet-to-Neumann* map, re-defines the spectral fractional Laplacian $(-\Delta)^\alpha u(x)$ in $\mathbb{R}^d$ in terms of the solution to a degenerate integer-order elliptic boundary value problem defined on $\mathbb{R}^d \times [0, \infty)$ with boundary values $u(x)$. This approach has been known since at least as early as 1968 [65], when it was used for the study of symmetric stable processes, and it was recently revived in the fractional literature by Caffarelli and Silvestre [15]. The elliptic extension is discussed in detail in other sections of this book, but we briefly reproduce the main result below for our discussion.

The extension method can be written as follows [15]. Given a function $u(x)$ on $\mathbb{R}^d$, consider the extension $\tilde{u}(x, t)$ on $\mathbb{R}^d \times [0, \infty)$ that solves the integer-order equation

$$\begin{aligned} \Delta_x \tilde{u}(x, t) + \frac{1 - \alpha}{t} \partial_t \tilde{u}(x, t) + \partial_t^2 \tilde{u}(x, t) = 0, \\ \tilde{u}(x, 0) = u(x). \end{aligned} \tag{4.20}$$

Then,

$$(-\Delta)^\alpha u(x) = c(d, \alpha) \lim_{t \to 0} \frac{\tilde{u}(x, t) - \tilde{u}(x, 0)}{t^\alpha} \tag{4.21}$$

for some constant c that depends on the dimension d and the fractional order α.

This approach has been derived for the fractional Laplacian in bounded domains in [80]. In this case, the elliptic problem (4.20) is defined on the cylinder over Ω, i. e., $\Omega \times [0, \infty)$, and a trace similar to (4.21) is used. Therefore, to discretize the fractional Laplacian in Ω, one must solve an integer-order equation in the cylinder, which can be done using classical SEMs. Other types of numerical methods have been applied to the elliptic extension, as reported in [67, 4, 14]. Notably, Gatto and Hesthaven [33] developed an *hp*-adaptive finite element method to solve the integer elliptic problem in an appropriately weighted Sobolev space. We emphasize that the elliptic extension approach for fractional diffusion has been used in the development of numerical methods for approximating the spectral definition with *zero Dirichlet* boundary conditions, but no analogous approach for non-zero Dirichlet or any type of Neumann condition using the elliptic extension has been reported.

4.5 Integral fractional Laplacian

In a bounded domain $\Omega \subset \mathbb{R}^d$, to define the *integral* fractional Laplacian, we must first prescribe an *exterior* condition $u(x) = g(x)$ for $x \in \mathbb{R}^d \setminus \Omega$. For such functions $u(x)$, the integral fractional Laplacian is given by

$$\begin{aligned}(-\Delta)^\alpha u(x) &= C(d,\alpha)\,\text{p. v.}\int_{\mathbb{R}^d} \frac{u(x)-u(y)}{|x-y|^{d+\alpha}}\,dy \\ &= C(d,\alpha)\left[\text{p. v.}\int_{\Omega} \frac{u(x)-u(y)}{|x-y|^{d+\alpha}}\,dy + \int_{\mathbb{R}^d\setminus\Omega} \frac{u(x)-g(y)}{|x-y|^{d+\alpha}}\,dy\right], \end{aligned} \tag{4.22}$$

where the coefficient $C(d,\alpha)$ is defined as $C(d,\alpha) = \frac{2^{2\alpha}\Gamma(\alpha+\frac{d}{2})}{\pi^{d/2}|\Gamma(-\alpha)|}$ and "p. v." denotes the *principal value*, which is defined as

$$\text{p. v.}\int_{\mathbb{R}^d} \frac{u(x)-u(y)}{|x-y|^{d+\alpha}}\,dy = \lim_{\varepsilon\to 0}\int_{\mathbb{R}^d\setminus B_\varepsilon} \frac{u(x)-u(y)}{|x-y|^{d+\alpha}}\,dy, \quad B_\varepsilon = \{x \mid |x| < \varepsilon\}. \tag{4.23}$$

This definition is isotropic in x and depends directly on the boundary Ω and exterior condition $g(x)$. As the exterior condition is defined in the Dirichlet sense, we refer to this definition as the *Dirichlet integral fractional Laplacian*. However, other exterior conditions could also be prescribed, which would lead to different operators. Currently, the Neumann condition for the integral fractional Laplacian is the subject of some debate [28, 27, 12], and there is no consensus in the literature for how such a (non-local) Neumann condition should be formulated. Recently, one fractional Neumann operator was proposed in [12] based on mass conservation. Another formulation based on the derivation of a non-local flux was proposed in [28].

At the time of this writing, no SEMs have been developed for the integral definition in multiple dimensions. However, finite element methods are well developed [2, 3, 5], and a discussion of these methods can be found in the chapter on finite element methods in this handbook.

The following theorem describes the Sobolev regularity properties of the integral fractional Poisson equation (4.9).

Theorem 4.1 ([3, 34, 1]). *Let $\partial\Omega \in C^\infty$, $f \in H^s(\Omega)$ for $s \geq -\alpha$, and $u \in \mathbb{H}^\alpha(\Omega)$ be the solution of the fractional Poisson problem* (4.9). *Then*

$$u \in \begin{cases} H^{2\alpha+s}(\Omega) & \text{if } 0 < \alpha + s < 1/2, \\ H^{\alpha+1/2-\varepsilon}(\Omega)\ \forall \varepsilon > 0 & \text{if } 1/2 \leq \alpha + s. \end{cases} \tag{4.24}$$

We remark that while the above theorem is posed for $\partial\Omega \in C^\infty$, this condition can be relaxed to $\partial\Omega \in C^{1,1}$ or $\partial\Omega \in C^2(\Omega)$. We note that if $\alpha + s \geq 1/2$ and the regularity of f (which is represented by s) is increased, the global regularity of u does not necessarily improve. Indeed, increasing s merely improves the interior regularity of u. In [34], the author showed that for a smooth boundary $\partial\Omega$, $f \in H^s(\Omega)$ implies that $u \in H^{\alpha+s}_{\text{loc}}(\Omega)$, instead of $H^{\alpha+s}(\Omega)$ as one might expect.

5 Space-time FPDEs in high dimensions

We consider a $(1 + d)$-dimensional linear reaction–diffusion equation of the form

$$\partial_t^\beta u = \sum_{j=1}^{d} \kappa_j \partial_{x_j}^{\alpha_j} u + \gamma u + f, \quad (t, x) \in \Omega, \tag{5.1}$$

where $\partial_t^\beta \equiv {}_0^C\mathcal{D}_t^\beta$ is the Caputo derivative of order $\beta \in (0, 1]$ and $\partial_{x_j}^{\alpha_j} u \equiv ({}_{a_j}\mathcal{D}_{x_j}^{\alpha_j} + {}_{x_j}\mathcal{D}_{b_j}^{\alpha_j})u$ denotes a variant of a fractional Laplacian with two-sided Riemann–Liouville derivatives of order $\alpha_j \in (1, 2]$ in a $(1+d)$-dimensional *space-time* domain $\Omega = (0, T] \times (a_1, b_1) \times \cdots (a_d, b_d)$, subject to Dirichlet initial and boundary conditions. In bounded domains, the FPDE in (5.1) can be represented by the probability density function of a multi-dimensional *killed non-Markovian* stochastic processes [13].

5.1 The method of lines

As in numerical methods for integer-order equations, we can discretize in time first and then apply spectral and/or spectral element methods for (5.1); see, e. g., [52, 92] for a particular case of (5.1).

Since ${}_0^C\mathcal{D}_t^\beta u = {}_0^{\mathrm{RL}}\mathcal{D}_t^\beta v$ where $v = u - u_0$ is continuous when $0 < \beta < 1$, we can discretize the time-fractional derivative by discretizing the derivative first and then approximating the time-fractional integral. For example, we can use the backward Euler scheme and obtain $\int_0^{t_{n+1}} (t_{n+1} - s)^{-\beta} v(s)\, ds = \int_0^{t_n} (t_n - s)^{-\beta} v(s)\, ds + \Delta t \Gamma(1-\beta)(Lu^{n+1} + f^{n+1})$, where $Lu^{n+1} = \sum_{j=1}^d \kappa_j \partial_{x_j}^{\alpha_j} u^{n+1} + \gamma u^{n+1}$. In this formula, the left-hand side is the only integral to be evaluated since the integral in the right-hand side has been computed in the last time step. Then the integral in the left-hand side is separated into two pieces: $\int_{t_n}^{t_{n+1}} (t_{n+1} - s)^{-\beta} u(s)\, ds + \int_0^{t_n} (t_{n+1} - s)^{-\beta} u(s)\, ds$. The first term is approximated by approximating $u(s)$ with its piece-wise interpolation, i. e.,

$$\int_{t_n}^{t_{n+1}} (t_{n+1} - s)^{-\beta} u(s)\, ds \approx \phi_1 u(t_n) + \phi_2 \frac{u(t_{n+1}) - u(t_n)}{\Delta t}, \tag{5.2}$$

where $\phi_1 = \int_0^{\Delta t} (\Delta t - s)^{-\beta}\, ds$, $\phi_2 = \int_0^{\Delta t} s(\Delta t - s)^{-\beta}\, ds$. Approximating the integral in the left-hand side by (5.2) plus $\int_0^{t_n} (t_{n+1} - s)^{-\beta} u(s)\, ds$ and rewriting the above formula, we have

$$\begin{aligned} \frac{\phi_2}{\Delta t}(u^{n+1} - u^n) = -\phi_1 u^n - \int_0^{t_n} (t_{n+1} - s)^{-\beta} v(s)\, ds + \int_0^{t_n} (t_n - s)^{-\beta} v(s)\, ds \\ + \Delta t \Gamma(1 - \beta)(Lu^{n+1} + f^{n+1}). \end{aligned} \tag{5.3}$$

The time consuming part in (5.3) is the computation of $\int_0^{t_n} (t_{n+1} - s)^{-\beta} v(s)\, ds$. We can apply the fast convolution method from [57] in time to calculate this fractional integral with quasi-linear complexity and logarithm storage requirement. The resulting linear system of (5.3) can be readily solved by deploying any stable and accurate numerical methods, such as the spectral and spectral element methods described in Section 3.

Remark 5.1. It is possible that a solution has weak singularity around $t = 0$. Usually, the solution will quickly become smooth after a short time. If one needs higher-order accuracy, one can apply Lubich's correction method [56] when t is close to zero. The corrections are only needed in several time steps while no corrections are needed for t far away from 0. The details of implementation and practical instructions of the correction method have been presented in [95].

The method of lines also include semi-discretization in space first and then treating the resulting equations as fractional ordinary differential equations. For example, a discretization in space is performed in [26] using a local discontinuous Galerkin method before discretization in time.

5.2 Space-time spectral methods

Another approach to solving (1+d)-dimensional linear reaction–diffusion equations in space-time is to apply spectral methods both in time and space. For example, Legendre spectral methods in both time and space are applied for a space-time equation with one time-fractional derivative in [100] and an equation with multi-term time-fractional derivatives in [99]; see also a space-time spectral method for a time-fractional diffusion equation in [51]. Here we present the space-time spectral method from [84, 73].

For equation (5.1), we consider the following weak formulation:

$$a(u,v) = ({}_0\mathcal{D}_t^{\frac{\beta}{2}} u, {}_t\mathcal{D}_T^{\frac{\beta}{2}} v)_\Omega - \sum_{j=1}^{d} k_j [({}_{a_j}\mathcal{D}_{x_j}^{\frac{\alpha_j}{2}} u, {}_{x_j}\mathcal{D}_{b_j}^{\frac{\alpha_j}{2}} v)_\Omega$$
$$+ ({}_{a_j}\mathcal{D}_{x_j}^{\frac{\alpha_j}{2}} v, {}_{x_j}\mathcal{D}_{b_j}^{\frac{\alpha_j}{2}} u)_\Omega] + \gamma(u,v)_\Omega = (f,v)_\Omega.$$

The Petrov–Galerkin spectral method is to find $u_N \in U_N$ such that $a(u_N, v) = (f,v)_\Omega$, $\forall v \in V_N$, where

$$U_N = \operatorname{span}\left\{(\psi_n^\tau \circ \eta)(t) \prod_{j=1}^{d} (\phi_{m_j} \circ \xi_j)(x_j) : n = 1,\dots,\mathcal{N},\ m_j = 1,\dots,\mathcal{M}_j\right\},$$
$$V_N = \operatorname{span}\left\{(\Psi_r^\tau \circ \eta)(t) \prod_{j=1}^{d} (\phi_{k_j} \circ \xi_j)(x_j) : r = 1,\dots,\mathcal{N},\ k_j = 1,\dots,\mathcal{M}_j\right\}.$$

Here, we let $\beta = 2\tau$, $\alpha_j = 2\nu_j$, $\eta(t) = 2t/T - 1$, and $\xi_j(s) = 2\frac{s-a_j}{b_j-a_j} - 1$, and

$$\phi_m(\xi) = (P_{m+1}(\xi) - P_{m-1}(\xi)), \quad m = 1,2,\dots \text{ and } \xi \in [-1,1],$$

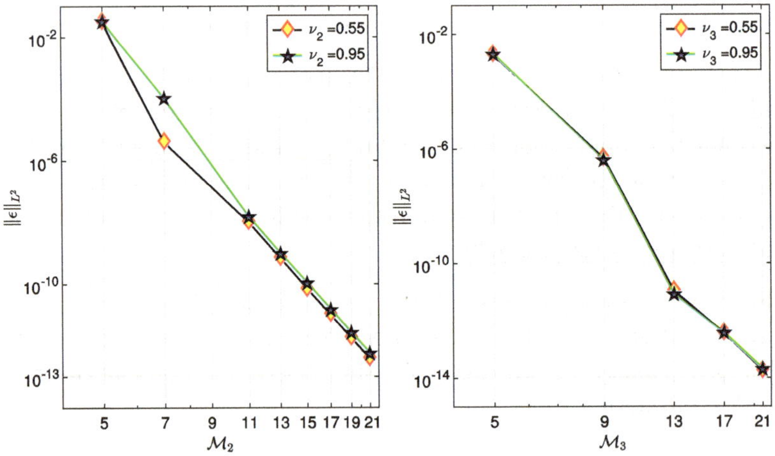

Figure 4: Spatial *p-refinement*: log-log scale L^∞-error versus spatial expansion orders $\mathcal{M}_2$, $\mathcal{M}_3$ for the limit orders $\nu_j = \alpha_j/2$, $j = 2,3$.

$$\psi_n^\tau(\eta) = \sigma_n(1+\eta)^\tau P_{n-1}^{-\tau,\tau}(\eta), \quad n = 1,2,\ldots \text{ and } \eta \in [-1,1],$$
$$\Psi_r^\tau(\eta) = \tilde{\sigma}_r(1-\eta)^\tau P_{r-1}^{\tau,-\tau}(\eta), \quad r = 1,2,\ldots \text{ and } \eta \in [-1,1].$$

It is shown in [84, 73] that this Petrov–Galerkin spectral method is stable. The resulting linear system can be made symmetric and solved with a complexity $\mathcal{O}(N^d)$ in [72].

Now let us consider a numerical example where the solution has weak boundary singularity: $u = t^{p_1} \times \prod_{i=1}^{d=3}((1+x_i)^{p_{2i}} - \epsilon_i(1+x)^{p_{2i+1}})$, where $p_1 = 7\frac{2}{3}$, $p_2 = 6\frac{1}{3}$, $p_3 = 6\frac{2}{7}$, $p_4 = 7\frac{4}{5}$, $p_5 = 7\frac{1}{7}$, $p_6 = 7\frac{3}{5}$, $p_7 = 7\frac{1}{7}$ and $\epsilon_1 = 2^{p_2-p_3}$, $\epsilon_2 = 2^{p_4-p_5}$, $\epsilon_3 = 2^{p_6-p_7}$ in the hypercube domain as $[0,1] \times [-1,1] \times [-1,1] \times [-1,1]$. Higher accuracy is seen in Figure 4 when the polynomial orders are increased in space.

Bibliography

[1] G. Acosta, F. M. Bersetche, and J. P. Borthagaray, A short FE implementation for a 2D homogeneous Dirichlet problem of a fractional Laplacian, *Comput. Math. Appl.*, **74**(4) (Aug. 2017), 784–816.

[2] G. Acosta and J. Borthagaray, A fractional Laplace equation: Regularity of solutions and finite element approximations, *SIAM J. Numer. Anal.*, **55** (2017), 472–495.

[3] M. Ainsworth and C. Glusa, Aspects of an adaptive finite element method for the fractional Laplacian: A priori and a posteriori error estimates, efficient implementation and multigrid solver, *Comput. Methods Appl. Mech. Eng.*, **327** (2017), 4–35.

[4] M. Ainsworth and C. Glusa, Hybrid finite element – spectral method for the fractional Laplacian: Approximation theory and efficient solver, arXiv:1709:01639, Sep. 2017.

[5] M. Ainsworth and C. Glusa, Towards an efficient finite element method for the integral fractional Laplacian on polygonal domains, in *Festschrift for 80th Birthday of Ian Sloan*, 2017.

[6] M. Ainsworth and Z. Mao, Analysis and approximation of a fractional Cahn–Hilliard equation, *SIAM J. Numer. Anal.*, **55**(4) (2017), 1689–1718.

[7] G. Akagi, G. Schimperna, and A. Segatti, Fractional Cahn–Hilliard, Allen–Cahn and porous medium equations, arXiv:1502.06383, 2015.

[8] H. Antil, J. Pfefferer, and S. Rogovs, Fractional operators with inhomogeneous boundary conditions: Analysis, control, and discretization, arXiv:1703.05256, 2017.

[9] R. Askey, *Orthogonal Polynomials and Special Functions*, SIAM, Philadelphia, PA, 1975.

[10] T. M. Atanackovic, M. Budincevic, and S. Pilipovic, On a fractional distributed-order oscillator, *J. Phys. A, Math. Gen.*, **38** (2005), 6703–6713.

[11] I. Babuska and B. Guo, Direct and inverse approximation theorems for the p-version of the finite element method in the framework of weighted Besov spaces, part I: Approximability of functions in the weighted Besov spaces, *SIAM J. Numer. Anal.*, **39**(5) (2001/02), 1512–1538.

[12] B. Baeumer, M. Kovács, M. M. Meerschaert, and H. Sankaranarayanan, Boundary conditions for fractional diffusion, arXiv:1706.07991, 2017.

[13] B. Baeumer, T. Luks, and M. M. Meerschaert, Space-time fractional Dirichlet problems, arXiv:1604.06421, 2016.

[14] A. Bonito, J. P. Borthagaray, R. H. Nochetto, E. Ota-Rola, and A. J. Salgado, Numerical methods for fractional diffusion, arXiv:1707.01566, 2017.

[15] L. Caffarelli and L. Silvestre, An extension problem related to the fractional Laplacian, *Commun. Partial Differ. Equ.*, **32**(8) (2007), 1245–1260.

[16] Y. Cao, T. Herdman, and Y. Xu, A hybrid collocation method for Volterra integral equations with weakly singular kernels, *SIAM J. Numer. Anal.*, **41**(1) (2003), 364–381.
[17] M. Caputo, Distributed order differential equations modelling dielectric induction and diffusion, *Fract. Calc. Appl. Anal.*, **4** (2001), 421–442.
[18] M. Caputo, Diffusion with space memory modelled with distributed order space fractional differential equations, *Ann. Geophys.*, **46** (2003), 223–234.
[19] F. Chen, Q. Xu, and J. S. Hesthaven, A multi-domain spectral method for time-fractional differential equations, *J. Comput. Phys.*, **293**(C) (July 2015), 157–172.
[20] S. Chen, J. Shen, and L. L. Wang, Generalized Jacobi functions and their applications to fractional differential equations, *Math. Comput.*, **85**(300) (2016), 1603–1638.
[21] S. Chen, J. Shen, and L.-L. Wang, Laguerre functions and their applications to tempered fractional differential equations on infinite intervals, *J. Sci. Comput.*, **74**(3) (Mar. 2018), 1286–1313.
[22] W. Chen and S. Holm, Fractional Laplacian time-space models for linear and nonlinear lossy media exhibiting arbitrary frequency power-law dependency, *J. Acoust. Soc. Am.*, **115**(4) (2004), 1424–1430.
[23] P. Constantin and J. Wu, Behavior of solutions of 2D quasi-geostrophic equations, *SIAM J. Math. Anal.*, **30**(5) (1999), 937–948.
[24] N. Cusimano, F. D. Teso, L. Gerardo-Giorda, and G. Pagnini, Discretisations of the spectral fractional Laplacian on general domains with Dirichlet, Neumann, and Robin boundary conditions, arXiv:1708.03602v1, 2017.
[25] O. Defterli, M. D'Elia, Q. Du, M. Gunzburger, R. Lehoucq, and M. M. Meerschaert, Fractional diffusion on bounded domains, *Fract. Calc. Appl. Anal.*, **18**(2) (2015), 342–360.
[26] W. Deng and J. Hesthaven, Local discontinuous Galerkin methods for fractional diffusion equations, *ESAIM: M2AN*, **47**(6) (2013), 1845–1864.
[27] S. Dipierro, X. Ros-Oton, and E. Valdinoci, Nonlocal problems with Neumann boundary conditions, *Rev. Mat. Iberoam.*, **33** (2017), 377–416.
[28] Q. Du, M. Gunzburger, R. Lehoucq, and K. Zhou, Analysis and approximation of nonlocal diffusion problems with volume constraints, *SIAM Rev.*, **54** (2012), 667–696.
[29] V. Ervin, N. Heuer, and J. Roop, Regularity of the solution to 1-D fractional order diffusion equations, *Math. Comput.*, (2018).
[30] S. Esmaeili, M. Shamsi, and Y. Luchko, Numerical solution of fractional differential equations with a collocation method based on Müntz polynomials, *Comput. Math. Appl.*, **62**(3) (2011), 918–929.
[31] L. Fatone and D. Funaro, Optimal collocation nodes for fractional derivative operators, *SIAM J. Sci. Comput.*, **37**(3) (2015), A1504–A1524.
[32] N. J. Ford, M. L. Morgado, and M. Rebelo, Nonpolynomial collocation approximation of solutions to fractional differential equations, *Fract. Calc. Appl. Anal.*, **16**(4) (2013), 874–891.
[33] P. Gatto and J. S. Hesthaven, Numerical approximation of the fractional Laplacian via hp-finite elements, with an application to image denoising, *J. Sci. Comput.*, **65**(1) (Oct. 2015), 249–270.
[34] G. Grubb, Fractional Laplacians on domains, a development of Hörmander's theory of μ-transmission pseudodifferential operators, *Adv. Math.*, **268** (2015), 478–528.
[35] G. Grubb, Regularity of spectral fractional Dirichlet and Neumann problems, *Math. Nachr.*, **289**(7) (2016), 831–844.
[36] Q. Guan and M. Gunzburger, θ schemes for finite element discretization of the space–time fractional diffusion equations, *J. Comput. Appl. Math.*, **288** (2015), 264–273.
[37] B.-y. Guo and L.-l. Wang, Jacobi approximations in non-uniformly Jacobi-weighted Sobolev spaces, *J. Approx. Theory*, **128**(1) (2004), 1–41.

[38] Z. Hao, G. Lin, and Z. Zhang, Regularity and spectral methods for two-sided fractional diffusion equations with a low-order term, arXiv:1705.07209, 2017.

[39] Z. Hao and Z. Zhang, Regularity in weighted Sobolev spaces and spectral methods for fractional advection-diffusion-reaction equations, *preprint*, 2018.

[40] D. Hou and C. Xu, A fractional spectral method with applications to some singular problems, *Adv. Comput. Math.*, **43**(5) (2017), 911–944.

[41] C. Huang, Y. Jiao, L.-L. Wang, and Z. Zhang, Optimal fractional integration preconditioning and error analysis of fractional collocation method using nodal generalized Jacobi functions, *SIAM J. Numer. Anal.*, **54**(6) (2016), 3357–3387.

[42] C. Huang and Z. Zhang, Convergence of a *p*-version/*hp*-version method for fractional differential equations, *J. Comput. Phys.*, **286** (2015), 118–127.

[43] G. E. Karniadakis and S. J. Sherwin, *Spectral/hp Element Methods for CFD*, 2nd edn., Oxford University Press, 2005.

[44] J. F. Kelly and R. J. McGough, Approximate analytical time-domain Green's functions for the Caputo fractional wave equation, *J. Acoust. Soc. Am.*, **140**(2) (2016), 1039–1047.

[45] J. F. Kelly and M. M. Meerschaert, Space-time duality for the fractional advection-dispersion equation, *Water Resour. Res.*, **53**(4) (2017), 3464–3475.

[46] M. M. Khader, T. S. E. Danaf, and A. S. Hendy, Efficient spectral collocation method for solving multi-term fractional differential equations based on the generalized Laguerre polynomials, *J. Fract. Calc. Appl.*, **3**(13) (2012), 1–14.

[47] E. Kharazmi, M. Zayernouri, and G. Karniadakis, A Petrov–Galerkin spectral element method for fractional elliptic problems, *J. Comput. Methods Appl. Mech. Eng.*, **324** (2017), 512–536.

[48] M. Kwaśnicki, Ten equivalent definitions of the fractional Laplace operator, *Fract. Calc. Appl. Anal.*, **20**(1) (Feb 2017), 7–51.

[49] N. Laskin, Fractional quantum mechanics and Lévy path integrals, *Phys. Lett. A*, **268**(4) (2000), 298–305.

[50] C. Li, F. Zeng, and F. Liu, Spectral approximations to the fractional integral and derivative, *Fract. Calc. Appl. Anal.*, **15**(3) (2012), 383–406.

[51] X. Li and C. Xu, A space-time spectral method for the time fractional diffusion equation, *SIAM J. Numer. Anal.*, **47**(3) (2009), 2108–2131.

[52] Y. Lin and C. Xu, Finite difference/spectral approximations for the time-fractional diffusion equation, *J. Comput. Phys.*, **225**(2) (2007), 1533–1552.

[53] J.-L. Lions and E. Magenes, *Problèmes aux limites non homogènes et applications, vol. 2*, Travaux et Recherches Mathématiques, vol. 18, Dunod, Paris, 1968.

[54] A. Lischke, G. Pang, M. Gulian, F. Song, C. Glusa, X. Zheng, Z. Mao, W. Cai, M. M. Meerschaert, M. Ainsworth, and G. E. Karniadakis, What is the fractional Laplacian? arXiv:1801.09767, 2018.

[55] A. Lischke, M. Zayernouri, and G. E. Karniadakis, A Petrov–Galerkin spectral method of linear complexity for fractional multiterm ODEs on the half line, *SIAM J. Sci. Comput.*, **39**(3) (2017), A922–A946.

[56] C. Lubich, Discretized fractional calculus, *SIAM J. Math. Anal.*, **17**(3) (1986), 704–719.

[57] C. Lubich and A. Schädle, Fast convolution for nonreflecting boundary conditions, *SIAM J. Sci. Comput.*, **24**(1) (2002), 161–182.

[58] F. Mainardi and R. Gorenflo, Time-fractional derivatives in relaxation processes: a tutorial survey, arXiv:0801.4914, 2008.

[59] Z. Mao, S. Chen, and J. Shen, Efficient and accurate spectral method using generalized Jacobi functions for solving Riesz fractional differential equations, *Appl. Numer. Math.*, **106** (2016), 165–181.

[60] Z. Mao and G. E. Karniadakis, A spectral method (of exponential convergence) for singular solutions of the diffusion equation with general two-sided fractional derivative, *SIAM J. Numer. Anal.*, **56**(1) (2018), 24–49.
[61] Z. Mao and J. Shen, Efficient spectral–Galerkin methods for fractional partial differential equations with variable coefficients, *J. Comput. Phys.*, **307** (2016), 243–261.
[62] Z. Mao and J. Shen, Spectral element method with geometric mesh for two-sided fractional differential equations, *Adv. Comput. Math.*, (2017), 1–27.
[63] S. Mashayekhi, P. Miles, M. Y. Hussaini, and W. S. Oates, Fractional viscoelasticity in fractal and non-fractal media: Theory, experimental validation, and uncertainty analysis, *J. Mech. Phys. Solids*, **111** (2018), 134–156.
[64] R. Metzler and J. Klafter, The restaurant at the end of the random walk: recent developments in the description of anomalous transport by fractional dynamics, *J. Phys. A, Math. Gen.*, **37** (2004), R161–R208.
[65] S. Molchanov and E. Ostrovskii, Symmetric stable processes as traces of degenerate diffusion processes, *Theory Probab. Appl.*, **14**(1) (1968), 128–131.
[66] M. Naghibolhosseini, *Estimation of Outer-Middle Ear Transmission Using DPOAEs and Fractional-Order Modeling of Human Middle Ear*, Ph. D. thesis, City University of New York, NY, 2015.
[67] R. H. Nochetto, E. Otàrola, and A. J. Salgado, A PDE approach to fractional diffusion in general domains: A priori error analysis, *Found. Comput. Math.*, **15**(3) (2015), 733–791.
[68] I. Podlubny, *Fractional Differential Equations*, Academic Press, Inc., San Diego, CA, 1999.
[69] C. Pozrikidis, *The Fractional Laplacian*, CRC Press, Boca Raton, FL, 2016.
[70] M. Reed and B. Simon, *Methods of Modern Mathematical Physics I: Functional Analysis*, Revised and enlarged edition, Academic Press, 1980.
[71] W. Rudin, *Functional Analysis*, 2nd edn., McGraw-Hill, 1991.
[72] M. Samiee, M. Zayernouri, and M. M Meerschaert, A unified spectral method for FPDEs with two-sided derivatives; part I: A fast solver, *J. Comput. Phys.*, (2018), https://doi.org/10.1016/j.jcp.2018.02.014.
[73] M. Samiee, M. Zayernouri, and M. M. Meerschaert, A unified spectral method for FPDEs with two-sided derivatives; part II: Stability, and error analysis, *J. Comput. Phys.*, (2018), https://doi.org/10.1016/j.jcp.2018.07.041.
[74] S. G. Samko, A. A. Kilbas, and O. I. Marichev, *Fractional Integrals and Derivatives: Theory and Applications*, Gordon and Breach Science Publishers, Yverdon, 1993.
[75] A. Schaedle, M. Lopez-Fernandez, and C. Lubich, Fast and oblivious convolution quadrature, *SIAM J. Sci. Comput.*, **28**(2) (2006), 421–438.
[76] C. Sheng and J. Shen, A hybrid spectral element method for fractional two-point boundary value problems, *Numer. Math., Theory Methods Appl.*, **10**(2) (2017), 437–464.
[77] I. Sokolov, A. Chechkin, and J. Klafter, Distributed order fractional kinetics, *Acta Phys. Pol. B*, **35** (2004), 1323–1341.
[78] F. Song, C. Xu, and G. E. Karniadakis, Computing fractional Laplacians on complex-geometry domains: Algorithms and simulations, *SIAM J. Sci. Comput.*, **39**(4) (2017), A1320–A1344.
[79] P. Stinga, *Fractional Powers of Second Order Partial Differential Operators: Extension Problem and Regularity Theory*, Ph. D. thesis, Universidad Autonoma de Madrid, 2010.
[80] P. R. Stinga and J. L. Torrea, Extension problem and Harnack's inequality for some fractional operators, *Commun. Partial Differ. Equ.*, **35**(11) (2010), 2092–2122.
[81] B. E. Treeby and B. T. Cox, Modeling power law absorption and dispersion for acoustic propagation using the fractional Laplacian, *J. Acoust. Soc. Am.*, **127**(5) (2010), 2741–2748.

[82] H. Wang and X. Zhang, A high-accuracy preserving spectral Galerkin method for the Dirichlet boundary-value problem of variable-coefficient conservative fractional diffusion equations, *J. Comput. Phys.*, **281** (2015), 67–81.
[83] M. Yamamoto, Asymptotic expansion of solutions to the dissipative equation with fractional Laplacian, *SIAM J. Math. Anal.*, **44**(6) (2012), 3786–3805.
[84] M. Zayernouri, M. Ainsworth, and G. Em Karniadakis, A unified Petrov–Galerkin spectral method for fractional PDEs, *Comput. Methods Appl. Mech. Eng.*, **283** (2015), 1545–1569.
[85] M. Zayernouri, M. Ainsworth, and G. E. Karniadakis, Tempered fractional Sturm–Liouville eigenproblems, *SIAM J. Sci. Comput.*, **37**(4) (2015), A1777–A1800.
[86] M. Zayernouri, W. Cao, Z. Zhang, and G. E. Karniadakis, Spectral and discontinuous spectral element methods for fractional delay equations, *SIAM J. Sci. Comput.*, **36**(6) (2014), B904–B929.
[87] M. Zayernouri and G. E Karniadakis, Exponentially accurate spectral and spectral element methods for fractional odes, *J. Comput. Phys.*, **257** (2014), 460–480.
[88] M. Zayernouri and G. Karniadakis, Discontinuous spectral element methods for time- and space-fractional advection equations, *SIAM J. Sci. Comput.*, **36**(4) (2014), B684–B707.
[89] M. Zayernouri and G. E. Karniadakis, Fractional Sturm–Liouville eigen-problems: Theory and numerical approximation, *J. Comput. Phys.*, (2013), 495–517.
[90] M. Zayernouri and G. E. Karniadakis, Exponentially accurate spectral and spectral element methods for fractional ODEs, *J. Comput. Phys.*, **257** (2014), 460–480.
[91] M. Zayernouri and G. E. Karniadakis, Fractional spectral collocation method, *SIAM J. Sci. Comput.*, **36**(1) (2014), A40–A62.
[92] F. Zeng, F. Liu, C. Li, K. Burrage, I. Turner, and V. Anh, A Crank–Nicolson ADI spectral method for a two-dimensional Riesz space fractional nonlinear reaction-diffusion equation, *SIAM J. Numer. Anal.*, **52**(6) (2014), 2599–2622.
[93] F. Zeng, Z. Mao, and G. E. Karniadakis, A generalized spectral collocation method with tunable accuracy for fractional differential equations with end-point singularities, *SIAM J. Sci. Comput.*, **39**(1) (2017), A360–A383.
[94] F. Zeng, Z. Zhang, and G. E. Karniadakis, A generalized spectral collocation method with tunable accuracy for variable-order fractional differential equations, *SIAM J. Sci. Comput.*, **37**(6) (2015), A2710–A2732.
[95] F. Zeng, Z. Zhang, and G. E. Karniadakis, Second-order numerical methods for multi-term fractional differential equations: Smooth and non-smooth solutions, *Comput. Methods Appl. Mech. Eng.*, **327**(1) (2017), 478–502.
[96] Z. Zhang, F. Zeng, and G. E. Karniadakis, Optimal error estimates of spectral Petrov–Galerkin and collocation methods for initial value problems of fractional differential equations, *SIAM J. Numer. Anal.*, **53**(4) (2015), 2074–2096.
[97] L. Zhao, W. Deng, and J. S. Hesthaven, Spectral methods for tempered fractional differential equations, arXiv:1603.06511, 2016.
[98] X. Zhao and Z. Zhang, Superconvergence points of fractional spectral interpolation, *SIAM J. Sci. Comput.*, **38**(1) (2016), A598–A613.
[99] M. Zheng, F. Liu, V. Anh, and I. Turner, A high-order spectral method for the multi-term time-fractional diffusion equations, *Appl. Math. Model.*, **40**(7–8) (2016), 4970–4985.
[100] M. Zheng, F. Liu, I. Turner, and V. Anh, A novel high order space-time spectral method for the time fractional Fokker–Planck equation, *SIAM J. Sci. Comput.*, **37**(2) (2015), A701–A724.

Weihua Deng and Xudong Wang

Discontinuous Galerkin and finite element methods

Abstract: This chapter is a short essay on discontinuous Galerkin methods for fractional (convection–) diffusion equations in one and two dimensions. The method is based on the local discontinuous Galerkin methods for the classical parabolic equation, i. e., decomposing the high-order derivative and rewriting the equation into a first-order system. Depending on the properties of fractional operators, we decompose it into several first-order derivatives and one fractional integral. Then we propose the corresponding numerical schemes and discuss their stability and convergence. Some algorithms in two dimensions are provided.

Keywords: Discontinuous Galerkin methods, fractional calculus, fractional diffusion equation, algorithm

MSC 2010: 26A33 , 35R11 , 65M60, 65M12

1 Introduction

In 1973, Reed and Hill [15] introduced the first discontinuous Galerkin (DG) method for hyperbolic equations, and since that time there has been an active development of DG methods for hyperbolic and nearly hyperbolic problems, resulting in a variety of different schemes. Especially, DG methods for the numerical treatment of non-linear hyperbolic systems experienced a vigorous development due to a strong interaction with the ideas of finite volume methods for hyperbolic problems [6]. But the development of the DG methods did not stop there. Some authors extended the DG methods to elliptic problems with a convection-dominated part. In 1997, Bassi and Rebay [1] introduced a DG method for the Navier–Stokes equations and in 1998, Cockburn and Shu [9] introduced the so-called local discontinuous Galerkin (LDG) methods by generalizing the original DG method of Bassi and Rebay. This scheme was in turn an extension of the Runge–Kutta discontinuous Galerkin (RKDG) method developed by Cockburn and Shu [7, 8] for non-linear hyperbolic systems.

Acknowledgement: This work was partially supported by the National Natural Science Foundation of China under Grant No. 11671182 and the Fundamental Research Funds for the Central Universities under Grant No. lzujbky-2018-ot03.

Weihua Deng, Xudong Wang, School of Mathematics and Statistics, Gansu Key Laboratory of Applied Mathematics and Complex Systems, Lanzhou University, Lanzhou 730000, P. R. China, e-mails: dengwh@lzu.edu.cn, xdwang14@lzu.edu.cn

https://doi.org/10.1515/9783110571684-007

A careless application of the DG method to a problem with high-order derivatives could yield an inconsistent scheme [20]. The idea of LDG methods [9] for time-dependent partial differential equations (PDEs) with high-order derivatives is to rewrite the equation into a first-order system and then apply the DG method to the system. A key ingredient for the success of these methods is the reasonable design of interface numerical fluxes. These fluxes must be designed to guarantee stability and local solvability of all the auxiliary variables introduced to approximate the derivatives of the solution.

Here we review the LDG methods extended to solve problems containing spatial fractional derivatives (FDs). It can be noted that almost all the advantages or characteristics [12] of DG methods when used to solve classical PDEs carry over to fractional PDEs, i. e., the numerical solution does not have to satisfy any inter-element continuity constraint which renders it ideal for *hp*-adaptivity; the mass matrix is local and easily invertible, leading to an explicit formulation for time-dependent problems. We shall also discuss that the choice of the numerical fluxes is essential to ensure the accuracy and stability of the LDG schemes. Taking the classical heat equation as an example, in this section, we introduce the basic idea of the LDG method. In Section 2, we review the properties of fractional calculus (FC) and the appropriate functional setting. The LDG schemes and their stability when used to solve the fractional diffusion equations in one and two dimensions are provided in Section 3 and Section 4, respectively. In Section 5, we deal with the fractional convection–diffusion equations with two kinds of schemes and Section 6 concludes the chapter with some remarks.

We show the idea of the LDG scheme by solving a simple heat equation in $\Omega = (a, b)$,

$$u_t = u_{xx}, \quad \text{in } \Omega \times (0, T), \tag{1.1}$$

with the initial and Dirichlet boundary conditions

$$\begin{aligned} u(x, 0) &= u_0(x), && \text{in } \Omega, \\ u(a, t) &= u(b, t) = 0, && \text{on } (0, T), \end{aligned} \tag{1.2}$$

which can be written into a first-order system, i. e.,

$$u_t - q_x = 0, \quad q - u_x = 0. \tag{1.3}$$

We can then formally use the same DG scheme for the convection equation to solve (1.3). To do that, we denote the mesh by $\mathcal{T} = \{I_j = [x_{j-1}, x_j], j = 1, \dots, N\}$, and set $h_j = x_j - x_{j-1}$, $h = \max_{j=1}^N h_j$. Then we look for $u \in V_k$, where

$$V_k = \{v : v \text{ is a polynomial of degree at most } k \text{ for } x \in I_j,\ j = 1, \dots, N\}, \tag{1.4}$$

such that, for all test functions $v, p \in V_k$,

$$\begin{aligned}\int_{I_j} u_t v dx + \int_{I_j} q v_x dx - \hat{q}_j v_j^- + \hat{q}_{j-1} v_{j-1}^+ &= 0,\\ \int_{I_j} q p dx + \int_{I_j} u p_x dx - \hat{u}_j p_j^- + \hat{u}_{j-1} p_{j-1}^+ &= 0,\end{aligned} \tag{1.5}$$

where $v^{\pm}(x) = \lim_{\epsilon \to 0} v(x \pm \epsilon)$ and the so-called numerical fluxes $(\hat{u}, \hat{q})$, affecting the consistency, stability, and even accuracy, remain to be defined to complete the definition of the LDG method. Using the notations

$$\{v\} = \frac{1}{2}(v^- + v^+), \quad [v] = v^- - v^+,$$

we then choose the numerical fluxes as [3]

$$\begin{pmatrix} \hat{q} \\ \hat{u} \end{pmatrix} = \begin{pmatrix} \{q\} \\ \{u\} \end{pmatrix} - \begin{pmatrix} c_{11} & c_{12} \\ -c_{12} & 0 \end{pmatrix} \begin{pmatrix} [u] \\ [q] \end{pmatrix}, \tag{1.6}$$

which also holds at the boundary if we let

$$(u, q)(a^-) = (0, q(a^+)), \quad (u, q)(b^+) = (0, q(b^-)),$$

reflecting a homogeneous Dirichlet boundary condition.

There are several important points concerning these numerical fluxes [3]:

- The coefficient c_{11} must be a positive number, which aims to enhance the stability of the scheme and improve the accuracy.
- We take $c_{21} = -c_{12}$ to ensure the stability of the LDG scheme.
- Note that $c_{22} = 0$. This is so because we want to solve the auxiliary variable q in terms of u element by element. This local solvability gives the name to the LDG method.
- The main purpose of the coefficient c_{12} is also to enhance the accuracy of the method, i. e., if taking $c_{12} = 0$ [1], the rate of convergence of the energy norm is of order h^k for smooth functions, while taking $c_{12} = 1/2$ [2] leads to the optimal rate of order h^{k+1}.

Taking $v = u$ and $p = q$ in (1.5), integrating by parts and adding over j, we have

$$\begin{aligned}&\frac{1}{2}\frac{d}{dt}\|u(t)\|^2_{L^2(\Omega)} + \|q\|^2_{L^2(\Omega)} + \sum_{j=1}^{N-1}([u_j q_j] - \hat{u}_j[q_j] - \hat{q}_j[u_j]) + u(b^-)q(b^-)\\ &\quad - u(a^+)q(a^+) + \hat{q}(a)u(a^+) - \hat{q}(b)u(b^-) + \hat{u}(a)q(a^+) - \hat{u}(b)q(b^-) = 0.\end{aligned} \tag{1.7}$$

Substituting the numerical fluxes $(\hat{u}, \hat{q})$ (1.6) into (1.7) and using the formula

$$[v_j w_j] = \{v_j\}[w_j] + [w_j]\{v_j\}, \quad \text{for } j = 1, \ldots, N-1,$$

we obtain

$$\frac{1}{2}\frac{d}{dt}\|u(t)\|^2_{L^2(\Omega)} + \|q\|^2_{L^2(\Omega)} + \sum_{j=1}^{N-1} c_{11}[u_j]^2 = 0. \tag{1.8}$$

Then integrating it over $(0, T)$, we have the following.

Proposition 1.1 (L^2 **stability**). *The semi-discrete scheme* (1.5) *is stable, and for any* $T > 0$,

$$\|u(x,T)\|^2_{L^2(\Omega)} + 2\int_0^T \|q(x,t)\|^2_{L^2(\Omega)} dt + 2\int_0^T \sum_{j=1}^{N-1} c_{11}[u_j]^2 dt = \|u_0(x)\|^2_{L^2(\Omega)}.$$

2 Fractional calculus

A fundamental difference between problems in classical calculus and FC is the global nature of the latter formulations. The global nature of such models often leads to the high expectation on advanced computational techniques which are substantially more important than those associated with classical problems. On the other hand, the solutions of fractional diffusion problems are generally smooth when the boundary conditions and the source term are specified properly. All this suggests that it may be worth considering higher-order accurate formulations, such as DG methods.

Before solving the fractional PDEs, we offer the functional settings associated with FDs. Let $n-1 < \alpha < n$. The Riemann–Liouville FD ${}^{\mathrm{RL}}_{-\infty}D^{\alpha}_{x}({}^{\mathrm{RL}}_{x}D^{\alpha}_{+\infty})$ and the Caputo FD ${}^{\mathrm{C}}_{-\infty}D^{\alpha}_{x}({}^{\mathrm{C}}_{x}D^{\alpha}_{+\infty})$ are naturally related. They are equivalent provided all derivatives of order less than n of a specified function $u(x)$ disappear when $x \to \pm\infty$. This result remains true if a bounded domain is considered, i. e., "$-\infty(+\infty)$" is replaced by "a(b)".

Lemma 2.1 ([14]). *Suppose* $u^{(j)}(a) = u^{(j)}(b) = 0$, $\forall 0 \le j \le n-1$ $(n-1 < \alpha < n)$. *Then*

$${}^{\mathrm{RL}}_{a}D^{\alpha}_{x}u(x) = {}^{\mathrm{C}}_{a}D^{\alpha}_{x}u(x), \quad {}^{\mathrm{RL}}_{x}D^{\alpha}_{b}u(x) = {}^{\mathrm{C}}_{x}D^{\alpha}_{b}u(x). \tag{2.1}$$

From Lemma 2.1, we have, for $1 < \alpha < 2$ and a continuous function u with $u(a) = u(b) = 0$,

$${}^{\mathrm{RL}}_{a}D^{\alpha-1}_{x}u(x) = {}^{\mathrm{C}}_{a}D^{\alpha-1}_{x}u(x), \quad {}^{\mathrm{RL}}_{x}D^{\alpha-1}_{b}u(x) = {}^{\mathrm{C}}_{x}D^{\alpha-1}_{b}u(x), \tag{2.2}$$

i. e.,

$${}^{\mathrm{RL}}_{a}D^{\alpha}_{x}u(x) = \frac{d}{dx}{}_{a}I^{2-\alpha}_{x}\frac{d}{dx}u(x), \quad {}^{\mathrm{RL}}_{x}D^{\alpha}_{b}u(x) = \frac{d}{dx}{}_{x}I^{2-\alpha}_{b}\frac{d}{dx}u(x). \tag{2.3}$$

These equalities will be used in the LDG scheme in the subsequent sections.

To carry out the analysis, we introduce the negative fractional norms and fractional integral (FI) spaces [10], and we define the norms of FI spaces.

Definition 2.1 ([10]). Let $\alpha > 0$. Define the norm

$$\|v\|_{J_L^{-\alpha}(\mathbb{R})} := \|{}_{-\infty}I_x^{\alpha} v\|_{L^2(\mathbb{R})}, \quad \|v\|_{J_R^{-\alpha}(\mathbb{R})} := \|{}_x I_{+\infty}^{\alpha} v\|_{L^2(\mathbb{R})} \tag{2.4}$$

and let $J_L^{-\alpha}(\mathbb{R})$ (or $J_R^{-\alpha}(\mathbb{R})$) denote the closure of $C_0^{\infty}(\mathbb{R})$ w. r. t. $\|\cdot\|_{J_L^{-\alpha}(\mathbb{R})}$ (or $\|\cdot\|_{J_R^{-\alpha}(\mathbb{R})}$).

Recall the Fourier transform of $v(x)$ defined as $\mathscr{F}[v(x)] = \int_{-\infty}^{+\infty} e^{-ix\omega} v(x)dx = \hat{v}(\omega)$. As in classical Fourier analysis, we define a higher norm for functions in $H^{-\alpha}(\mathbb{R})$ in terms of the Fourier transform.

Definition 2.2 ([10]). Let $\alpha \in (0, \frac{1}{2})$. Define the norm

$$\|v\|_{H^{-\alpha}(\mathbb{R})} := \||\omega|^{-\alpha}\hat{v}\|_{L^2(\mathbb{R})} \tag{2.5}$$

and let $H^{-\alpha}(\mathbb{R})$ denote the closure of $C_0^{\infty}(\mathbb{R})$ w. r. t. $\|\cdot\|_{H^{-\alpha}(\mathbb{R})}$.

The three norms are closely related as stated in the following results.

Theorem 2.3 ([10]). *If $\alpha \in (0, \frac{1}{2})$, the three spaces $H^{-\alpha}(\mathbb{R})$, $J_L^{-\alpha}(\mathbb{R})$, and $J_R^{-\alpha}(\mathbb{R})$ are equal with equivalent norms.*

Lemma 2.2 ([10]). *If $\alpha \in (0, \frac{1}{2})$, then*

$$({}_{-\infty}I_x^{\alpha} v, {}_x I_{+\infty}^{\alpha} v) = \cos(\alpha\pi)\|{}_x I_{+\infty}^{\alpha} v\|_{L^2(\mathbb{R})}^2 = \cos(\alpha\pi)\|v\|_{H^{-\alpha}(\mathbb{R})}^2. \tag{2.6}$$

Let us now restrict our attention to the case in which $\mathrm{supp}(v) \subset \Omega = (a, b)$. Then ${}_{-\infty}I_x^{\alpha} v = {}_a I_x^{\alpha} v$ and ${}_x I_{+\infty}^{\alpha} v = {}_x I_b^{\alpha} v$. Straightforward extension of the definitions given above yields the following.

Definition 2.4 ([10]). Define the spaces $H_0^{-\alpha}(\Omega)$, $J_{L,0}^{-\alpha}(\Omega)$, $J_{R,0}^{-\alpha}(\Omega)$ as the closures of $C_0^{\infty}(\Omega)$ under their respective norms.

The following theorem gives the relations among the FI spaces with different α.

Theorem 2.5 ([10]). *If $-\alpha_2 < -\alpha_1 < 0$, then $J_{L,0}^{-\alpha_1}(\Omega)(H_0^{-\alpha_1}(\Omega)$ or $J_{R,0}^{-\alpha_1}(\Omega))$ is embedded into $J_{L,0}^{-\alpha_2}(\Omega)(H_0^{-\alpha_2}(\Omega)$ or $J_{R,0}^{-\alpha_2}(\Omega))$, and $L^2(\Omega)$ is embedded into both of them.*

3 Fractional diffusion equation in one dimension

Let us consider the fractional diffusion equation

$$\partial_t u(x,t) = d\, {}_{-\infty}^{\mathrm{RL}}D_x^{\alpha} u(x,t) + f(x,t) \quad \text{in } \Omega \times (0, T], \tag{3.1}$$

with the initial and Dirichlet boundary condition

$$\begin{aligned} u(x,0) &= u_0(x) && \text{in } \Omega, \\ u(x,t) &= 0 && \text{in } \mathbb{R}\backslash\Omega \times (0,T], \end{aligned} \tag{3.2}$$

where $1 < \alpha < 2$ and $\Omega = (a, b)$ is bounded. In fact, when $\alpha = 1$ and 2, ${}^{\mathrm{RL}}_{-\infty}D^{\alpha}_{x}u$ still makes sense and recovers exactly the first-order and second-order classical derivatives, respectively. At this moment, the equation becomes a pure convection equation or a pure diffusion equation.

Here, we specify the generalized boundary conditions on the whole supplement of Ω, i. e., $\mathbb{R}\backslash\Omega$ [11]. In fact, the fractional diffusion equations are the macroscopic manifestation of anomalous diffusion and can be derived by continuous time random walk (CTRW) model [13], governed by the waiting time and jump length; generally the first moment of the waiting time or/and the second moment of the jump length diverge(s). From the process of derivation, one can see that the issue of initial condition can be easily/reasonably fixed, as classical ones, just specifying the value of $u(x, 0)$ in the domain Ω. The boundary condition, however, is much different from the classical ones. Actually, for Lévy processes, except Brownian motion, all others have discontinuous paths. As a result, the boundary $\partial\Omega$ itself cannot be hit by the majority of discontinuous sample trajectories, which implies that the generalized boundary conditions must be introduced when solving fractional diffusion equations.

3.1 Weak formulation and LDG scheme

Considering the Dirichlet boundary condition in (3.2), we have ${}^{\mathrm{RL}}_{-\infty}D^{\alpha}_{x}u = {}^{\mathrm{RL}}_{a}D^{\alpha}_{x}u$. Furthermore, since $1 < \alpha < 2$, the fractional operator is composed of a second-order derivative and an FI of order $2 - \alpha$. Then by (2.3), there is

$$ {}^{\mathrm{RL}}_{a}D^{\alpha}_{x}u(x,t) = \frac{\partial}{\partial x}\,{}_{a}I^{2-\alpha}_{x}\frac{\partial}{\partial x}u(x,t). $$

To obtain a consistent and high-accuracy scheme for this problem, we rewrite the fractional diffusion equation (3.1) into a system of low-order equations, i. e.,

$$ \begin{aligned} u_t &= \sqrt{d}q_x + f && \text{in } \Omega \times (0,T], \\ q &= {}_{a}I^{2-\alpha}_{x}p && \text{in } \Omega \times (0,T], \\ p &= \sqrt{d}u_x && \text{in } \Omega \times (0,T]. \end{aligned} \tag{3.3} $$

Associated with the mesh $\mathcal{T} = \{I_j = [x_{j-1}, x_j], j = 1, \ldots, N\}$, we define the broken Sobolev spaces

$$ L^2(\Omega, \mathcal{T}) := \{v : \Omega \to \mathbb{R}\ v|_{I_j} \in L^2(I_j), j = 1, \ldots, N\} $$

and

$$ H^1(\Omega, \mathcal{T}) := \{v : \Omega \to \mathbb{R}\ v|_{I_j} \in H^1(I_j), j = 1, \ldots, N\}. $$

Then the weak form is to find (u,q,p) of (3.3) belonging to $H^1(0,T;H^1(\Omega,\mathcal{T}))\times L^2(0,T;H^1(\Omega,\mathcal{T}))\times L^2(0,T;L^2(\Omega,\mathcal{T}))$, such that, for $j=1,\dots,N$ and all test functions $w\in L^2(0,T;L^2(\Omega,\mathcal{T}))$ and $v,z\in L^2(0,T;H^1(\Omega,\mathcal{T}))$,

$$\begin{aligned}
&\int_{I_j} u_t v dx + \sqrt{d}\int_{I_j} q v_x dx - \sqrt{d} q v\Big|_{x_{j-1}^+}^{x_j^-} = \int_{I_j} f v dx,\\
&\int_{I_j} q w dx - \int_{I_j} {}_aI_x^{2-\alpha} p\, w dx = 0,\\
&\int_{I_j} p z dx + \sqrt{d}\int_{I_j} u z_x dx - \sqrt{d} u z\Big|_{x_{j-1}^+}^{x_j^-} = 0.
\end{aligned} \tag{3.4}$$

Next, we shall propose the numerical schemes. The trial and test functions should be restricted to the finite-dimensional subspaces $V_k\in H^1(\Omega,\mathcal{T})$, where V_k is defined by (1.4). Furthermore, we define U, Q, and P as the approximations of u, q, and p in the weak form, respectively. We seek $(U,Q,P)\in H^1(0,T;V_k)\times L^2(0,T;V_k)\times L^2(0,T;V_k)$ such that for all $(v,w,z)\in L^2(0,T;V_k)\times L^2(0,T;V_k)\times L^2(0,T;V_k)$, and for $j=1,\dots,N$,

$$\begin{aligned}
&\int_{I_j} U_t v dx + \sqrt{d}\int_{I_j} Q v_x dx - \sqrt{d}\hat{Q} v\Big|_{x_{j-1}^+}^{x_j^-} = \int_{I_j} f v dx,\\
&\int_{I_j} Q w dx - \int_{I_j} {}_aI_x^{2-\alpha} P\, w dx = 0,\\
&\int_{I_j} P z dx + \sqrt{d}\int_{I_j} U z_x dx - \sqrt{d}\hat{U} z\Big|_{x_{j-1}^+}^{x_j^-} = 0.
\end{aligned} \tag{3.5}$$

To complete the formulation of the numerical schemes, we must define the numerical fluxes $(\hat{U},\hat{Q})$. Inspired by traditional LDG schemes of heat equations in Section 1, alternating fluxes should be a good choice. We choose

$$\hat{U}_j = U_j^-,\quad \hat{Q}_j = Q_j^+,\quad 0\le j\le N-1,$$

and

$$\hat{U}_N = 0,\quad \hat{Q}_N = Q(b^-) - \frac{\beta}{h}U(b^-),$$

where β is a positive constant. Adding the term $-\frac{\beta}{h}U(b^-)$ is equivalent to taking c_{11} to be $\frac{\beta}{h}$ on the right boundary of (1.6), which improves the accuracy of the scheme.

Remark 3.1. Another choice of the alternating fluxes $(\hat{U},\hat{Q})$ is

$$\hat{U}_j = U_j^+,\quad \hat{Q}_j = Q_j^-,\quad 1\le j\le N,$$

and

$$\hat{U}_0 = 0,\quad \hat{Q}_0 = Q(a^+) + \frac{\beta}{h}U(a^+).$$

There are some other cases, where the LDG scheme will be a little different.

The proposed LDG schemes can be extended directly to the model including two-sided (mixed) FDs, that is, the model

$$\partial_t u(x,t) = \theta_1 \cdot {}^{\mathrm{RL}}_{a}D^{\beta}_{x} u(x,t) + \theta_2 \cdot {}^{\mathrm{RL}}_{x}D^{\beta}_{b} u(x,t) + f(x,t), \quad \theta_1, \theta_2 > 0, 1 < \beta < 2,$$

in $\Omega \times (0, T]$, with appropriate initial and Dirichlet boundary conditions. In this case, we can introduce the auxiliary variables q and p and write the LDG scheme as

$$\begin{cases} u_t - q_x = f & \text{in } \Omega \times (0, T], \\ q = (\theta_1 \cdot {}_aI_x^{2-\beta} + \theta_2 \cdot {}_xI_b^{2-\beta})p & \text{in } \Omega \times (0, T], \\ p - u_x = 0 & \text{in } \Omega \times (0, T]. \end{cases}$$

3.2 Stability analysis

Now, we perform the numerical stability analysis for the proposed scheme.

Lemma 3.1 (Continuous Grönwall inequality [16]). *Let f, g, h be piece-wise continuous non-negative functions defined on (a, b). Assume that g is non-decreasing and there is a positive constant C independent of t such that*

$$\forall t \in (a, b), \quad f(t) + h(t) \le g(t) + C \int_a^t f(s)ds.$$

Then

$$\forall t \in (a, b), \quad f(t) + h(t) \le e^{C(t-a)} g(t).$$

The continuous Grönwall inequality is an important tool for analyzing time-dependent problems, especially in the stability and convergence analyses.

In this subsection, we will show the stability of the LDG scheme (3.5) for fractional diffusion problems. Integrating (3.5) w. r. t. t from 0 to T and summing over all elements, it turns out that the scheme can be expressed as follows. Find $(U, Q, P) \subset H^1(0, T; V_k) \times L^2(0, T; V_k) \times L^2(0, T; V_k)$ such that for all $(v, w, z) \in H^1(0, T; V_k) \times L^2(0, T; V_k) \times L^2(0, T; V_k)$, we have

$$B(U, Q, P; v, w, z) = \mathcal{L}(v, w, z), \tag{3.6}$$

where the discrete bilinear form B is defined by

$$B(U, Q, P; v, w, z) = \int_0^T (U_t, v)dt - \int_0^T ({}_aI_x^{2-\alpha} P, w)dt + \sqrt{d} \int_0^T ((Q, v_x) + (U, z_x))dt$$

$$+\int_0^T ((Q,w)+(P,z))dt+\sqrt{d}\int_0^T\bigg(Q(a^+)v(a^+)-Q(b^-)v(b^-)$$
$$+\frac{\beta}{h}U(b^-)v(b^-)-\sum_{j=1}^{N-1}(Q_j^+[v_j]+U_j^-[z_j])\bigg)dt, \tag{3.7}$$

and the discrete linear form $\mathcal{L}$ is given by

$$\mathcal{L}(v,w,z)=\int_0^T (f,v)dt. \tag{3.8}$$

Theorem 3.1 (L^2 **stability**). *Scheme (3.6) is L^2 stable, i. e., for all $T>0$, its solution satisfies*

$$\|U(x,T)\|^2_{L^2(\Omega)}+2\cos((\alpha/2-1)\pi)\int_0^T \|P(x,t)\|^2_{H^{\alpha/2-1}(\Omega)}dt+\int_0^T \frac{2\sqrt{d}\beta}{h}U(b^-)^2dt$$
$$\le C\bigg(\int_0^T \|f(x,t)\|^2_{L^2(\Omega)}dt+\int_0^T \|u_0(x)\|^2_{L^2(\Omega)}dt\bigg). \tag{3.9}$$

Proof. Taking $(v,w,z)=(U,-P,Q)$ and using formula (2.6), we have

$$B(U,Q,P;U,-P,Q)=\frac{1}{2}\int_0^T \frac{\partial}{\partial t}\|U\|^2_{L^2(\Omega)}dt+\cos((\alpha/2-1)\pi)\int_0^T \|P\|^2_{H^{\alpha/2-1}(\Omega)}dt$$
$$+\sqrt{d}\int_0^T\int_a^b \frac{\partial(UQ)}{\partial x}dxdt-\sqrt{d}\int_0^T\sum_{j=1}^{N-1}(Q_j^+[U_j]+U_j^-[Q_j])dt$$
$$+\sqrt{d}\int_0^T\bigg(Q(a^+)U(a^+)-Q(b^-)U(b^-)+\frac{\beta}{h}U(b^-)^2\bigg)dt. \tag{3.10}$$

Substituting the numerical fluxes $(\hat{U},\hat{Q})$ and using the formula

$$\int_0^T\int_a^b \frac{\partial(UQ)}{\partial x}dxdt=\int_0^T\sum_{j=1}^{N-1}[U_jQ_j]dt-\int_0^T (Q(a^+)U(a^+)-Q(b^-)U(b^-))dt$$
$$=\int_0^T\sum_{j=1}^{N-1}(U_j^-[Q_j]+Q_j^+[U_j])dt-\int_0^T (Q(a^+)U(a^+)-Q(b^-)U(b^-))dt,$$

we have

$$B(U,Q,P;U,-P,Q) = \frac{1}{2}\int_0^T \frac{\partial}{\partial t}\|U\|^2_{L^2(\Omega)}dt + \cos((\alpha/2-1)\pi)\int_0^T \|P\|^2_{H^{\alpha/2-1}(\Omega)}dt$$
$$+ \int_0^T \frac{\sqrt{d}\beta}{h}U(b^-)^2 dt. \tag{3.11}$$

On the other hand, with Hölder's inequality,

$$\mathcal{L}(U,-P,Q) = \int_0^T (f,U)dt \le \frac{1}{2}\int_0^T \|f\|^2_{L^2(\Omega)}dt + \frac{1}{2}\int_0^T \|U\|^2_{L^2(\Omega)}dt. \tag{3.12}$$

Combining (3.11) with (3.12), and using the continuous Grönwall inequality, we obtain (3.9). □

3.3 Error estimates

For the error analysis, we define the projection operators $\pi^\pm$, π from $H^1(\Omega,\mathcal{T})$ to V_k. For intervals $I_j = (x_{j-1}, x_j)$, $j = 1, 2, \ldots, N$, $\pi^\pm$ are defined by the following $k+1$ conditions:

$$\begin{aligned}(\pi^\pm u - u, v)_{I_j} &= 0 \quad \forall v \in P^{k-1}(I_j),\ \text{if } k > 0,\\ \pi^- u(x_j) &= u(x_j^-), \quad \pi^+ u(x_{j-1}) = u(x_{j-1}^+),\end{aligned} \tag{3.13}$$

where $P^{k-1}(I_j)$ denotes the polynomials of degree less than or equal to $k-1$ on element I_j. Operator π is the standard L^2-projection, which is defined as

$$(\pi u - u, v)_{I_j} = 0 \quad \forall v \in P^k(I_j). \tag{3.14}$$

We denote

$$e_u = u - U, \quad e_q = q - Q, \quad e_p = p - P.$$

Since the scheme (3.5) is consistent with (3.3), the exact solution (u, q, p) of (3.3) satisfies

$$B(u,q,p;v,w,z) = \mathcal{L}(v,w,z) \tag{3.15}$$

for all $(v,w,z) \in H^1(0,T;V_k) \times L^2(0,T;V_k) \times L^2(0,T;V_k)$. Combining (3.6) and (3.15), we have the error equation

$$B(e_u, e_q, e_p; v, w, z) = 0. \tag{3.16}$$

Taking

$$v = \pi^- u - U, \quad w = P - \pi p, \quad z = \pi^+ q - Q,$$

we obtain

$$B(v,z,-w;v,w,z) = B(v^e,z^e,-w^e;v,w,z), \tag{3.17}$$

where

$$v^e = \pi^- u - u, \quad w^e = p - \pi p, \quad z^e = \pi^+ q - q.$$

The left-hand side of (3.17) has been considered in (3.11). We have

$$B(v,z,-w;v,w,z) \\ = \frac{1}{2}\int_0^T \frac{\partial}{\partial t}\|v\|^2_{L^2(\Omega)}dt + \cos((\alpha/2-1)\pi)\int_0^T \|w\|^2_{H^{\alpha/2-1}(\Omega)}dt + \int_0^T \frac{\sqrt{d}\beta}{h}v(b^-)^2 dt. \tag{3.18}$$

It only remains to obtain an upper bound of the right-hand side of (3.17), which can be expressed as

$$B(v^e,z^e,-w^e;v,w,z) = \mathcal{I} + \mathcal{II} + \mathcal{III} + \mathcal{IV} + \mathcal{V},$$

where

$$\mathcal{I} = \int_0^T \left(\frac{\partial v^e}{\partial t}, v\right)dt,$$

$$\mathcal{II} = \sqrt{d}\int_0^T \left(z^e, \frac{\partial v}{\partial x}\right)dt + \sqrt{d}\int_0^T \left(v^e, \frac{\partial z}{\partial x}\right)dt - \int_0^T (w^e, z)dt,$$

$$\mathcal{III} = -\sqrt{d}\int_0^T \sum_{j=1}^{N-1}(z^e)_j^+[v_j]dt - \sqrt{d}\int_0^T \sum_{j=1}^{N-1}(v^e)_j^-[z_j]dt,$$

$$\mathcal{IV} = \sqrt{d}\int_0^T \left(z^e(a^+)v(a^+) - z^e(b^-)v(b^-) + \frac{\beta}{h}v^e(b^-)v(b^-)\right)dt,$$

$$\mathcal{V} = \int_0^T (z^e, w)dt + \int_0^T ({}_aI_x^{2-\alpha}p, w)dt.$$

Using the standard approximation theory [5], we obtain

$$\begin{aligned}\mathcal{I} &\le \frac{1}{2}\int_0^T\int_a^b \left(\frac{\partial v^e}{\partial t}\right)^2 dxdt + \frac{1}{2}\int_0^T\int_a^b v^2 dxdt \\ &\le c_u h^{2k+2} + \frac{1}{2}\int_0^T \|v\|^2_{L^2(\Omega)}dt,\end{aligned}$$

where c_u depends on $|u|_{H^{k+1}(\Omega)}$.

All the terms in $\mathcal{II}$ and $\mathcal{III}$ vanish due to the definition of projections $(\pi^{\pm}, \pi)$ and the fluxes $(\hat{u}, \hat{q})$. Similarly, the first term and third term in $\mathcal{IV}$ vanish. Therefore,

$$\begin{aligned}\mathcal{IV} &\leq \frac{\sqrt{d}h}{4\beta}\int_0^T z^e(b^-)^2 dt + \frac{\sqrt{d}\beta}{h}\int_0^T v(b^-)^2 dt \\ &\leq c_q h^{2k+2} + \frac{\sqrt{d}\beta}{h}\int_0^T v(b^-)^2 dt.\end{aligned}$$

Finally, using Theorem 2.5 and inverse inequality, we have

$$\begin{aligned}\mathcal{V} &\leq \int_0^T\int_a^b \frac{h^{\alpha-2}}{2\epsilon}(z^e)^2 dxdt + \frac{\epsilon}{2}\int_0^T h^{2-\alpha}\|w\|^2_{L^2(\Omega)}dt \\ &\quad + \int_0^T\int_a^b \frac{(w^e)^2}{2\epsilon}dxdt + \frac{\epsilon}{2}\int_0^T \|w\|^2_{H^{\alpha-2}(\Omega)}dt \\ &\leq (c_p/\epsilon)h^{2k+\alpha} + c\epsilon\int_0^T \|w\|^2_{H^{\alpha/2-1}(\Omega)}dt.\end{aligned}$$

Combining the above estimates leads to

$$\begin{aligned}B(v^e, z^e, -w^e; v, w, z) &\leq c_u h^{2k+\alpha} + \frac{1}{2}\int_0^T \|v\|^2_{L^2(\Omega)}dt + \frac{\sqrt{d}\beta}{h}\int_0^T v(b^-)^2 dt \\ &\quad + c\epsilon\int_0^T \|w\|^2_{H^{\alpha/2-1}(\Omega)}dt, \end{aligned} \tag{3.19}$$

where c_u depends on $|u|_{H^{k+2}(\Omega)}$ due to $p = \sqrt{d}u_x$ in (3.3). Then, choosing a sufficiently small ϵ such that $c\epsilon < \cos((\alpha/2-1)\pi)$, from (3.17), (3.18), and (3.19), we have

$$\begin{aligned}\|v(T)\|^2_{L^2(\Omega)} &\leq \|v(0)\|^2_{L^2(\Omega)} + c_u h^{2k+\alpha} + \int_0^T \|v\|^2_{L^2(\Omega)}dt \\ &\leq c_u h^{2k+\alpha} + \int_0^T \|v\|^2_{L^2(\Omega)}dt.\end{aligned}$$

According to the continuous Grönwall inequality and the standard approximation on $v^e = P^-u - u$, we obtain the following.

Theorem 3.2 (Error estimate). *The error for the scheme* (3.6) *satisfies*

$$\|u(x,t) - U(x,t)\|_{L^2(\Omega)} \leq c_u h^{k+\alpha/2}. \tag{3.20}$$

Since $\|w\|_{L^2(\Omega)}$ cannot be bounded by $\|w\|_{H^{\alpha/2-1}(\Omega)}$ in the estimation of $\mathcal{V}$, we do not get the optimal convergence order. Theorem 3.2 shows that the convergence order decreases as $\alpha \to 1$. At the same time, the numerical dissipation term in (3.9) vanishes. To overcome this, we add a penalty term $[P(x,t)]$, as a local dissipation, to enhance the stability of the LDG scheme (3.5). Adding it to the second equation of (3.5) makes

$$\int_{I_j} Qwdx - \int_{I_j} {}_aI_x^{2-\alpha}P\,wdx - L(h,\alpha)\cdot([P]w)\Big|_{x_{j-1}^+}^{x_j^-} = 0. \tag{3.21}$$

Here $L(h,\alpha)$ is a positive constant depending on h and α; we take it to be h^α in agreement with the scaling of the global operator. In the numerical examples of the next subsection, this term helps to recover the optimal $k+1$ order of convergence for any $\alpha \in [1,2]$.

3.4 Numerical results

In this subsection, we offer some numerical results of the LDG scheme with alternating fluxes for fractional diffusion problems to validate our theoretical results. We use a fourth-order explicit Runge–Kutta method to solve the method-of-line fractional PDEs, i. e., the classical ODE system. To ensure that the overall error is dominated by the space error, small time steps are used.

We solve the problem

$$\frac{\partial u(x,t)}{\partial t} = \frac{\Gamma(6-\beta)}{\Gamma(6)}\,{}^{\mathrm{RL}}D_{0+}^{\alpha}u(x,t) - \mathrm{e}^{-t}(x^5 + x^{5-\alpha}) \tag{3.22}$$

on the computational domain $\Omega = (0,1)$, and the initial condition

$$u(x,0) = x^5, \quad x \in \Omega, \tag{3.23}$$

and the Dirichlet boundary conditions

$$u(0,t) = 0, \quad u(1,t) = \mathrm{e}^{-t}, \tag{3.24}$$

where $\alpha \in [1,2]$. The exact solution is given by $\mathrm{e}^{-t}x^5$.

The errors are measured in the discrete L^2 norm at $t = 1$. Tables 1 and 2 for $k = 1,2$ demonstrate the errors and the orders of convergence. When $\alpha \to 1$, the last two rows recover the optimal orders of convergence only with the stabilization included.

4 Fractional diffusion equation in two dimensions

Let us consider the fractional diffusion equation in two dimensions

$$\partial_t u(x,y,t) = {}_{-\infty}^{\mathrm{RL}}D_x^{\alpha}u + {}_{-\infty}^{\mathrm{RL}}D_y^{\beta}u + f \quad \text{in } \Omega \times (0,T] \tag{4.1}$$

Table 1: Errors and orders of convergence for first-order polynomial approximation ($k = 1$) with M being the number of elements.

α	$M = 2^6$	$M = 2^7$		$M = 2^8$		$M = 2^9$	
	error	error	order	error	order	error	order
2.00	7.82e−05	2.04e−05	1.94	5.20e−06	1.97	1.31e−06	1.99
1.80	9.37e−05	2.44e−05	1.94	6.14e−06	1.99	1.52e−06	2.01
1.50	1.27e−04	3.23e−05	1.98	7.93e−06	2.03	1.95e−06	2.02
1.20	3.17e−04	8.32e−05	1.93	2.02e−05	2.05	5.09e−06	1.99
1.10	1.90e−04	5.22e−05	1.86	1.32e−05	1.98	3.31e−06	2.00
1.09	2.09e−04	5.96e−05	1.81	1.53e−05	1.96	3.86e−06	1.99
1.08	2.31e−04	6.85e−05	1.76	1.78e−05	1.94	4.53e−06	1.98
1.07	2.56e−04	7.94e−05	1.69	2.12e−05	1.90	5.40e−06	1.98
1.06	2.82e−04	9.26e−05	1.61	2.57e−05	1.85	6.52e−06	1.98
1.05	3.11e−04	1.09e−04	1.52	3.17e−05	1.78	8.11e−06	1.97
1.04	3.43e−04	1.29e−04	1.41	4.01e−05	1.68	1.04e−05	1.95
1.03	3.75e−04	1.52e−04	1.30	5.18e−05	1.55	1.45e−05	1.84
1.02	4.10e−04	1.80e−04	1.19	6.89e−05	1.39	2.16e−05	1.67
1.01	4.46e−04	2.12e−04	1.07	9.31e−05	1.19	3.44e−05	1.44
1.00	4.83e−04	2.49e−04	0.95	1.27e−04	0.97	6.38e−05	0.99
1.01	5.38e−04	1.43e−04	1.91	3.71e−05	1.95	9.45e−06	1.97
1.00	5.38e−04	1.43e−04	1.91	3.71e−05	1.95	9.45e−06	1.97

Table 2: Errors and orders of convergence for second-order polynomial approximation ($k = 2$) with M being the number of elements.

α	$M = 2^3$	$M = 2^4$		$M = 2^5$		$M = 2^6$	
	error	error	order	error	order	error	order
2.00	4.92e−04	6.62e−05	2.89	8.61e−06	2.94	1.09e−06	2.98
1.80	5.19e−04	6.76e−05	2.94	8.50e−06	2.99	1.07e−06	2.99
1.50	5.53e−04	7.27e−05	2.93	9.08e−06	3.00	1.12e−06	3.02
1.20	6.04e−04	7.02e−05	3.11	8.73e−06	3.01	1.10e−06	2.99
1.10	3.04e−04	8.28e−05	1.88	9.80e−06	3.08	9.74e−07	3.33
1.09	2.98e−04	8.20e−05	1.86	1.01e−05	3.01	8.97e−07	3.50
1.08	2.92e−04	8.10e−05	1.85	1.05e−05	2.95	7.99e−07	3.71
1.07	2.86e−04	7.99e−05	1.84	1.08e−05	2.88	6.80e−07	4.00
1.06	2.80e−04	7.87e−05	1.83	1.12e−05	2.82	5.51e−07	4.34
1.05	2.74e−04	7.75e−05	1.82	1.15e−05	2.75	4.36e−07	4.72
1.04	2.69e−04	7.62e−05	1.82	1.18e−05	2.69	3.63e−07	5.02
1.03	2.63e−04	7.48e−05	1.81	1.21e−05	2.63	3.69e−07	5.03
1.02	2.57e−04	7.34e−05	1.81	1.23e−05	2.58	4.28e−07	4.85
1.01	2.51e−04	7.19e−05	1.81	1.25e−05	2.53	5.08e−07	4.62
1.00	2.46e−04	7.04e−05	1.81	1.26e−05	2.48	5.79e−07	4.45
1.01	1.50e−05	1.43e−06	3.41	1.10e−07	3.70	1.07e−08	3.36
1.00	1.52e−05	1.43e−06	3.39	1.10e−07	3.71	1.06e−08	3.36

with the initial and Dirichlet boundary conditions

$$\begin{aligned} u(x,y,0) &= u_0(x,y) \quad && \text{in } \Omega, \\ u(x,y,t) &= 0 && \text{in } \mathbb{R}\backslash\Omega \times (0,T], \end{aligned} \tag{4.2}$$

where $1 < \alpha, \beta < 2$ and $\Omega = (a,b) \times (c,d)$ is bounded.

4.1 Weak formulation and LDG scheme

Similar to the one-dimensional case, by (2.3) and the boundary conditions (4.2), we have

$$ {}_a^{\mathrm{RL}}D_x^{\alpha} = \frac{\partial}{\partial x} {}_aI_x^{2-\alpha} \frac{\partial}{\partial x}, \quad {}_c^{\mathrm{RL}}D_y^{\beta} = \frac{\partial}{\partial y} {}_cI_y^{2-\beta} \frac{\partial}{\partial y}. $$

Following the standard approach for the development of the LDG method, we introduce the auxiliary variables $\boldsymbol{p} = (p_1, p_2)$ and $\boldsymbol{q} = (q_1, q_2)$ and rewrite the problem into a first-order system, i. e.,

$$\begin{aligned} u_t &= \nabla \cdot \boldsymbol{q} + f && \text{in } \Omega \times (0,T], \\ \boldsymbol{q} &= ({}_aI_x^{2-\alpha} p_1, {}_cI_y^{2-\beta} p_2) && \text{in } \Omega \times (0,T], \\ \boldsymbol{p} &= \nabla u && \text{in } \Omega \times (0,T]. \end{aligned} \tag{4.3}$$

Then we introduce some notations for the LDG scheme. The domain is subdivided into non-overlapping elements E, which can be triangles or rectangles. Here we take elements E to be triangles. The mesh $\mathcal{T}$ is called regular if $\forall E \in \mathcal{T}$, $h_E/\rho_E \le C$, where C is a positive constant, h_E the diameter of the element E, and ρ_E the diameter of the inscribed circle in element E. The set of edges of the subdivision $\mathcal{T}$ is denoted by Γ. Let Γ_i denote the set of interior edges and let $\Gamma_b = \Gamma \backslash \Gamma_i$ denote the set of edges on $\partial\Omega$. With each edge e, the unit normal vector is $\mathbf{n}_e$. If e is on the boundary $\partial\Omega$, then $\mathbf{n}_e$ is taken to be the unit outward vector normal to $\partial\Omega$.

Associated with the mesh $\mathcal{T}$, the broken Sobolev spaces $L^2(\Omega, \mathcal{T})$ and $H^1(\Omega, \mathcal{T})$ can be defined similarly as the ones in Section 3. If v belongs to $H^1(\Omega, \mathcal{T})$, then the trace of v along any side of one element E is well defined. If two elements E^+ and E^- are neighbors and share one common side e, i. e., $e = \overline{\partial E^+} \cap \overline{\partial E^-}$, let $\boldsymbol{x}$ be a point on e and $\mathbf{n}^{\pm}$ be the unit outward normal to $\partial E^{\pm}$ at the point $\boldsymbol{x}$. Let $v^{\pm}(\boldsymbol{x})$ denote the value $\lim_{\epsilon \to 0} v(\boldsymbol{x} - \epsilon \mathbf{n}^{\pm})$ and set

$$ \{v\} = \frac{1}{2}(v^+ + v^-), \quad [v] = (v^- \mathbf{n}^- + v^+ \mathbf{n}^+). $$

The weak form is to find

$$ (u, \boldsymbol{q}, \boldsymbol{p}) \in H^1(0,T; H^1(\Omega, \mathcal{T})) \times (L^2(0,T; H^1(\Omega, \mathcal{T})))^2 \times (L^2(0,T; L^2(\Omega, \mathcal{T})))^2 $$

such that

$$\begin{aligned} (u_t, v)_{E_j} &= (\mathbf{n}\cdot \boldsymbol{q}, v)_{\partial E_j} - (\boldsymbol{q}, \nabla v)_{E_j} + (f, v)_{E_j}, \\ (\boldsymbol{q}, \boldsymbol{\phi})_{E_j} &= (({}_aI_x^{2-\alpha} p_1, {}_cI_y^{2-\beta} p_2), \boldsymbol{\phi})_{E_j}, \\ (\boldsymbol{p}, \boldsymbol{\pi})_{E_j} &= (u, \mathbf{n}\cdot \boldsymbol{\pi})_{\partial E_j} - (u, \nabla\cdot \boldsymbol{\pi})_{E_j}, \end{aligned} \tag{4.4}$$

for all test functions $v \in L^2(0,T;H^1(\Omega,\mathcal{T}))$, $\boldsymbol{\phi} \in (L^2(0,T;L^2(\Omega,\mathcal{T})))^2$, and $\boldsymbol{\pi} \in (L^2(0,T;H^1(\Omega,\mathcal{T})))^2$. Then the LDG scheme is to find $(u_h, \boldsymbol{q}_h, \boldsymbol{p}_h) \in H^1(0,T;V_k)\times(L^2(0,T;V_k))^2\times(L^2(0,T;V_k))^2$ such that, for all test functions $(v, \boldsymbol{\phi}, \boldsymbol{\pi}) \in L^2(0,T;V_k)\times(L^2(0,T;V_k))^2\times(L^2(0,T;V_k))^2$ and all elements E_j, there exist

$$\begin{aligned} (u_{ht}, v)_{E_j} &= (\mathbf{n}\cdot \hat{\boldsymbol{q}}_h, v)_{\partial E_j} - (\boldsymbol{q}_h, \nabla v)_{E_j} + (f, v)_{E_j}, \\ (\boldsymbol{q}_h, \boldsymbol{\phi})_{E_j} &= (({}_aI_x^{2-\alpha} p_{1h}, {}_cI_y^{2-\beta} p_{2h}), \boldsymbol{\phi})_{E_j}, \\ (\boldsymbol{p}_h, \boldsymbol{\pi})_{E_j} &= (\hat{u}_h, \mathbf{n}\cdot \boldsymbol{\pi})_{\partial E_j} - (u_h, \nabla\cdot \boldsymbol{\pi})_{E_j}. \end{aligned} \tag{4.5}$$

In the two-dimensional case, we adopt the central fluxes, defined as

$$\hat{u}_h = \frac{1}{2}(u_h^+ + u_h^-), \quad \hat{\boldsymbol{q}}_h = \frac{1}{2}(\boldsymbol{q}_h^+ + \boldsymbol{q}_h^-)$$

at all internal edges and

$$\hat{u}_h = 0, \quad \hat{\boldsymbol{q}}_h = \boldsymbol{q}_h^+ = \boldsymbol{q}_h^-$$

at the external edges.

4.2 Numerical algorithm

In this subsection, we illustrate the numerical performance of the proposed schemes by the numerical simulations of the problems in two dimensions. For the computational part, we introduce the strong form, recovered by integration by parts from the weak form (4.5), as

$$\begin{aligned} (u_t, v)_{E_j} &= (\nabla\cdot \boldsymbol{q}, v)_{E_j} - (\mathbf{n}\cdot(\boldsymbol{q} - \hat{\boldsymbol{q}}), v)_{\partial E_j}, \\ (\boldsymbol{q}, \boldsymbol{\phi})_{E_j} &= (({}_aI_x^{2-\alpha} p_1, {}_cI_y^{2-\beta} p_2), \boldsymbol{\phi})_{E_j}, \\ (\boldsymbol{p}, \boldsymbol{\pi})_{E_j} &= (\nabla u, \boldsymbol{\pi})_{E_j} - (u - \hat{u}, \mathbf{n}\cdot \pi)_{\partial E_j}. \end{aligned} \tag{4.6}$$

Let N_e denote the number of the triangle elements and $N_k = \frac{1}{2}(k+1)(k+2)$ denote the degree of freedom of each element. We introduce some local and global vector and matrix notations to form the statement. We have

$$\boldsymbol{u}_{h,m} = [u_{1,m}, u_{2,m}, \ldots, u_{N_k,m}]^T, \qquad \boldsymbol{u}_h = [\boldsymbol{u}_{h,1}, \boldsymbol{u}_{h,2}, \ldots, \boldsymbol{u}_{h,N_e}]^T,$$
$$\boldsymbol{p}_{1(h,m)} = [p_{1(1,m)}, p_{1(2,m)}, \ldots, p_{1(N_k,m)}]^T,$$
$$\boldsymbol{p}_{1h} = [\boldsymbol{p}_{1(h,1)}, \boldsymbol{p}_{1(h,2)}, \ldots, \boldsymbol{p}_{1(h,N_e)}]^T,$$
$$\boldsymbol{q}_{1(h,m)} = [q_{1(1,m)}, q_{1(2,m)}, \ldots, q_{1(N_k,m)}]^T,$$
$$\boldsymbol{q}_{1h} = [\boldsymbol{q}_{1(h,1)}, \boldsymbol{q}_{1(h,2)}, \ldots, \boldsymbol{q}_{1(h,N_e)}]^T,$$
$$\boldsymbol{f}_{h,m} = [f_{1,m}, f_{2,m}, \ldots, f_{N_k,m}]^T, \qquad \boldsymbol{f}_h = [\boldsymbol{f}_{h,1}, \boldsymbol{f}_{h,2}, \ldots, \boldsymbol{f}_{h,N_e}]^T,$$

which are, respectively, the local variables for u, p_1, p_2, f on the mth element and the global ones by assembling all the local ones. We also have the local mass matrix M^m with

$$M_{ij}^m = (l_i^m(\boldsymbol{x}), l_j^m(\boldsymbol{x}))_{E_m}$$

and the local spatial stiffness matrix S_x^m, S_y^m with the entries

$$(S_x^m)_{ij} = \left(\frac{\partial l_j^m(\boldsymbol{x})}{\partial x}, l_i^m(\boldsymbol{x})\right)_{E_m}, \quad (S_y^m)_{ij} = \left(\frac{\partial l_j^m(\boldsymbol{x})}{\partial y}, l_i^m(\boldsymbol{x})\right)_{E_m}.$$

It is a little bit complex to compute the fractional spatial stiffness matrices of (4.6). The non-local property of the fractional operators makes each element of the matrices depend on the quantities of its affected regions. As in Figure 1, due to the left Riemann–Liouville FD in the x or y direction, the value on one Gauss quadrature point in an element (denoted by a black square) is affected by all the values on the solid line, which crosses over the triangles on the left side or the downside and yields some intersection points (denoted by a small black spot).

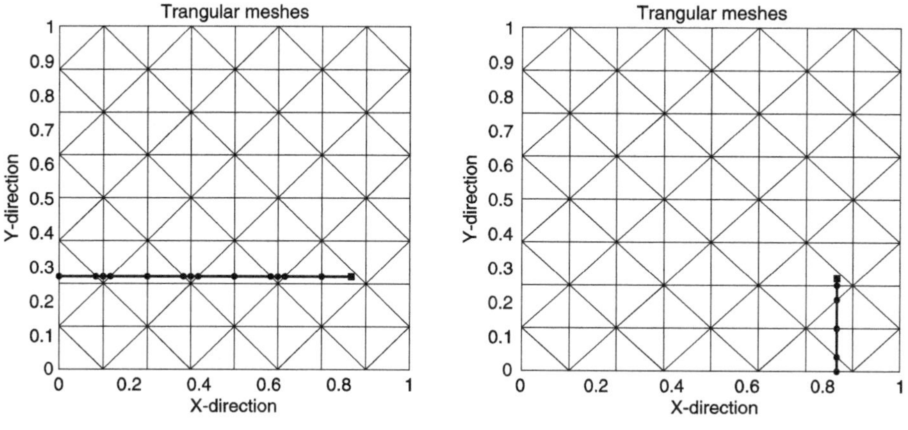

Figure 1: The intersection points (denoted by a small black spot) on triangles in the x or y direction related to the Gauss points (denoted by a black square).

Next, we discuss how to form the global fractional spatial stiffness matrix in detail. Denote $\widetilde{S}_x$ or $\widetilde{S}_y$ as the global fractional spatial stiffness matrices in the x or y direction, respectively; $\widetilde{S}_x$ and $\widetilde{S}_y$ are $N_e \times N_e$ block matrices and every element is an $N_k \times N_k$ matrix. Here, we use numerical quadrature on a triangle element to compute these fractional stiffness matrices. We only describe the specific procedure of forming the global fractional spatial stiffness matrix $\widetilde{S}_x$ for simplicity; $\widetilde{S}_y$ is obtained in the same fashion.

The integral of a function u defined on the physical element E can be approximated by using the Gauss quadrature rule on the triangle, i. e.,

$$\int_E u(\boldsymbol{x})d\boldsymbol{x} = J\int_I u(\boldsymbol{x}(\boldsymbol{r}))d\boldsymbol{r} \simeq J\sum_{j=1}^{Q_E} u(\boldsymbol{x}(\boldsymbol{r}_j))w_j = J\sum_{j=1}^{Q_E} u(\boldsymbol{x}_j)w_j,$$

where Q_E is the total number of Gauss quadrature weights or points and J is the transformation Jacobian between the physical element E and the reference element I. The sets $\{\boldsymbol{r}_j\}_{j=1}^{Q_E} \in I$ and $\{\boldsymbol{x}_j\}_{j=1}^{Q_E} \in E$ correspond one-to-one by an affine map. Thus, for functions u and v, with u^m denoting the restriction of u on the mth element, we have

$$\begin{aligned}({}_aI_x^{2-\alpha}u, v^m(\boldsymbol{x}))_{E_m} &\simeq J^m \sum_{j=1}^{Q_{E_m}} {}_aI_x^{2-\alpha}u\, v^m(\boldsymbol{x}_j)w_j \\ &= J^m \sum_{j=1}^{Q_{E_m}} \frac{1}{\Gamma(2-\alpha)}\Bigg(\sum_{n\in A}\int_{x_{n-1}}^{x_n}(x_j-\xi)^{1-\alpha}u^n(\xi,y_j)d\xi \\ &\quad + \int_{x_m}^{x_j}(x_j-\xi)^{1-\alpha}u^m(\xi,y_j)d\xi\Bigg)v^m(\boldsymbol{x}_j)w_j, \end{aligned} \tag{4.7}$$

where A is the set of elements related to each quadrature point on every triangle element, and x_n denotes the intersection point on the horizontal line passing through the elements in A to the quadrature point x_j, as sketched in Figure 1. Here we decompose the integral interval of the FI into different parts due to the different approximations of function u on different elements. By a simple linear transformation,

$$\begin{aligned}&\int_{x_m}^{x_j}(x_j-\xi)^{1-\alpha}u^m(\xi,y_j)d\xi \\ &= \left(\frac{x_j-x_m}{2}\right)^{2-\alpha}\int_{-1}^{1}(1-\eta)^{1-\alpha}u^m\left(\frac{x_j+x_m}{2}+\frac{x_j-x_m}{2}\eta, y_j\right)d\eta, \end{aligned} \tag{4.8}$$

the integral of (4.7) can be accurately approximated by the Gauss–Jacobi quadrature formulas with weight functions $(1-\eta)^{\alpha-1}$. On the other hand,

$$\int_{x_{n-1}}^{x_n} (x_j - \xi)^{1-\alpha} u^n(\xi, y_j) d\xi$$
$$= \int_{x_{n-1}}^{x_j} (x_j - \xi)^{1-\alpha} u^n(\xi, y_j) d\xi - \int_{x_n}^{x_j} (x_j - \xi)^{1-\alpha} u^n(\xi, y_j) d\xi,$$

with the right-hand side being handled similarly as (4.8). We recover the global semi-discrete form (4.6) as

$$\begin{aligned}
M\frac{\partial \boldsymbol{u}_h}{\partial t} &= S_x \boldsymbol{q}_{1h} + S_y \boldsymbol{q}_{2h} - \bigcup_{m=1}^{N_e} \int_{\partial E_m} \mathbf{n} \cdot ((\boldsymbol{q}_{1h}, \boldsymbol{q}_{2h}) - \hat{\boldsymbol{q}}_h) l(\boldsymbol{x}) ds, \\
M\boldsymbol{q}_{1h} &= {}_l\widetilde{S}_x \boldsymbol{p}_{1h}, \quad M\boldsymbol{q}_{2h} = {}_l\widetilde{S}_y \boldsymbol{p}_{2h}, \\
M\boldsymbol{p}_{1h} &= S_x \boldsymbol{u}_h - \bigcup_{m=1}^{N_e} \int_{\partial E_m} n_x (\boldsymbol{u}_h - \hat{\boldsymbol{u}}_h) l(\boldsymbol{x}) ds, \\
M\boldsymbol{p}_{2h} &= S_y \boldsymbol{u}_h - \bigcup_{m=1}^{N_e} \int_{\partial E_m} n_y (\boldsymbol{u}_h - \hat{\boldsymbol{u}}_h) l(\boldsymbol{x}) ds,
\end{aligned}$$

where M, S_x, S_y are global mass and stiffness matrices and their non-zero diagonal blocks are constructed by M^m, S_x^m, S_y^m, respectively, and $\bigcup_{m=1}^{N_e}$ refers to the construction of the equations simultaneously by numerical computations.

5 Fractional convection–diffusion equation

We consider the fractional convection–diffusion equation

$$\partial_t u(x,t) + \partial_x f(u) = {}_{-\infty}^{\mathrm{RL}}D_x^\alpha u(x,t) \quad \text{in } \Omega \times (0,T], \tag{5.1}$$

with the initial and Dirichlet boundary conditions

$$\begin{aligned}
u(x,0) &= u_0(x) \quad && \text{in } \Omega, \\
u(x,t) &= 0 && \text{in } \mathbb{R}\backslash\Omega \times (0,T],
\end{aligned}$$

where $1 < \alpha < 2$, $\Omega = (a,b)$ and f is assumed to be Lipschitz continuous. With the special choice of $f(u) = u^2/2$, it is recognized as a fractional version of the viscous Burgers equation. The main difference of (5.1) from (3.1) in Section 3 is the convection term $\partial_x f(u)$; we show two schemes in the following. The proofs of stability and convergence can be found in [18] and [17], respectively.

5.1 Monotone flux for the convection term

The basic idea for constructing the LDG schemes is also to suitably rewrite the convection–diffusion equation into a first-order system. Similar to the fractional diffusion equation, with the introduced new variables q and p, the system becomes to

$$\begin{aligned} u_t + \partial_x f(u) &= q_x \quad \text{in } \Omega \times (0,T], \\ q - {}_aI_x^{2-\alpha} p &= 0 \quad \text{in } \Omega \times (0,T], \\ p - u_x &= 0 \quad \text{in } \Omega \times (0,T]. \end{aligned} \tag{5.2}$$

Then the numerical scheme is to find $(u,q,p) \in H^1(0,T;V_k) \times L^2(0,T;V_k) \times L^2(0,T;V_k)$ such that for all $(v,w,z) \in L^2(0,T;V_k) \times L^2(0,T;V_k) \times L^2(0,T;V_k)$, and for $j = 1,\ldots,N$,

$$\begin{aligned} &\int_{I_j} u_t v dx + \int_{I_j} (q-f)v_x dx + (\hat{f} - \hat{q})v\Big|_{x_{j-1}^+}^{x_j^-} = 0, \\ &\int_{I_j} qw dx - \int_{I_j} {}_aI_x^{2-\alpha} p\, w dx = 0, \\ &\int_{I_j} pz dx + \int_{I_j} u z_x dx - \hat{u}z\Big|_{x_{j-1}^+}^{x_j^-} = 0. \end{aligned} \tag{5.3}$$

To complete the formulation of the numerical schemes, we define the numerical fluxes $\hat{q}$ and $\hat{u}$ as

$$\hat{u} = u^-, \quad \hat{q} = q^+$$

at all interior boundaries and

$$\hat{u} = 0, \quad \hat{q} = q^+ = q^-$$

at the external boundaries. As to the non-linear part $\hat{f}$, we take

$$\hat{f} = \hat{f}(u^-, u^+),$$

where $\hat{f}(u^-,u^+)$ is a monotone flux for $f(u)$, namely, $\hat{f}(u^-,u^+)$ is consistent with $f(u)$ in the sense that $\hat{f}(u,u) = f(u)$, and is non-decreasing in u^- and non-increasing in u^+ [18]. More examples of monotone fluxes which are suitable for DG schemes can be found in [19]. We could, for example, use the simple Lax–Friedrichs flux

$$\hat{f}(u^-,u^+) = \frac{1}{2}(f(u^-) + f(u^+) - \gamma(u^+ - u^-)), \quad \gamma = \max_u |f'(u)|,$$

where the maximum is taken over a relevant range of u.

5.2 Characteristic method for the convection term

If $f(u)$ in (5.1) reduces to a linear case, the convection term can be handled by characteristic methods [4, 17]. Here we consider the fractional convection–diffusion equation

$$\partial_t u + d(x,t)\partial_x u = {}_a^{\mathrm{RL}}D_x^{\alpha} u. \tag{5.4}$$

Let $\psi(x,t) = (1 + d(x,t)^2)^{1/2}$ and the characteristic direction associated with $\partial_t u + d\partial_x u$ be denoted by $\partial_\tau = \frac{1}{\psi}(\partial_t + d \cdot \partial_x)$. Then we can rewrite (5.4) as a lower-order system, i. e.,

$$\begin{aligned} \psi\partial_\tau u - q_x &= 0, \\ q - {}_aI_x^{2-\alpha} p &= 0, \\ p - u_x &= 0; \end{aligned} \tag{5.5}$$

the numerical scheme is to find $(u, q, p) \in H^1(0,T;V_k) \times L^2(0,T;V_k) \times L^2(0,T;V_k)$ such that for all $(v, w, z) \in L^2(0,T;V_k) \times L^2(0,T;V_k) \times L^2(0,T;V_k)$, and for $j = 1, \dots, N$,

$$\begin{aligned} &\int_{I_j} \psi\partial_\tau uv dx + \int_{I_j} qv_x dx - \hat{q}v\Big|_{x_{j-1}^+}^{x_j^-} = 0, \\ &\int_{I_j} qw dx - \int_{I_j} {}_aI_x^{2-\alpha} p\, w dx = 0, \\ &\int_{I_j} pz dx + \int_{I_j} uz_x dx - \hat{u}z\Big|_{x_{j-1}^+}^{x_j^-} = 0. \end{aligned} \tag{5.6}$$

The numerical fluxes $\hat{q}$ and $\hat{u}$ are chosen to be the alternating fluxes like the ones in Section 5.1.

Next, we discretize the time derivative with the characteristic method. For a given positive integer M, let $0 = t^0 < t^1 < \cdots < t^M = T$ be a partition of $(0,T]$ into subintervals $J^n = (t^{n-1}, t^n]$ with uniform mesh and the interval length denoted by τ. The characteristic tracing back along the field $d(x,t)$ of a point x at time t^n to t^{n-1} is approximated by [4]

$$\check{x}(x, t^{n-1}) = x - d(x, t^n)\tau.$$

Therefore, the hyperbolic part of (5.4) at time t^n can be approximated as

$$\psi^n \partial_\tau u^n \approx \frac{1}{\tau}(u^n - \check{u}^{n-1}),$$

where $u^n = u(x, t^n)$, $\check{u}^{n-1} = u(\check{x}(x, t^{n-1}), t^{n-1})$, and $\check{u}^0 = u^0(x)$. Thus, the fully discrete scheme is to find $(u^n, q^n, p^n) \in H^1(0,T;V_k) \times L^2(0,T;V_k) \times L^2(0,T;V_k)$, for all test functions $(v,w,z) \in L^2(0,T;V_k) \times L^2(0,T;V_k) \times L^2(0,T;V_k)$, such that

$$\begin{aligned} &\int_{I_j} \frac{1}{\tau}(u^n - \breve{u}^{n-1})v dx + \int_{I_j} q^n v_x dx - \hat{q}^n v\Big|_{x_{j-1}^+}^{x_j^-} = 0, \\ &\int_{I_j} q^n w dx - \int_{I_j} {}_aI_x^{2-\alpha} p^n \, w dx = 0, \\ &\int_{I_j} p^n z dx + \int_{I_j} u^n z_x dx - \hat{u}^n z\Big|_{x_{j-1}^+}^{x_j^-} = 0. \end{aligned} \tag{5.7}$$

6 Concluding remarks

In this chapter, we propose the LDG methods for the fractional (convection–) diffusion equations. To obtain a high-order accuracy, we rewrite the Riemann–Liouville FD operator as a composite of several first-order derivatives and one FI, transforming the problem to a lower-order system. LDG methods have significant potential for problems with spatial FDs due to their flexibility and ability to support high-order accuracy while maintaining a large degree of locality in the formulation. With the smoothness associated with solutions to fractional problems, the high-order accuracy allows for a reduced number of degrees of freedom without impacting the accuracy. This is a substantial advantage with the global nature of the fractional operators, achieving a reduced computational cost over alternative methods, in particular for problems in higher dimensions.

Bibliography

[1] F. Bassi and S. Rebay, A high-order accurate discontinuous finite element method for the numerical solution of the compressible Navier–Stokes equations, *J. Comput. Phys.*, **131** (1997), 267–279.

[2] P. Castillo, An optimal estimate for the local discontinuous Galerkin method, in B. Cockburn, G. E. Karniadakis, and C.-W. Shu (eds.), *Discontinuous Galerkin Methods: Theory, Computation and Applications*, Lecture Notes in Computational Science and Engineering, vol. 11, pp. 285–290, Springer-Verlag, 2000.

[3] P. Castillo, B. Cockburn, D. Schotzau, and C. Schwab, Optimal a priori error estimates for the hp-version of the local discontinuous Galerkin method for convection-diffusion problems, *Math. Comp.*, **71** (2002), 455–478.

[4] Z. X. Chen, Characteristic mixed discontinuous finite element methods for advection-dominated diffusion problems, *Comput. Methods Appl. Mech. Eng.*, **191** (2002), 2509–2538.

[5] P. Ciarlet, *The Finite Element Method for Elliptic Problems*, North-Holland, Amsterdam, 1978.

[6] B. Cockburn, G. E. Karniadakis, and C.-W. Shu, The development of discontinuous Galerkin methods, in B. Cockburn, G. E. Karniadakis, and C.-W. Shu (eds.), *Discontinuous Galerkin Methods: Theory, Computation and Applications*, Lecture Notes in Computational Science and Engineering, vol. 11, pp. 3–50, Springer-Verlag, 2000.
[7] B. Cockburn, S. Y. Lin, and C.-W. Shu, TVB Runge–Kutta local projection discontinuous Galerkin finite element method for conservation laws III: One dimensional systems, *J. Comput. Phys.*, **84** (1989), 90–113.
[8] B. Cockburn and C.-W. Shu, TVB Runge–Kutta local projection discontinuous Galerkin finite element method for scalar conservation laws II: General framework, *Math. Comp.*, **52** (1989), 411–435.
[9] B. Cockburn and C.-W. Shu, The local discontinuous Galerkin finite element method for the time dependent convection-diffusion systems, *SIAM J. Numer. Anal.*, **35** (1998), 2440–2463.
[10] W. H. Deng and J. S. Hesthaven, Local discontinuous Galerkin methods for fractional diffusion equations, *ESAIM: Math. Model. Numer. Anal.*, **47** (2013), 1845–1864.
[11] W. H. Deng, B. Y. Li, W. Y. Tian, and P. W. Zhang, Boundary problems for the fractional and tempered fractional operators, *Multiscale Model. Simul.*, **16** (2018), 125–149.
[12] J. S. Hesthaven and T. Warburton, *Nodal Discontinuous Galerkin Methods: Algorithms, Analysis, and Applications*, Springer Verlag, New York, 2008.
[13] R. Metzler and J. Klafter, The random walk's guide to anomalous diffusion: A fractional dynamics approach, *Phys. Rep.*, **339** (2000), 1–77.
[14] I. Podlubny, *Fractional Differential Equations*, Academic Press, New York, 1999.
[15] W. H. Reed and T. R. Hill, Triangular Mesh methods for the neutron transport equation, in *Tech. Report LA-UR-73-479*, 1973.
[16] B. Rivière, *Discontinuous Galerkin Methods for Solving Elliptic and Parabolic Equations: Theory and Implementation*, SIAM, Philadelphia, 2008.
[17] S. Q. Wang, J. Y. Yuan, W. H. Deng, and Y. J. Wu, A hybridized discontinuous Galerkin method for 2D fractional convection-diffusion equations, *J. Sci. Comput.*, **68** (2016), 826–847.
[18] Q. Xu and J. S. Hesthaven, Discontinuous Galerkin method for fractional convection-diffusion equations, *SIAM J. Numer. Anal.*, **52** (2014), 405–423.
[19] J. Yan and C.-W. Shu, Local discontinuous Galerkin methods for partial differential equations with higher order derivatives, *J. Sci. Comput.*, **17** (2002), 27–47.
[20] M. P. Zhang and C.-W. Shu, An analysis of three different formulations of the discontinuous Galerkin method for diffusion equations, *Math. Models Methods Appl. Sci.*, **13** (2008), 395–413.

Fawang Liu and Ian Turner

Numerical methods for time-space fractional partial differential equations

Abstract: In this chapter, numerical methods for time-space fractional partial differential equations are presented. Firstly, some preliminary knowledge for fractional operators is introduced. Secondly, finite difference methods for the one-dimensional time-space fractional advection–dispersion equation and time-space Caputo–Reisz fractional diffusion equation in two dimensions are considered, respectively. Thirdly, an unstructured mesh finite element method for the two-dimensional time-space Riesz fractional diffusion equation on an irregular convex domain is presented. Finally, a two-dimensional time-space fractional diffusion equation based on the fractional Laplacian operator is investigated.

Keywords: Finite difference methods, unstructured mesh finite difference/finite element methods, irregular convex domains, Riesz fractional operator, fractional Laplacian operator

MSC 2010: 65M12, 26A33

1 Introduction

Due to the non-local property of fractional-order operators, fractional calculus has had growing success in the description of the physical processes that evolve in complex systems associated with temporal memory and spatial heterogeneity. In many cases, these operators have not only shown a great potential of providing more accurate mathematical models, but they also uncover extra details neglected by their conventional integer-order counterparts. Since a fractional derivative is a generalization of an ordinary derivative, it lose many of its basic properties; for example, it loses a clear geometric or physical interpretation, the index law is only valid when working in specific functional spaces, the derivative of the product of two functions is difficult to compute, and the chain rule cannot be straightforwardly applied. Models based on partial differential equations (PDEs) containing derivatives of fractional order in time, space, or both, have attracted considerable interest over the last decade because of

Fawang Liu, School of Mathematical Sciences, Queensland University of Technology, Brisbane, QLD 4001, Australia; and College of Mathematics and Computer Science, Fuzhou University, Fujian 350116, China, e-mail: f.liu@qut.edu.au
Ian Turner, School of Mathematical Sciences, Queensland University of Technology, Brisbane, QLD 4001, Australia; and Australian Research Council Centre of Excellence for Mathematical and Statistical Frontiers (ACEMS), Queensland University of Technology, Brisbane, QLD, Australia, e-mail: i.turner@qut.edu.au

https://doi.org/10.1515/9783110571684-008

their ability to model anomalous transport phenomena [6, 10]. These phenomena are strongly connected to the interactions within complex and non-homogeneous media exhibiting spatial heterogeneity.

In this chapter, numerical methods for solving time- and space-fractional PDEs are presented. The objective is to provide a survey of our recent work for deriving and analyzing time-space fractional PDEs. In particular, we highlight a range of computational models based on the finite difference method, unstructured mesh finite element method, and unstructured mesh finite volume method. The outline of the chapter is as follows. In Section 2, some preliminary knowledge for fractional operators is introduced. In Section 3, finite difference methods for the one-dimensional time-space fractional advection–dispersion equation and the time-space Caputo–Reisz fractional diffusion equation in two-dimensions are considered. In Section 4, an unstructured mesh finite element method for the 2D time-space Riesz fractional diffusion equation on an irregular convex domain is discussed. In Section 5, a two-dimensional time-space fractional diffusion equation involving the fractional Laplacian operator is considered.

2 Preliminary knowledge

Definition 2.1 ([6], Caputo fractional derivatives on a finite interval). Let $[a, b]$ be a finite interval of the real axis, $\alpha > 0$, $n - 1 < \alpha < n$, $n \in \mathbb{N}$. Then the operators

$$ {}^{C}D_{a+}^{\alpha} f(t) = \frac{1}{\Gamma(n-\alpha)} \int_{a}^{t} \frac{f^{(n)}(\xi)d\xi}{(t-\xi)^{\alpha-n+1}}, \tag{2.1} $$

$$ {}^{C}D_{b-}^{\alpha} f(t) = \frac{(-1)^n}{\Gamma(n-\alpha)} \int_{t}^{b} \frac{f^{(n)}(\xi)d\xi}{(\xi-t)^{\alpha-n+1}} \tag{2.2} $$

are called the left- and right-sided Caputo fractional derivatives with α order, respectively. Here $\Gamma(z) = \int_0^{+\infty} x^{z-1}e^{-x}dx$, $\mathrm{Re}(z) > 0$ is the *Gamma* function, where $z \in C$ and $\mathrm{Re}(z)$ is its real part.

Definition 2.2 ([6], Riemann–Liouville fractional derivatives). Let $[a, b]$ $(-\infty < a < b < \infty)$ be a finite interval of the real axis, $f(x) \in AC^{n-1}([a, b])$, and $n - 1 < \alpha < n$, $n \in \mathbb{N}$. Then the operators

$$ {}^{RL}D_{a+}^{\alpha} f(x) = \frac{1}{\Gamma(n-\alpha)} \frac{d^n}{dx^n} \int_{a}^{x} \frac{f(\xi)}{(x-\xi)^{\alpha-n+1}} d\xi, \tag{2.3} $$

$$ {}^{RL}D_{b-}^{\alpha} f(x) = \frac{(-1)^n}{\Gamma(n-\alpha)} \frac{d^n}{dx^n} \int_{x}^{b} \frac{f(\xi)}{(\xi-x)^{\alpha-n+1}} d\xi \tag{2.4} $$

are called the left-sided and right-sided Riemann–Liouville fractional derivatives of order α, respectively.

Definition 2.3 ([6], Riesz fractional derivative on a finite interval). Let $[a,b]$ $(-\infty < a < b < \infty)$ be a finite interval of the real axis, $f(x) \in AC^{n-1}([a,b])$, and $n-1 < \alpha < n$. Then the operator

$$\frac{\partial^\alpha}{\partial|x|^\alpha} f(x) = -\frac{1}{2\cos(\frac{\alpha\pi}{2})}\{{}^{\mathrm{RL}}D^\alpha_{a+} f(x) + {}^{\mathrm{RL}}D^\alpha_{b-} f(x)\}$$
$$= -c_\alpha\left\{\frac{\partial^\alpha f(x)}{\partial x^\alpha} + \frac{\partial^\alpha f(x)}{\partial(-x)^\alpha}\right\} \tag{2.5}$$

is called the Riesz fractional derivative (potential operation) with α $(\alpha > 0)$ order, $c_\alpha = \frac{1}{2\cos(\frac{\alpha\pi}{2})}$.

Theorem 2.1 ([6], The relationship between Caputo and Riemann–Liouville fractional derivatives). *Let $n-1 < \alpha < n$, $n \in \mathbb{N}$, $f(t) \in AC^n[a,b]$. Then the Caputo derivatives ${}^C D^\alpha_{a+} f(t)$ and ${}^C D^\alpha_{b-} f(t)$ exist almost everywhere on $[a,b]$ and can be expressed as*

$${}^C D^\alpha_{a+} f(t) = {}^{\mathrm{RL}}D^\alpha_{a+} f(t) - \sum_{j=0}^{n-1} \frac{f^{(j)}(a)}{\Gamma(j+1-\alpha)}(t-a)^{j-\alpha}, \tag{2.6}$$

$${}^C D^\alpha_{b-} f(t) = {}^{\mathrm{RL}}D^\alpha_{b-} f(t) - (-1)^n \sum_{j=0}^{n-1} \frac{f^{(j)}(b)}{\Gamma(j+1-\alpha)}(b-t)^{j-\alpha}. \tag{2.7}$$

Lemma 2.1 ([6]). *The L1 approximation for discretization of the Caputo time-fractional derivative* $(0 < \gamma < 1)$

$${}^C D^\gamma_{0+} f(t_{k+1}) = \frac{\tau^{-\gamma}}{\Gamma(2-\gamma)} \sum_{j=0}^{k} b_j^{(\gamma)}[f(t_{k+1-j}) - f(x, t_{k-j})] + R^{(C)}_{k+1}, \tag{2.8}$$

or it can be expressed in the form

$${}^C D^\gamma_{0+} f(t_{k+1}) = \frac{\tau^{-\gamma}}{\Gamma(2-\gamma)}\left[b_0^{(\gamma)} f(t_{k+1}) - b_k^{(\gamma)} f(t_0) - \sum_{j=1}^{k}(b_{j-1}^{(\gamma)} - b_j^{(\gamma)}) f(t_{k+1-j})\right] + R^{(C)}_{k+1},$$

where $b_j^{(\gamma)} = (j+1)^{1-\gamma} - j^{1-\gamma}, j = 0,1,2,\ldots,N$, $|R^{(C)}_{k+1}| \le C\tau^{2-\gamma}$.

Lemma 2.2 ([6]). *If* $0 < \gamma < 1$, *we have*

(i) $b_j^{(\gamma)} > b_{j+1}^{(\gamma)}, j = 0,1,2,\ldots;$

(ii) $b_0^{(\gamma)} = 1$, $b_j^{(\gamma)} > 0, j = 0,1,2,\ldots;$

(iii) *there exist constants* $C_1 > 0$ *and* $C_2 > 0$, *such that*

$$C_1 k^\gamma \le (b_k^{(\gamma)})^{-1} \le C_2 k^\gamma.$$

Definition 2.4 ([6], Left- and right-sided Grünwald–Letnikov fractional derivatives). The terms

$$^{\mathrm{GL}}D_{a+}^{\alpha}f(x) = \lim_{h\longrightarrow 0}\frac{1}{h^{\alpha}}\sum_{j=0}^{\infty} g_j^{(\alpha)} f(x-jh), \tag{2.9}$$

$$^{\mathrm{GL}}D_{b-}^{\alpha}f(x) = \lim_{h\longrightarrow 0}\frac{1}{h^{\alpha}}\sum_{j=0}^{\infty} g_j^{(\alpha)} f(x+jh) \tag{2.10}$$

are called the left- and right-sided Grünwald–Letnikov fractional derivatives, respectively. Here

$$g_0^{(\alpha)} = 1, \quad g_j^{(\alpha)} = \left(1-\frac{\alpha+1}{j}\right)g_{j-1}^{(\alpha)}, \quad j = 1,2,\dots. \tag{2.11}$$

Lemma 2.3 ([6]). *When $0 < \alpha < 1$, the coefficients $g_j^{(\alpha)}$ satisfy*
(1) $g_0^{(\alpha)} = 1, g_1^{(\alpha)} = -\alpha, g_j^{(\alpha)} < 0, j = 2,3,\dots;$
(2) $\sum_{j=0}^{\infty} g_j^{(\alpha)} = 0;$
(3) *for any positive integer N, the inequality $\sum_{j=0}^{N} g_j^{(\alpha)} > 0$ holds.*

Lemma 2.4 ([6]). *When $1 < \alpha < 2$, the coefficients $g_j^{(\alpha)}$ satisfy:*
(1) $g_0^{(\alpha)} = 1, g_1^{(\alpha)} = -\alpha < 0, g_j^{(\alpha)} > 0\ (j = 2,3,\dots);$
(2) $\sum_{j=0}^{\infty} g_j^{(\alpha)} = 0;$
(3) *for any positive integer N, the inequality $\sum_{j=0}^{N} g_j^{(\alpha)} < 0$ holds.*

Lemma 2.5 ([6]). *Let $f(x) \in C^n[a,b], \alpha \geq 0, N = (b-a)/h$. Then the finite left Grünwald–Letnikov formula is given by*

$$^{\mathrm{GL}}D_{a+}^{\alpha}f(x) = \frac{1}{h^{\alpha}}\sum_{j=0}^{\lceil (x-a)/h\rceil} g_j^{(\alpha)} f(x-jh), \tag{2.12}$$

i. e.,

$$^{\mathrm{RL}}D_{a+}^{\alpha}f(x) = \frac{1}{h^{\alpha}}\sum_{j=0}^{\lceil (x-a)/h\rceil} g_j^{(\alpha)} f(x-jh) + O(h), \tag{2.13}$$

where $\lceil\cdot\rceil$ denotes the ceiling function that rounds up to the nearest integer.

The Grünwald–Letnikov and Riemann–Liouville fractional derivatives have the following properties.

Property 2.1 ([6]). Let $0 < n-1 < \alpha < n, n \in \mathbb{N}$. Then for functions $f(x)$ having n continuous derivatives the left (right) Grünwald–Letnikov definition is equivalent to the left (right) Riemann–Liouville definition.

Property 2.2 ([6]). The relationship between the Riemann–Liouville and Grünwald–Letnikov definitions is important for the numerical approximation of fractional-order differential equations because it allows the use of the Riemann–Liouville definitions during problem formulation, and then the use of the Grünwald–Letnikov definitions for obtaining the numerical solution.

Property 2.3 ([6]). When considering a well-defined function $f(x)$ on the bounded interval $[a, b]$, if $f(a) = 0$ or $f(b) = 0$, the function $f(x)$ can be zero extended for $x < a$ or $x > b$, and the α-order left and right Riemann–Liouville fractional derivatives of $f(x)$ at each point x can be approximated by finite terms (i. e., replacing a finite number of terms as an infinite number of terms; see the formulas in Lemma 2.5 and Lemma 2.6).

Lemma 2.6 ([6]). *Let $f(x) \in C^n[a, b]$, $\alpha \geq 0$, $N = (b - a)/h$. Then the finite right Grünwald–Letnikov formula given by*

$$ {}^{\mathrm{GL}}D_{b-}^{\alpha} f(x) = \frac{1}{h^{\alpha}} \sum_{j=0}^{\lceil (b-x)/h \rceil} g_j^{(\alpha)} f(x + jh) \tag{2.14}$$

yields a first-order approximation for the right Riemann–Liouville differential operator, i. e.,

$$ {}^{\mathrm{RL}}D_{b-}^{\alpha} f(x) = \frac{1}{h^{\alpha}} \sum_{j=0}^{\lceil (b-x)/h \rceil} g_j^{(\alpha)} f(x + jh) + O(h). \tag{2.15}$$

Remark 2.1. Meerschaert et al. [9] found that, when solving a space-fractional diffusion equation using the standard Grünwald–Letnikov derivatives for the fractional diffusion term, the explicit Euler method, implicit Euler method, and Crank–Nicolson method are unconditionally unstable. Thus, Meerschaert et al. suggested the use of the shifted Grünwald–Letnikov formulas to approximate the fractional diffusion term.

Lemma 2.7 ([6]). *The finite left/right shifted Grünwald–Letnikov formulas are given by*

$$\begin{aligned} {}^{\mathrm{RL}}D_{a+,1}^{\alpha} f(x) &= {}^{\mathrm{GL}}D_{a+,1}^{\alpha} f(x) + O(h) \\ &= \frac{1}{h^{\alpha}} \sum_{j=0}^{\lceil (x-a)/h+1 \rceil} g_j^{(\alpha)} f(x - (j-1)h) + O(h), \end{aligned} \tag{2.16}$$

$$\begin{aligned} {}^{\mathrm{RL}}D_{b-,1}^{\alpha} f(x) &= {}^{\mathrm{GL}}D_{b-,1}^{\alpha} f(x) + O(h) \\ &= \frac{1}{h^{\alpha}} \sum_{j=0}^{\lceil (b-x)/h+1 \rceil} g_j^{(\alpha)} f(x + (j-1)h) + O(h). \end{aligned} \tag{2.17}$$

Remark 2.2 ([6]). Using the weighted and shifted Grünwald–Letnikov formulas, we can obtain some higher-order approximations.

3 Finite difference methods for the TSFPDE

3.1 Time-space fractional advection–dispersion equations

Fractional advection–dispersion equations are used, for example, in groundwater hydrology to model the transport of passive tracers carried by fluid flow in a porous medium. Fractional space derivatives are used to model anomalous diffusion or dispersion, where a particle plume spreads at a rate inconsistent with the classical Brownian motion model. When a fractional derivative replaces the second derivative in a diffusion or dispersion model, it leads to enhanced diffusion (also called superdiffusion [6]). For a one-dimensional advection–dispersion model with constant coefficients, analytical solutions are available using Fourier transform methods. However, many practical problems require a model with variable coefficients.

Consider the following one-dimensional time-space fractional advection–dispersion equation (TSFADE) [5, 6]:

$$ {}^{C}D^{\gamma}_{0+}u(x,t) = -v(x,t)^{\mathrm{RL}}D^{\beta}_{0+}u(x,t) + K(x,t)^{\mathrm{RL}}D^{\alpha}_{0+}u(x,t) + f(x,t), \tag{3.1} $$

with the initial and time-dependent Dirichlet boundary conditions given by

$$ u(x,0) = \psi(x), \tag{3.2} $$

$$ u(0,t) = \varphi_1(t) = 0, u(L,t) = \varphi_2(t), \tag{3.3} $$

where $0 < \gamma < 1$, $0 < \beta < 1$, and $1 < \alpha < 2$.

We introduce an implicit finite difference method for solving the TSFPDE.

Define $t_k = k\tau$, $k = 0, 1, 2, \dots, N$, $x_i = ih$, $i = 0, 1, 2, \dots, M$, where $\tau = \frac{T}{N}$ and $h = \frac{L}{M}$ are the time and space steps, respectively. Let u_i^k be the numerical solution of the difference method.

To solve (3.1) the L1 approximation is used to discretize the time-fractional derivative, the Grünwald–Letnikov formula is used for the advection term and the shifted Grünwald–Letnikov formula is used for the diffusion term.

Let $r^{(1)}_{i,k} = \frac{v(x_i,t_k)\tau^{\gamma}\Gamma(2-\gamma)}{h^{\beta}}$, $r^{(2)}_{i,k} = \frac{K(x_i,t_k)\tau^{\gamma}\Gamma(2-\gamma)}{h^{\alpha}}$, $v_{i,k} = v(x_i, t_k)$, $K_{i,k} = K(x_i, t_k)$, $f_i^k = f(x_i, t_k)$, $d_i^k = \tau^{\gamma}\Gamma(2-\gamma)f_i^k$. Then the following implicit fractional difference method (IFDM) can be obtained:

$$ \begin{aligned} &u_i^{k+1} - b_k^{(\gamma)}u_i^0 - \sum_{j=1}^{k}(b_{j-1}^{(\gamma)} - b_j^{(\gamma)})u_i^{k+1-j} \\ &\quad = -r^{(1)}_{i,k+1}\sum_{j=0}^{i} g_j^{(\beta)}u_{i-j}^{k+1} + r^{(2)}_{i,k+1}\sum_{j=0}^{i+1} g_j^{(\alpha)}u_{i-j+1}^{k+1} + d_i^{k+1}, \end{aligned} \tag{3.4} $$

which can be expressed in the following form:

$$ u_i^{k+1} + r^{(1)}_{i,k+1}\sum_{j=0}^{i} g_j^{(\beta)}u_{i-j}^{k+1} - r^{(2)}_{i,k+1}\sum_{j=0}^{i+1} g_j^{(\alpha)}u_{i-j+1}^{k+1} $$

$$= b_k^{(\gamma)} u_i^0 + \sum_{j=1}^{k} (b_{j-1}^{(\gamma)} - b_j^{(\gamma)}) u_i^{k+1-j} + d_i^{k+1}. \tag{3.5}$$

The explicit fractional difference method (EFDM) for solving (3.1)–(3.3) is given by

$$u_i^{k+1} = b_k^{(\gamma)} u_i^0 + \sum_{j=1}^{k} (b_{j-1}^{(\gamma)} - b_j^{(\gamma)}) u_i^{k+1-j}$$
$$- r_{i,k}^{(1)} \sum_{j=0}^{i} g_j^{(\beta)} u_{i-j}^k + r_{i,k}^{(2)} \sum_{j=0}^{i+1} g_j^{(\alpha)} u_{i-j+1}^k + d_i^k. \tag{3.6}$$

The boundary and initial conditions are discretized as follows:

$$u_i^0 = \psi(ih), \tag{3.7}$$
$$u_0^k = 0, \ u_M^k = \varphi_2(k\tau). \tag{3.8}$$

Theorem 3.1 ([5, 6]). *The IFDM is unconditionally stable.*

Theorem 3.2 ([5, 6]). *Let u_i^k and $u(x_i, t_k)$ be the numerical solution of the IFDM and exact solution of the TSFPDE, respectively. Then there is a positive constant C, such that*

$$|u_i^k - u(x_i, t_k)| \le C(\tau^{2-\gamma} + h), \quad i = 1, 2, \dots, m-1, \ k = 1, 2, \dots, n.$$

Theorem 3.3 ([5, 6]). *If $r_{i,k}^{(1)} + \beta r_{i,k}^{(2)} < 2 - 2^{1-\alpha}$, the fractional explicit difference scheme is stable.*

Theorem 3.4 ([5, 6]). *Let u_i^k and $u(x_i, t_k)$ be the numerical solution of the FEDM and the exact solution of the TSFADE, respectively. If $r_{i,k}^{(1)} + \beta r_{i,k}^{(2)} < 2 - 2^{1-\alpha}$, there is a positive constant C, such that*

$$|u_i^k - u(x_i, t_k)| \le C(\tau^{2-\gamma} + h), \quad i = 1, 2, \dots, m-1, \ k = 1, 2, \dots, n.$$

Remark 3.1. By using the weighted Grünwald–Letnikov formula or extrapolation technique, higher-order finite difference schemes can be derived [6].

Example 3.1. Consider the following space-time fractional advection–dispersion equation (STFADE):

$$ {}^{C}D_{0+}^{\gamma} u(x,t) = -v\, {}^{RL}D_{0+}^{\beta} u(x,t) + K\, {}^{\mathrm{RL}}D_{0+}^{\alpha} u(x,t) + f(x,t),$$
$$0 \le x \le 1, \ 0 < t \le T, \tag{3.9}$$

with the initial and boundary conditions given by

$$u(x,0) = \psi(x), \quad (0 \le x \le 1), \tag{3.10}$$
$$u(0,t) = 0, u(1,t) = 0, \quad (0 \le t \le T), \tag{3.11}$$

where $0 < \gamma < 1$, $0 < \beta < 1$, $1 < \alpha < 2$, $\nu > 0$, and $K > 0$ are constants. To illustrate the effectiveness of the proposed numerical scheme the function $f(x,t)$ is chosen so that the exact solution of (3.9)–(3.11) is given by

$$u(x,t) = (t^{\gamma+\beta} + t^{\alpha})x^2(1-x)^2.$$

To obtain the exact form of $f(x,t)$, the following formulas are used:

$$\begin{aligned}
{}^{C}D^{\gamma}_{0+}(t-a)^{\nu} &= \frac{\Gamma(\nu+1)(t-a)^{\nu-\gamma}}{\Gamma(\nu-\gamma+1)},\\
{}^{RL}D^{\alpha}_{0+}(x-a)^{\nu} &= \frac{\Gamma(\nu+1)(x-a)^{\nu-\alpha}}{\Gamma(\nu-\alpha+1)},
\end{aligned}$$

so that

$$\begin{aligned}
{}^{C}D^{\gamma}_{0+}u(x,t) &= \left(\frac{\Gamma(\gamma+\beta+1)}{\Gamma(\beta+1)}t^{\beta} + \frac{\Gamma(\alpha+1)}{\Gamma(\alpha+1-\gamma)}t^{\alpha-\gamma}\right)x^2(1-x)^2,\\
{}^{RL}D^{\beta}_{0+}u(x,t) &= (t^{\gamma+\beta}+t^{\alpha})\left(\frac{2}{\Gamma(3-\beta)}x^{2-\beta} - \frac{12}{\Gamma(4-\beta)}x^{3-\beta} + \frac{24}{\Gamma(5-\beta)}x^{4-\beta}\right),\\
{}^{RL}D^{\alpha}_{0+}u(x,t) &= (t^{\gamma+\beta}+t^{\alpha})\left(\frac{2}{\Gamma(3-\alpha)}x^{2-\alpha} - \frac{12}{\Gamma(4-\alpha)}x^{3-\alpha} + \frac{24}{\Gamma(5-\alpha)}x^{4-\alpha}\right),
\end{aligned}$$

and therefore

$$\begin{aligned}
f(x,t) &= {}^{C}D^{\gamma}_{0+}u(x,t) + K\,{}^{RL}D^{\beta}_{0+}u(x,t) - \nu\,{}^{RL}D^{\alpha}_{0+}u(x,t)\\
&= \left(\frac{\Gamma(\gamma+\beta+1)}{\Gamma(\beta+1)}t^{\beta} + \frac{\Gamma(\alpha+1)}{\Gamma(\alpha+1-\gamma)}t^{\alpha-\gamma}\right)x^2(1-x)^2\\
&\quad + \nu(t^{\gamma+\beta}+t^{\alpha})\left(\frac{2}{\Gamma(3-\beta)}x^{2-\beta} - \frac{12}{\Gamma(4-\beta)}x^{3-\beta} + \frac{24}{\Gamma(5-\beta)}x^{4-\beta}\right)\\
&\quad - K(t^{\gamma+\beta}+t^{\alpha})\left(\frac{2}{\Gamma(3-\alpha)}x^{2-\alpha} - \frac{12}{\Gamma(4-\alpha)}x^{3-\alpha} + \frac{24}{\Gamma(5-\alpha)}x^{4-\alpha}\right).
\end{aligned}$$

An implicit difference method (IDM) is now established for solving the STFADE (3.9)–(3.11), i. e.,

$$\begin{aligned}
&u_i^{k+1} + r^{(1)}\sum_{j=0}^{i} g_j^{(\beta)}u_{i-j}^{k+1} - r^{(2)}\sum_{j=0}^{i+1} g_j^{(\alpha)}u_{i-j+1}^{k+1}\\
&\quad = b_k^{(\gamma)}u_i^0 + \sum_{j=1}^{k}(b_{j-1}^{(\gamma)} - b_j^{(\gamma)})u_i^{k+1-j} + d_i^{k+1},
\end{aligned} \tag{3.12}$$

where $r^{(1)} = \frac{\nu\tau^{\gamma}\Gamma(2-\gamma)}{h^{\beta}}$, $r^{(2)} = \frac{K\tau^{\gamma}\Gamma(2-\gamma)}{h^{\alpha}}$, $f_i^k = f(x_i,t_k)$, $d_i^k = \tau^{\gamma}\Gamma(2-\gamma)f_i^k$.

We take $\nu = 0.5$, $K = 0.5$, $\gamma = 0.7$, $\beta = 0.3$, $\alpha = 1.6$, $T = 1.5$, $h \approx \tau^{2-\gamma}$. The error and convergence rate are exhibited in Table 1. It can be seen from the table that the expected order of time is $2-\gamma$, which shows the stability and convergence of the IDM.

Table 1: The error and convergence order of τ for $h \approx \tau^{2-\gamma}$ with $v = 0.5$, $K = 0.5$, $\gamma = 0.7$, $\beta = 0.3$, $\alpha = 1.6$ at $T = 1.5$.

τ	$h \approx \tau^{2-\gamma}$	Error	Order
1/10	5.0119E−02	7.1151E−03	
1/20	2.0355E−02	2.2979E−03	1.63
1/40	8.2665E−03	9.8542E−04	1.22
1/80	3.3572E−03	3.8951E−04	1.34
1/160	1.3635E−03	1.5628E−04	1.32

3.2 Time-space Caputo–Reisz fractional diffusion equation in 2D

Fractional-order dynamics in physics, particularly when applied to diffusion, leads to an extension of the concept of Brownian motion through a generalization of the Gaussian probability function to what is termed anomalous diffusion. As magnetic resonance imaging (MRI) is applied with increasing temporal and spatial resolution, the spin dynamics are being examined more closely; such examinations extend our knowledge of biological materials through a detailed analysis of relaxation time distribution and water diffusion heterogeneity. In MRI, it is tissue water that is diffusing in the complex tissue environment of cells, membranes, and connective tissue. Fractional-order models are used to fit the observed magnetization signal decay on a pixel-by-pixel basis, hence providing a new contrast mechanism that has the potential to improve our understanding of disease. Here the dynamical models become more complex as they attempt to correlate new data with a multiplicity of tissue compartments where processes are often anisotropic. Anomalous diffusion in the human brain using fractional-order calculus has been investigated in [7]. Recently, a new diffusion model was proposed by solving the Bloch–Torrey equation using fractional-order calculus with respect to time and space [8]. However, effective numerical methods and supporting error analyses for the fractional Bloch–Torrey equation are still limited [11, 12].

In this section, the space-time fractional Bloch–Torrey equation (ST-FBTE) with Riesz derivative form is considered [16]. The time and space derivatives in the ST-FBTE are replaced by the Caputo and the sequential Riesz fractional derivatives, respectively. The ST-FBTE can be written in the following form:

$$K_\alpha \, {}_0^C D_t^\alpha M_{xy}(\mathbf{r}, t) = \lambda M_{xy}(\mathbf{r}, t) + K_\beta \mathbf{R}^\beta M_{xy}(\mathbf{r}, t), \tag{3.13}$$

where $\lambda = -i\gamma(\mathbf{r} \cdot \mathbf{G}(t))$, $\mathbf{G}(t)$ is the magnetic field gradient, γ is the gyromagnetic ratio, and $\mathbf{r} = (x, y, z)$. Here $\mathbf{R}^\beta = (R_x^\beta + R_y^\beta + R_z^\beta)$ is a sequential Riesz fractional-order Laplacian operator in space; $M_{xy}(\mathbf{r}, t) = M_x(\mathbf{r}, t) + iM_y(\mathbf{r}, t)$, where $i = \sqrt{-1}$, comprises the transverse components of the magnetization ($0 < \alpha \le 1$ and $1 < \beta \le 2$).

For the numerical solution of the ST-FBTE in Riesz form, the real and imaginary components are equated to express equation (3.13) as a coupled system of PDEs for the components M_x and M_y, namely,

$$K_\alpha\, {}_0^C D_t^\alpha M_x(\mathbf{r},t) = \lambda_G M_y(\mathbf{r},t) + K_\beta\left(\frac{\partial^\beta}{\partial|x|^\beta} + \frac{\partial^\beta}{\partial|y|^\beta} + + \frac{\partial^\beta}{\partial|z|^\beta}\right) M_x(\mathbf{r},t), \tag{3.14}$$

$$K_\alpha\, {}_0^C D_t^\alpha M_y(\mathbf{r},t) = -\lambda_G M_x(\mathbf{r},t) + K_\beta\left(\frac{\partial^\beta}{\partial|x|^\beta} + \frac{\partial^\beta}{\partial|y|^\beta} + + \frac{\partial^\beta}{\partial|z|^\beta}\right) M_y(\mathbf{r},t), \tag{3.15}$$

where $\lambda_G = \gamma(\mathbf{r}\cdot\mathbf{G}(t))$.

For convenience, we only consider the two-dimensional case here, however the following methods and techniques can be directly extended to the three-dimensional time-space Caputo–Reisz fractional diffusion equation (3D-TS-CR-FDE) case. The 2D-ST-FBTE versions of (3.14) and (3.15) are decoupled, which leads to solving a 2D-TS-CR-FDE of the following form [6]:

$$ {}^C D_{0+}^\gamma u(x,y,t) = K_\alpha\left(\frac{\partial^\alpha}{\partial|x|^\alpha} + \frac{\partial^\alpha}{\partial|y|^\alpha}\right) u(x,y,t) + f(x,y,t), \tag{3.16}$$

$$u(x,y,0) = \psi(x,y), \tag{3.17}$$

$$u(x,y,t)|_\Gamma = 0, \tag{3.18}$$

where $0<\gamma<1$, $1<\alpha<2$, $0\le t\le T$, and $(x,y)\in\Omega$, Ω is the finite region $[0,L_x]\times[0,L_y]$, and Γ is the boundary of Ω.

3.2.1 2D implicit fractional difference method for the 2D-TS-CR-FDE

Let $h_x = L_x/M_1$, $h_y = L_y/M_2$, and $\tau = T/N$ be the spatial and time steps, respectively.

Using an L1 approximation to discretize the time-fractional derivative, and the shifted Grünwald–Letnikov formulas for the Riesz fractional derivatives, the following implicit numerical scheme (INM) can be derived [16]:

$$\frac{\tau^{-\gamma}}{\Gamma(2-\gamma)}\sum_{l=0}^{n} b_l[u_{i,j}^{n+1-l} - u_{i,j}^{n-l}] = -c_\alpha K_\alpha\left[\frac{1}{h_x^\alpha}\left(\sum_{p=0}^{i+1} g_p u_{i-p+1,j}^{n+1} + \sum_{p=0}^{M_1-i+1} g_p u_{i+p-1,j}^{n+1}\right) + \frac{1}{h_y^\alpha}\left(\sum_{q=0}^{j+1} g_q u_{i,j-q+1}^{n+1} + \sum_{q=0}^{M_2-j+1} g_q u_{i,j+q-1}^{n+1}\right)\right] + f_{i,j}^{n+1}. \tag{3.19}$$

Thus, the following two-dimensional IFDM (2D-IFDM) is obtained:

$$u_{i,j}^{n+1} + \mu_1\left(\sum_{p=0}^{i+1} g_p u_{i-p+1,j}^{n+1} + \sum_{p=0}^{M_1-i+1} g_p u_{i+p-1,j}^{n+1}\right)$$

$$
\begin{aligned}
&+\mu_2\left(\sum_{q=0}^{j+1} g_q u_{i,j-q+1}^{n+1}+\sum_{q=0}^{M_2-j+1} g_q u_{i,j+q-1}^{n+1}\right)\\
&=\sum_{l=0}^{n-1}(b_l-b_{l+1})u_{i,j}^{n-l}+b_n u_{i,j}^0+\mu_0 f_{i,j}^{n+1}, \qquad (3.20)\\
&i=1,2,\dots,M_1-1, j=1,2,\dots,M_2-1,
\end{aligned}
$$

with $(i=0,1,\dots,M_1, j=0,1,\dots,M_2)$

$$u_{i,j}^0=\psi_{i,j}=\psi(x_i,y_j), \qquad (3.21)$$

$$u_{0,j}^{n+1}=u_{M_1,j}^{n+1}=u_{i,0}^{n+1}=u_{i,M_2}^{n+1}=0, \qquad (3.22)$$

where $\mu_0=\tau^\gamma\Gamma(2-\gamma)$, $\mu_1=\frac{c_\alpha K_\alpha\tau^\gamma\Gamma(2-\gamma)}{h_x^\alpha}$, and $\mu_2=\frac{c_\alpha K_\alpha\tau^\gamma\Gamma(2-\gamma)}{h_y^\alpha}$ and the coefficients $\mu_0>0$, $\mu_1,\mu_2<0$ for $0<\gamma\le 1$ and $1<\alpha\le 2$.

Theorem 3.5 ([16]). *The 2D-IFDM for the 2D-TS-CR-FDE is unconditionally stable.*

Theorem 3.6 ([16]). *Let $u_{i,j}^n$ and $u(x_i,y_j,t_n)$ be the numerical solution of the 2D-IFDM and the exact solution of the 2D-TS-CR-FDE, respectively. Then there is a positive constant C, such that*

$$|u_{i,j}^n-u(x_i,y_j,t_n)|\le C(\tau^{2-\gamma}+h_x+h_y), \quad n=1,2,\dots.$$

Example 3.2. The following space-time fractional Bloch–Torrey equation with initial and boundary conditions on a finite domain is considered:

$$K_\alpha\, {}_0^C D_t^\alpha M(\mathbf{r},t)=K_\beta\left(\frac{\partial^\beta}{\partial|x|^\beta}+\frac{\partial^\beta}{\partial|y|^\beta}\right)M(\mathbf{r},t)+f(\mathbf{r},t), \qquad (3.23)$$

$$M(\mathbf{r},0)=0, \qquad (3.24)$$

$$M(\mathbf{r},t)|_\Gamma=0, \qquad (3.25)$$

where

$$
\begin{aligned}
f(\mathbf{r},t)=&\frac{K_\beta t^{\alpha+\beta}}{2\cos(\beta\pi/2)}\Bigg(\bigg(\frac{2}{\Gamma(3-\beta)}[x^{2-\beta}+(1-x)^{2-\beta}]\\
&-\frac{12}{\Gamma(4-\beta)}[x^{3-\beta}+(1-x)^{3-\beta}]+\frac{24}{\Gamma(5-\beta)}[x^{4-\beta}\\
&+(1-x)^{4-\beta}]\bigg)y^2(1-y)^2+\bigg(\frac{2}{\Gamma(3-\beta)}[y^{2-\beta}+(1-y)^{2-\beta}]\\
&-\frac{12}{\Gamma(4-\beta)}[y^{3-\beta}+(1-y)^{3-\beta}]\\
&+\frac{24}{\Gamma(5-\beta)}[y^{4-\beta}+(1-y)^{4-\beta}]\bigg)x^2(1-x)^2\Bigg)\\
&+\frac{K_\alpha\Gamma(\alpha+\beta+1)}{\Gamma(\beta+1)}t^\beta x^2(1-x)^2y^2(1-y)^2, \qquad (3.26)
\end{aligned}
$$

and $0 < \alpha \leq 1$, $1 < \beta \leq 2$, $t > 0$, $\mathbf{r} = (x, y) \in \Omega$, Ω is the finite rectangular region $[0, 1] \times [0, 1]$, and Γ is the boundary of Ω.

The exact solution of this problem is $M(\mathbf{r}, t) = t^{\alpha+\beta}x^2(1-x)^2y^2(1-y)^2$, which can be verified by substituting directly into (3.23). The maximum absolute error between the exact solution and the numerical solutions obtained by INM, with spatial and temporal steps $\tau^{2-\alpha} \approx h_x = h_y = 1/8, 1/16, 1/32$ at time $t = 1.0$ when $K_\alpha = 1.0$, $K_\beta = 0.5$, $\alpha = 0.8$, $\beta = 1.8$, is listed in Table 2.

Table 2: Comparison of maximum error for INM at time $t = 1.0$ with $\alpha = 0.8$, $\beta = 1.8$, $K_\alpha = 1.0$, $K_\beta = 0.5$.

τ	$h_x = h_y (\approx \tau^{2-\alpha})$	Maximum computed error	Error rate
$\frac{1}{6}$	$\frac{1}{8}$	0.00049160	–
$\frac{1}{10}$	$\frac{1}{16}$	0.00027001	$1.82 \approx 2$
$\frac{1}{18}$	$\frac{1}{32}$	0.00013504	$1.9995 \approx 2$

From Table 2, it can be seen that the Error rate $= \frac{\text{error}_1}{\text{error}_2} \approx \frac{h_1}{h_2} = 2$. Here, h_1 and h_2 denote different spatial step sizes. Table 3 shows the maximum absolute error between the exact solution and the numerical solutions obtained by INM, with spatial steps $h_x = h_y = 1/32$ at time $t = 1.0$ when $K_\alpha = 1.0$, $K_\beta = 0.5$, $\alpha = 0.8$, $\beta = 1.8$. From Table 3, it can be seen that the Error rate $= \frac{\text{error}_1}{\text{error}_2} \approx (\frac{\tau_1}{\tau_2})^{2-\alpha} = 2.3$. This is in good agreement with our theoretical analysis, namely that the convergence order of the numerical method INM for this problem is $O(\tau^{2-\alpha} + h_x + h_y)$.

Table 3: Comparison of maximum error for INM with $h_x = h_y = 1/32$ at time $t = 1.0$ when $\alpha = 0.8$, $\beta = 1.8$, $K_\alpha = 1.0$, $K_\beta = 0.5$.

τ	Maximum computed error	Error rate
$\frac{1}{4}$	0.00078984	–
$\frac{1}{8}$	0.00035091	$2.25 \approx 2.30$
$\frac{1}{16}$	0.00015520	$2.26 \approx 2.30$

3.2.2 Two-dimensional fractional alternating direction implicit method for the 2D-TS-CR-FDE

A two-dimensional fractional alternating direction implicit method (2D-FADDIM) is now proposed for solving the 2D-TS-CR-FDE [15]. We consider the following fractional partial differential discrete operators:

$$\delta_x^\alpha u_{i,j}^{n+1} = \sum_{p=0}^{i+1} g_p u_{i-p+1,j}^{n+1} + \sum_{p=0}^{M_1-i+1} g_p u_{i+p-1,j}^{n+1}, \tag{3.27}$$

$$\delta_y^\alpha u_{i,j}^{n+1} = \sum_{q=0}^{j+1} g_q u_{i,j-q+1}^{n+1} + \sum_{q=0}^{M_2-j+1} g_q u_{i,j+q-1}^{n+1}, \tag{3.28}$$

which are $O(h_x)$ and $O(h_y)$ approximations of the Riesz fractional derivatives by the shifted Grünwald–Letnikov schemes, respectively.

Thus, the implicit fractional difference scheme (3.20) may be rearranged into the following form involving δ_x^α and δ_y^α:

$$(1 + \mu_1 \delta_x^\alpha + \mu_2 \delta_y^\alpha) u_{i,j}^{n+1} = \sum_{l=0}^{n-1} (b_l - b_{l+1}) u_{i,j}^{n-l} + b_n u_{i,j}^0 + \mu_0 f_{i,j}^{n+1}. \tag{3.29}$$

The implicit fractional difference scheme (3.20) for the 2D-TS-CR-FDE has a local truncation error of the form $O(\tau^{2-\gamma} + h_x + h_y)$ and is unconditionally stable. Unfortunately, (3.20) requires a linear system of equations to be solved for calculating the difference solution $u_{i,j}^{n+1}$ that does not have the desirable property of the coefficient matrix being sparse and band structured as for the classical case. That is to say, at each time step, the implicit fractional difference scheme (3.20) requires the solution of a very large dense linear system of equations with $(M_1 - 1) \times (M_2 - 1)$ unknowns, which is computationally expensive to solve. It is therefore necessary to construct other numerical methods that are unconditionally stable, with less computational overheads.

For this purpose, the alternating direction implicit method is adopted. The aim is to divide the calculation into three steps with reduced calculation. In the first step, the problem is solved in the x direction, and in the second step, the problem is solved in the y direction. To do this an additional negligible term is introduced, i. e.,

$$\mu_1 \mu_2 \delta_x^\alpha \delta_y^\alpha u_{i,j}^{n+1}, \tag{3.30}$$

to the left side of (3.29) that has no impact on the convergence of the scheme. This leads to the following equation:

$$(1 + \mu_1 \delta_x^\alpha)(1 + \mu_2 \delta_y^\alpha) u_{i,j}^{n+1} = \sum_{l=0}^{n-1} (b_l - b_{l+1}) u_{i,j}^{n-l} + b_n u_{i,j}^0 + \mu_0 f_{i,j}^{n+1}. \tag{3.31}$$

Now the IDM defined by equation (3.20) can be solved by using the following iterative scheme with LU factorization (LU factorization applied on the smaller linear systems obtained from the discretization of the corresponding one-dimensional systems), and the fractional alternating direction implicit scheme (FADIS) at time $t = t_{n+1}$ is then given as follows.

1. For each fixed $y = y(j)$, obtain an intermediate solution $u^*_{i,j}$ by solving the problem in the x direction $1 \le i \le M_1 - 1$:

$$(1 - \mu_1 \delta_x^\alpha) u^*_{i,j} = \sum_{l=0}^{n-1} (b_l - b_{l+1}) u^{n-l}_{i,j} + b_n u^0_{i,j} + \mu_0 f^{n+1}_{i,j}. \tag{3.32}$$

2. For each fixed $x = x(i)$, then solve the following linear system in the y direction $1 \le j \le M_2 - 1$:

$$(1 - \mu_2 \delta_y^\alpha) u^{n+1}_{i,j} = u^*_{i,j}. \tag{3.33}$$

Prior to solving equation (3.32), the boundary conditions should be accommodated for the additional intermediate solution $u^*_{i,j}$. Specifically, for homogeneous Dirichlet boundary conditions at each fixed $y = y(j)$ $(1 \le j \le M_2)$, we have

$$u^n_{x_0,j} = 0, \quad u^n_{x_{M_1},j} = 0. \tag{3.34}$$

Therefore, the boundary values for the intermediate solution can be computed from

$$u^*_{x_0,j} = (1 - \mu_2 \delta_y^\beta) u^n_{x_0,j}, \quad u^*_{x_{M_2},j} = (1 - \mu_2 \delta_y^\beta) u^n_{x_{M_2},j}. \tag{3.35}$$

As can be seen from the operator $\mu_2 \delta_y^\alpha$ defined in equation (3.28), the boundary values of $u^*_{0,j}$ and $u^*_{M_1,j}$ are computed by using both the node values on and inside the boundary. Therefore, the boundary values of the intermediate solution are not equal to zero.

Theorem 3.7 ([15]). *The 2D-FADDIM for the 2D-TS-CR-FDE is unconditionally stable.*

Theorem 3.8 ([15]). *The 2D-FADDIM is convergent, and there is a positive constant C^*, such that*

$$\|e^{n+1}\|_\infty \le C^* (\tau^{2-\alpha} + h_x + h_y), \tag{3.36}$$

where $e^n_{i,j} = u(x_i, y_j, t_n) - u^n_{i,j}$ and $e^n = (e^n_{1,1}, e^n_{2,1}, \ldots, e_{M_1-1,M_2-1})^T$.

Table 4: Comparison of CPU time (seconds) for the numerical schemes 2D-FADIS, 2D-INS, and N2D-INM with temporal step $\tau = 1/100$ at time $t = 1.0$.

$h_x = h_y$	**2D-FADIS** CPU time	**2D-INS** CPU time
$\frac{1}{8}$	0.1476875	5.515625
$\frac{1}{16}$	0.3196875	28.06201
$\frac{1}{32}$	2.5780217	1356.328

3.2.3 Numerical results

In order to better present the efficiency of the FADIS, Example 3.2 is used for comparison. The numerical tests presented in this section are performed on a laptop Lenovo Y430. Table 4 lists the CPU time calculated by the two-dimensional FADIS (2D-FADIS) and the two-dimensional implicit numerical scheme (2D-INM) that is first order in space at time $t = 1$. The temporal step is $\tau = 1/100$ and the spatial steps are $h_x = h_y = 1/8, 1/16, 1/32$.

From Table 4 it can be seen that the traditional implicit numerical schemes are extremely time consuming while the FADIS is computationally efficient, with the CPU time greatly reduced for all mesh sizes. Note in particular that the smaller the mesh size is, the more significant the reduction in CPU time is.

4 2D time-space Riesz fractional diffusion equation on irregular convex domains

As mentioned in the introduction, fractional differential equations are powerful tools for modeling the non-locality and spatial heterogeneity evident in many real-world problems. Although numerous numerical methods have been proposed, most of them are limited to regular domains and uniform meshes. In fact, many problems from science and engineering involve mathematical models that must be computed on irregular domains and therefore seeking effective numerical methods to solve the FDE on such domains is important. For irregular convex domains, the treatment of the space-fractional derivative becomes more challenging and the general methods such as the ones discussed in Sections 2 and 3 are no longer feasible. In this section, a novel numerical technique based on the Galerkin finite element method (FEM) with an unstructured mesh is derived to deal with the space-fractional derivative on arbitrarily shaped convex and non-convex domains [3]. The theory has been extended and a computational model for the case of a multiply connected domain is discussed in [3].

4.1 2D time-space Riesz fractional diffusion equation

Consider the following two-dimensional time-space Riesz fractional diffusion equation (2D-TSRFDE) on an irregular convex domain:

$$ {}_0^C D_t^{\gamma} u(x,y,t) = K_1 \frac{\partial^{2\alpha} u(x,y,t)}{\partial |x|^{2\alpha}} + K_2 \frac{\partial^{2\beta} u(x,y,t)}{\partial |y|^{2\beta}} + f(x,y,t), \qquad (4.1) $$
$$ (x,y,t) \in \Omega \times (0,T], $$

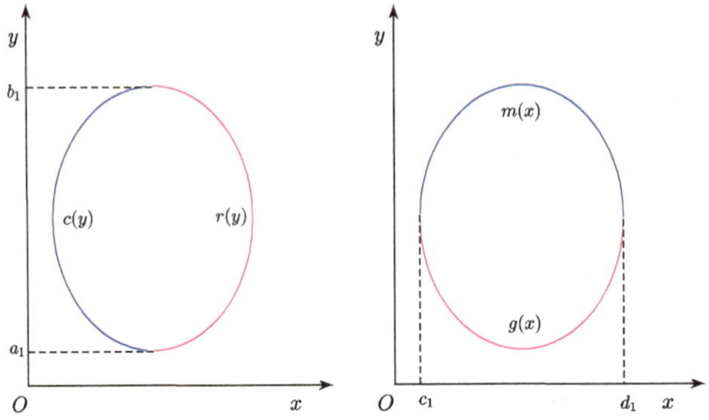

Figure 1: The boundaries of convex domain Ω.

with the initial condition

$$u(x,y,0)=\phi(x,y),\quad (x,y)\in\Omega \tag{4.2}$$

and boundary condition

$$u(x,y,t)=0,\quad (x,y,t)\in\partial\Omega\times(0,T], \tag{4.3}$$

where $0<\gamma<1$, $\frac{1}{2}<\alpha,\beta\leq 1$, $K_1>0$, $K_2>0$, and $f(x,y,t)$ and $\phi(x,y)$ are two known smooth functions. The irregular convex domain Ω is defined as $\Omega=\{(x,y)\,|\,c(y)<x<r(y), a_1<y<b_1\}$ or $\Omega=\{(x,y)\,|\,g(x)<y<m(x), c_1<x<d_1\}$, where $a_1=\min_{(x,y)\in\Omega} g(x)$, $b_1=\max_{(x,y)\in\Omega} m(x)$, $c_1=\min_{(x,y)\in\Omega} c(y)$, and $d_1=\max_{(x,y)\in\Omega} r(y)$. Here the method is illustrated for the convex domain with a curved boundary, as shown in Figure 1.

We define the Riesz fractional derivatives $\frac{\partial^{2\alpha}u}{\partial|x|^{2\alpha}}$ and $\frac{\partial^{2\beta}u}{\partial|y|^{2\beta}}$ as

$$\begin{aligned}\frac{\partial^{2\alpha}u(x,y,t)}{\partial|x|^{2\alpha}}&=-\frac{1}{2\cos(\alpha\pi)}\big({}_{c(y)}\tilde{D}_x^{2\alpha}u(x,y,t)+{}_x\tilde{D}_{r(y)}^{2\alpha}u(x,y,t)\big),\\ \frac{\partial^{2\beta}u(x,y,t)}{\partial|y|^{2\beta}}&=-\frac{1}{2\cos(\beta\pi)}\big({}_{g(x)}\tilde{D}_y^{2\beta}u(x,y,t)+{}_y\tilde{D}_{m(x)}^{2\beta}u(x,y,t)\big),\end{aligned}$$

and the Riemann–Liouville fractional derivatives with varying boundary ${}_{c(y)}\tilde{D}_x^{2\alpha}u(x,y,t)$, ${}_x\tilde{D}_{r(y)}^{2\alpha}u(x,y,t)$, ${}_{g(x)}\tilde{D}_y^{2\beta}u(x,y,t)$, and ${}_y\tilde{D}_{m(x)}^{2\beta}u(x,y,t)$ $(n-1<2\alpha, 2\beta<n)$ are given by

$$\begin{aligned}{}_{c(y)}\tilde{D}_x^{2\alpha}u(x,y,t)&:=\frac{1}{\Gamma(n-2\alpha)}\frac{\partial^n}{\partial x^n}\int_{c(y)}^{x}(x-s)^{n-2\alpha-1}u(s,y,t)\,ds,\\ {}_x\tilde{D}_{r(y)}^{2\alpha}u(x,y,t)&:=\frac{(-1)^n}{\Gamma(n-2\alpha)}\frac{\partial^n}{\partial x^n}\int_{x}^{r(y)}(s-x)^{n-2\alpha-1}u(s,y,t)\,ds,\end{aligned}$$

$$
{}_{g(x)}\tilde{D}_y^{2\beta}u(x,y,t) := \frac{1}{\Gamma(n-2\beta)}\frac{\partial^n}{\partial y^n}\int_{g(x)}^{y}(y-s)^{n-2\beta-1}u(x,s,t)\,ds,
$$

$$
{}_y\tilde{D}_{m(x)}^{2\beta}u(x,y,t) := \frac{(-1)^n}{\Gamma(n-2\beta)}\frac{\partial^n}{\partial y^n}\int_{y}^{m(x)}(s-y)^{n-2\beta-1}u(x,s,t)\,ds.
$$

4.2 Preliminary knowledge

At first some definitions and lemmas are introduced on the fractional derivative space, which were first established by Roop and Ervin [1, 6, 13] in the one-dimensional case. For two-dimensional rectangular domains, these results are also applicable [2]. Here, we extend them to convex domains. Referring to Figure 1, we define

$$
(u,v)_\Omega := \int_{a_1}^{b_1}\int_{c(y)}^{r(y)} u(x,y)v(x,y)dxdy = \int_{c_1}^{d_1}\int_{g(x)}^{m(x)} u(x,y)v(x,y)dydx
$$

and $\|u\|_{L_2(\Omega)} := (u,u)_\Omega^{1/2}$.

Definition 4.1 ([2], Left fractional derivative space). For $\mu > 0$, denote the semi-norm and the norm, respectively, as

$$
\begin{aligned}
|u|_{\tilde{J}_L^\mu(\Omega)} &:= \left(\|{}_{c(y)}\tilde{D}_x^\mu u\|_{L_2(\Omega)}^2 + \|{}_{g(x)}\tilde{D}_y^\mu u\|_{L_2(\Omega)}^2\right)^{1/2},\\
\|u\|_{\tilde{J}_L^\mu(\Omega)} &:= \left(\|u\|_{L_2(\Omega)}^2 + |u|_{\tilde{J}_L^\mu(\Omega)}^2\right)^{1/2}
\end{aligned}
$$

and define $\tilde{J}_L^\mu(\Omega)$ ($\tilde{J}_{L,0}^\mu(\Omega)$) as the closure of $C^\infty(\Omega)$ ($C_0^\infty(\Omega)$) with respect to $\|\cdot\|_{\tilde{J}_L^\mu(\Omega)}$.

Definition 4.2 ([3], Right fractional derivative space). For $\mu > 0$, denote the semi-norm and the norm, respectively, as

$$
\begin{aligned}
|u|_{\tilde{J}_R^\mu(\Omega)} &:= \left(\|{}_x\tilde{D}_{r(y)}^\mu u\|_{L_2(\Omega)}^2 + \|{}_y\tilde{D}_{m(x)}^\mu u\|_{L_2(\Omega)}^2\right)^{1/2},\\
\|u\|_{\tilde{J}_R^\mu(\Omega)} &:= \left(\|u\|_{L_2(\Omega)}^2 + |u|_{\tilde{J}_R^\mu(\Omega)}^2\right)^{1/2}
\end{aligned}
$$

and define $\tilde{J}_R^\mu(\Omega)$ ($\tilde{J}_{R,0}^\mu(\Omega)$) as the closure of $C^\infty(\Omega)$ ($C_0^\infty(\Omega)$) with respect to $\|\cdot\|_{\tilde{J}_R^\mu(\Omega)}$.

Definition 4.3 ([3], Symmetric fractional derivative space). For $\mu > 0, \mu \neq n-\frac{1}{2}, n \in \mathbb{N}$, denote the semi-norm and the norm, respectively, as

$$
\begin{aligned}
|u|_{\tilde{J}_S^\mu(\Omega)} &:= \left(|({}_{c(y)}\tilde{D}_x^\mu u, {}_x\tilde{D}_{r(y)}^\mu u)_\Omega| + |({}_{g(x)}\tilde{D}_y^\mu u, {}_y\tilde{D}_{m(x)}^\mu u)_\Omega|\right)^{1/2},\\
\|u\|_{\tilde{J}_S^\mu(\Omega)} &:= \left(\|u\|_{L_2(\Omega)}^2 + |u|_{\tilde{J}_S^\mu(\Omega)}^2\right)^{1/2}
\end{aligned}
$$

and define $\tilde{J}_S^\mu(\Omega)$ ($\tilde{J}_{S,0}^\mu(\Omega)$) as the closure of $C^\infty(\Omega)$ ($C_0^\infty(\Omega)$) with respect to $\|\cdot\|_{\tilde{J}_S^\mu(\Omega)}$.

Definition 4.4 ([3], Fractional Sobolev space). For $\mu > 0$, denote the semi-norm and the norm, respectively, as

$$|u|_{H^\mu(\Omega)} := \| \, |\xi|^\mu \mathcal{F}(\hat{u})(\xi) \|_{L_2(\mathbb{R}^2)},$$
$$\|u\|_{H^\mu(\Omega)} := \left(\|u\|^2_{L_2(\Omega)} + |u|^2_{H^\mu(\Omega)} \right)^{1/2},$$

where $\mathcal{F}(\hat{u})(\xi)$ is the Fourier transform of $\hat{u}$, which is the zero extension of u outside Ω, and define $H^\mu(\Omega)$ ($H^\mu_0(\Omega)$) as the closure of $C^\infty(\Omega)$ ($C^\infty_0(\Omega)$) with respect to $\| \cdot \|_{H^\mu(\Omega)}$.

Define the following fractional derivative and integral operators:

$$_{-\infty}D^\mu_x u(x,y,t) = \frac{1}{\Gamma(n-\mu)} \frac{\partial^n}{\partial x^n} \int_{-\infty}^{x} (x-s)^{n-\mu-1} u(s,y,t)\, ds,$$

$$_x D^\mu_{+\infty} u(x,y,t) = \frac{(-1)^n}{\Gamma(n-\mu)} \frac{\partial^n}{\partial x^n} \int_{x}^{+\infty} (s-x)^{n-\mu-1} u(s,y,t)\, ds,$$

$$_{-\infty}I^{-\mu}_x u(x,y,t) = \frac{1}{\Gamma(\mu)} \int_{-\infty}^{x} (x-s)^{\mu-1} u(s,y,t)\, ds,$$

$$_x I^{-\mu}_{+\infty} u(x,y,t) = \frac{(-1)^n}{\Gamma(\mu)} \int_{x}^{+\infty} (s-x)^{\mu-1} u(s,y,t)\, ds,$$

$$_{c(y)}\tilde{I}^{-\mu}_x u(x,y,t) = \frac{1}{\Gamma(\mu)} \int_{c(y)}^{x} (x-s)^{\mu-1} u(s,y,t)\, ds,$$

$$_x \tilde{I}^{-\mu}_{r(y)} u(x,y,t) = \frac{(-1)^n}{\Gamma(\mu)} \int_{x}^{r(y)} (s-x)^{\mu-1} u(s,y,t)\, ds.$$

The definitions of the operators in the y direction are similar.

Remark 4.1. If $\mathrm{supp}(u) \subset \Omega$, then $_{-\infty}D^\mu_x u = {}_{c(y)}\tilde{D}^\mu_x u$, $_x D^\mu_{+\infty} u = {}_x\tilde{D}^\mu_{r(y)} u$, $_{-\infty}I^{-\mu}_x u = {}_{c(y)}\tilde{I}^{-\mu}_x u$, and $_x I^{-\mu}_{+\infty} u = {}_x\tilde{I}^{-\mu}_{r(y)}$.

Lemma 4.1 ([2]). *Let $\mu > 0$ and define the operators*
(i) $_{-\infty}I^{-\mu}_x : L_2(\Omega) \to L_2(\Omega)$;
(ii) $_{-\infty}D^\mu_x : \tilde{J}^\mu_L(\Omega) \to L_2(\Omega)$;
(iii) $_x I^{-\mu}_{+\infty} : L_2(\Omega) \to L_2(\Omega)$;
(iv) $_x D^\mu_{+\infty} : \tilde{J}^\mu_R(\Omega) \to L_2(\Omega)$.

Then all the operators are bounded operators.

Lemma 4.2 ([2]). *For $u \in \tilde{J}^\mu_{L,0}(\Omega) \cap \tilde{J}^\mu_{R,0}(\Omega)$ and $0 < s < \mu$, we have*

$$\|u\|_{L_2(\Omega)} \le C_1 \|{}_{c(y)}\tilde{D}^s_x u\|_{L_2(\Omega)} \le C_2 \|{}_{c(y)}\tilde{D}^\mu_x u\|_{L_2(\Omega)},$$

$$\|u\|_{L_2(\Omega)} \le C_3 \|_{g(x)}\tilde{D}_y^s u\|_{L_2(\Omega)} \le C_4 \|_{g(x)}\tilde{D}_y^\mu u\|_{L_2(\Omega)},$$

where C_1, C_2, C_3, and C_4 are some positive constants independent of u.

Lemma 4.3 ([2]). *If $\mu > 0$, then $\tilde{J}_L^\mu(\Omega)$, $\tilde{J}_R^\mu(\Omega)$, and $H^\mu(\Omega)$ are equivalent with equivalent norms and semi-norms; if $\mu > 0$, $\mu \neq n - \frac{1}{2}$, $n \in \mathbb{N}$, then $\tilde{J}_{L,0}^\mu(\Omega)$, $\tilde{J}_{R,0}^\mu(\Omega)$, $\tilde{J}_{S,0}^\mu(\Omega)$, and $H_0^\mu(\Omega)$ have equivalent norms and semi-norms.*

Lemma 4.4 ([2]). *If $\mu \in (1/2, 1)$, $u, v \in \tilde{J}_{L,0}^{2\mu}(\Omega) \cap \tilde{J}_{R,0}^{2\mu}(\Omega)$, then*

$$\big({}_{c(y)}\tilde{D}_x^{2\mu} u(x,y), v(x,y)\big)_\Omega = \big({}_{c(y)}\tilde{D}_x^\mu u(x,y), {}_x\tilde{D}_{r(y)}^\mu v(x,y)\big)_\Omega,$$
$$\big({}_x\tilde{D}_{r(y)}^{2\mu} u(x,y), v(x,y)\big)_\Omega = \big({}_x\tilde{D}_{r(y)}^\mu u(x,y), {}_{c(y)}\tilde{D}_x^\mu v(x,y)\big)_\Omega.$$

4.3 Variational formulation

4.3.1 The fully discrete variational formulation

For convenience, it is supposed that $C, C_1, C_2, \ldots$ are positive constants, which may take distinct values according to the different contexts discussed throughout this section.

Let $\tau = \frac{T}{N}$ be the time step and $t_n = n\tau$, $n = 0, 1, 2, \ldots, N$. Using the L1 approximation, the following approximation is obtained [3]:

$${}_0^C D_t^\gamma u(x,y,t_n) = \frac{\tau^{-\gamma}}{\Gamma(2-\gamma)} \sum_{k=0}^{n} b_k^n u(x,y,t_k) + R_t^n,$$

where $b_n^n = 1$, $b_0^n = (n-1)^{1-\gamma} - n^{1-\gamma} < 0$, $b_k^n = (n-k+1)^{1-\gamma} - 2(n-k)^{1-\gamma} + (n-k-1)^{1-\gamma} < 0$, $k = 1, 2, \ldots, n-1$. Denote

$$\nabla_t^\gamma u(x,y,t_n) = \frac{\tau^{-\gamma}}{\Gamma(2-\gamma)} \sum_{k=0}^{n} b_k^n u(x,y,t_k), \quad n = 1, 2, \ldots, N.$$

Then

$$\|R_t^n\|_0 = \|{}_0^C D_t^\gamma u(x,y,t_n) - \nabla_t^\gamma u(x,y,t_n)\|_0 \le C\tau^{2-\gamma}.$$

Denote $V = H_0^\alpha(\Omega) \cap H_0^\beta(\Omega)$. The domain Ω is divided into a number of regular triangular regions. Let $\mathcal{T}_h$ be this triangulation and h be the maximum diameter of the triangular elements. Define the finite element subspace as

$$V_h := \{v_h \mid v_h \in C(\Omega) \cap V,\ v_h|_K \text{ is linear for all } K \in \mathcal{T}_h \text{ and } v_h|_{\partial\Omega} = 0\}.$$

Assume that $u_h^n \in V_h$ is the approximation of $u(x,y,t)$ at $t = t_n$. Then, by Lemma 4.4, the fully discrete scheme associated with the variational form of equation (4.1) is obtained as follows. Find $u_h^n \in V_h$ for each $t = t_n$ ($n = 1, 2, \ldots, N$) such that

$$\begin{cases} (\nabla_t^\gamma u_h^n, v_h)_\Omega + A(u_h^n, v_h)_\Omega = (f^n, v_h)_\Omega, & \forall v_h \in V_h, \\ u_h^0 = u_{0h}, \end{cases} \tag{4.4}$$

where $u_{0h} \in V_h$ is a reasonable approximation of u^0, $K_x = \frac{K_1}{2\cos(\alpha\pi)}$, $K_y = \frac{K_2}{2\cos(\beta\pi)}$, and

$$A(u,v)_\Omega := K_x\{({}_{c(y)}\tilde{D}^{\alpha}_{x}u, {}_{x}\tilde{D}^{\alpha}_{r(y)}v)_\Omega + ({}_{x}\tilde{D}^{\alpha}_{r(y)}u, {}_{c(y)}\tilde{D}^{\alpha}_{x}v)_\Omega\}$$
$$+ K_y\{({}_{g(x)}\tilde{D}^{\beta}_{y}u, {}_{y}\tilde{D}^{\beta}_{m(x)}v)_\Omega + ({}_{y}\tilde{D}^{\beta}_{m(x)}u, {}_{g(x)}\tilde{D}^{\beta}_{y}v)_\Omega\}.$$

4.3.2 The implementation of FEM with an unstructured mesh

The computation process is now considered for piece-wise linear polynomials on the triangular element e_p, $p = 1, 2, \ldots, N_e$, where N_e is the total number of triangles. Then, within element e_p, the field function $u^p(x,y)$ can be written as

$$u^p(x,y) = \sum_{j_0=1}^{3} u_{j_0}\varphi_{j_0}(x,y),$$

where the triangle vertices are numbered in a counter-clockwise order as 1, 2, 3 and the basis function $\varphi_{j_0}(x,y)$ is defined by

$$\varphi_{j_0}(x,y)|_{(x,y)\in e_p} = \frac{1}{2\Delta_{e_p}}(a_{j_0}x + b_{j_0}y + c_{j_0}), \quad \varphi_{j_0}(x,y)|_{(x,y)\notin e_p} = 0,$$
$$a_1 = y_2 - y_3, \quad a_2 = y_3 - y_1, \quad a_3 = y_1 - y_2,$$
$$b_1 = x_3 - x_2, \quad b_2 = x_1 - x_3, \quad b_3 = x_2 - x_1,$$
$$c_1 = x_2y_3 - x_3y_2, \quad c_2 = x_3y_1 - x_1y_3, \quad c_3 = x_1y_2 - x_2y_1,$$

where Δ_{e_p} is the area of triangle element p. It is well known that

$$\varphi_{j_0}(x_{i_0}, y_{i_0}) = \delta_{i_0 j_0}, \quad i_0, j_0 = 1, 2, 3,$$

and δ is the Kronecker function. With these local field functions and basis functions, the formulation of $u(x,y)$ for the whole triangulation is defined as

$$u(x,y) = \sum_{i=1}^{N_p} u_i l_i(x,y),$$

where $l_i(x,y)$ is the new basis function whose support domain is Ω_{e_l} (Figure 2) and N_p is the total number of vertices on the convex domain Ω.

Now, rewrite u_h^n in the form

$$u_h^n = \sum_{i=1}^{N_p} u_i^n l_i(x,y), \tag{4.5}$$

where u_i^n are the coefficients that are to be solved for. Substituting equation (4.5) into equation (4.4) with $v_h = l_j(x,y)$, $j = 1, 2, \ldots, N_p$, gives

$$\sum_{i=1}^{N_p} u_i^n[(l_i, l_j)_\Omega + \omega_0 A(l_i, l_j)_\Omega] = -\sum_{i=1}^{N_p}\sum_{k=1}^{n-1} b_k^n u_i^k (l_k, l_j)_\Omega - b_0^n(u^0, l_j)_\Omega + \omega_0(f^n, l_j)_\Omega,$$

where $\omega_0 = \tau^{\gamma}\Gamma(2-\gamma)$. The above equation can be expressed in matrix form as

$$(M + \omega_0 B)U^n = -M \sum_{k=1}^{n-1} b_k^n U^k - G^0 + \omega_0 F_1^n,$$

where M is the mass matrix with elements $M_{ij} = (l_j, l_i)_\Omega = \sum_{p=1}^{N_e}(l_j, l_i)_{e_p}$ and B is the stiffness matrix with elements $B_{ij} = A(l_j, l_i)_\Omega$ and $U^n = [u_1^n, u_2^n, \dots, u_{N_p}^n]^T$. The vectors G^0 and F_1^n are given, respectively, by

$$\begin{aligned} G^0 &= b_0^n[(u^0, l_1)_\Omega, (u^0, l_2)_\Omega, \dots, (u^0, l_{N_p})_\Omega]^T \\ &= \sum_{p=1}^{N_e} b_0^n[(u^0, l_1)_{e_p}, (u^0, l_2)_{e_p}, \dots, (u^0, l_{N_p})_{e_p}]^T, \\ F_1^n &= [(f^n, l_1)_\Omega, (f^n, l_2)_\Omega, \dots, (f^n, l_{N_p})_\Omega]^T \\ &= \sum_{p=1}^{N_e} [(f^n, l_1)_{e_p}, (f^n, l_2)_{e_p}, \dots, (f^n, l_{N_p})_{e_p}]^T. \end{aligned}$$

Due to the non-local property of the fractional derivative, matrix B is the most difficult part to calculate. For matrix B, the (i, j) entry is given by

$$\begin{aligned} B_{ij} = A(l_j, l_i)_\Omega = {} & K_x\big({}_{c(y)}\tilde{D}_x^\alpha l_j(x,y), {}_x\tilde{D}_{r(y)}^\alpha l_i(x,y)\big)_\Omega \\ & + K_x\big({}_x\tilde{D}_{r(y)}^\alpha l_j(x,y), {}_{c(y)}\tilde{D}_x^\alpha l_i(x,y)\big)_\Omega \\ & + K_y\big({}_{g(x)}\tilde{D}_y^\beta l_j(x,y), {}_y\tilde{D}_{m(x)}^\beta l_i(x,y)\big)_\Omega \\ & + K_y\big({}_y\tilde{D}_{m(x)}^\beta l_j(x,y), {}_{g(x)}\tilde{D}_y^\beta l_i(x,y)\big)_\Omega. \end{aligned} \tag{4.6}$$

The support domains of the four fractional derivatives ${}_{c(y)}\tilde{D}_x^\alpha l(x,y)$, ${}_x\tilde{D}_{r(y)}^\alpha l(x,y)$, ${}_{g(x)}\tilde{D}_y^\beta l(x,y)$, ${}_y\tilde{D}_{m(x)}^\beta l(x,y)$ are denoted as $\Omega_{e_i}^L$, $\Omega_{e_i}^R$, $\Omega_{e_i}^D$, $\Omega_{e_i}^U$, respectively. The notation

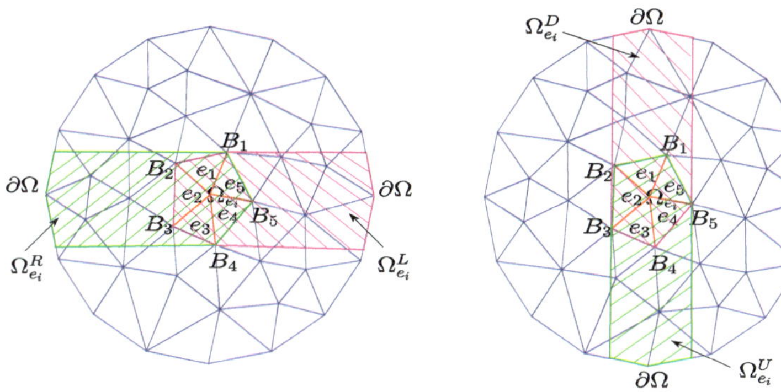

Figure 2: An illustration of support domains $\Omega_{e_i}^L$, $\Omega_{e_i}^R$, $\Omega_{e_i}^D$, $\Omega_{e_i}^U$.

that L, R, D, U correspond to the “left”, “right”, “down”, and “up” directions for the four fractional derivatives is adopted. They do not correspond to the actual location in the domain Ω. From Figure 2, it can be observed that node i is surrounded by the elements e_1, e_2, e_3, e_4, e_5 and the boundary $\partial\Omega_{e_i}$ is constituted by $\partial\Omega^l_{e_i}$ and $\partial\Omega^r_{e_i}$, where $\partial\Omega^l_{e_i}$ is made up of line segments B_1B_2, B_2B_3, B_3B_4 and $\partial\Omega^r_{e_i}$ is connected by the line segments B_4B_5, B_5B_1, respectively. Then $\partial\Omega^l_{e_i}$, $y = y_{B_4}$ $(x \geq x_{B_4})$, $y = y_{B_1}$ $(x \geq x_{B_1})$, and $\partial\Omega$ form the support domain $\Omega^L_{e_i}$. In a similar manner, the support domains can be defined for $\Omega^R_{e_i}$, $\Omega^D_{e_i}$, and $\Omega^U_{e_i}$ (Figure 2).

In view of the similarity of the four terms in the right-hand side of equation (4.6), the computation of $({}_{c(y)}\tilde{D}^{\alpha}_{x}l_j, {}_{x}\tilde{D}^{\alpha}_{r(y)}l_i)_{\Omega}$ is illustrated as an example (the interested reader is referred to [3] for further details). By applying Gauss quadrature, the following equation is obtained:

$$\begin{aligned}({}_{c(y)}\tilde{D}^{\alpha}_{x}l_j, {}_{x}\tilde{D}^{\alpha}_{r(y)}l_i)_{\Omega} &= \sum_{p=1}^{N_e}({}_{c(y)}\tilde{D}^{\alpha}_{x}l_j, {}_{x}\tilde{D}^{\alpha}_{r(y)}l_i)_{e_p}\\ &= \sum_{p=1}^{N_e}\int_{e_p} {}_{c(y)}\tilde{D}^{\alpha}_{x}l_j\,{}_{x}\tilde{D}^{\alpha}_{r(y)}l_i\,dxdy\\ &\approx \sum_{p=1}^{N_e}\sum_{(\hat{x_q},\hat{y_q})\in G_K} \omega_q\,{}_{c(y)}\tilde{D}^{\alpha}_{x}l_j|_{(\hat{x_q},\hat{y_q})}\,{}_{x}\tilde{D}^{\alpha}_{r(y)}l_i|_{(\hat{x_q},\hat{y_q})},\end{aligned}$$

where G_K denotes the set of all Gauss points in element e_p and ω_q are the weights associated with the Gauss point $P(\hat{x_q},\hat{y_q})$ (Figure 2). When point $P(\hat{x_q},\hat{y_q})$ is out of the support domains $\Omega^L_{e_i}$ and $\Omega^R_{e_i}$, then ${}_{c(y)}\tilde{D}^{\alpha}_{x}l_j(x,y) = 0$ and ${}_{x}\tilde{D}^{\alpha}_{r(y)}l_i(x,y) = 0$. To evaluate ${}_{c(y)}\tilde{D}^{\alpha}_{x}l_j|_{(\hat{x_q},\hat{y_q})}$, suppose that segment $y = \hat{y_q}$, $c(\hat{y_q}) \leq x \leq \hat{x_q}$ intersects n_q points with the triangular element of Ω_{e_j}, and these points are numbered as $x^0_q < x^1_q < x^2_q < \cdots x^{n_q}_q$. Then the definition of the fractional derivative is given by

$$\begin{aligned}{}_{c(y)}\tilde{D}^{\alpha}_{x}l_j|_{(\hat{x_q},\hat{y_q})} &= {}_{c(\hat{y_q})}\tilde{D}^{\alpha}_{x}l_j(x,\hat{y_q})|_{x=\hat{x_q}}\\ &= \left(\frac{1}{\Gamma(1-\alpha)}\frac{\partial}{\partial x}\int_{c(\hat{y_q})}^{x}(x-\xi)^{-\alpha}l_j(\xi,\hat{y_q})d\xi\right)|_{x=\hat{x_q}}\\ &= \sum_{k=1}^{n_i}\left(\frac{1}{\Gamma(1-\alpha)}\frac{\partial}{\partial x}\int_{x_i^{k-1}}^{x_i^k}(x-\xi)^{-\alpha}l_j(\xi,\hat{y_q})d\xi\right)|_{x=\hat{x_q}}.\end{aligned} \tag{4.7}$$

Then, there are three different cases that need to be discussed (Figure 3). In case I, the

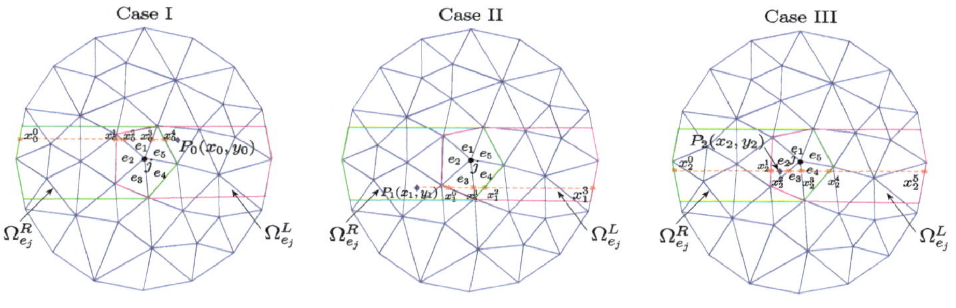

Figure 3: The points of intersection by $y = \hat{y_q}$ with the triangle element of Ω_{e_j} and $\partial\Omega$.

Gauss point $P_0(x_0, y_0)$ is only located in $\Omega_{e_j}^L$, and

$$l_j(x, y_0) = \begin{cases} 0, & x_0^0 \le x \le x_0^1, \\ \varphi_{j2}(x, y_0), & x_0^1 \le x \le x_0^2, \\ \varphi_{j1}(x, y_0), & x_0^2 \le x \le x_0^3, \\ \varphi_{j5}(x, y_0), & x_0^3 \le x \le x_0^4, \\ 0, & x_0^4 \le x, \end{cases}$$

where $\varphi_{jp}(x, y)$ is the basis function of node j on element p and $c(y_0) = x_0^0$. In case II, the Gauss point $P_1(x_1, y_1)$ is only located in the $\Omega_{e_j}^R \backslash \Omega_{e_j}^L$. Then we have ${}_{c(y)}\tilde{D}_x^\alpha l_j|_{(x_1,y_1)} = 0$. In case III, the Gauss point $P_2(x_2, y_2)$ is located in $\Omega_{e_j}^L \cap \Omega_{e_j}^R = \Omega_{e_j}$. Then

$$l_j(x, y_2) = \begin{cases} 0, & x_2^0 \le x \le x_2^1, \\ \varphi_{j2}(x, y_2), & x_2^1 \le x \le x_2^2, \\ \varphi_{j3}(x, y_2), & x_2^2 \le x \le x_2^3, \\ \varphi_{j4}(x, y_2), & x_2^3 \le x \le x_2^4. \end{cases}$$

Similarly, to evaluate ${}_x\tilde{D}_{r(y)}^\alpha l_j|_{(\hat{x_q},\hat{y_q})}$, the following three cases are considered. In case I, ${}_x\tilde{D}_{r(y)}^\alpha l_j|_{(x_0,y_0)} = 0$. In case II,

$$l_j(x, y_1) = \begin{cases} 0, & x \le x_1^0, \\ \varphi_{j3}(x, y_1), & x_1^0 \le x \le x_1^1, \\ \varphi_{j4}(x, y_1), & x_1^1 \le x \le x_1^2, \\ 0, & x_1^2 \le x \le x_1^3. \end{cases}$$

In case III,

$$l_j(x, y_2) = \begin{cases} \varphi_{j2}(x, y_2), & x_2^1 \le x \le x_2^2, \\ \varphi_{j3}(x, y_2), & x_2^2 \le x \le x_2^3, \\ \varphi_{j4}(x, y_2), & x_2^3 \le x \le x_2^4, \\ 0, & x_2^4 \le x \le x_2^5. \end{cases}$$

As $l_j(\xi, \hat{y_q})$ is a linear function on $[x_i^{k-1}, x_i^k]$, $k = 1, 2, \ldots, n_i$, equation (4.7) can be evaluated using integration by parts.

4.3.3 Stability and convergence of the fully discrete scheme

At first the notations are simplified for $(\cdot,\cdot)_\Omega$, $\|\cdot\|_{L_2(\Omega)}$ and $\|\cdot\|_{H^s(\Omega)}$ as $(\cdot,\cdot)$, $\|\cdot\|_0$, $\|\cdot\|_s$, respectively. Let $\sigma = \max\{\alpha,\beta\}$ and define the semi-norm $|\cdot|_{(\alpha,\beta)}$ and norm $|||\cdot|||_{(\alpha,\beta)}$ as follows:

$$|u|_{(\alpha,\beta)} := (K_1\|_{c(y)}\tilde{D}_x^\alpha u\|_0^2 + K_2\|_{g(x)}\tilde{D}_y^\beta u\|_0^2)^{\frac{1}{2}}, \quad |||u|||_{(\alpha,\beta)} := (\|u\|_0^2 + |u|_{(\alpha,\beta)}^2)^{\frac{1}{2}}.$$

Theorem 4.1. *The fully discrete variational scheme* (4.4) *is unconditionally stable* [3].

Theorem 4.2. *Suppose that $u(t_n)$, u_h^n are the exact solution and numerical solution of problem* (4.1)–(4.3) *at $t = t_n$, respectively, and u, u_{tt}, ${}_0^CD_t^\gamma u \in L^\infty(0,T;H^{s+1}(\Omega))$. Then for $n = 1, 2, \ldots, N$, the following error estimate holds* [3]*:*

$$\begin{aligned}|||u_h^n - u(t_n)|||_{(\alpha,\beta)}^2 &\le C\{\tau^{4-2\gamma} + \|u_h^0 - u(t_0)\|_\sigma^2\} \\ &\quad + C\Big\{h^{2s+2-2\sigma}\Big(\|u(t_0)\|_{s+1}^2 + \|u(t_n)\|_{s+1}^2 + \max_{0\le t\le T}\|{}_0^CD_t^\gamma u(t)\|_{s+1}^2\Big)\Big\}.\end{aligned}$$

Remark 4.2. It can be deduced from Theorem 4.2 that when triangular linear basis functions ($s = 1$) are used, the error satisfies [3]

$$|||u_h^n - u(t_n)|||_{(\alpha,\beta)} \le C(\tau^{2-\gamma} + h^{2-\sigma}).$$

4.3.4 Numerical example

In this section, some numerical examples are presented to verify the effectiveness of the theoretical analysis presented in Section 4.3.3. Linear polynomials on triangles are adopted, where h is the maximum length of the triangles and N_e is the number of triangles in $\mathcal{T}_h$. By Theorem 4.2, it is expected that $\|u(t_n)-u_h^n\|_0 \sim O(h^2)$, $|||u(t_n)-u_h^n|||_{(\alpha,\beta)} \sim O(h^{2-\sigma})$, $\sigma = \max(\alpha,\beta)$ in the spatial direction and $\|u(t_n) - u_h^n\|_0 \sim O(\tau^{2-\gamma})$ in the temporal direction. Here, the following formulation is used to calculate the convergence order:

$$\text{Order} = \begin{cases} \dfrac{\log(\|E(h_1)\|_0/\|E(h_2)\|_0)}{\log(h_1/h_2)}, & \text{in space,} \\ \dfrac{\log(\|E(\tau_1)\|_0/\|E(\tau_2)\|_0)}{\log(\tau_1/\tau_2)}, & \text{in time.} \end{cases}$$

Example 4.1. The following 2D-TSRFDE is considered on an elliptical domain:

$$\begin{cases} {}_0^CD_t^\gamma u(x,y,t) = K_1\frac{\partial^{2\alpha}u(x,y,t)}{\partial|x|^{2\alpha}} + K_2\frac{\partial^{2\beta}u(x,y,t)}{\partial|y|^{2\beta}} + f(x,y,t), & (x,y,t) \in \Omega\times(0,T], \\ u(x,y,0) = \frac{1}{10}(4x^2 + y^2 - 1)^2, & (x,y) \in \Omega, \\ u(x,y,t) = 0, & (x,y,t) \in \partial\Omega\times(0,T], \end{cases}$$

where $\Omega = \{(x,y) \mid 4x^2 + y^2 < 1\}$, $K_1 = 1$, $K_2 = 1$, $T = 1$,

$$\begin{aligned}
f(x,y,t) = {} & \frac{t^{2-\gamma}}{5\Gamma(3-\gamma)}(4x^2+y^2-1)^2 + \frac{t^2+1}{20\cos(\alpha\pi)}\{16[f_1(x,a_0,2\alpha)+g_1(x,b_0,2\alpha)] \\
& + 8(y^2-1)[f_2(x,a_0,2\alpha)+g_2(x,b_0,2\alpha)] \\
& + (y^4-2y^2+1)[f_3(x,a_0,2\alpha)+g_3(x,b_0,2\alpha)]\} \\
& + \frac{t^2+1}{20\cos(\beta\pi)}\{[f_1(y,c_0,2\beta)+g_1(y,d_0,2\beta)] \\
& + 8(x^2-1)[f_2(y,c_0,2\beta)+g_2(y,d_0,2\beta)] \\
& + (16x^4-8x^2+1)[f_3(y,c_0,2\beta)+g_3(y,d_0,2\beta)]\}, \\
& a_0 = -\frac{1}{2}\sqrt{1-y^2}, \quad b_0 = \frac{1}{2}\sqrt{1-y^2}, \quad c_0 = -\sqrt{1-4x^2}, \quad d_0 = \sqrt{1-4x^2}, \\
& f_1(x,a,\alpha) = {}_aD_x^\alpha(x^4), \quad f_2(x,a,\alpha) = {}_aD_x^\alpha(x^2), \quad f_3(x,a,\alpha) = {}_aD_x^\alpha(1), \\
& g_1(x,b,\alpha) = {}_xD_b^\alpha(x^4), \quad g_2(x,b,\alpha) = {}_xD_b^\alpha(x^2), \quad g_3(x,b,\alpha) = {}_xD_b^\alpha(1).
\end{aligned}$$

The exact solution is $u(x,y,t) = \frac{t^2+1}{10}(4x^2+y^2-1)^2$.

The numerical results for the simulations are given in Table 5, which illustrate the $L_{(\alpha,\beta)}$ error, L_2 error, and corresponding convergence order of h. Table 6 displays the L_2 error and the convergence order of τ. From these two tables it can be seen that the expected convergence orders $O(h^{2-\sigma})$, $O(h^2)$, and $O(\tau^{2-\gamma})$ are attained. Table 7 shows the L_2 error and the convergence order of $\tau = h$ for the second-order temporal numerical scheme. The numerical results appear to be in excellent agreement with the exact solution and second-order accuracy is achieved, which demonstrates the effectiveness of the numerical method.

Table 5: The $L_{(\alpha,\beta)}$ error, L_2 error, and convergence order of h for different α, β at $t = 1$ with $\gamma = 0.7$, $\tau = 1/1000$.

$\gamma = 0.7$	N_e	h	$L_{(\alpha,\beta)}$ error	Order	L_2 error	Order
	70	3.0312E−01	8.9507E−02	–	9.1296E−03	–
$\alpha = 0.75$	468	1.2558E−01	3.5831E−02	1.04	1.7483E−03	1.88
$\beta = 0.95$	1142	8.3913E−02	2.1668E−02	1.25	6.9743E−04	2.28
	4324	4.5308E−02	1.0303E−02	1.21	1.8504E−04	2.15
	70	3.0312E−01	9.0145E−02	–	8.6868E−03	–
$\alpha = 0.8$	468	1.2558E−01	3.1969E−02	1.18	1.4612E−03	2.02
$\beta = 0.8$	1142	8.3913E−02	1.8311E−02	1.38	5.5341E−04	2.41
	4324	4.5308E−02	8.1488E−03	1.31	1.3987E−04	2.23

5 2D time-space fractional diffusion equation with fractional Laplacian operator

In this section, we consider the following two-dimensional time-space fractional diffusion equation involving the fractional Laplacian operator (2D-TSFDE-FL), with the following given initial and boundary conditions:

$$ {}_tD_*^{\gamma}u(x,y,t) = -K_{\alpha}(-\Delta)^{\alpha/2}u(x,y,t), \tag{5.1} $$

$$ u(0,y,t) = u(a,y,t) = 0, \tag{5.2} $$

$$ u(x,0,t) = u(x,b,t) = 0, \tag{5.3} $$

$$ u(x,y,0) = u_0(x,y), \tag{5.4} $$

where the Laplacian operator is defined as $-\Delta = -\frac{\partial^2}{\partial x^2} - \frac{\partial^2}{\partial y^2}$.

5.1 Analytical solution of the 2D-TSFDE-FL

Definition 5.1 ([4]). Suppose the two-dimensional Laplacian $(-\Delta)$ has a complete set of orthonormal eigenfunctions $\varphi_{n,m}$ corresponding to eigenvalues $\lambda_{n,m}$ in a rectangular region $\mathcal{D} = \{(x,y)|0 \le x \le a, 0 \le y \le b\}$, i. e., $(-\Delta)\varphi_{n,m} = \lambda_{n,m}\varphi_{n,m}$; $\mathcal{B}(\varphi) = 0$ on $\partial\mathcal{D}$, where $\mathcal{B}(\varphi)$ is the homogeneous Dirichlet boundary condition. Let

$$ \mathcal{F}_{\eta} = \left\{ f = \sum_{n=1}^{\infty}\sum_{m=1}^{\infty} c_{n,m}\varphi_{n,m},\ c_{n,m} = \langle f, \varphi_{n,m}\rangle,\ \sum_{n=1}^{\infty} |c_{n,m}|^2|\lambda_{n,m}|^{\alpha} < \infty,\ 1 < \alpha \le 2 \right\}. \tag{5.5} $$

Then for any $f \in \mathcal{F}_{\eta}$, the two-dimensional fractional Laplacian $(-\Delta)^{\alpha/2}$ is defined by

$$ (-\Delta)^{\alpha/2}f = \sum_{n=1}^{\infty}\sum_{m=1}^{\infty} c_{n,m}(\lambda_{n,m})^{\alpha/2}\varphi_{n,m}, \tag{5.6} $$

where $\lambda_{n,m} = \frac{n^2\pi^2}{a^2} + \frac{m^2\pi^2}{b^2}$ and $\varphi_{n,m} = \frac{2}{\sqrt{ab}} \sin\frac{n\pi x}{a} \sin\frac{m\pi y}{b}$ for $n, m = 1, 2, \ldots$.

Lemma 5.1 ([6]). *The initial value problem* $(0 < \alpha < 1)$

$$ \begin{cases} {}^{C}D_{0+}^{\gamma}y(t) + \bar{\lambda}y(t) = 0, & 0 < \alpha < 1,\ \bar{\lambda} \in \mathcal{R}, \\ y(0) = c_0 \in \mathcal{R} \end{cases} \tag{5.7} $$

Table 6: The L_2 error and convergence order of τ for $\gamma = 0.7$ at $t = 1$ with $\alpha = \beta = 0.8$ and $h^2 \approx \tau^{2-\gamma}$.

N_e	τ	h	L_2 error	Order
276	$\frac{1}{14}$	1.8428E−01	2.2122E−03	–
1142	$\frac{1}{46}$	8.3913E−02	5.5444E−04	1.16
1738	$\frac{1}{61}$	6.9134E−02	3.7308E−04	1.40

Table 7: The L_2 error and convergence order of $\tau = h$ for the second-order numerical scheme with $\gamma = 0.7$, $\alpha = \beta = 0.8$ at $t = 1$.

N_e	h	L_2 error	Order
276	1.8428E−01	2.2182E−03	–
1142	8.3913E−02	5.7249E−04	1.72
1738	6.9134E−02	3.9132E−04	1.96
4324	4.5308E−02	1.6187E−04	2.09

has a unique solution

$$y(t) = c_0 E_\gamma(-\bar{\lambda} t^\gamma). \tag{5.8}$$

We derive the analytical solution of the two-dimensional time-space fractional diffusion equation with fractional Laplacian operator. First set

$$u(x,y,t) = \sum_{n=1}^{\infty}\sum_{m=1}^{\infty} c_{n,m}(t)\varphi_{n,m},$$

where $\varphi_{n,m}$ are orthonormal eigenfunctions. Using Definition 5.1 and substituting $u(x,y,t)$ into (5.1), we have

$$\sum_{n=1}^{\infty}\sum_{m=1}^{\infty} {}^C D_{0+}^{\gamma} c_{n,m}(t)\varphi_{n,m} = -K_\alpha \sum_{n=1}^{\infty}\sum_{m=1}^{\infty} c_{n,m}(t)(\lambda_{n,m})^{\alpha/2}\varphi_{n,m}, \tag{5.9}$$

i. e.,

$${}^C D_{0+}^{\gamma} c_{n,m}(t) = -K_\alpha c_{n,m}(t)(\lambda_{n,m})^{\alpha/2}. \tag{5.10}$$

Since $u(x,y,t)$ must also satisfy the initial condition (5.4)

$$\sum_{n=1}^{\infty}\sum_{m=1}^{\infty} c_{n,m}(0)\varphi_{n,m} = u_0(x,y), \quad 0 \le x \le a,\ 0 \le y \le b, \tag{5.11}$$

we obtain

$$c_{n,m}(0) = \int_0^a\int_0^b u_0(x,y)\,\varphi_{n,m}\,dy\,dx. \tag{5.12}$$

Using Lemma 5.1, we obtain

$$c_{n,m}(t) = E_\gamma(-K_\alpha(\lambda_{n,m})^{\alpha/2} t^\gamma) c_{n,m}(0)\,. \tag{5.13}$$

Hence, the analytic solution of the 2D-TSFDE-FL (5.1)–(5.4) is given by

$$u(x,y,t) = \sum_{n=1}^{\infty}\sum_{m=1}^{\infty} c_{n,m}(t)\varphi_{n,m}$$

$$= \sum_{n=1}^{\infty}\sum_{m=1}^{\infty} E_\gamma(-K_\alpha(\lambda_{n,m})^{\alpha/2}t^\gamma)c_{n,m}(0)\,\varphi_{n,m}. \tag{5.14}$$

This solution can be used to verify our numerical solution of the 2D-TSFDE-FL discussed throughout the following subsections.

5.2 Numerical solution of the 2D-TSFDE-FL

The matrix transfer technique [4] is now used to obtain the discretization in space for the 2D-TSFDE-FL. The fundamental idea of this technique is to find the approximate matrix representation of the Laplacian $(-\triangle)$ in the standard diffusion equation using the finite difference, finite element, or finite volume methods. These methods can be executed directly in the numerical approximations.

By introducing a mesh and denoting the value of $u(x, y, t)$ at the ith node by $u_i(t)$, and the vector of such values by $\mathbf{u}(t)$, the matrix transfer technique [4] for solving the 2D-TSFDE-FL (5.1)–(5.4) proceeds by first considering the non-fractional discrete equation

$$\frac{\mathrm{d}\mathbf{u}}{\mathrm{d}t} = -\mathbf{A}\mathbf{u}, \tag{5.15}$$

where $\mathbf{A}$ is the approximate matrix representation of the standard Laplacian $(-\Delta)$ under homogeneous Dirichlet boundary conditions, obtained by using either finite difference or finite element methods. Under the matrix transfer technique, the fractional Laplacian operator is approximated as

$$-(-\Delta)^{\alpha/2}\mathbf{u} \approx -\mathbf{A}^{\alpha/2}\mathbf{u}. \tag{5.16}$$

In summary, the matrix transfer technique transforms the 2D-TSFDE-FL (5.1)–(5.4) into the following time-fractional differential system:

$${}^{C}D^{\gamma}_{0+}\mathbf{u} = -K_\alpha \mathbf{A}^{\alpha/2}\mathbf{u}. \tag{5.17}$$

5.2.1 Finite difference method in space

The standard five-point finite difference stencil with equal grid spacing in both the x and y directions, i. e., $h = \frac{a}{M_1} = \frac{b}{M_2}$, will result in the following block tridiagonal approximate matrix representation of the Laplacian:

$$\mathbf{A} = \frac{1}{h^2}\begin{bmatrix} \mathbf{B} & -\mathbf{I} & & & \\ -\mathbf{I} & \mathbf{B} & -\mathbf{I} & & \\ & \ddots & \ddots & \ddots & \\ & & -\mathbf{I} & \mathbf{B} & -\mathbf{I} \\ & & & -\mathbf{I} & \mathbf{B} \end{bmatrix}, \tag{5.18}$$

where $\mathbf{A} \in \mathbb{R}^{(M_1-1)(M_2-1)\times(M_1-1)(M_2-1)}$, $\mathbf{I} \in \mathbb{R}^{(M_1-1)\times(M_1-1)}$, and the tridiagonal matrix $\mathbf{B} = \text{tridiag}(-1, 4, -1) \in \mathbb{R}^{(M_1-1)\times(M_1-1)}$.

5.2.2 Finite element method in space

In the case of the finite element method, we begin with the following non-fractional governing equation:

$$\frac{\partial u}{\partial t} = \frac{\partial^2 u}{\partial x^2} + \frac{\partial^2 u}{\partial y^2} \equiv -(-\Delta)u \quad \text{in } \Omega, \tag{5.19}$$

for a two-dimensional domain $\Omega = [0, a] \times [0, b]$. The boundary conditions are

$$u = 0, \quad \text{on } \partial\Omega. \tag{5.20}$$

Multiplying (5.1) by a test function v and integrating over the computational domain Ω gives

$$\int_\Omega \frac{\partial u}{\partial t} v \, d\Omega = -\int_\Omega (-\Delta)u \, v \, d\Omega. \tag{5.21}$$

In order to develop the weak form of (5.21), integration by parts is applied to the right-hand side to reduce the order of differentiation within the integral. Requiring that the test function v vanishes on $\partial\Omega$, we obtain the weak form of (5.19) as

$$\int_\Omega \frac{\partial u}{\partial t} v \, d\Omega = -\int_\Omega \nabla u \cdot \nabla v \, d\Omega. \tag{5.22}$$

Discretization of the domain in (5.22) is performed using three-node triangular elements, which is discussed in Section 4.

Expanding $u(x, y, t)$ in terms of the shape functions $\{\phi_i(x, y)\}_{i=1}^{\mathcal{M}}$, we obtain

$$u(x, y, t) = \sum_{i=1}^{\mathcal{M}} u_i(t)\, \phi_i(x, y), \tag{5.23}$$

where $\mathcal{M}$ is the number of free nodes in the mesh. We note from (5.23) that the shape functions are used to interpolate the spatial variation, while the temporal variation is related with the nodal variables.

We take $v = \phi_j, j = 1, \ldots \mathcal{M}$ in (5.22), and we obtain the discrete formulation

$$\int_\Omega \sum_{i=1}^{\mathcal{N}} \frac{\mathrm{d}u_i}{\mathrm{d}t} \phi_i \, \phi_j \, d\Omega = -\int_\Omega \sum_{i=1}^{\mathcal{N}} u_i \nabla\phi_i \cdot \nabla\phi_j \, d\Omega, \quad j = 1, \ldots \mathcal{N}, \tag{5.24}$$

or, upon interchanging summation and integration,

$$\sum_{i=1}^{\mathcal{N}} \frac{du_i}{dt} \int_{\Omega} \phi_i \phi_j \, d\Omega = -\sum_{i=1}^{\mathcal{N}} u_i \int_{\Omega} \nabla\phi_i \cdot \nabla\phi_j \, d\Omega \,, \quad j = 1, \ldots \mathcal{N}. \tag{5.25}$$

Introducing the "mass" matrix $(\mathbf{M})_{ij} = \int_{\Omega} \phi_i \phi_j \, d\Omega$ and the "stiffness" matrix $(\mathbf{K})_{ij} = \int_{\Omega} \nabla\phi_i \cdot \nabla\phi_j \, d\Omega$, (5.25) can be written as

$$\frac{d\mathbf{u}}{dt} = -\mathbf{M}^{-1}\mathbf{K}\mathbf{u}. \tag{5.26}$$

Thus, we have derived that the approximate matrix representation of the standard Laplacian for the finite element method is

$$\mathbf{A} = \mathbf{M}^{-1}\mathbf{K}. \tag{5.27}$$

Interestingly, although both **M** and **K** are symmetric positive definite and sparse, **A** is non-symmetric and dense. Nevertheless, a standard argument in linear algebra shows that **A** is similar to the symmetric positive matrix $\tilde{\mathbf{A}} = \mathbf{M}^{-\frac{1}{2}}\mathbf{K}\mathbf{M}^{-\frac{1}{2}}$ and therefore its eigenvalues are positive and real, and hence **A** itself is positive definite; see [14].

5.2.3 Unstructured mesh finite volume method in space

We consider the following two-dimensional time-space fractional diffusion equation with fractional Laplacian operator (2D-TSFDE-FL) and given initial condition on an arbitrarily shaped domain Ω:

$$_tD_*^{\gamma} u(x, y, t) = -K_{\alpha}(-\Delta)^{\frac{\alpha}{2}} u, \tag{5.28}$$

$$u(x, y, 0) = u_0(x, y), \quad (x, y) \in \Omega, \tag{5.29}$$

$$u(x, y, t) = 0, \quad (x, y) \in \partial\Omega, \quad t > 0. \tag{5.30}$$

The finite volume discretization of equation (5.28) is now derived from the non-fractional equation

$$\frac{\partial u}{\partial t} = -K(-\Delta)u, \tag{5.31}$$

where K is a positive constant. The above equation can also be rewritten as

$$\frac{\partial u}{\partial t} = -K(-\nabla) \cdot \nabla u. \tag{5.32}$$

In the solution domain, each node (p_i) is associated with one control volume (denoted by $V_i, i = 1, \ldots, n_p$) (the domain surrounded with blue lines in Figure 4). Each surface of

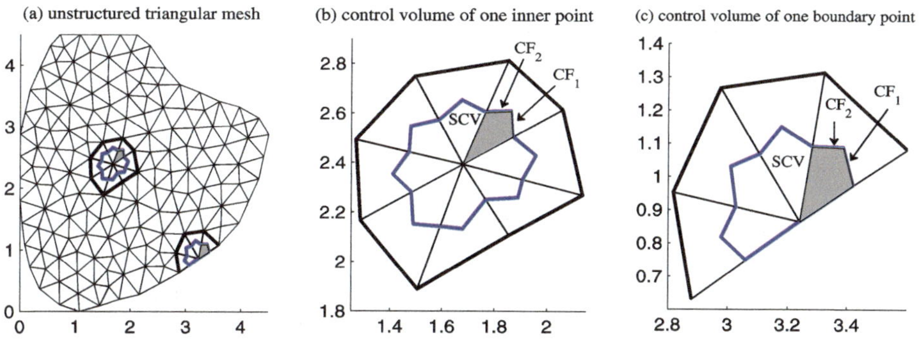

Figure 4: Construction of a control volume from the triangular finite element.

the control volume is defined as the vector that joins the centroid of the element to the mid point of one of its sides, as shown in Figure 4, denoted by CF1 and CF2 for example. Consequently, each of the triangular elements is divided into three subdomains by these control surfaces. These quadrilateral shapes are called subcontrol volumes (SCVs) and are illustrated in Figure 4 (for example, the gray quadrilateral). Thus, a control volume consists of the sum of all neighboring SCVs that surround any given node. The control volume can be assembled in a straightforward and efficient manner at the element level. The flow across each control surface must be determined by an integral. The finite volume unstructured mesh method (FVUM) discretization process is initiated by utilizing the integrated form of equation (5.32). Integrating equation (5.32) over an arbitrary control volume V_i yields

$$\int_{V_i} \frac{\partial u}{\partial t} dV_i = \int_{V_i} (-K(-\nabla) \cdot \nabla u) dV_i. \tag{5.33}$$

Applying the Gauss divergence theorem to the right-hand side of equation (5.33), we obtain

$$\frac{\partial}{\partial t} \int_{V_i} u dV_i = -K \int_{\Gamma_i} (-\nabla u) \cdot \mathbf{n}_i d\Gamma_i, \tag{5.34}$$

and using a lumped mass approach for the time derivative term gives

$$\Delta V_i \frac{\partial u_i}{\partial t} = -K \int_{\Gamma_i} (-\nabla u) \cdot \mathbf{n}_i d\Gamma_i, \tag{5.35}$$

where $\mathbf{n}_i$ represents the outward unit normal surface vector to the control surface (CF) and an anti-clockwise traversal of the finite volume integration is assumed. In the discrete sense, we use the approximation $\mathbf{n}_i \Delta\Gamma_i = \Delta yi - \Delta xj$; Δx and Δy represent the x and y components of the SCV face, and ΔV_i and ΔV_{ij} are used to denote the area of the control volume and the SCV surrounding the point p_i; they are evaluated for the vertex case as

$$\Delta V_i = \sum_{j=1}^{m_i} \Delta V_{ij}, \tag{5.36}$$

where m_i is the total number of SCVs that make up the control volume associated with node p_i. The integral term on the right-hand side of equation (5.35) is a line integral. It will be approximated by the mid point approximation for each control surface. To effect this mid point approximation, the value of the integrand is required at the mid point of the control surface and it is for these surfaces that the outward normal vector will be specified.

The integral term in equation (5.35) can then be rewritten as

$$\begin{aligned} -K \int_{\Gamma_i} (-\nabla u) \cdot \mathbf{n}_i d\Gamma_i &= -K \sum_{j=1}^{m_i} \int_{\Gamma_{ij}^1+\Gamma_{ij}^2} (-\nabla u^j) \cdot \mathbf{n}_{ij} d\Gamma_{ij} \\ &= -K \sum_{j=1}^{m_i} \sum_{r=1}^{2} -\left(\frac{\partial u^j}{\partial x} \Delta y_{ij}^r - \frac{\partial u^j}{\partial y} \Delta x_{ij}^r \right). \end{aligned} \tag{5.37}$$

Here, i is the global nodal number, m_i denotes the number of SCVs associated with the ith control volume. To evaluate the terms in equation (5.37), one of the triangular elements, ω_j, is considered, where node i is one of its vertices. For simplicity, the considered element ω_j is denoted as $\Delta_{1,2,3}$, where 1, 2, 3 is the local nodal number of the considered element in the counter-clockwise direction. The coordinates and the values of $u(x,y)$ at the three vertices of the element are denoted (x_l, y_l) $(l = 1,2,3)$, and u_l^j $(l = 1,2,3)$, which are the local ordering of the numerical solution of u at the vertices of the considered element. The linear interpolation function within the element is given by

$$\varphi_{\omega_j} = a_1 + a_2 x + a_3 y, \tag{5.38}$$

satisfying $\varphi_{\omega_j}(x_l, y_l) = u_l^j$. By using the above linear interpolation, the shape function $N_l^j(x,y)$ can be expressed as functions of u_l^j. The integral in equation (5.37) is then expressed as

$$-K \int_{\Gamma_i} (-\nabla u) \cdot \mathbf{n}_i d\Gamma_i = -K \sum_{j=1}^{m_i} \sum_{r=1}^{2} \sum_{l=1}^{3} -\left(\frac{\partial N_l^j}{\partial x} \Delta y_{ij}^r - \frac{\partial N_l^j}{\partial y} \Delta x_{ij}^r \right) u_l^j. \tag{5.39}$$

Thus, the discrete form of (5.39) is

$$\Delta V_i \frac{\partial u_i}{\partial t} = -K \sum_{j=1}^{m_i} \sum_{r=1}^{2} \sum_{l=1}^{3} -\left(\frac{\partial N_l^j}{\partial x} \Delta y_{ij}^r - \frac{\partial N_l^j}{\partial y} \Delta x_{ij}^r \right) u_l^j. \tag{5.40}$$

Using the same method for all nodes and their control volumes, and using a global ordering u_i in place of the local ordering u_l^j, the above discretized equation can be written in the following matrix form:

$$\mathbf{D}\frac{d\mathbf{u}}{dt} = -K\mathbf{G}\mathbf{u}, \tag{5.41}$$

where $\mathbf{u} = [u_1, u_2, \ldots, u_N]^T$ is the numerical solution approximating $u(x_i, y_i, t)$ at each mesh node (x_i, y_i). The matrix $\mathbf{D} = \operatorname{diag}(\triangle V_i)$ is a diagonal matrix with the control volume areas along its diagonal. The matrix $\mathbf{G}$ is sparse, symmetric, and positive semi-definite and represents the contributions from each node towards the total flux through each control volume.

By setting $\mathbf{F} = \mathbf{D}^{-1}\mathbf{G}$, the above equation could be rewritten as

$$\frac{d\mathbf{u}}{dt} = -K\mathbf{F}\mathbf{u}, \tag{5.42}$$

where $\mathbf{F}$ is the finite volume matrix representation of the negative Laplacian $(-\Delta)$.

According to [4] the matrix transfer technique uses the representation of the fractional Laplacian $-(-\Delta)^{\alpha/2}$ as $-\mathbf{F}^{\alpha/2}$ since matrix $\mathbf{F}$ represents the operator Laplacian $(-\Delta)$ with homogeneous Dirichlet boundary conditions under the finite volume discretization.

5.2.4 Finite difference method with matrix transfer technique for the 2D-TSFDE-FL

Define $t_n := n\tau$, $n = 0, 1, 2, \ldots$, where τ is the time step. We adopt the L1-approximation to discretize the Caputo time-fractional derivative [6], i. e.,

$${}_tD_*^{\gamma} u^n = \frac{1}{\mu_0}\sum_{j=0}^{n-1} b_j[u^{n-j} - u^{n-1-j}] + O(\tau^{2-\gamma}), \tag{5.43}$$

where $\mu_0 = \tau^{\gamma}\Gamma(2-\gamma)$, $b_j = (j+1)^{1-\gamma} - j^{1-\gamma}$, $j = 0, 1, 2, \ldots, n-1$. Combining the approximation for the Caputo time-fractional derivative with the approximation of the fractional Laplacian (5.16), we obtain the following numerical approximation of the 2D-TSFDE-FL (5.1)–(5.4):

$$\frac{1}{\mu_0}\sum_{j=0}^{n-1} b_j[\mathbf{u}^{n-j} - \mathbf{u}^{n-1-j}] = -K_\alpha \mathbf{A}^{\alpha/2}\mathbf{u}^n. \tag{5.44}$$

After some further manipulations, (5.44) can be written in the form

$$\mathbf{u}^n = [\mathbf{I} + \mu_0 K_\alpha \mathbf{A}^{\alpha/2}]^{-1}\left[\sum_{j=0}^{n-2}(b_j - b_{j+1})\mathbf{u}^{n-1-j} + b_{n-1}\mathbf{u}^0\right]. \tag{5.45}$$

Defining the scalar function $\bar{G}_1(\mathbf{A}) = [I + \mu_0 K_\alpha \mathbf{A}^{\alpha/2}]^{-1}$, we obtain the first numerical scheme (Scheme 1) for approximating the 2D-TSFDE-FL (5.1)–(5.4) as

$$\mathbf{u}^n = \bar{G}_1(\mathbf{A})\mathbf{B}_1^n, \quad \text{with } \mathbf{B}_1^n = \sum_{j=0}^{n-2}(b_j - b_{j+1})\mathbf{u}^{n-1-j} + b_{n-1}\mathbf{u}^0, \tag{5.46}$$

where $\mathbf{A}$ is either generated using the finite difference method (5.18) or the FEM (5.27), or the unstructured mesh finite volume method (5.42).

These numerical schemes can be implemented using either the Lanczos method ($\mathbf{A}$ is generated from the finite difference method), or the M-Lanczos method ($\mathbf{A}$ is generated from the FEM or the unstructured mesh finite volume method). See [14] for further details.

5.2.5 Exact-in-time method with matrix transfer technique for the 2D-TSFDE-FL

We now consider an alternative strategy for approximating the fractional differential system (5.16) associated with the 2D-TSFDE-FL (5.1)–(5.4). Taking the Laplace transform of (5.1) with $\tilde{\mathbf{u}}^n = \mathcal{L}\{\mathbf{u}^n(t)\}$ yields

$$s^\gamma \tilde{\mathbf{u}}^n - s^{\gamma-1}\mathbf{u}^0 = -K_\alpha \mathbf{A}^{\alpha/2}\tilde{\mathbf{u}}^n, \tag{5.47}$$

i. e.,

$$\tilde{\mathbf{u}}^n = \left[s\mathbf{I} + s^{1-\gamma}K_\alpha \mathbf{A}^{\alpha/2}\right]^{-1}\mathbf{u}^0. \tag{5.48}$$

If $\mathbf{A}$ can be diagonalized as

$$\mathbf{A} = \mathbf{P}\mathbf{\Lambda}\mathbf{P}^{-1}, \tag{5.49}$$

where $\mathbf{\Lambda} = \operatorname{diag}(\lambda_i,\ i = 1,\dots,m)$, λ_i being the eigenvalues of $\mathbf{A}$, then from (5.47), we obtain

$$\begin{aligned} \mathbf{u}^n &= \mathcal{L}^{-1}\{[s\mathbf{I} + s^{1-\gamma}K_\alpha \mathbf{A}^{\alpha/2}]^{-1}\mathbf{u}^0\} \\ &= \mathbf{P}\operatorname{diag}\left\{\mathcal{L}^{-1}\left\{\frac{1}{s + s^{1-\gamma}K_\alpha \lambda_i^{\alpha/2}}\right\},\ i = 1,\dots,m\right\}\mathbf{P}^{-1}\mathbf{u}^0. \end{aligned} \tag{5.50}$$

To perform the required inversion we require the Mittag-Leffler function $E_\gamma(z)$ [10]. We have

$$E_\gamma(z) = \sum_{k=0}^{\infty}\frac{z^k}{\Gamma(\gamma k + 1)}, \tag{5.51}$$

which is a generalization of the exponential function, with Laplace transform given by

$$\mathcal{L}\{E_\gamma(-\omega t^\gamma)\} = \frac{1}{s + s^{1-\gamma}\omega}, \quad \Re(s) > |\omega|^{1/\gamma}. \tag{5.52}$$

Using (5.52), we obtain

$$\begin{aligned}\mathbf{u}^n &= \mathbf{P}\,\mathrm{diag}\{E_\gamma(-K_\alpha\lambda_i^{\alpha/2}t_n^\gamma),\quad i=1,\ldots,m\}\mathbf{P}^{-1}\mathbf{u}^0\\ &= E_\gamma(-K_\alpha\mathbf{A}^{\alpha/2}t_n^\gamma)\mathbf{u}^0.\end{aligned} \tag{5.53}$$

Defining the scalar function $\bar{G}_2(\mathbf{A}) = E_\gamma(-K_\alpha\mathbf{A}^{\alpha/2}t_n^\gamma)$, we obtain the second numerical scheme (Scheme 2) for approximating the 2D-TSFDE-FL (5.1)–(5.4) as

$$\mathbf{u}^n = \bar{G}_2(\mathbf{A})\mathbf{B}_2, \quad \text{with } \mathbf{B}_2 = \mathbf{u}^0, \tag{5.54}$$

where **A** can again be either generated using the finite difference method (5.18) or the FEM (5.27), or the unstructured mesh finite volume method (5.42).

5.2.6 Numerical examples

Example 5.1. Consider the 2D-TSFDE-FL (5.1)–(5.4) on the domain $[0,1]\times[0,1]$ with $K_\alpha = 1$ and initial condition $u_0(x,y) = xy(1-x)(1-y)$ (see [14]).

The analytical solution of the 2D-TSFDE-FL (5.1)–(5.4) is given by

$$u(x,y,t) = \sum_{n=1}^{\infty}\sum_{m=1}^{\infty} E_\gamma(-\lambda_{n,m}^{\alpha/2}t^\gamma)c_{n,m}(0)\varphi_{n,m}, \tag{5.55}$$

where

$$\lambda_{n,m} = n^2\pi^2 + m^2\pi^2, \tag{5.56}$$

$$\varphi_{n,m} = 2\sin(n\pi x)\sin(m\pi y), \tag{5.57}$$

$$c_{n,m}(0) = \int_0^1\int_0^1 xy(1-x)(1-y)\varphi_{n,m}\,dy\,dx. \tag{5.58}$$

The scheme outlined in Section 5.2.5 is exact in time, all of the error in this scheme is associated with the spatial discretization, by either the finite difference method or the FEM. To identify the order of convergence in space, we compute the error in the numerical solution at $t = 0.01$ with $\alpha = 1.3$, and $\gamma = 0.5$ for a sequence of refined meshes. In Table 8, we present the ℓ^∞-norm error for both the finite difference method and the

Table 8: Spatial errors at $t = 0.01$ using Scheme 2 with $\alpha = 1.3$ and $\gamma = 0.5$.

h	FDM	FEM
0.1	6.4294e−005	2.2014e−004
0.05	1.6289e−005	6.3283e−005
0.025	4.0930e−006	1.5994e−005
0.0125	1.0264e−006	4.2674e−006
0.00625	2.6016e−007	1.0331e−006
Order	2.0	2.0

Table 9: Temporal errors at $t = 0.01$ using Scheme 1 with $h = 0.00625$, $\alpha = 1.3$, and $\gamma = 0.5$.

τ	FDM
0.001	4.8872e−004
0.0005	2.3559e−004
0.00025	1.1354e−004
0.000125	5.6336e−005
0.0000625	2.7265e−005
Order	1.0

FEM. For the finite difference method, uniform grids with grid spacing h were used. For the FEM, unstructured triangular meshes with maximum element edge length h were used. The order of convergence in space is estimated to be $O(h^2)$ for the finite difference method and $O(h^2)$ for the FEM.

To identify the order of convergence in time for the scheme given in Section 5.2.4, we compute the error in the numerical solution at $t = 0.01$ using the finite difference method on the finest mesh with $h = 0.00625$, $\alpha = 1.3$, and $\gamma = 0.5$. In Table 9, we present the ℓ^∞-norm error. The order of convergence in time for Scheme 1 is estimated to be $O(\tau)$.

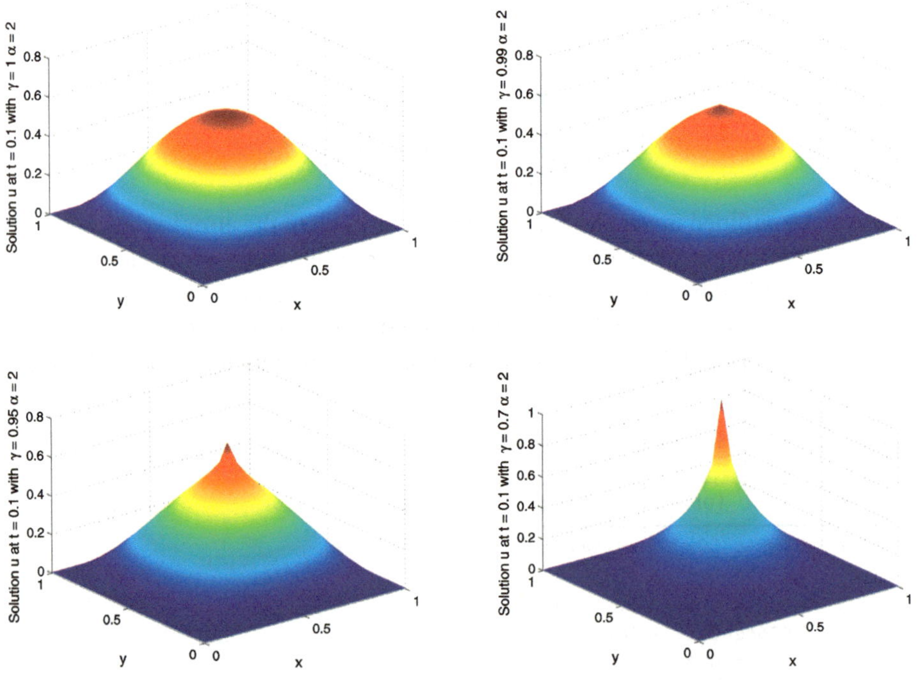

Figure 5: Effect of fractional order in time.

To further illustrate the effect of the fractional order in time and space, we present another example with the Delta function as the initial condition.

Example 5.2. Consider the 2D-TSFDE-FL (5.1)–(5.4) on the domain $[0,1] \times [0,1]$ with $K_\alpha = 1$ and $u_0(x,y) = \delta(x - \frac{1}{2}, y - \frac{1}{2})$. In this example, we use the scheme from Section 5.2.4 to compute the numerical solution.

In Figure 5 we illustrate the effect of the fractional order in time for this problem, with α fixed at 2. The diffusion process with $0 < \gamma < 1$ is called a subdiffusion process. It is very interesting to see the appearance of cusps for the different choices of fractional order in time γ, as compared to the standard diffusion with $\gamma = 1$.

In Figure 6 we illustrate the effect of the fractional order in space for this problem, with γ fixed at 1. The diffusion process with $1 < \alpha < 2$ is called a Lévy flight. We observe the slower rate of diffusion associated with the Lévy flight, as compared to the standard diffusion.

The effect of the fractional order in both time and space is illustrated in Figure 7. The feature of the anomalous diffusion process is characterized by the different combinations of the fractional order in time γ and the fractional order in space α.

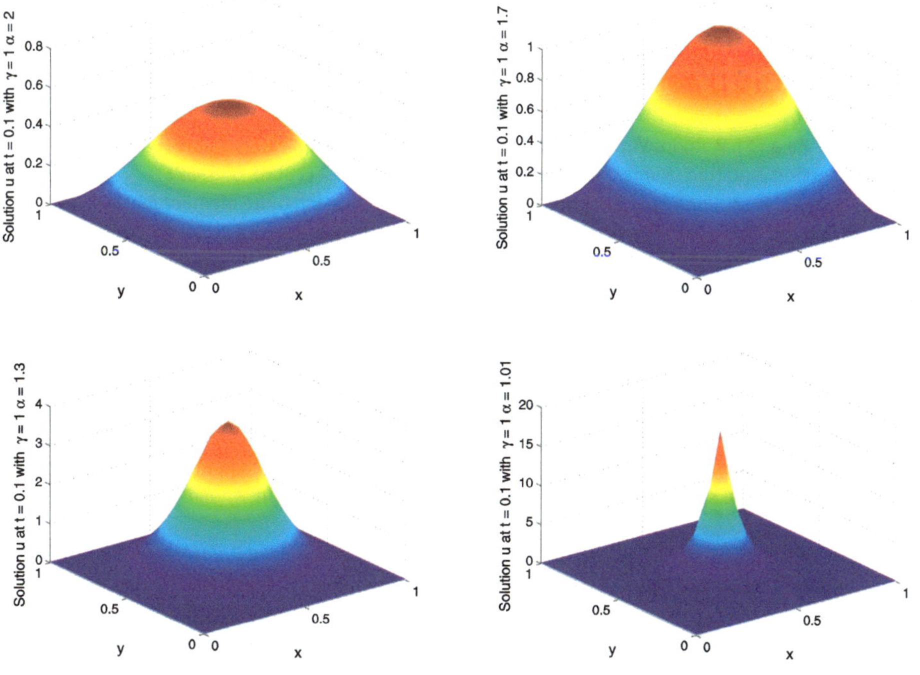

Figure 6: Effect of fractional order in space.

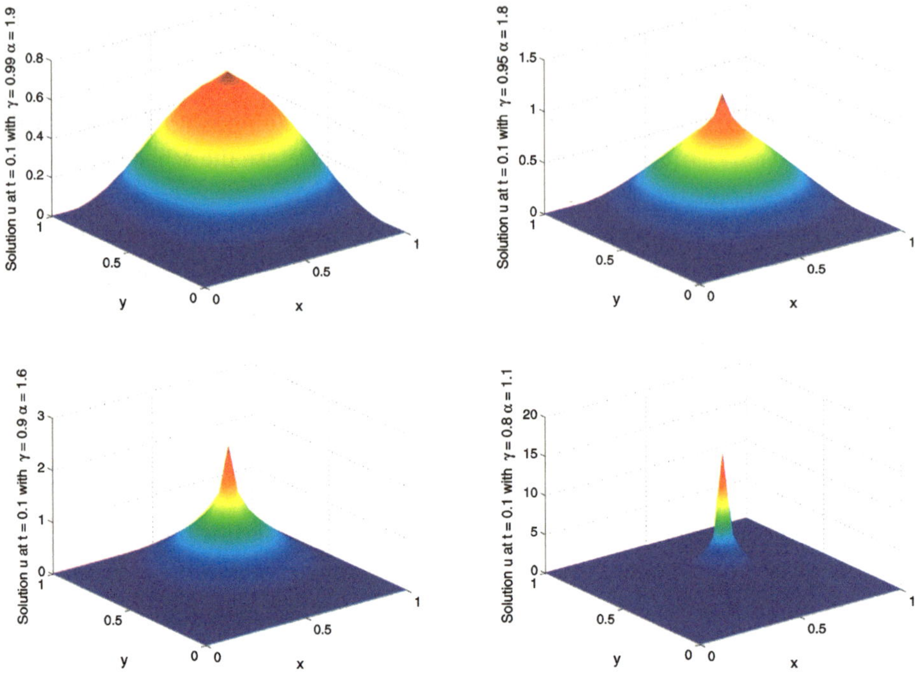

Figure 7: Effect of fractional order in both time and space.

Bibliography

[1] V. J. Ervin and J. P. Roop, Variational formulation for the stationary fractional advection dispersion equation, *Numer. Methods Partial Differ. Equ.*, **22** (2006), 558–576.

[2] W. Fan, F. Liu, X. Jiang, and I. Turner, A novel unstructured mesh finite element method for solving the time-space fractional wave equation on a two-dimensional irregular convex domain, *Fract. Calc. Appl. Anal.*, **672** (2017), 352–383.

[3] L. Feng, F. Liu, I. Turner, Q. Yang, and P. Zhuang, Unstructured mesh finite difference/finite element method for the 2D time-space Riesz fractional diffusion equation on irregular convex domains, *Appl. Math. Model.*, **59** (2018), 441–463.

[4] M. Ilić, F. Liu, I. Turner, and V. Anh, Numerical approximation of a fractional-in-space diffusion equation (II) – with nonhomogeneous boundary conditions, *Fract. Calc. Appl. Anal.*, **9** (2006), 333–349.

[5] F. Liu, P. Zhuang, V. Anh, I. Turner, and K. Burrage, Stability and convergence of the difference methods for the space-time fractional advection-diffusion equation, *Appl. Math. Comput.*, **191** (2007), 12–20.

[6] F. Liu, P. Zhuang, and Q. Liu, *Numerical Methods of Fractional Partial Differential Equations and Applications*, Science Press, China, November 2015, ISBN 978-7-03-046335-7 (Chinese).

[7] R. Magin, O. Abdullah, D. Baleanu, and X. Zhou, Anomalous diffusion expressed through fractional order differential operators in the Bloch–Torrey equation, *J. Magn. Res.*, **190**(2) (2008), 255–270.

[8] R. Magin, B. Akpa, T. Neuberger, and A. Webb, Fractional order analysis of Sephadex gel structures: NMR measurements reflecting anomalous diffusion, *Commun. Nonlinear Sci. Numer. Simul.*, **16**(12) (2011), 4581–4587.
[9] M. M. Meerschaert and C. Tadjeran, Finite difference approximations for fractional advection–dispersion flow equations, *Appl. Math. Comput.*, **172** (2004), 65–77.
[10] I. Podlubny, *Fractional Differential Equations*, Academic Press, San Diego, 1999.
[11] S. Qin, F. Liu, and I. Turner, A two-dimensional multi-term time and space fractional Bloch–Torrey model based on bilinear rectangular finite elements, *Commun. Nonlinear Sci. Numer. Simul.*, **56** (2018), 270–286.
[12] S. Qin, F. Liu, I. Turner, V. Vegh, Q. Yu, and Q. Yang, Multi-term time fractional Bloch equations and application in magnetic resonance imaging, *J. Comput. Appl. Math.*, **319** (2017), 308–319.
[13] J. P. Roop, Variational Solution of the Fractional Advection-Dispersion Equation, Ph. D. thesis, Clemson University, 2004.
[14] Q. Yang, I. Turner, F. Liu, and M. Ilis, Novel numerical methods for solving the time-space fractional diffusion equation in 2D, *SIAM J. Sci. Comput.*, **33** (2011), 1159–1180.
[15] Q. Yu, F. Liu, I. Turner, and K. Burrage, A computationally effective alternating direction method for the space and time fractional Bloch–Torrey equation in 3-D, *Appl. Math. Comput.*, **219** (2012), 4082–4095.
[16] Q. Yu, F. Liu, I. Turner, and K. Burrage, Stability and convergence of an implicit numerical method for the space and time fractional Bloch–Torrey equation, in *The Special Issue of Fractional Calculus and Its Applications*, *Philos. Trans. R. Soc. A, Math. Phys. Eng. Sci.*, **A371** (2013), 20120150.

Guofei Pang and Wen Chen

Comparison of two radial basis collocation methods for Poisson problems with fractional Laplacian

Abstract: Research on meshless methods for partial differential equations with fractional Laplacian, especially in high spatial dimensions, is rare. This chapter compares two existing radial basis collocation methods on 1D, 2D, and 3D fractional Poisson problems with zero non-local boundary condition. Benchmark solutions over unit balls in 1D, 2D, and 3D spaces are employed to validate the methods. The two methods are compared in terms of solution accuracy, computational cost, and flexibility. Numerical results show that the two methods have comparable solution accuracy and same-order time complexity, but varied flexibilities. Additionally, the influences of fractional order on the solution accuracy differ from each other.

Keywords: Three-dimensional fractional Laplacian, meshless method, radial basis collocation, space-fractional derivative, fractional Poisson problem

MSC 2010: 35J67, 35J05 , 65N35

1 Introduction

One of the motivations to develop mesh-free numerical methods is to ease the simulations where creating an effective mesh from the geometry of a complex 3D object is rather difficult. Another motivation is to remedy the problems where the mesh may be re-created, such as cracking and bending simulations in computational solid mechanics.

The past decade has seen much progress in applications of meshless methods to fractional partial differential equations (PDEs). Most of the research concentrates on time-fractional PDEs, but research on space-fractional PDEs is relatively rare. The reason why the latter is rare is that the meshless approaches for the time-fractional case are just the same as or similar to the counterparts for the integer-order derivative case, whereas the meshless approaches for the space-fractional case need re-

Acknowledgement: This work was supported by National Natural Science Foundation of China (11701025).

Guofei Pang, Division of Applied Mathematics, Brown University, Providence, RI 02912, USA, e-mail: guofei_pang@brown.edu
Wen Chen, College of Mechanics and Materials, Hohai University, Nanjing, Jiangsu 211100, China, e-mail: chenwen@hhu.edu.cn

https://doi.org/10.1515/9783110571684-009

formulation due to the non-locality the fractional operator exhibits. More importantly, the re-formulation strategy varies with different types of space-fractional derivatives. Among them are the one-sided derivative, two-sided derivative, fractional Laplacian [16], (generalized) multi-dimensional derivative [18], multi-scaling derivative [27], and even variable-order derivative [31].

The research on meshless methods for space-fractional PDEs includes, but is not limited to, the following. In [17], the meshless approach based on the point interpolation method for 1D diffusion equations with the left-sided fractional derivative is developed. In [9] and [10], the meshless approaches based on improved moving least-squares approximation for 1D and 2D wave equations with the two-sided fractional derivatives, respectively, are presented. In [11], the elementary-free Galerkin method to 2D diffusion equations with the two-sided fractional derivatives is extended. Considering a more complex space-fractional operator, i. e., the fractional Laplacian, in [8] and [20] the boundary element-free and the radial basis function (RBF) collocation methods for 2D and/or 3D problems with irregular computation domains, respectively, are proposed. In [30], the RBF collocation methods proposed in [20] are combined with the analytical method in the time domain in order to solve the time-space fractional PDEs. Very recently, in [16] another RBF collocation method has been presented for the Poisson problem with fractional Laplacian. The new RBF collocation method can handle the problems which the RBF collocation method presented in [20] cannot. Comparison of the two RBF methods proposed in [16] and [20] in terms of solution accuracy, computational cost, and flexibility will be the topic of the present chapter.

The chapter only concerns the meshless methods for PDEs with fractional Laplacian. For PDEs with other types of space-fractional operators, e. g., one-sided and two-sided ones, the readers are referred to the relevant aforementioned papers. Actually, the fractional Laplacian can be considered the extension of the one-sided or two-sided derivative through generalizing the discretized differentiation directions to the continuous ones. When restricting the differentiation directions to the coordinate axis directions, the fractional Laplacian degenerates to the one-sided or two-sided fractional derivative. The main reason why we exclusively consider the fractional Laplacian is the operator's widespread applications in modeling the varied physical problems which exhibit anomalous behaviors not following classical exponential or local laws. Among these are solute transport in fractured and porous media [18, 2, 31], acoustic wave propagation with frequency-dependent dissipation [6, 33, 14, 36, 4], and laminar and turbulent flows [32, 28, 34, 3, 12], just to mention a few.

It is still an open problem to develop effective numerical methods for PDEs with fractional Laplacian because of three primary difficulties. First, there are varied definitions for the fractional Laplacian. These definitions are equivalent for the whole space $\mathbb{R}^d$ but could be different for a bounded domain. For instance, on a bounded domain, as [16] illustrates, the spectral definition differs from the Riesz definition (or the directional definition). Thus, one cannot find a unified numerical method that works well for all definitions. Usually, each definition has its preferable method. For example, the spectral element method works well for the spectral definition [29], the adaptive finite element method (FEM) is good for the Riesz definition [1], and the RBF collocation method is promising for the directional definition [20]. The second difficulty comes from the strong non-locality of the fractional Laplacian compared to the weak non-locality of the one-sided or two-sided derivative. This difficulty is more obvious for the directional definition of the fractional Laplacian. In numerical discretization, the discretization matrix is actually the superposition of the matrices for different differentiation directions. The fractional Laplacian has much more differentiation directions than the one-sided or two-sided derivatives, leading to a discretization matrix having a more complicated structure. It is thus more difficult to develop fast matrix-vector product methods via exploiting the matrix structure. Development of the corresponding preconditioning techniques is also challenging because of the complicated matrix structure. The third difficulty manifests itself when considering high-dimensional, irregular-domain problems. Mesh-based methods, such as adaptive FEM, could require much more meshing effort. Additionally, the non-locality of the fractional Laplacian makes these essentially local methods non-local. When the directional definition of the fractional Laplacian is considered, the FEM needs to evaluate the non-local interaction between two arbitrary separate elements, and the non-local interaction intensity is characterized by the length of the line segment that is the intersection part between the current element and the ray ejected from another element [25, 22]. For 3D problems, the calculation of the line segment length could be troublesome.

RBF collocation methods were developed by the authors to lower the third difficulty [20]. Being one of the typical meshless methods, the RBF collocation methods [5] can easily handle high-dimensional, irregular-domain fractional Laplacian problems and are rather easy to implement compared to the mesh-based methods such as FEM. Due to the strong-form collocation, the solution accuracy of the RBF collocation methods is often higher than that of the weak-form methods, say, FEM. In this chapter, we specifically investigate two RBF collocation methods, presented in [20] and [16]. The two methods are both designed for the directional definition of the fractional Laplacian. Particularly, one method is based on Gauss–Jacobi quadrature, and the other relies on the vector Grünwald formula [19]. We abbreviate the quadrature-based method as Q-RBF and the Grünwald formula-based method as G-RBF. According to the definition of the directional fractional derivative in the fractional Laplacian, the Q-RBF and the G-RBF methods correspond to the Caputo and the Riemann–Liouville directional derivatives, respectively. Since the two types of directional derivatives can be different, the two RBF methods can tackle different problems. More importantly, because of the dimension-independent merit of the RBF, it is straightforward to extend the two existing methods to 3D problems, which is another contribution of the chapter apart from the comparison.

The chapter is organized as follows. Section 2 introduces the problem we consider and elaborates on the directional definition of the fractional Laplacian as well as two types of directional fractional derivatives. The equivalence condition of the two directional derivatives is given. Section 3 details the implementations of the two RBF collocation methods: Q-RBF and G-RBF. Section 4 shows the numerical results. The performance of the current RBF methods is tested on the benchmark solutions in 1D, 2D, and 3D spaces. Conclusions of the comparison are drawn in Section 5.

2 Fractional Poisson problem

Without loss of generality, we simply consider the numerical solution to the time-independent fractional Poisson problem. It is straightforward to obtain the solution to the time-dependent problem using standard numerical approaches for temporal discretization. For instance, in [20] we applied the finite difference scheme to the temporal discretization.

The fractional Poisson problem defined on a bounded domain Ω with zero non-local boundary condition is [16]

$$\begin{aligned} (-\Delta)^{\alpha/2}u(\boldsymbol{x}) &= f(\boldsymbol{x}), \quad \boldsymbol{x} \in \Omega \subseteq \mathbb{R}^d,\ \alpha \in (0,1) \cup (1,2), \\ u(\boldsymbol{x}) &= 0, \quad \boldsymbol{x} \in \mathbb{R}^d \setminus \Omega. \end{aligned} \tag{1}$$

Restricting the solution both on the boundary $\partial\Omega$ and over the exterior $\mathbb{R}^d \setminus \overline{\Omega}$ guarantees the solution uniqueness. The directional definition of the fractional Laplacian is adopted here (equation (26.24) of [26], [21]), i. e.,

$$(-\Delta)^{\alpha/2}u(\boldsymbol{x}) := C_{\alpha,d} \int\limits_{\|\boldsymbol{\theta}\|_2=1} D^{\alpha}_{\boldsymbol{\theta}} u(\boldsymbol{x}) d\boldsymbol{\theta}, \quad \boldsymbol{\theta} \in \mathbb{R}^d, \tag{2}$$

where $\boldsymbol{\theta}$ is the differentiation direction represented by a unit vector and the integration domain $\|\boldsymbol{\theta}\|_2 = 1$ is a unit sphere ($\|\cdot\|_2$ is the L_2 norm). The scaling constant before the integral is $C_{\alpha,d} = \frac{\Gamma(\frac{1-\alpha}{2})\Gamma(\frac{d+\alpha}{2})}{2\pi^{\frac{d+1}{2}}}$, where d is the spatial dimension.

The Riemann–Liouville directional derivative is defined by[1]

$$D^{\alpha}_{\boldsymbol{\theta}} u(\boldsymbol{x}) := \begin{cases} \frac{1}{\Gamma(1-\alpha)} D^1_{\boldsymbol{\theta}} \int_0^{d(\boldsymbol{x},\boldsymbol{\theta})} \zeta^{-\alpha} u(\boldsymbol{x} - \zeta\boldsymbol{\theta}) d\zeta, & \alpha \in (0,1), \\ \frac{1}{\Gamma(2-\alpha)} D^2_{\boldsymbol{\theta}} \int_0^{d(\boldsymbol{x},\boldsymbol{\theta})} \zeta^{1-\alpha} u(\boldsymbol{x} - \zeta\boldsymbol{\theta}) d\zeta, & \alpha \in (1,2), \end{cases} \tag{3}$$

where $d(\boldsymbol{x}, \boldsymbol{\theta})$ is the distance from the point $\boldsymbol{x}$ to the boundary $\partial\Omega$ along the direction "$-\boldsymbol{\theta}$". Let $P(\boldsymbol{x}) := \boldsymbol{x} - d(\boldsymbol{x}, \boldsymbol{\theta})\boldsymbol{\theta}$, and then $P(\boldsymbol{x}) \in \partial\Omega$, which we call the backward boundary point of $\boldsymbol{x}$. We also call $d(\boldsymbol{x}, \boldsymbol{\theta})$ the backward distance. The non-locality of the

1 The derivative was first defined in [26] (see the equation below equation (24.39) of [26]).

fractional Laplacian exists in the whole space, and the upper integration limit $d(\boldsymbol{x}, \boldsymbol{\theta})$ should have been $+\infty$. But, because the solution vanishes outside the bounded domain, the contribution of the solutions in the exterior is simply zero! We thus derive a finite distance $d(\cdot, \cdot)$ that depends on the domain shape. The integer-order directional derivative is defined by [35]

$$D^n_{\boldsymbol{\theta}} u(\boldsymbol{x}) := (\boldsymbol{\theta} \cdot \nabla)^n u(\boldsymbol{x}), \quad n \in \mathbb{Z}, \tag{4}$$

where ∇ is the gradient operator with respect to $\boldsymbol{x}$ and "$\cdot$" represents the inner product of two vectors. In the 2D case, the directional derivative can be written as $(\cos\theta \frac{\partial}{\partial x_1} + \sin\theta \frac{\partial}{\partial x_2})^n u(x_1, x_2)$, where $\boldsymbol{\theta} = [\cos\theta, \sin\theta]^T$.

We also consider the Caputo directional derivative defined by [20]

$$D^\alpha_{*\boldsymbol{\theta}} u(\boldsymbol{x}) := \begin{cases} \frac{1}{\Gamma(1-\alpha)} \int_0^{d(\boldsymbol{x},\boldsymbol{\theta})} \zeta^{-\alpha} D^1_{\boldsymbol{\theta}} u(\boldsymbol{x} - \zeta\boldsymbol{\theta}) d\zeta, & \alpha \in (0,1), \\ \frac{1}{\Gamma(2-\alpha)} \int_0^{d(\boldsymbol{x},\boldsymbol{\theta})} \zeta^{1-\alpha} D^2_{\boldsymbol{\theta}} u(\boldsymbol{x} - \zeta\boldsymbol{\theta}) d\zeta, & \alpha \in (1,2). \end{cases} \tag{5}$$

The corresponding fractional Laplacian is

$$(-\Delta)^{\alpha/2}_* u(\boldsymbol{x}) := \int\limits_{\|\boldsymbol{\theta}\|_2 = 1} D^\alpha_{*\boldsymbol{\theta}} u(\boldsymbol{x}) d\boldsymbol{\theta}, \quad \boldsymbol{\theta} \in \mathbb{R}^d. \tag{6}$$

The directional fractional derivative is not a new fractional derivative, and it is essentially the copy of the left-sided fractional derivative after the rotation of the coordinate axes. Take the 2D case for an example. Denote in the polar coordinate $x_1 = r\cos\theta$, $x_2 = r\sin\theta$, where θ is the angle between the vector $\overrightarrow{Ox_1}$ and the positive direction of the axis x_1. Implementing the coordinate counter-clockwise rotation with an angle θ from the original coordinate axes (x_1, x_2) to the new axes (x_1', x_2') yields

$$\begin{cases} x_1' = x_1 \cos\theta + x_2 \sin\theta = r, \\ x_2' - -x_1 \sin\theta + x_2 \cos\theta = 0. \end{cases} \tag{7}$$

We note that the differentiation direction $\boldsymbol{\theta}$ is also dependent on the coordinate axes, and after the coordinate transformation the differentiation direction changes from $\boldsymbol{\theta} = [\cos\theta, \sin\theta]^T$ to $\boldsymbol{\theta} = [1, 0]^T$. We thus rewrite the directional fractional derivative in form of the left-sided derivative, i. e.,

$$\begin{aligned} D^\alpha_{\boldsymbol{\theta}} u(x_1, x_2) &= D^\alpha_{\boldsymbol{\theta}} u(x_1(x_1'), x_2(x_2')) = D^\alpha_{a+} u'(r, 0), \\ D^\alpha_{*\boldsymbol{\theta}} u(x_1, x_2) &= D^\alpha_{*\boldsymbol{\theta}} u(x_1(x_1'), x_2(x_2')) = {}^C D^\alpha_{a+} u'(r, 0), \end{aligned} \tag{8}$$

where $D^\alpha_{a+}(\cdot)$ and ${}^C D^\alpha_{a+}(\cdot)$ are the left-sided Riemann–Liouville and Caputo fractional derivatives with respect to the first coordinate direction, x_1', of $u'(x_1', x_2') := u(x_1(x_1'), x_2(x_2'))$, respectively. The new coordinate of the backward boundary point $P([x_1, x_2]^T)$ is $[a, 0]^T$.

Analogous to the correlation between the two types of left-sided fractional derivatives, the Riemann–Liouville and Caputo directional differential operators are equivalent at the given point $\boldsymbol{x}$, if the operand and/or its directional derivative vanish at the backward boundary point of $\boldsymbol{x}$. We give the following two theorems. The proof is omitted here because it is, after the coordinate rotation, the same as the proof for the left-sided fractional derivatives (namely, performing repeatedly integration by parts and differentiation [24]).

Theorem 2.1. *For $v(\boldsymbol{x}) \in C^1(\Omega)$ which vanishes outside the open set Ω, $\alpha \in (0,1)$ and the fixed $\boldsymbol{x} \in \Omega$, we have $D_{\boldsymbol{\theta}}^{\alpha} v(\boldsymbol{x}) = D_{*\boldsymbol{\theta}}^{\alpha} v(\boldsymbol{x})$.*

Theorem 2.2. *For $v(\boldsymbol{x}) \in C^2(\Omega) \cap C^1(\partial\Omega)$ which vanishes outside the open set Ω, $\alpha \in (1,2)$ and the fixed $\boldsymbol{x} \in \Omega$, we have $D_{\boldsymbol{\theta}}^{\alpha} v(\boldsymbol{x}) = D_{*\boldsymbol{\theta}}^{\alpha} v(\boldsymbol{x})$ if and only if $D_{\boldsymbol{\theta}}^{1}(v(P(\boldsymbol{x}))) = 0$, where the backward boundary point $P(\boldsymbol{x}) = \boldsymbol{x} - d(\boldsymbol{x}, \boldsymbol{\theta})\boldsymbol{\theta} \in \partial\Omega$.*

From the definition of the directional fractional Laplacian, we further have the following corollaries.

Corollary 2.2.1. *For $u(\boldsymbol{x}) \in C^1(\Omega)$ which vanishes outside the open set Ω, $\alpha \in (0,1)$, and any $\boldsymbol{x} \in \Omega$, we have $(-\Delta)^{\alpha/2} u(\boldsymbol{x}) = (-\Delta)_*^{\alpha/2} u(\boldsymbol{x})$.*

Corollary 2.2.2. *For $u(\boldsymbol{x}) \in C^2(\Omega) \cap C^1(\partial\Omega)$ which vanishes outside the open set Ω, $\alpha \in (1,2)$, and any $\boldsymbol{x} \in \Omega$, we have $(-\Delta)^{\alpha/2} u(\boldsymbol{x}) = (-\Delta)_*^{\alpha/2} u(\boldsymbol{x})$ if and only if $\nabla u(\boldsymbol{y}) = \boldsymbol{0}$, $\forall \boldsymbol{y} \in \partial\Omega$.*

Note that the Q-RBF and the G-RBF methods are employed to discretize $(-\Delta)_*^{\alpha/2}(\cdot)$ and $(-\Delta)^{\alpha/2}(\cdot)$, respectively. It can be learned from the above corollaries that for $\alpha \in (0,1)$, or $\alpha \in (1,2)$ with $\nabla u(\partial\Omega) = 0$, the equations $(-\Delta)^{\alpha/2} u = f$ and $(-\Delta)_*^{\alpha/2} u = f$ are equivalent, and thus the two RBF methods can be employed to solve the same problem. However, for $\alpha \in (1,2)$ and $\nabla u(\partial\Omega) \neq 0$, the two equations differ from each other, and therefore the two RBF methods will produce completely different solutions. In this case, the equation model $(-\Delta)_*^{\alpha/2} u = f$ is an independent model, and because the Caputo derivative is a regularized form of the Riemann–Liouville derivative, the model features lower singularity.

3 RBF collocation methods

The RBF collocation method represents the approximate solution as a weighted sum of the RBFs $\phi_j(\boldsymbol{x}) = \phi(\|\boldsymbol{x} - \boldsymbol{x}_j\|_2)$, i. e.,

$$u(\boldsymbol{x}) \approx \sum_{j=1}^{M+N} \lambda_j \phi_j(\boldsymbol{x}), \tag{9}$$

where the expansion coefficients λ_j are unknown and will be determined through solving a linear system. The RBF $\phi_j(\boldsymbol{x})$ is globally defined over the whole domain $\boldsymbol{x} \in \Omega$, but it vanishes in Ω's exterior.

We collocate M domain points in Ω and N boundary points on $\partial\Omega$. The sequence of the collocation points are rearranged so that the first M points are the domain points. Substituting the above RBF expansion in the fractional Poisson equation and collocating M points in the domain yields

$$\sum_{j=1}^{M+N} \lambda_j (-\Delta)^{\alpha/2} \phi_j(\boldsymbol{x}_i) = f(\boldsymbol{x}_i), \quad i = 1, 2, \ldots, M. \tag{10}$$

Noticing that the non-local boundary condition $u(\boldsymbol{x}) = 0$ for $\boldsymbol{x} \in \mathbb{R}^d \setminus \Omega$, it is impossible to collocate points in the whole exterior of Ω. To collocate points in an exterior neighborhood of Ω is a candidate strategy. To simplify the RBF method, without sacrificing the solution accuracy, we restrict the collocation points merely on the boundary, i. e.,

$$\sum_{j=1}^{M+N} \lambda_j \phi_j(\boldsymbol{x}_i) = 0, \quad i = M+1, M+2, \ldots, M+N. \tag{11}$$

Note that the strategy to collocate points on the boundary rather than the exterior neighborhood of Ω simply works for zero non-local boundary condition. For non-zero boundary condition, the exterior neighborhood strategy has to be taken, as was done in [16]. The linear system formed by equations (10) and (11) is finally solved for the expansion coefficient λ_j. Similar implementation can be made for the fractional Laplacian in the Caputo sense $(-\Delta)_*^{\alpha/2}$.

Given the approximation of the directional fractional derivative, the fractional Laplacian of the RBFs can be approximated by using the quadrature rule with respect to the differentiation direction $\boldsymbol{\theta}$. Note that for 1D space, we do not need the quadrature rule ($C_{\alpha,1} = \frac{1}{2\cos(\alpha\pi/2)}$). We have

$$\begin{aligned}
(-\Delta)^{\alpha/2}\phi_j(x_1) &= \frac{1}{2\cos(\alpha\pi/2)}(D_0^\alpha + D_\pi^\alpha)\phi_j(x_1) = -\frac{\partial^\alpha \phi_j(x_1)}{\partial|x_1|^\alpha} \quad \text{(1D case)},\\
(-\Delta)^{\alpha/2}\phi_j(x_1,x_2) &= C_{\alpha,2}\int_0^{2\pi} D_\theta^\alpha \phi_j(x_1,x_2)d\theta \approx C_{\alpha,2}\sum_{k=1}^{N_\theta} w_k D_{\theta_k}^\alpha \phi_j(x_1,x_2) \quad \text{(2D case)},\\
(-\Delta)^{\alpha/2}\phi_j(x_1,x_2,x_3) &= C_{\alpha,3}\int_0^{\pi}\int_0^{2\pi} D_{\theta,\psi}^\alpha \phi_j(x_1,x_2,x_3)\sin\psi d\theta d\psi\\
&\approx C_{\alpha,3}\sum_{k=1}^{N_\theta}\sum_{l=1}^{N_\psi} w_k v_l \sin\psi_l D_{\theta_k,\psi_l}^\alpha \phi_j(x_1,x_2,x_3) \quad \text{(3D case)},
\end{aligned} \tag{12}$$

where N_θ-point Gauss–Legendre quadrature with the point–weight pairs (θ_k, w_k) is applied to the integral with respect to θ and N_ψ-point Gauss–Legendre quadrature with the point–weight pairs (ψ_l, v_l) is applied to the integral with respect to ψ. For simplicity, here we write as a slight abuse of notation D_θ^α in place of $D^\alpha_{[\cos\theta,\sin\theta]^T}$ for the 2D case and $D^\alpha_{\theta,\psi}$ in place of $D^\alpha_{[\sin\psi\cos\theta,\sin\psi\sin\theta,\cos\psi]^T}$ for the 3D case. For the 3D case, the $\sin\psi$ in the integrand appears from the spherical coordinate transformation. Additionally, it is seen that in the 1D case, the fractional Laplacian is reduced to the Riesz space derivative, $\frac{\partial^\alpha(\cdot)}{\partial|x_1|^\alpha}$, up to a minus sign.

Similar implementation can be made for the fractional Laplacian in the Caputo sense $(-\Delta)_*^{\alpha/2}$. G-RBF and Q-RBF exploit varied strategies to approximate the directional fractional derivatives $D_\theta^\alpha(\cdot)$ (or $D^\alpha_{\theta,\psi}(\cdot)$) and $D^\alpha_{*\theta}(\cdot)$ (or $D^\alpha_{*\theta,\psi}(\cdot)$), respectively.

3.1 Quadrature-based RBF method (Q-RBF)

The method approximates the Caputo directional derivative using the Gauss–Jacobi quadrature rule [20, 21].

1D case (The differentiation direction $\boldsymbol{\theta} = \pm 1$): We have

$$
\begin{aligned}
D^\alpha_{*+1}\phi_j(x_1) &\approx \frac{(d(x_1,+1)/2)^{1-\alpha}}{\Gamma(1-\alpha)}\sum_{k=1}^{N_g} w_k \frac{\partial\phi_j(x_1 - d(x_1,+1)(1+\xi_k)/2)}{\partial x_1}, && \alpha\in(0,1),\\
D^\alpha_{*+1}\phi_j(x_1) &\approx \frac{(d(x_1,+1)/2)^{2-\alpha}}{\Gamma(2-\alpha)}\sum_{k=1}^{N_g} w_k \frac{\partial^2\phi_j(x_1 - d(x_1,+1)(1+\xi_k)/2)}{\partial x_1^2}, && \alpha\in(1,2),\\
D^\alpha_{*-1}\phi_j(x_1) &\approx \frac{(d(x_1,-1)/2)^{1-\alpha}}{\Gamma(1-\alpha)}\sum_{k=1}^{N_g} w_k \frac{\partial\phi_j(x_1 + d(x_1,-1)(1+\xi_k)/2)}{\partial x_1}, && \alpha\in(0,1),\\
D^\alpha_{*-1}\phi_j(x_1) &\approx \frac{(d(x_1,-1)/2)^{2-\alpha}}{\Gamma(2-\alpha)}\sum_{k=1}^{N_g} w_k \frac{\partial^2\phi_j(x_1 + d(x_1,-1)(1+\xi_k)/2)}{\partial x_1^2}, && \alpha\in(1,2),
\end{aligned}
\tag{13}
$$

where w_k and ξ_k are the Gauss–Jacobi quadrature weights and points, respectively [20, 21]. Note that $\xi_k \in (-1,1)$.

2D case (The differentiation direction $\boldsymbol{\theta} = [\cos(\theta),\sin(\theta)]^T, \theta\in[0,2\pi)$; d is the abbreviation of $d(x_1,x_2,\boldsymbol{\theta})$): We have

$$
\begin{aligned}
&D^\alpha_{*\boldsymbol{\theta}}\phi_j(x_1,x_2)\\
&\quad\approx \frac{(d/2)^{1-\alpha}}{\Gamma(1-\alpha)}\sum_{k=1}^{N_g} w_k D^1_{\boldsymbol{\theta}}\phi_j\left(x_1 - \frac{d(1+\xi_k)\cos\theta}{2}, x_2 - \frac{d(1+\xi_k)\sin\theta}{2}\right), \quad \alpha\in(0,1),\\
&D^\alpha_{*\boldsymbol{\theta}}\phi_j(x_1,x_2)\\
&\quad\approx \frac{(d/2)^{2-\alpha}}{\Gamma(2-\alpha)}\sum_{k=1}^{N_g} w_k D^2_{\boldsymbol{\theta}}\phi_j\left(x_1 - \frac{d(1+\xi_k)\cos\theta}{2}, x_2 - \frac{d(1+\xi_k)\sin\theta}{2}\right), \quad \alpha\in(1,2).
\end{aligned}
\tag{14}
$$

3D case (The differentiation direction $\boldsymbol{\theta} = [\sin\psi\cos\theta, \sin\psi\sin\theta, \cos\psi]^T$, $\theta\in[0,2\pi)$, $\psi\in[0,\pi]$; d is the abbreviation of $d(x_1, x_2, x_3, \boldsymbol{\theta})$): We have

$$\begin{aligned}
D^{\alpha}_{*\boldsymbol{\theta}}\phi_j(x_1,x_2,x_3) &\approx \frac{(d/2)^{1-\alpha}}{\Gamma(1-\alpha)}\sum_{k=1}^{N_g} w_k D^1_{\boldsymbol{\theta}}\phi_j\Big(x_1 - \frac{d(1+\xi_k)\sin\psi\cos\theta}{2},\\
&\quad x_2 - \frac{d(1+\xi_k)\sin\psi\sin\theta}{2}, x_3 - \frac{d(1+\xi_k)\cos\psi}{2}\Big), \quad \alpha\in(0,1),\\
D^{\alpha}_{*\boldsymbol{\theta}}\phi_j(x_1,x_2,x_3) &\approx \frac{(d/2)^{2-\alpha}}{\Gamma(2-\alpha)}\sum_{k=1}^{N_g} w_k D^2_{\boldsymbol{\theta}}\phi_j\Big(x_1 - \frac{d(1+\xi_k)\sin\psi\cos\theta}{2},\\
&\quad x_2 - \frac{d(1+\xi_k)\sin\psi\sin\theta}{2}, x_3 - \frac{d(1+\xi_k)\cos\psi}{2}\Big), \quad \alpha\in(1,2).
\end{aligned} \tag{15}$$

The compact form for the 1D, 2D, and 3D cases is

$$D^{\alpha}_{*\boldsymbol{\theta}}\phi_j(\boldsymbol{x}) \approx \frac{(d(\boldsymbol{x},\boldsymbol{\theta})/2)^{n-\alpha}}{\Gamma(n-\alpha)}\sum_{k=1}^{N_g} w_k D^n_{\boldsymbol{\theta}}\phi_j\Big(\boldsymbol{x} - \frac{d(\boldsymbol{x},\boldsymbol{\theta})(1+\xi_k)\boldsymbol{\theta}}{2}\Big), \quad \alpha\in(n-1,n). \tag{16}$$

The number of Gauss–Jacobi quadrature points, N_g, controls the approximation accuracy. Throughout the chapter, we fix the RBF to be the frequently used multi-quadratic function

$$\phi_j(\boldsymbol{x}) = \sqrt{\|\boldsymbol{x}-\boldsymbol{x}_j\|_2^2 + C^2},$$

where C is the *ad hoc* shape parameter. It is straightforward to consider other RBFs, say, thin plate spline [20]. The corresponding first-order and second-order directional derivative of the RBF are given as

$$\begin{aligned}
D^1_{\boldsymbol{\theta}}\phi_j\Big(\boldsymbol{x} - \frac{d(\boldsymbol{x},\boldsymbol{\theta})(1+\xi_k)\boldsymbol{\theta}}{2}\Big) &= \frac{A_0}{\sqrt{R_0^2+C^2}},\\
D^2_{\boldsymbol{\theta}}\phi_j\Big(\boldsymbol{x} - \frac{d(\boldsymbol{x},\boldsymbol{\theta})(1+\xi_k)\boldsymbol{\theta}}{2}\Big) &= \frac{1}{\sqrt{R_0^2+C^2}} - \frac{A_0^2}{(R_0^2+C^2)^{1.5}},
\end{aligned} \tag{17}$$

where

$$\begin{aligned}
A_0 &= (\boldsymbol{x}-\boldsymbol{x}_j)^T\cdot\boldsymbol{\theta} - \frac{d(\boldsymbol{x},\boldsymbol{\theta})}{2}(1+\xi_k),\\
R_0 &= \Big\|\boldsymbol{x} - \frac{d(\boldsymbol{x},\boldsymbol{\theta})(1+\xi_k)\boldsymbol{\theta}}{2} - \boldsymbol{x}_j\Big\|_2.
\end{aligned} \tag{18}$$

We want to mention that the dimension-independent merit of the RBF makes the above integer-order directional derivatives look the same for the varied dimensions $d = 1, 2$, and 3.

3.2 Grünwald formula-based RBF method (G-RBF)

The method approximates the Riemann–Liouville directional derivative using the vector Grünwald formula [19], i. e.,

$$D_{\boldsymbol{\theta}}^{\alpha}\phi_j(\boldsymbol{x}) = h(\boldsymbol{\theta})^{-\alpha} \sum_{k=1}^{[\kappa d(\boldsymbol{x},\boldsymbol{\theta})]} (-1)^k \binom{\alpha}{k} \phi_j(\boldsymbol{x} - kh(\boldsymbol{\theta})\boldsymbol{\theta}) + O(h(\boldsymbol{\theta})). \tag{19}$$

The spatial step $h(\boldsymbol{\theta})$ is defined by $h(\boldsymbol{\theta}) := d(\boldsymbol{x},\boldsymbol{\theta})/[\kappa d(\boldsymbol{x},\boldsymbol{\theta})] \approx 1/\kappa$, where the operation $[\cdot]$ takes the nearest integer. The constant parameter κ controls the approximation error of the vector Grünwald formula.

Compared to the Gauss–Jacobi quadrature approximation for the Caputo derivative, the current approximation has a simpler form and does not employ the derivatives of the RBF. Thus, the current approximation is easier to program.

4 Numerical results

This section adopts the benchmark solutions [13] defined over unit interval, unit disk, and unit ball. For the unit interval, three exact solutions having varied smoothness are considered, i. e., u_1, u_2, and u_3 (Table 1). For the unit disk, two exact solutions are considered, i. e., u_4 and u_5. For the unit ball, the solution u_5 is considered. All the solutions vanish outside computational domains. Note that these solutions are derived for the Riesz fractional Laplacian. Since the directional fractional Laplacian $(-\Delta)^{\alpha/2}(\cdot)$ considered here is equivalent to the Riesz definition when considering zero non-local boundary condition [16], the benchmark solutions are still valid for the present cases.

Table 1: Benchmark solutions over the unit interval $(-1,1)$, unit disk, and unit ball.

Exact solution $u(x_1)$ (1D case) or $u(\boldsymbol{x})$ (2D and 3D)	**Source term $f(x_1)$ or $f(\boldsymbol{x})$ ($d = 2, 3$ for the dimensionality)**
$u_1(x_1) = x_1(1 - x_1^2)^{\alpha/2}$	$f_1(x_1) = \Gamma(\alpha + 2)x_1$
$u_2(x_1) = x_1(1 - x_1^2)^{1+\alpha/2}$	$f_2(x_1) = \frac{\Gamma(\alpha+3)}{6}(3 - (3 + \alpha)x_1^2)x_1$
$u_3(x_1) = x_1(1 - x_1^2)^{2+\alpha/2}$	$f_3(x_1) = \frac{\Gamma(\alpha+3)(\alpha+4)}{120}(15 - (10\alpha + 30)x_1^2 + (\alpha + 3)(\alpha + 5)x_1^4)x_1$
$u_4(\boldsymbol{x}) = (1 - \lVert\boldsymbol{x}\rVert_2^2)^{\alpha/2}$	$f_4(\boldsymbol{x}) = 2^{\alpha}\Gamma(\frac{\alpha}{2} + 1)\Gamma(\frac{d+\alpha}{2})\Gamma(\frac{d}{2})^{-1}$
$u_5(\boldsymbol{x}) = (1 - \lVert\boldsymbol{x}\rVert_2^2)^{1+\alpha/2}$	$f_5(\boldsymbol{x}) = 2^{\alpha}\Gamma(\frac{\alpha}{2} + 2)\Gamma(\frac{d+\alpha}{2})\Gamma(\frac{d}{2})^{-1}(1 - (1 + \frac{\alpha}{d})\lVert\boldsymbol{x}\rVert_2^2)$

It should be emphasized that from Corollary 2.2.1, for $\alpha \in (0,1)$, the two fractional Laplacians in the Caputo and the Riemann–Liouville senses are the same, and thus the benchmark solutions work for both the Q-RBF and the G-RBF methods. However, from Corollary 2.2.2, we learn that for $\alpha \in (1,2)$, if the solution gradient does not vanish on the boundary, say, $u_1(\cdot)$ and $u_4(\cdot)$, the two fractional Laplacians will not be equivalent. In this case, the benchmark solutions do not work for Q-RBF.

In order to show the solution accuracy, we define the average relative error

$$\epsilon =: \frac{\|\boldsymbol{u} - \bar{\boldsymbol{u}}\|_2}{\|\boldsymbol{u}\|_2}, \tag{20}$$

where $\boldsymbol{u}$ and $\bar{\boldsymbol{u}}$ are the vectors formed by the exact and the approximate solutions over the test points, respectively.

The calculation of the backward distance for the unit interval is trivial: $d(x_1, +1) = x_1 + 1$ and $d(x_1, -1) = 1 - x_1$. For the unit disk and ball, it becomes somewhat complicated, and the corresponding Matlab code is given in the appendix.

4.1 One-dimensional problem

For simplicity, the collocation points for the 1D problem are directly taken as the equispaced nodes in $[-1, 1]$, while for the 2D problem, we will consider non-uniform nodes. The 201 test points are evenly distributed in $[-1, 1]$ with the spatial step 0.01.

The shape parameter C of the multi-quadratic RBF $\phi_j(\boldsymbol{x}) = \sqrt{\|\boldsymbol{x} - \boldsymbol{x}_j\|_2^2 + C^2}$ plays a significant role in affecting the solution accuracy of the RBF collocation method. As Figure 1 shows, there exists an optimal shape parameter given the distribution of the collocation points as well as the fractional order α. Note that these two factors both affect the form of the collocation matrix

$$A_{ij} = \begin{bmatrix} (-\Delta)^{\alpha/2}\phi_j(\boldsymbol{x}_i) \\ \phi_j(\boldsymbol{x}_i) \end{bmatrix}.$$

The optimal parameter, however, varies with different collocation point distributions and fractional orders. A larger shape parameter lead to a larger condition number, and thus the rule of thumb is to select a smaller shape parameter when the number of collocation points is larger. Although there are a variety of strategies to optimize the shape parameter (see [7] and the references therein), none of them can work well for all the cases. In most cases, one just takes a relatively small but fixed shape parameter in order to balance the solution accuracy and the ill-posedness of the collocation matrix.

For 1D, 2D, and 3D numerical examples we consider here, we take the three fixed shape parameters $C = 0.01$, $C = 0.3$, and $C = 0.1$, respectively. As for how to select the fixed shape parameter, one straightforward strategy is trial and error. This strategy will become computationally expensive when solving the forward problem once takes much time. One possible alternative strategy is to regard the shape parameter as the extra equation parameter in addition to the fractional order and then discover these two equation parameters together from the experimental or field data using machine learning [23]. The implementation of this strategy is beyond the scope of this chapter.

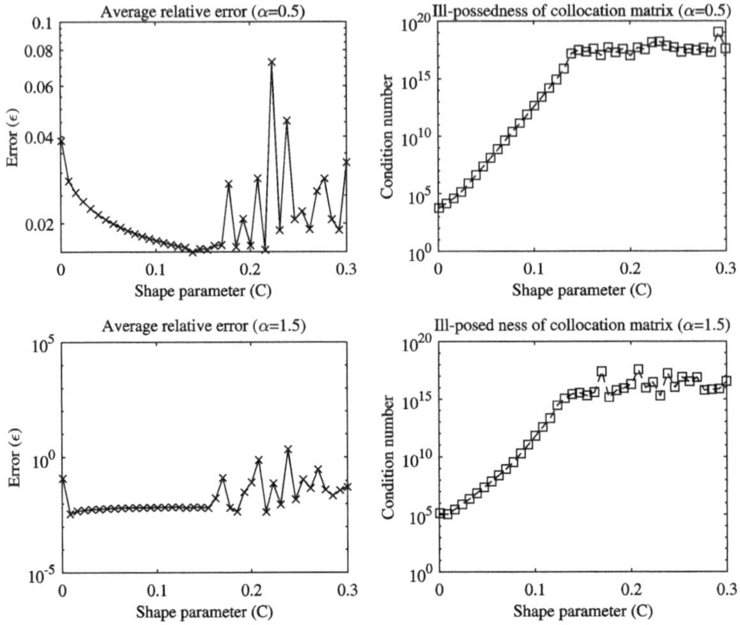

Figure 1: One-dimensional problem: Influence of the shape parameter of the multi-quadratic RBF on the solution accuracy and on the condition number of the collocation matrix. The problem $(-\Delta)^{\alpha/2}u_1 = f_1$ is solved with 150-collocation-point G-RBF where the controlling parameter $\kappa = 500$.

The solution accuracies of the G-RBF and Q-RBF methods are compared in Figures 2, 3, and 4 on the benchmark solutions u_1, u_2, and u_3, respectively. It is observed that the less smooth the benchmark solution is, the less accurate the Q-RBF method will become. Differently, the G-RBF method is less sensitive to the solution smoothness. Note that u_1 is least smooth due to the abrupt variation near the boundary (Figure 5). Furthermore, in Figure 2, for $\alpha = 1.8$, the Q-RBF fails in that in this case $(-\Delta)^{\alpha/2}$ and $(-\Delta)_*^{\alpha/2}$ are not equivalent and thus u_1 is not the solution of $(-\Delta)_*^{\alpha/2}u = f_1$ that is solved by Q-RBF.

4.2 Two-dimensional problem

For the 2D case, we consider uniform and non-uniform distributions of the collocation points in order to show the flexibility of the RBF collocation method in selecting collocation points. The uniform distribution is obtained through letting the domain collocation points align with the Cartesian grid, and the non-uniform distribution generates the M domain collocation points from the first M points of a quasi-random number sequence – the Halton sequence [15]. Both distributions generate the domain points within a smaller disk with a radius of 0.98 in order to avoid the overlap between

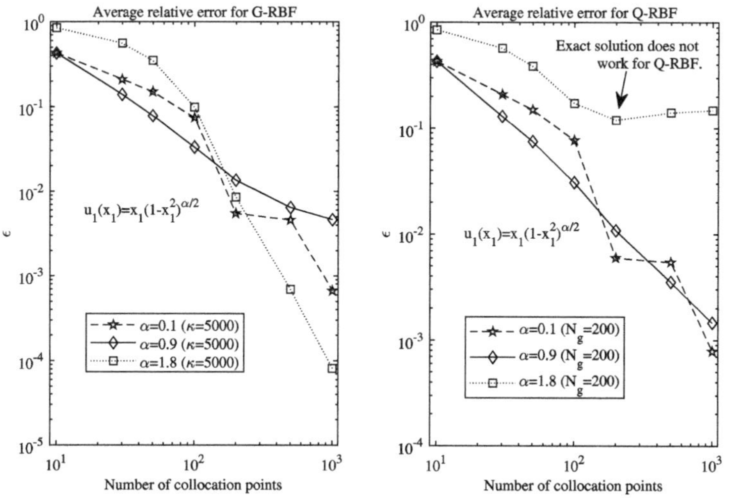

Figure 2: One-dimensional problem: Comparison of the G-RBF and Q-RBF methods for the benchmark solution $u_1(x_1) = x_1(1 - x_1^2)^{\alpha/2}$. Note that the shape parameter of the RBF is fixed to be $C = 0.01$ for all the 1D examples; κ is the parameter controlling the accuracy of approximating the Riemann–Liouville directional fractional derivative in the G-RBF method, and N_g controls the approximation accuracy for the Caputo directional fractional derivative in the Q-RBF method. Here we let the two parameters κ and N_g be large in order to achieve highly accurate solutions when fewer collocation points are used. Actually, the use of smaller parameters is also acceptable, say, $\kappa = 500$ and $N_g = 20$, achieving a same-order error as when the number of collocation points is large.

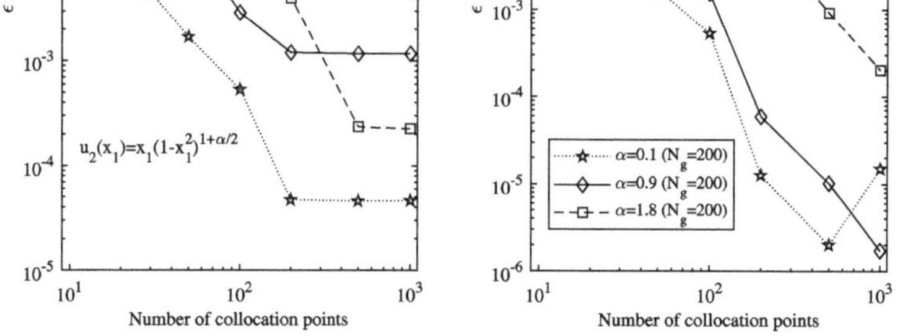

Figure 3: One-dimensional problem: Comparison of the G-RBF and Q-RBF methods for the benchmark solution $u_2(x_1) = x_1(1 - x_1^2)^{1+\alpha/2}$.

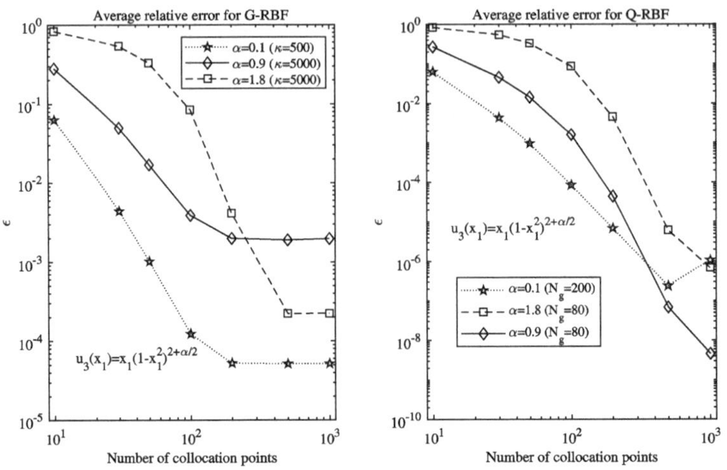

Figure 4: One-dimensional problem: Comparison of the G-RBF and Q-RBF methods for the benchmark solution $u_3(x_1) = x_1(1 - x_1^2)^{2+\alpha/2}$.

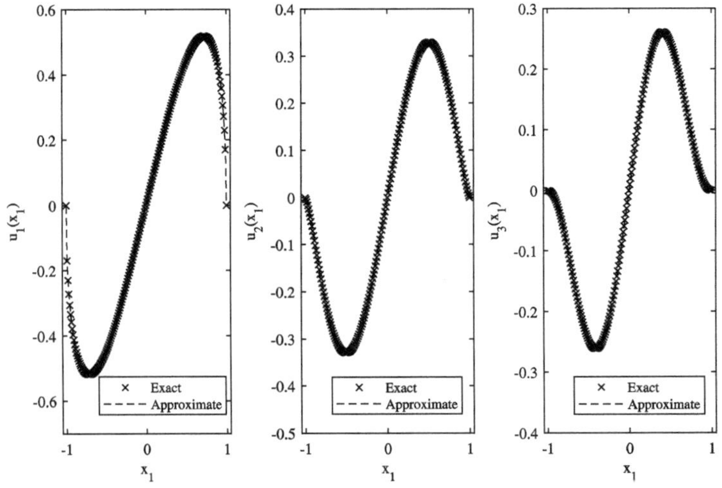

Figure 5: One-dimensional problem: three benchmark solutions u_1, u_2, and u_3 with $\alpha = 0.9$ and their approximations computed from RBF method. The approximate solutions are produced by the G-RBF methods with 1000 collocation points and $\kappa = 500$; u_1 is least smooth since near the boundary its gradient is much larger (gradient is actually infinite at the boundary).

the domain and the boundary collocation points, and moreover, both distributions set the boundary collocation points to be uniformly distributed on the perimeter. See Figure 6 for the distributions.

Figures 7 and 8 display the RBF solution accuracy for u_4 with the uniform and the non-uniform distributions of the collocation points, respectively. It is observed that

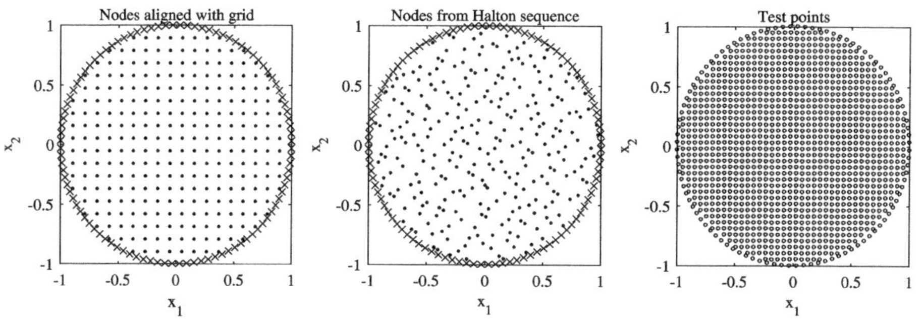

Figure 6: Two-dimensional problem: Uniform (left) and non-uniform (center) distributions of collocation points. In total 376 collocation points are shown with 276 domain and 100 boundary points. The right subplot shows the distribution of the 1252 test points.

the two distributions yield solution accuracy of the same order, but the uniform distribution produces slightly more accurate solutions. Figures 9 and 10 further show the RBF solution accuracy for u_5 with the uniform and the non-uniform distributions of the collocation points, respectively. Similar to the 1D case, due to the sharp change of u_4 near the boundary, the solution accuracy for u_4 is obviously lower than that for u_5. Comparison of the G-RBF and Q-RBF methods indicates that the two methods can both achieve an average relative error below 5 % with around 1000 collocation points for two distributions of collocation points.

Figures 11 and 12 are the contour plots of the numerical solutions for $\alpha = 0.9$ and $\alpha = 0.1$ in comparison with the exact solutions, respectively. It can be seen that the maximum error appears near the boundary, which is caused by the inaccuracy of the RBF interpolation on the boundary. Also, the average relative error of $\epsilon = 1.8\,\%$ indicates sufficiently accurate solutions although some obvious errors appear near the boundary.

It should be noted that the influence of the fractional order on the solution accuracy cannot be ignored, especially as the fractional order approaches one. Tables 2 and 3 show the influence of the fractional order for the G-RBF and Q-RBF methods, respectively. For the G-RBF method, when $\alpha = 0.99$ and $\alpha = 1.01$, one needs to increase the controlling parameter, κ, in order to preserve the solution accuracy. For the Q-RBF method, it is safe to approach $\alpha = 1$ from the left, say $\alpha = 0.99$, but the solution becomes extremely inaccurate when approaching $\alpha = 1$ from the right, even though more collocation points are used. Note that for $\alpha = 1.01$, increasing N_g does not remedy the problem. Another observation is that compared to the G-RBF method, the Q-RBF method needs much more collocation points to preserve the solution accuracy as the fractional order approaches one from the right.

In addition, the two RBF collocation methods have same-order time complexity for computing the collocation matrix: $O(N_\theta M(M + N)\kappa + N(M + N))$ for G-RBF and $O(N_\theta M(M+N)N_g + N(M+N))$ for Q-RBF in the 2D case. For the 3D case, the time com-

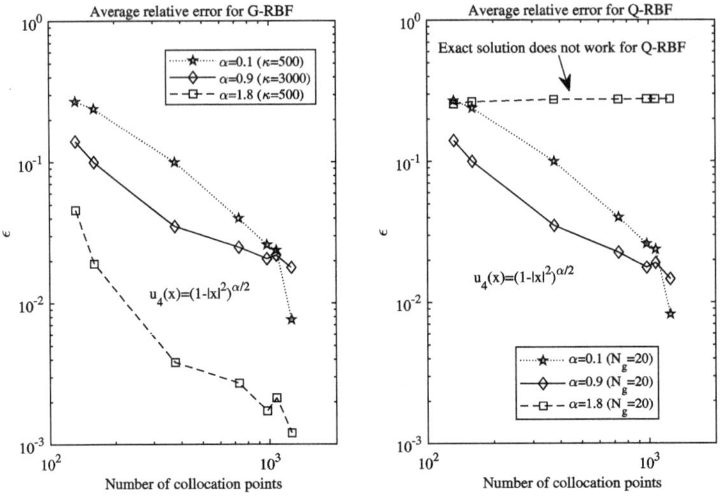

Figure 7: Two-dimensional problem (uniform collocation points): Comparison of the G-RBF and Q-RBF methods for the benchmark solution $u_4(x_1, x_2) = (1 - x_1^2 - x_2^2)^{\alpha/2}$. The 20-point Gauss–Legendre quadrature rule is used, namely, $N_\theta = 20$ in (12). Twenty quadrature points are sufficient to preserve the high accuracy for approximating the integral with respect to θ. Note that the shape parameter of the RBF is fixed to be $C = 0.3$ for all the 2D examples. Also, the solution u_4 does not work for the Q-RBF when $\alpha \in (1, 2]$ since u_4 simply solves the equation $(-\Delta)^{\alpha/2} u = f_4$ but does not solve $(-\Delta)_*^{\alpha/2} u = f_4$ that is solved by Q-RBF.

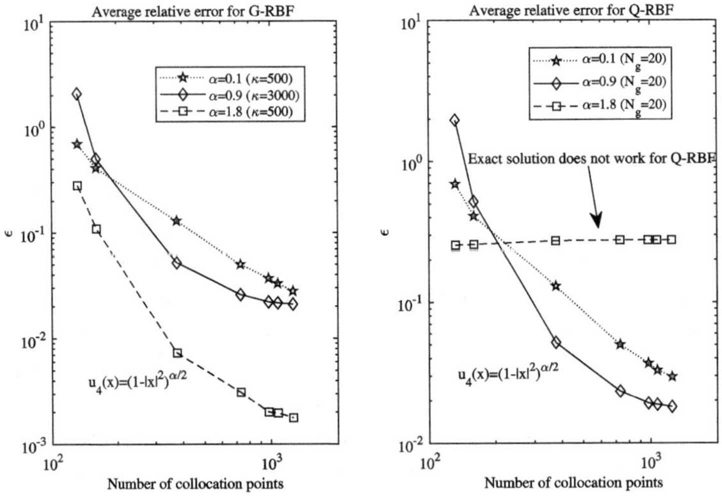

Figure 8: Two-dimensional problem (Non-uniform collocation points): Comparison of the G-RBF and Q-RBF methods for the benchmark solution $u_4(x_1, x_2) = (1 - x_1^2 - x_2^2)^{\alpha/2}$.

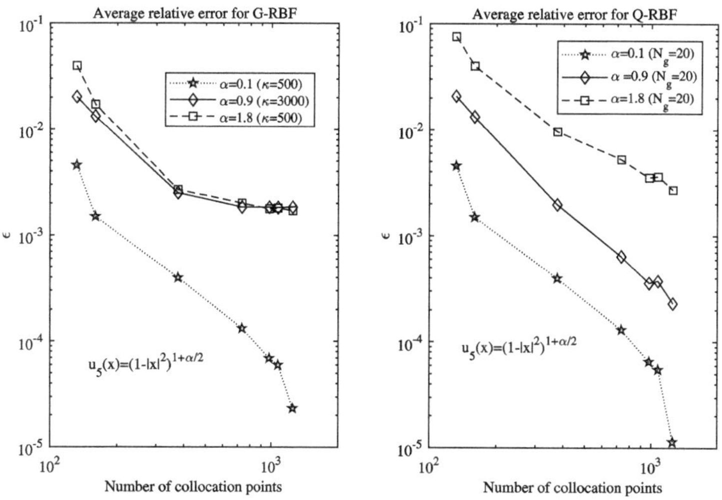

Figure 9: Two-dimensional problem (uniform collocation points): Comparison of the G-RBF and Q-RBF methods for the benchmark solution $u_5(x_1, x_2) = (1 - x_1^2 - x_2^2)^{1+\alpha/2}$.

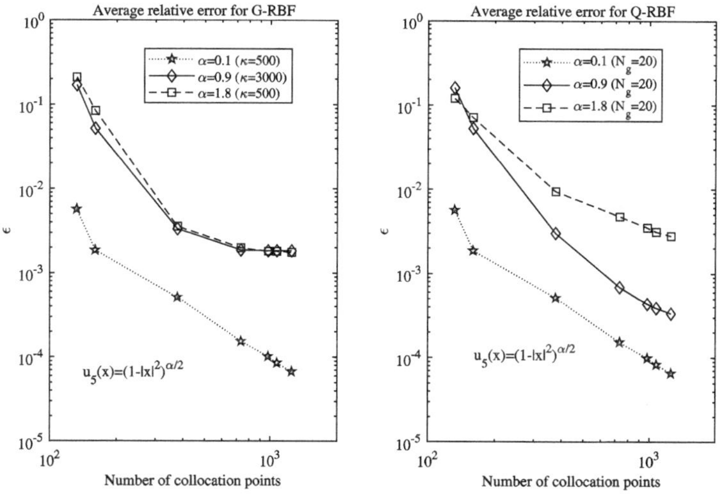

Figure 10: Two-dimensional problem (non-uniform collocation points): Comparison of the G-RBF and Q-RBF methods for the benchmark solution $u_5(x_1, x_2) = (1 - x_1^2 - x_2^2)^{1+\alpha/2}$.

plexity is replaced by $O(N_\theta N_\psi M(M+N)\kappa + N(M+N))$ for G-RBF and by $O(N_\theta N_\psi M(M+N)N_g + N(M+N))$ for Q-RBF. Note that the CPU time listed in Tables 2 and 3 is the total time of computing the collocation matrix and solving the resulting linear system. But, since the number of collocation points is small (less than 10000), solving the linear

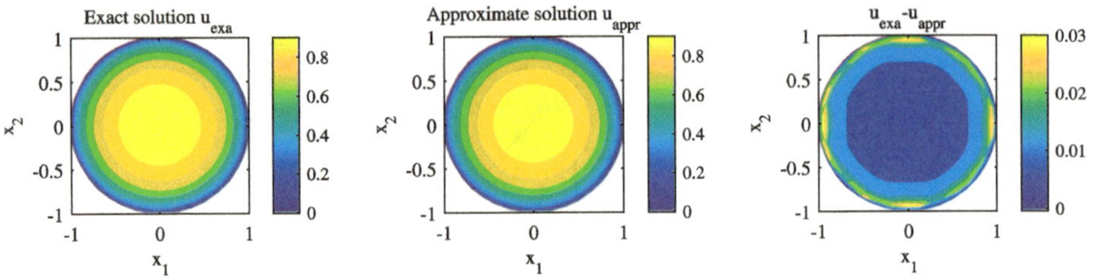

Figure 11: Two-dimensional problem: Comparison of the exact and the approximate solutions for u_4 with $\alpha = 0.9$. The numerical solutions are produced by G-RBF with 1200 uniformly distributed collocation points and $\kappa = 3000$. The average relative error $\epsilon = 1.8\,\%$.

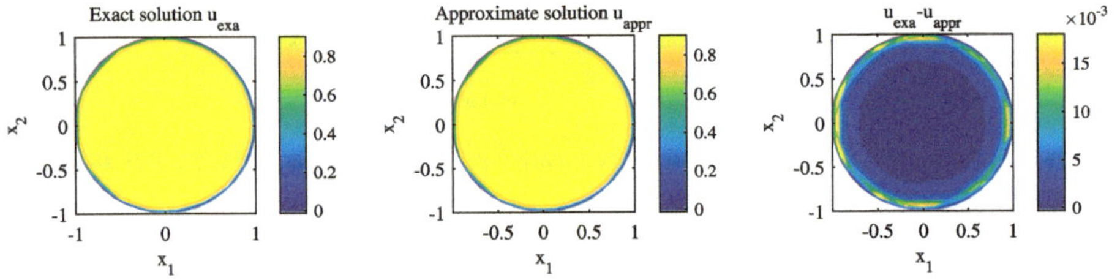

Figure 12: Two-dimensional problem: Comparison of the exact and the approximate solutions for u_4 with $\alpha = 0.1$. The numerical solutions are produced by G-RBF with 1200 uniformly distributed collocation points and $\kappa = 500$. The average relative error $\epsilon = 0.77\,\%$.

Table 2: Two-dimensional problem: Variation of solution accuracy and CPU time of the G-RBF method for varied fractional orders. The benchmark problem with the solution u_5 is solved using the uniformly distributed collocation points.

G-RBF	α **= 0.01**	**0.2**	**0.5**	**0.8**	**0.99**	**1.01**	**1.2**	**1.5**	**1.8**	**1.99**
κ	500	500	500	500	**8000**	**8000**	500	500	500	500
ϵ	7.4e−6	2.6e−4	9.3e−4	4.7e−3	8.0e−3	8.0e−3	8.7e−3	4.1e−3	2.1e−3	5.0e−4
Time (sec)	24	24	24	24	360	360	24	24	24	24
$M + N$	732	732	732	732	732	732	732	732	732	732

Table 3: Two-dimensional problem: Variation of solution accuracy and CPU time of the Q-RBF method for varied fractional orders. The benchmark problem with the solution u_5 is solved using the uniformly distributed collocation points.

Q-RBF	α **= 0.01**	**0.2**	**0.5**	**0.8**	**0.99**	**1.01**	**1.2**	**1.5**	**1.8**	**1.99**
N_g	20	20	20	20	20	20	20	20	20	20
ϵ	7.4e−6	2.4e−4	4.7e−4	6.1e−4	6.6e−4	**4.4e−1**	2.0e−2	9.4e−3	5.2e−3	2.9e−4
Time (sec)	17	17	17	17	17	475	274	283	31	31
$M + N$	732	732	732	732	732	**2732**	**2076**	**2076**	732	732

system (less than one second) is much faster than computing the collocation matrix. The CPU time was calculated on a PC with Intel i7-4770 CPU with 3.40 GHz and 16 GB memory.

4.3 Three-dimensional problem

The uniform collocation points, as is shown in Figure 13, are employed in the RBF methods for the 3D case. The domain collocation points are aligned with the Cartesian grid, and these points are located inside a smaller sphere with a radius of 0.98 in order to avoid the overlap between the domain and the boundary collocation points. The boundary collocation point are evenly distributed in the spherical coordinates.

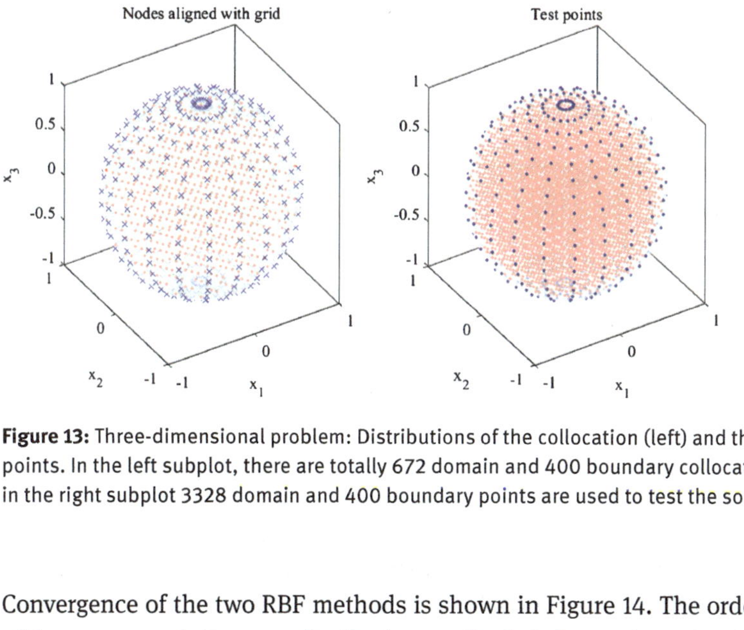

Figure 13: Three-dimensional problem: Distributions of the collocation (left) and the test (right) points. In the left subplot, there are totally 672 domain and 400 boundary collocation points, and in the right subplot 3328 domain and 400 boundary points are used to test the solution accuracy.

Convergence of the two RBF methods is shown in Figure 14. The order of magnitude of the average relative error for the two methods is from −4 to −2, with using at most 1757 collocation points. Similar to the 1D and 2D cases, the two RBF methods exhibit close solution accuracy. We want to mention that for Q-RBF and $\alpha = 1.8$ we changed the RBF's shape parameter from 0.1 to 0.5 in order to attain higher solution accuracy. This shows again the significant role the shape parameter plays for the accuracy.

Figures 15 and 16 display the numerical solutions in comparison with the exact solutions. We see again that an average relative error of around 1 % indicates a sufficiently good approximation. In addition, the CPU time in computing the collocation matrices for the two RBF methods is around 9 minutes for 1757 collocation points.

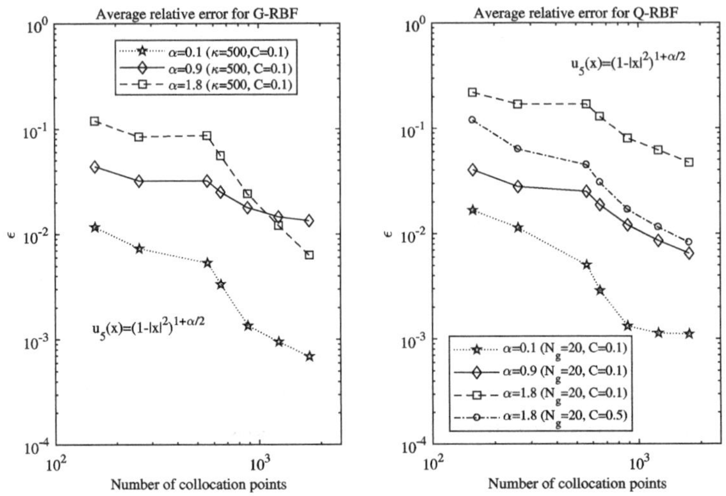

Figure 14: Three-dimensional problem: Comparison of the G-RBF and Q-RBF methods for the benchmark solution $u_5(x_1, x_2, x_3) = (1 - x_1^2 - x_2^2 - x_3^2)^{1+\alpha/2}$. The 10-point Gauss–Legendre quadrature rule is employed for approximating the integral with respect to θ or ψ in (12), namely, letting $N_\theta = N_\psi = 10$. Our numerical trials show that increasing N_θ and N_ψ to values larger than 10 will not significantly increase the solution accuracy.

5 Conclusions

In this chapter we compare the Q-RBF and G-RBF collocation methods on 1D, 2D, and 3D Poisson problems with directional fractional Laplacian. The two RBF collocation methods are shown to have sufficiently high accuracy, and their accuracies are comparable. As regards the computational cost, their time complexities in computing the collocation matrix are of the same order.

The influence of the fractional order on the solution accuracy for the two methods is, however, not the same. The G-RBF method needs a much smaller spatial step in approximating the directional fractional derivative (i. e., larger κ) in order to preserve the solution accuracy when the fractional order approaches one. Differently, the Q-RBF method works well as the fractional order tends toward one from the left, but the method's performance degrades as the fractional order approaches one from the right.

Other than the varied effects of the fractional order, the flexibilities of the two RBF collocation methods differ from each other as well. When $\alpha \in (1, 2]$ and the solution's gradient does not vanish on the boundary (namely, $\nabla u(\partial\Omega) \neq 0$), the fractional Laplacians in Riemann–Liouville and Caputo sense are not equivalent. In other words,

$$(-\Delta)^{\alpha/2} u \neq (-\Delta)_*^{\alpha/2} u.$$

In this case, the Riemann–Liouville fractional Poisson problem $(-\Delta)^{\alpha/2} u = f$ and the Caputo fractional Poisson problem $(-\Delta)_*^{\alpha/2} u = f$ with the same source term f will have

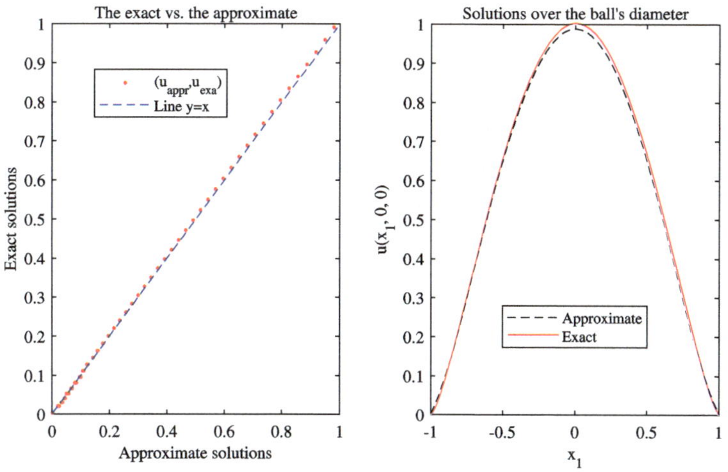

Figure 15: Three-dimensional problem: Comparison of the exact and the approximate solutions for u_5 with $\alpha = 0.9$. The numerical solutions are produced by G-RBF with 1757 collocation points and $\kappa = 500$. The average relative error $\epsilon = 1.4\,\%$.

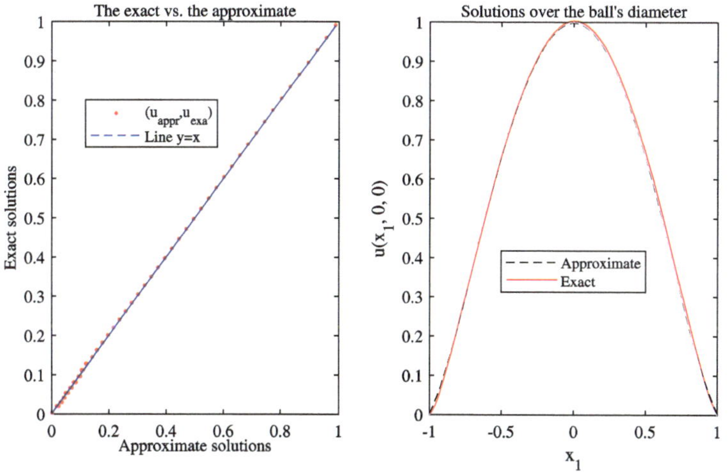

Figure 16: Three-dimensional problem: Comparison of the exact and the approximate solutions for u_5 with $\alpha = 0.9$. The numerical solutions are produced by Q-RBF with 1757 collocation points and $N_g = 20$. The average relative error $\epsilon = 0.76\,\%$.

different solutions. Note that the G-RBF and Q-RBF methods are designed to solve the two problems separately, and therefore, the two methods will definitely produce different numerical solutions.

Appendix

The Matlab code for computing the distance from the point (x_0, y_0) in a unit disk to the disk's boundary along the direction $[-\cos\theta, -\sin\theta]^T$ is given below. The input variable $\theta \in [0, 2\pi)$. The output is the computed backward distance.

```
function output=backward_distance_disk(x0,y0,theta)

  if theta>=0 && theta<pi/2 || theta>3*pi/2 && theta<2*pi
    k=tan(theta);
    xA1=(k*(k*x0-y0)-sqrt(1+k^2-(y0-k*x0)^2))/(1+k^2);
    output=abs(xA1-x0)*sqrt(1+k^2);

  elseif theta>pi/2 && theta<3*pi/2
    k=tan(theta);
    xA2=(k*(k*x0-y0)+sqrt(1+k^2-(y0-k*x0)^2))/(1+k^2);
    output=abs(xA2-x0)*sqrt(1+k^2);

  elseif theta==pi/2
    yA1=-(1-x0^2)^0.5;
    output=abs(yA1-y0);

  elseif theta==3*pi/2
    yA2=(1-x0^2)^0.5;
    output=abs(yA2-y0);
  end
```

The Matlab code for computing the distance from the point (x_0, y_0, z_0) in a unit ball to the ball's boundary along the direction $[-\sin\phi\cos\theta, -\sin\phi\sin\theta, -\cos\phi]^T$ is given below. The input variables $\phi \in [0, \pi]$ and $\theta \in [0, 2\pi)$. The output is the computed backward distance.

```
function output=backward_distance_ball(x0,y0,z0,phi,theta)

  t1=sin(phi)*cos(theta);
  t2=sin(phi)*sin(theta);
  t3=cos(phi);

  if abs(t1)<1e-10

      if t2>0
          k=t3/t2;
```

```
        xA1=(k*(k*y0-z0)-sqrt(1+k^2-(z0-k*y0)^2)-x0^2*k^2-
            x0^2)/(1+k^2);
        output=abs(xA1-y0)*sqrt(1+k^2);
    elseif t2<0
        k=t3/t2;
        xA2=(k*(k*y0-z0)+sqrt(1+k^2-(z0-k*y0)^2-x0^2*k^2-
            x0^2))/(1+k^2);
        output=abs(xA2-x0)*sqrt(1+k^2);
    elseif abs(t2)<1e-10 && t3>0
        yA1=-(1-x0^2-y0^2)^0.5;
        output=abs(yA1-z0);
    elseif abs(t2)<1e-10 && t3<0
        yA2=(1-x0^2-y0^2)^0.5;
        output=abs(yA2-z0);
    end

elseif abs(t1)>1e-10

    k21=t2/t1;
    k31=t3/t1;
    b=2*(-k21*x0+y0)*k21+2*(-k31*x0+z0)*k31;
    a=k21^2+k31^2+1;
    c=(-k21*x0+y0)^2+(-k31*x0+z0)^2-1;
    xplus=(-b+sqrt(b^2-4*a*c))/(2*a);
    xminus=(-b-sqrt(b^2-4*a*c))/(2*a);

    if (xplus-x0)*(-t1)>0
      x00=xplus;
    elseif (xminus-x0)*(-t1)>0
      x00=xminus;
    end

    output=abs((x00-x0)/t1);

end
```

Bibliography

[1] M. Ainsworth and C. Glusa, Aspects of an adaptive finite element method for the fractional Laplacian: a priori and a posteriori error estimates, efficient implementation and multigrid solver, *Comput. Methods Appl. Mech. Eng.*, **327** (2017), 4–35.

[2] D. A. Benson, S. W. Wheatcraft, and M. M. Meerschaert, Application of a fractional advection-dispersion equation, *Water Resour. Res.*, **36**(6) (2000), 1403–1412.

[3] W. Chen, A speculative study of 2/3-order fractional Laplacian modeling of turbulence: Some thoughts and conjectures, *Chaos, Interdiscip. J. Nonlinear Sci.*, **16**(2) (2006), 023126.

[4] W. Chen, J. Fang, G. Pang, and S. Holm, Fractional biharmonic operator equation model for arbitrary frequency-dependent scattering attenuation in acoustic wave propagation, *J. Acoust. Soc. Am.*, **141**(1) (2017), 244–253.

[5] W. Chen, Z.-J. Fu, and C. Ching-Shyang, *Recent Advances in Radial Basis Function Collocation Methods*, Springer, 2014.

[6] W. Chen and S. Holm, Fractional Laplacian time-space models for linear and nonlinear lossy media exhibiting arbitrary frequency power-law dependency, *J. Acoust. Soc. Am.*, **115**(4) (2004), 1424–1430.

[7] W. Chen, Y. Hong, and J. Lin, The sample solution approach for determination of the optimal shape parameter in the multiquadric function of the Kansa method, *Comput. Math. Appl.*, (2018).

[8] W. Chen and G. Pang, A new definition of fractional Laplacian with application to modeling three-dimensional nonlocal heat conduction, *J. Comput. Phys.*, **309** (2016), 350–367.

[9] R. Cheng, H. Ge, and Y. Wu, The numerical analysis of two-sided space-fractional wave equation with improved moving least-square Ritz method, *Math. Probl. Eng.*, **2016** (2016).

[10] R. Cheng, F. Sun, and J. Wang, Meshless analysis of two-dimensional two-sided space-fractional wave equation based on improved moving least-squares approximation, *Int. J. Comput. Math.*, **95**(3) (2018), 540–560.

[11] R. Cheng, F. Sun, Q. Wei, and J. Wang, Numerical treatment for solving two-dimensional space-fractional advection–dispersion equation using meshless method, *Mod. Phys. Lett. B*, **32**(06) (2018), 1850073.

[12] A. G. Churbanov and P. N. Vabishchevich, Numerical investigation of a space-fractional model of turbulent fluid flow in rectangular ducts, *J. Comput. Phys.*, **321** (2016), 846–859.

[13] B. Dyda, Fractional calculus for power functions and eigenvalues of the fractional Laplacian, *Fract. Calc. Appl. Anal.*, **15**(4) (2012), 536–555.

[14] S. Holm and S. P. Näsholm, Comparison of fractional wave equations for power law attenuation in ultrasound and elastography, *Ultrasound Med. Biol.*, **40**(4) (2014), 695–703.

[15] L. Kocis and W. J. Whiten, Computational investigations of low-discrepancy sequences, *ACM Trans. Math. Softw.*, **23**(2) (1997), 266–294.

[16] A. Lischke, G. Pang, M. Gulian, F. Song, C. Glusa, X. Zheng, Z. Mao, W. Cai, M. M. Meerschaert, M. Ainsworth, and G. E. Karniadakis, What is the fractional Laplacian? *ArXiv e-prints*, January 2018.

[17] Q. Liu, F. Liu, Y. T. Gu, P. Zhuang, J. Chen, and I. Turner, A meshless method based on point interpolation method (PIM) for the space fractional diffusion equation, *Appl. Math. Comput.*, **256** (2015), 930–938.

[18] M. M. Meerschaert, D. A. Benson, and B. Bäumer, Multidimensional advection and fractional dispersion, *Phys. Rev. E*, **59**(5) (1999), 5026.

[19] M. M. Meerschaert, J. Mortensen, and H.-P. Scheffler, Vector Grunwald formula for fractional derivatives, *Fract. Calc. Appl. Anal.*, **7**(1) (2004), 61–82.

[20] G. Pang, W. Chen, and Z. Fu, Space-fractional advection–dispersion equations by the Kansa method, *J. Comput. Phys.*, **293** (2015), 280–296.
[21] G. Pang, W. Chen, and K. Y. Sze, Gauss–Jacobi-type quadrature rules for fractional directional integrals, *Comput. Math. Appl.*, **66**(5) (2013), 597–607.
[22] G. Pang, W. Chen, and K. Y. Sze, A comparative study of finite element and finite difference methods for two-dimensional space-fractional advection-dispersion equation, *Adv. Appl. Math. Mech.*, **8**(1) (2016), 166–186.
[23] G. Pang, P. Perdikaris, W. Cai, and G. E. Karniadakis, Discovering variable fractional orders of advection–dispersion equations from field data using multi-fidelity Bayesian optimization, *J. Comput. Phys.*, **348** (2017), 694–714.
[24] I. Podlubny, *Fractional Differential Equations: An Introduction to Fractional Derivatives, Fractional Differential Equations, to Methods of Their Solution and Some of Their Applications*, vol. 198, Elsevier, 1998.
[25] J. P. Roop, Computational aspects of fem approximation of fractional advection dispersion equations on bounded domains in R2, *J. Comput. Appl. Math.*, **193**(1) (2006), 243–268.
[26] S. G. Samko, A. A. Kilbas, and O. I. Marichev et al., *Fractional Integrals and Derivatives. Theory and Applications*, Gordon and Breach, Yverdon, p. 44, 1993.
[27] R. Schumer, D. A. Benson, M. M. Meerschaert, and B. Baeumer, Multiscaling fractional advection-dispersion equations and their solutions, *Water Resour. Res.*, **39**(1) (2003).
[28] F. Song, C. Xu, and G. E. Karniadakis, A fractional phase-field model for two-phase flows with tunable sharpness: Algorithms and simulations, *Comput. Methods Appl. Mech. Eng.*, **305** (2016), 376–404.
[29] F. Song, C. Xu, and G. E. Karniadakis, Computing fractional Laplacians on complex-geometry domains: algorithms and simulations, *SIAM J. Sci. Comput.*, **39**(4) (2017), A1320–A1344.
[30] H. G. Sun, X. Liu, Y. Zhang, G. Pang, and R. Garrard, A fast semi-discrete Kansa method to solve the two-dimensional spatiotemporal fractional diffusion equation, *J. Comput. Phys.*, **345** (2017), 74–90.
[31] H. G. Sun, Y. Zhang, W. Chen, and D. M. Reeves, Use of a variable-index fractional-derivative model to capture transient dispersion in heterogeneous media, *J. Contam. Hydrol.*, **157** (2014), 47–58.
[32] B. J. Szekeres and J.-M. I. Ferenc, Fractional derivatives for vortex simulations, in *Proceedings of the Conference Algoritmy*, Podbanské, Slovakia, pp. 175–182, 2016.
[33] B. E. Treeby and B. T. Cox, Modeling power law absorption and dispersion for acoustic propagation using the fractional Laplacian, *J. Acoust. Soc. Am.*, **127**(5) (2010), 2741–2748.
[34] V. V. Uchaikin, *Fractional Derivatives for Physicists and Engineers, vol. 2*, Springer, 2013.
[35] E. W. Weisstein, Directional derivative, http://mathworld.wolfram.com/DirectionalDerivative.html.
[36] T. Zhu and J. M. Harris, Modeling acoustic wave propagation in heterogeneous attenuating media using decoupled fractional Laplacians, *Geophysics*, **79**(3) (2014), T105–T116.

Yong Zhang and Mark Meerschaert

Particle tracking solutions of vector fractional differential equations: A review

Abstract: This chapter reviews the Lagrangian solvers developed in the last two decades for fractional differential equations (FDEs). Both the Langevin approach and the fractional Lévy motion define dynamics of random walkers, whose density solves the FDE. For the vector FDEs, a multi-scaling compound Poisson process can track the trajectory of particles moving along arbitrary directions with direction-dependent scaling rates. Random walk particle tracking (RWPT) schemes, including streamline projection and flow subordination, are also needed to track particles whose mechanical dispersion follows streamlines. Particle paths affected by boundaries can also be modeled using RWPT, leading to a fully Lagrangian approximation for the vector spatiotemporal FDEs with streamline-dependent superdiffusion in domains of any size and boundary conditions, as required by real-world applications.

Keywords: Fractional differential equation, Lagrangian, random walk particle tracking

MSC 2010: 35R11, 60E07, 60H30

1 Introduction

Fractional differential equations (FDEs) built upon fractional calculus, a mathematical tool initiated by Leibniz and Liouville [21, 23], have been proposed by the physics community and have remained an active area of research simulating the ubiquitous anomalous diffusion for three decades [19, 20]. The FDEs use fractional derivatives in time or space to capture the extremely long-term correlation of individual particle waiting times and/or motions. Analytical solutions are not available for most FDEs, motivating the development of numerical methods. Among these numerous methods, the Lagrangian solver built upon random walk particle tracking (RWPT) is relatively new and exhibits promising applicability in the Earth sciences, where many geological

Acknowledgement: YZ was funded by the National Natural Science Foundation of China grants 41628202 and 41330632. MMM was funded by ARO MURI grant W911NF-15-1-0562 and National Science Foundation grants DMS-1462156 and EAR-1344280.

Yong Zhang, Department of Geological Sciences, University of Alabama, Tuscaloosa, Alabama, USA, e-mail: yzhang264@ua.edu
Mark Meerschaert, Department of Statistics and Probability, Michigan State University, East Lansing, Michigan, USA, e-mail: mcubed@stt.msu.edu

https://doi.org/10.1515/9783110571684-010

materials, such as water, chemicals, sediment, rocks, soil, and glacial, can be represented by "particles."

RWPT yields a grid-free, fully Lagrangian solution by simulating sample paths of the underlying stochastic process for the target material without modifying (such as truncating or discretizing) the governing equation. Solutions to constant coefficient FDEs are Lévy motions, whose α-stable densities [25] are the scaling limits of random walks with power-law transition probabilities [18]. Hence, by tracking the trajectory of random-walking particles whose jump sizes and/or waiting times follow heavy-tailed distributions, one solves the FDE. RWPT increases computational efficiency for large-scale flow systems with fine mesh [26], overcomes grid-average errors for simulating a sharp density front [12], retains minimal numerical dispersion [11], and has easy implementation and near-linear scaling on parallel computers, due to computational expense that depends on independent particle numbers rather than grid dimensions [9]. Most importantly, RWPT decomposes real-world dynamics into major components of particle jumps, which helps interpret the driving mechanisms, estimate effective parameters in a parsimonious model, and understand better the nature of the complex process. These advantages of the Lagrangian solver have motivated the recent development of particle-based approximations for the FDEs.

This chapter reviews the fundamental methodology of Lagrangian approximations for FDEs in the last two decades. Future research directions for the Lagrangian solver will also be identified. During the review, we will also focus on RWPT required by hydrological applications. The solid Earth system contains multi-scale heterogeneity, which motivates various heavy-tailed dynamics and provides an ideal base for the application of FDEs. Many moving targets in the Earth sciences can be regarded as particles, whose trajectories and dynamics can be described by the RWPT following specific FDEs [35].

2 Random walk particle tracking schemes

2.1 History and RWPT for 1D FDEs

The RWPT has a long history in solving the advection–dispersion equation (ADE) by hydrologists to describe pollutant transport in aquifers [1, 12, 27, 28]. Other random walk methods developed by computational physics and engineering communities include lattice gas and lattice Boltzmann approaches, the use of particles for numerical thermodynamics, stochastic properties of perfect gases, and molecular simulations in theoretical chemistry, which usually focus on different scales and will not be discussed here. Random walk solutions of the ADE typically depends on an analogy between the ADE and the forward equation for the associated Markov process. The forward equation, which provides a Lagrangian description of particles moving under specific Markov processes, is usually derived by the Langevin approach [24].

The Langevin approach is the first grid-free, formal Lagrangian solver developed to approximate the 1D space FDE with constant or continuous parameters [30], after the pioneer work of Chechkin and co-workers [4, 3], who modeled 1D fractional Lévy motion using random walkers, and Gorenflo et al. [6], who proposed a discrete random walk model after replacing the space/time FDEs by finite difference equations. Gorenflo et al. [8, 7] then extended their discrete random walk model to a 1D continuous time random walk (CTRW) model, by subordinating the random walk to a renewal process. Marseguerra and Zoia [14, 15, 16] developed a Monte Carlo approach to model subdiffusion across a discontinuity of parameters in 1D. Zhang et al. [32] developed an empirical reflection scheme to track particles undergoing superdiffusion across an interface with discontinuous parameters. Heinsalu et al. [10] developed a CTRW scheme for the 1D time fractional Fokker–Planck equation (FFPE). Magdziarz and Weron [13] developed a robust Monte Carlo approach to solve the 1D spatiotemporal FFPE. A brief review of the Langevin approach for 1D space and time FDEs is given in Appendix A.

2.2 Compound Poisson process for solving vector FDE

Many real-world transport problems require a vector FDE with variable parameters and multi-scaling spreading rates, motivating the multi-dimensional RWPT approach based on the compound Poisson process [29, 33]. The multi-scaling fractional ADE (FADE) takes the form [17]

$$\left(b\frac{\partial}{\partial t}+\beta\frac{\partial^{\gamma}}{\partial t^{\gamma}}\right)p(\vec{x},t)=-[\nabla\cdot(V(\vec{x})+\nabla_{M(d\theta)}^{H^{-1}-I}D(\vec{x}))]p(\vec{x},t)+\frac{\beta t^{-\gamma}}{\Gamma(1-\gamma)}p_0(\vec{x}),\tag{1}$$

where $b \geq 0$ and $\beta \geq 0$ are arbitrary parameters, $0 < \gamma < 1$ is the order of the Riemann–Liouville fractional time derivative, p is a PDF, V is the velocity vector, D is the dispersion coefficient, $p_0(x)$ denotes the initial condition, H^{-1} is the inverse of the scaling matrix providing the order and direction of the space fractional derivatives, I is the identity matrix, and $M(d\theta)$ is the mixing measure.

The RWPT approximation of model (1) contains two components: jumping in space and waiting in immobile zones. The individual jump vector of each particle at time step i, which is essentially a multi-scaling compound Poisson process, can be calculated by [29]

$$\vec{Z}(\tau)=\sum_{i=1}^{[\tau/d\tau]}\vec{x}_i=\sum_{i=1}^{[\tau/d\tau]}R_i^H\cdot\vec{\theta}_i,\tag{2}$$

where $\vec{Z}(\tau)$ represents the particle location at operational time τ, $[\tau/d\tau]$ is the number of random jumps by time τ using the step size $d\tau$, R_i is the random length of the ith jump with the jump direction $\vec{\theta}_i$ drawn independently from the mixing measure $M(d\theta)$ in equation (1). The matrix R^H is anisotropic to allow direction-dependent jump sizes. If the eigenvectors of H are orthogonal (representing perpendicular growth directions for particles), the random displacement along the kth eigenvector of H (with the jump length probabilities falling off as $P(R^{1/\alpha_k} > r) \propto r^{-\alpha_k}$) can be calculated by

$$R^{1/\alpha_k} = \left[D(x_k)\left|\cos\frac{\pi\alpha_k}{2}\right| d\tau\right]^{1/\alpha_k} dL_{\alpha_k} + \left[(\alpha_k - 1)\left|\cos\frac{\pi(\alpha_k-1)}{2}\right| d\tau\right]^{1/(\alpha_k-1)} \times \Theta\left|\frac{\partial D}{\partial x_k}\right|^{1/(\alpha_k-1)} dL_{\alpha_k-1}, \tag{3}$$

where $\Theta = 1, -1$, and 0 if $\partial D/\partial x_k > 0, < 0$, and 0, respectively; dL_{α_k} and dL_{α_k-1} denote the α_k- and $(\alpha_k - 1)$-order standard stable random variables with the maximum skewness, scale one, and zero shift.

If the scaling matrix H contains non-orthogonal eigenvectors such as plume growing in a fractured aquifer with non-orthogonal fracture orientations, the mixing measure and the scaling matrix would have the same directions, and the jump vectors R^{1/α_k} along each eigenvector are independent [33].

In the second step, the operational time can be simulated as the number of renewals by time $T > 0$ for a given waiting time distribution with power-law probability tails, i. e.,

$$dT = b\,d\tau + \left[\beta\left|\cos\frac{\pi\gamma}{2}\right| d\tau\right]^{1/\gamma} dS_\gamma, \tag{4}$$

where dS_γ is a γ-order, standard stable random variable. Note here the particle motion is not instantaneous, and thus we can distinguish the status (mobile or immobile) for each particle at any given time. This distinction is critical in modeling field-measured plumes, since the sampling process tends to collect mobile solutes preferentially.

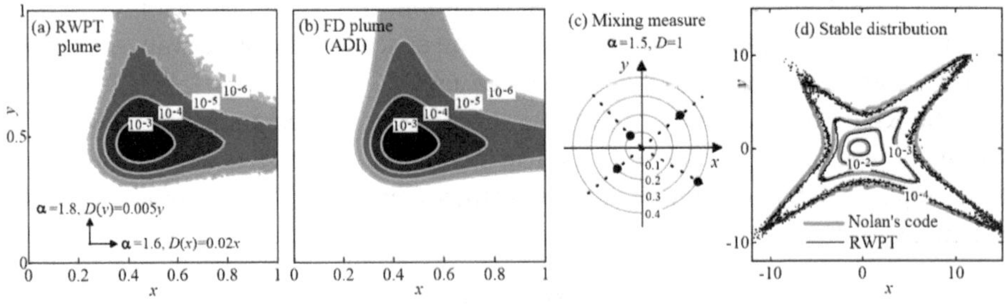

Figure 1: Comparison of numerical solutions for the FADE (1). Case 1 (modified from [29]): The RWPT solution (a) vs. the implicit Euler finite difference solution (b) for the FADE (1) with orthogonal eigenvectors in the scaling matrix H. Case 2: Polar plot of the discrete mixing measure showing four directions and weights (c) and the RWPT solution vs. Nolan's [22] multi-variate stable distribution (i. e., H with non-orthogonal eigenvectors).

Numerical examples of the above RWPT are shown in Figure 1, where the Lagrangian solutions generally match the other numerical solutions. It is also noteworthy that the mixing measure and scaling matrix defined in the vector FADE (1) provides a convenient way to capture complex diffusion in a system with limited information in local velocities. This feature can be useful in hydrologic sciences, since many field sites have only limited subsurface information and can only provide a coarse resolution of the actual velocity field.

2.3 Vector FDE with space-dependent mixing measure

The above RWPT considers a constant mixing measure $M(d\theta)$ in the FADE (1), while real-world diffusion may require a variable $M(d\theta)$. Mechanical dispersion in natural geological media represents the local variation of transport speed deviating from the mean velocity. The dispersion tensor in classical Fickian dispersion is usually aligned with the velocity vector, whose orientation may not remain constant but can change with the medium's internal architecture and/or external forcing. Generalizing to superdiffusion, the eigenvectors of the fractional derivative and the weights in the mixing measure will not be fixed in space, but may vary with streamlines. One example is the ancient, inter-connected braided river channels (i. e., the direction of channels fluctuates in space) in alluvial deposits, which form the major preferential flow paths for superdiffusive solutes. The other example is the regional-scale fractured rock masses whose orientation can change in space due to the change of stress fields and tectonic dynamics. Both media can motivate superdiffusion with a space-dependent mixing measure.

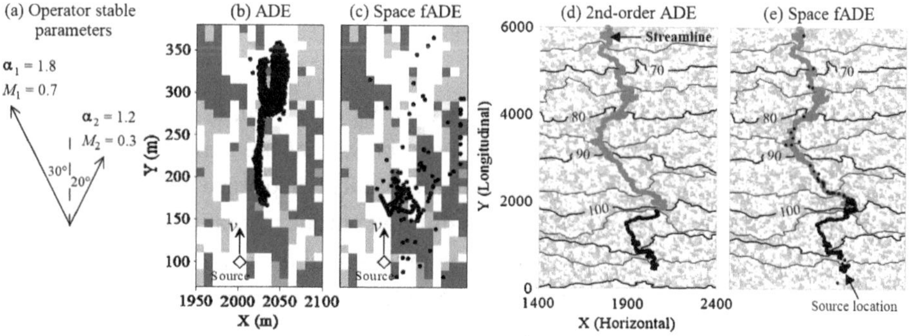

Figure 2: Particle plumes for the ADE (b) and the vector FADE (1) (c), given the streamline-dependent operator stable parameters defined in (a). The direction of V shows the mean flow direction; 10,000 particles were released at the point shown with a diamond. Plots (d) and (e) show the particle plumes of RWPT along streamlines, where the vector FADE (1) has parameters $\alpha = 1.8$, $\beta = 0$, and time $t = 5$.

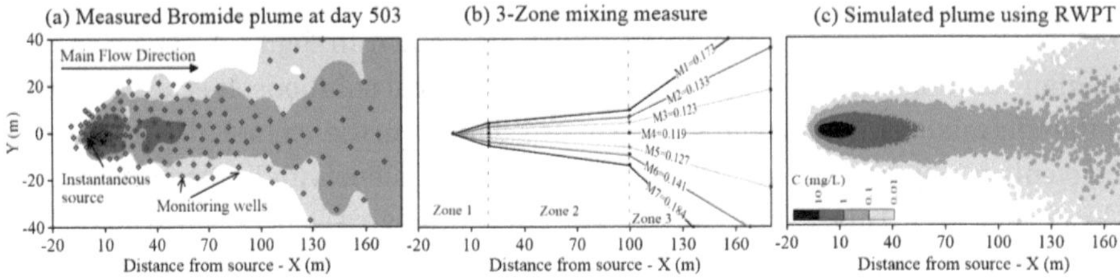

Figure 3: Application of the FADE (1) with a variable mixing measure for the MADE site: the measured bromide plume at day 503 (a), the best-fit three-zone mixing measure with seven discrete directions and weights (b), and the RWPT plume (c) (modified from [33]).

Here we define a streamline-dependent mixing measure, with the mean flow advected along streamlines. The streamline projection approach proposed by Zhang et al. [29] can track the resultant particle motion. One example is shown in Figure 2. During each jump, we assign a random jump length L, and a random direction θ according to the local velocity vector and the local mixing measure. Then the particle's displacement L is projected to adjacent streamlines, along the angle θ. The angle θ is then adjusted to the new direction of $v(x)$ and the particle is re-projected in the next jump. The FADE (1) with a variable mixing measure was found to capture the expanded, fan-shape plume observed at the well-known MADE test site, an alluvial aquifer in Mississippi (Figure 3) [33].

The above RWPT can be simplified if the mechanical dispersion is assumed to follow the stable distribution and particle superdiffusion follows exactly the streamlines. These two fundamental assumptions lead to the subordination to the general flow model [2], i. e.,

$$\left(b\frac{\partial}{\partial t}+\beta\frac{\partial^{\gamma}}{\partial t^{\gamma}}\right)p=-\nabla_{\vec{V}}p+\sigma^{*}(\nabla_{\vec{V}})^{\alpha}p+\nabla[D^{*}\nabla p]+\frac{\beta t^{-\gamma}}{\Gamma(1-\gamma)}p_{0}(\vec{x}), \tag{5}$$

where the advection operator $\nabla_{\vec{V}}$ is defined via $\nabla_{\vec{V}} = \nabla(\vec{V}p)$, σ^* is a scalar factor, and D^* is the molecular diffusivity. Model (5) shows that the density change is due to the advective flux $\nabla_{\vec{V}}p$, the subordinated mechanical dispersive flux $\sigma^*(\nabla_{\vec{V}})^{\alpha}p$, and the molecular diffusive flux $\nabla[D^*\nabla p]$. In saturated porous media, superdiffusion due to fast motion of dissolved chemicals along preferential flow paths does not deviate from $v(x)$ (with a certain angle) but follows exactly the streamlines. This subordinated flow model improves the computational efficiency of the RWPT scheme, allowing additional processes (such as bimolecular reactions) to be added to particle tracking. One example is shown in Figure 4, where the Lagrangian solver calculates multi-scale reactive transport with small-scale chemical reactions and large-scale non-Fickian diffusion [34].

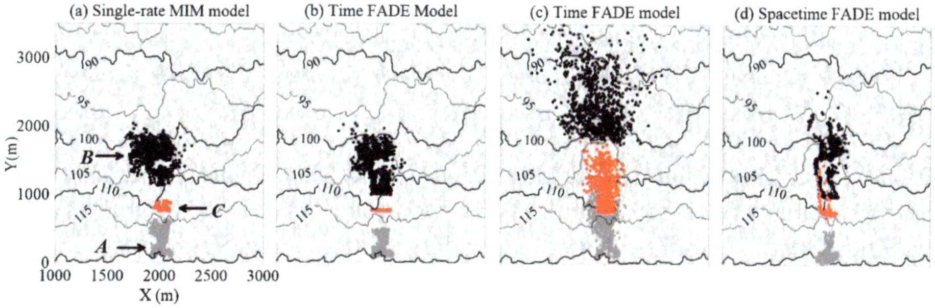

Figure 4: Particle plumes for bimolecular reaction $A + B \to C$ calculated by RWPT (snapshots at time $t = 5$): Single-rate mobile-immobile transport model, with rate coefficient 1 and capacity coefficient 1 (a). Time FADE (5) with factor $\sigma^* = 0$, capacity coefficient $\beta = 0.1$, and the time index $\gamma = 0.1$ (b) and 0.5 (c). The full FADE model (5) with the space index $\alpha = 1.6$, $\beta = 0.1$, and $\gamma = 0.1$ (d). Reactants A and B and the product C are represented by grey, black, and red particles, respectively.

2.4 Bounded fractional diffusion

Natural processes are usually bounded, motivating the FDE and its Lagrangian solver in bounded domains. Zhang et al. [31] defined non-local boundary conditions and then developed a Lagrangian solver to approximate bounded, 1D fractional diffusion, which can be extended to multiple dimensions using the RWPT schemes reviewed above. Zhang et al. [31] showed that, to define Neumann and mixed Robin boundary conditions, the sign of the Riemann–Liouville fractional derivative should remain consistent with the sign of the fractional-diffusive flux term in the FDE; otherwise the boundary value problem becomes ill-posed. Care is also required when approximating particle dynamics around the reflective boundary, where the exiting particles can be either reflected (symmetrically) back to the internal domain for (local and symmetric) Fickian diffusion, or relocated at the boundary for (the non-local and non-symmetric) fractional diffusion so that the reflected particles will not alter the overall dynamics of transport in the domain.

3 Future research directions

The low resolution in solution is one of the historical shortcomings for particle-based solvers. A large number of particles are required to reliably capture the large jump or long trapping for particles. To solve this issue, Allouch et al. (https://arxiv.org/pdf/1707.03871.pdf) developed particle-based, smooth particle approximations, which can obtain fine-resolution solutions for 1D space FDEs with constant parameters. This approach, similar to smooth particle hydrodynamics, requires collections of particles to discretize the domain and can therefore be computational demanding. Another approach is to assign variable weights for particles located at different positions, and hence the resolution at low-density regions may be improved without significant additional calculation.

RWPT algorithms for variable-order FDEs are also needed. Natural geological media can contain non-stationary heterogeneity, and water flow in aquifers and streams can change daily or seasonally, motivating the application of the FDEs with space- and time-dependent indices. The corresponding Lagrangian solver has not been fully developed. In addition, natural processes can involve multi-scale dynamics, motivating the rapid development of multi-scale physical models. The multi-scale FDE and its Lagrangian solver remain to be shown.

4 Conclusion

We reviewed the fully Lagrangian solver to approximate the FDEs. For the 1D FDE, both the Langevin approach and the fractional Lévy motion can define particle dynamics and guide the particle tracking schemes. For the vector FDEs with a constant mixing measure, the multi-scaling compound Poisson process can be used to track particles moving along arbitrary directions with direction-dependent scaling rates. Real-world transport, however, may require a space-dependent mixing measure in the FDE, which can be modeled by the streamline projection and flow subordination methods in RWPT. Particles' trajectories affected by Dirichlet, Neumann, or mixed Robin boundary conditions can also be tracked using RWPT algorithms, leading to a fully Lagrangian approximation for the vector spatiotemporal FDEs with superdiffusion along streamlines in domains with any size and boundary conditions, as required by hydrological applications.

Appendix A. Langevin approach for 1D FDEs

We start with the forward equation, derive the backward equation by taking adjoints, and then apply the general theory of Markov processes [5] to obtain the Langevin equation. For example, the Langevin approach contains three steps to solve the following FDE:

$$\frac{\partial p(x,t)}{\partial t} = -\frac{\partial}{\partial x}[v(x)p(x,t)] + \frac{\partial^\alpha}{\partial x^\alpha}[D(x)p(x,t)], \tag{6}$$

where $1 < \alpha \leq 2$ is the order of the Riemann–Liouville fractional space derivative. In step 1, using the fractional adjoint operator, one can derive the backward model of (6) [30], i. e.,

$$\frac{\partial p(x,t)}{\partial t} = v(x)\frac{\partial p(x,t)}{\partial x} + D(x)\frac{\partial^\alpha}{\partial(-x)^\alpha}p(x,t). \tag{7}$$

Step 2 builds the Markov process, containing the backward generator $Lu = v(x)\frac{\partial u}{\partial x} + \int [u(x+y) - u(x) - \frac{y}{1+y^2}\frac{\partial u(x)}{\partial x}]|\cos\frac{\pi\alpha}{2}|D(x)\phi(dy)$ and the Langevin equation [5, 30]

$$dX(t) = v(X(t))\,dt + \left[D(X(t))\left|\cos\frac{\pi\alpha}{2}\right|dt\right]^{1/\alpha} dL_{\gamma}. \tag{8}$$

The last step is to track particle dynamics defined by (8).

The Langevin approach can also solve the time FDE, i. e.,

$$\left(b\frac{\partial}{\partial t} + \beta\frac{\partial^{\gamma}}{\partial t^{\gamma}}\right)p(x,t) = A_x p(x,t) + \beta\frac{t^{-\gamma}}{\Gamma(1-\gamma)}p_0(x), \tag{9}$$

where A_x is the advection–diffusion operator such as that shown on the right-hand side of (6). Here we assume decoupled jump sizes and waiting times, and hence the density p in equation (9) can be calculated by subordinating the jump process against the waiting time process, i. e.,

$$p(x,t) = \int_0^{\infty} u(x,\tau)h(\tau,t)\,d\tau, \tag{10}$$

where τ denotes operational time. The first density $u(x,\tau)$ in (10) models particle motion in τ, which follows the Markov process (8) except that dt is replaced by $d\tau$. The second density $h(\tau,t)$ accounts for the waiting times after each jump, i. e.,

$$\frac{\partial}{\partial\tau}h(\tau,t) = -\left[b\frac{\partial h(\tau,t)}{\partial t} + \beta\frac{\partial^{\gamma}h(\tau,t)}{\partial t^{\gamma}}\right], \tag{11}$$

with initial condition $h(\tau = 0,t) = b\delta(t) + \beta t^{-\gamma}/\Gamma(1-\gamma)$. Equation (11) is analogous to (6), leading to the time Langevin equation (4) [33].

Bibliography

[1] S. W. Ahlstrom, H. P. Foote, R. C. Arnett, C. R. Cole, and R. J. Serne, *Multi-Component Mass Transport Model: Theory and Numerical Implementation*, Report BNWL-2127, Battelle Pacific Northwest Laboratory, Richlnd, WA, 1977.

[2] B. Baeumer, Y. Zhang, and R. Schumer, Incorporating super-diffusion due to sub-grid heterogeneity to capture non-Fickian transport, *Ground Water*, **53**(5) (2015), 699–708.

[3] A. Chechkin, V. Y. Gonchar, J. Klafter, R. Metzler, and L. V. Tanatarov, Lévy flights in a steep potential well, *J. Stat. Phys.*, **115**(516) (2004), 1505–1535.

[4] A. V. Chechkin and V. Y. Gonchar, A model for persistent Lévy motion, *Physica A*, **277**(3) (2000), 312–326.

[5] S. N. Ethier and T. G. Kurtz, *Markov Processes: Characterization and Convergence*, p. 380, John Wiley & Sons, New York, 1986.

[6] R. Gorenflo, F. Mainardi, D. Moretti, G. Pagnini, and P. Paradisi, Discrete random walk models for space–time fractional diffusion, *Chem. Phys.*, **284**(1) (2002), 521–541.

[7] R. Gorenflo, F. Mainardi, and A. Vivoli, Continuous-time random walk and parametric subordination in fractional diffusion, *Chaos Solitons Fractals*, **34**(1) (2007), 87–103.

[8] R. Gorenflo, A. Vivoli, and F. Mainardi, Discrete and continuous random walk models for space-time fractional diffusion, *Nonlinear Dyn.*, **38**(1) (2004), 101–116.

[9] C. T. Green, *Effects of Heterogeneity on Reactive Transport in Geologic Media*, Ph. D. Thesis, University of California, 2002.

[10] E. Heinsalu, M. Patriarca, I. Goychuk, G. Schmid, and P. Hänggi, Fractional Fokker–Planck dynamics: Numerical algorithm and simulations, *Phys. Rev. E*, **73** (2006), 046133.

[11] W. Kinzelbach, The random walk method in pollutant transport simulation, in E. Custodio, et al. (eds.), *Groundwater Flow and Quality Modeling*, pp. 227–245, Reidel Publishing Company, 1988.

[12] E. M. LaBolle, G. E. Fogg, and A. F. B. Tompson, Random-walk simulation of transport in heterogeneous porous media: Local mass-conservation problem and implementation methods, *Water Resour. Res.*, **32**(3) (1996), 583–593.

[13] M. Magdziarz, A. Weron, and K. Weron, Fractional Fokker–Planck dynamics: Stochastic representation and computer simulation, *Phys. Rev. E*, **75** (2007), 016708.

[14] M. Marseguerra and A. Zoia, The Monte Carlo and fractional kinetics approaches to the underground anomalous subdiffusion of contaminants, *Ann. Nucl. Energy*, **33** (2006), 223–235.

[15] M. Marseguerra and A. Zoia, Normal and anomalous transport across an interface: Monte Carlo and analytical approach, *Ann. Nucl. Energy*, **33** (2006), 1396–1407.

[16] M. Marseguerra and A. Zoia, Monte Carlo investigation of anomalous transport in presence of a discontinuity and of an advection field, *Physica A*, **377**(2) (2007), 448–464.

[17] M. M. Meerschaert, D. A. Benson, and B. Baeumer, Operator Lévy motion and multiscaling anomalous diffusion, *Phys. Rev. E*, **63**(2) (2001), 12–17.

[18] M. M. Meerschaert and A. Sikorskii, *Stochastic Models for Fractional Calculus*, De Gruyter Studies in Mathematics, vol. 43, De Gruyter, Berlin, 2012, ISBN 978-3-11-025869-1.

[19] R. Metzler and J. Klafter, The random walk's guide to anomalous diffusion: A fractional dynamics approach, *Phys. Rep.*, **339**(1) (2000), 1–77.

[20] R. Metzler and J. Klafter, The restaurant at the end of the random walk: Recent development in fractional dynamics of anomalous transport processes, *J. Phys. A*, **37** (2004), R161–R208.

[21] K. S. Miller and B. Ross, *An Introduction to Fractional Calculus and Fractional Differential Equations*, John Wiley, New York, 1993.

[22] J. P. Nolan, Multivariate stable distributions: Approximation, estimation, simulation and identification, in R. J. Adler, R. Feldman, and M Taqqu (eds.), *A Practical Guide to Heavy Tails: Statistical Techniques and Applications*, pp. 509–526, Birkhäuser Boston, Cambridge, MA, 1998.

[23] K. B. Oldham and J. Spanier, *The Fractional Calculus*, Academic Press, 1974.

[24] H. Risken, *The Fokker–Planck Equation*, Spinger-Verlag, p. 454, New York, 1984.

[25] G. Samorodnitsky and M. S. Taqqu, *Stable Non-Gaussian Random Processes: Stochastic Models with Infinite Variance*, Chapman and Hill, New York, 1994.

[26] A. F. B. Tompson, Numerical simulation of solute transport in three-dimensional randomly heterogeneous porous media, *Water Resour. Res.*, **29**(11) (1993), 3709–3726.

[27] A. F. B. Tompson and D. E. Dougherty, Particle-grid methods for reacting flows in porous media with application to Fisher's equation, *Appl. Math. Model.*, **16** (1992), 374–383.

[28] G. J. M. Uffink, A random walk method for the simulation of macrodispersion in a stratified aquifer, in *Relation of Groundwater Quality and Quantity*, IAHS Publication, vol. 146, pp. 103–114, 1985.

[29] Y. Zhang, D. A. Benson, M. M. Meerschaert, E. M. LaBolle, and H. P. Scheffler, Random walk approximation of fractional-order multiscaling anomalous diffusion, *Phys. Rev. E*, **74** (2006), 026706, doi:10.1103/PhysRevE.74.026706.

[30] Y. Zhang, D. A. Benson, M. M. Meerschaert, and H. P. Scheffler, On using random walks to solve the space-fractional advection-dispersion equations, *J. Stat. Phys.*, **123**(1) (2006), 89–110.

[31] Y. Zhang, C. T. Green, E. M. LaBolle, R. M. Neupauer, and H. G. Sun, Bounded fractional diffusion in geological media: Definition and Lagrangian approximation, *Water Resour. Res.*, **52** (2016), 8561–8577.

[32] Y. Zhang, E. M. LaBolle, and K. Pohlmann, Monte Carlo approximation of anomalous diffusion in macroscopic heterogeneous media, *Water Resour. Res.*, **45** (2009), W10417, doi:10.1029/2008WR007448.

[33] Y. Zhang, M. M. Meerschaert, and B. Baeumer, Particle tracking for time-fractional diffusion, *Phys. Rev. E*, **78** (2008), 036705.

[34] Y. Zhang and C. Papelis, Particle-tracking simulation of fractional diffusion-reaction processes, *Phys. Rev. E*, **84** (2011), 066704.

[35] Y. Zhang, H. G. Sun, H. H. Stowell, M. Zayernouri, and S. E. Hansen, A review of applications of fractional calculus in Earth system dynamics, *Chaos Solitons Fractals*, **102** (2017), 29–46.

Martin Stynes

Singularities

Abstract: First, it is demonstrated by simple examples that singularities are commonplace in the solutions of FD problems on bounded domains. Then for more general problems, bounds on derivatives of the solutions are presented to illustrate the exact nature of these singularities. It is shown that assuming more regularity than is generally true will restrict severely the class of problems under study. For numerical methods, singularities will often reduce the rate of convergence. Ways of addressing this deficiency are described – there are four main classes of methods designed specifically to handle singularities in the solutions of FD problems – and many references to the recent numerical research literature are given.

Keywords: Weak singularity, derivative bounds, regularity of solution, solution decomposition, special grid, special basis function, reformulation of problem

MSC 2010: 34A08, 65L05, 65L10, 65M12

1 Singularities in solutions are typical

In this introductory section, we show by simple examples that (weak) singularities are usually present in the solutions of fractional differential equations (FDEs). Here *singularity* means that the solution has only low-order regularity – even when the data of the problem are very smooth. This phenomenon causes a deterioration in the behavior of most numerical methods and renders their analysis more difficult.

1.1 Examples

Example 1 (Caputo example). Let $n \in \mathbb{N}$ and $n-1 < \alpha < n$. Consider the differential equation

$$ {}^{C}D^{\alpha}_{0+}u(x) = x^{\beta} \quad \text{for } 0 < x < 1, \text{ where } \beta \geq 0 \text{ is constant.} \tag{1} $$

By [7, p. 193] a particular solution is $\Gamma(\beta+1)x^{\alpha+\beta}/\Gamma(\alpha+\beta+1)$, and thus by [7, p. 55] the general solution of (1) is

$$ u(x) = \frac{\Gamma(\beta+1)x^{\alpha+\beta}}{\Gamma(\alpha+\beta+1)} + \sum_{k=0}^{n-1} c_k x^k \quad \text{for } 0 < x < 1, \tag{2} $$

for some constants c_k which are determined by initial or boundary conditions.

Martin Stynes, Beijing Computational Science Research Center, Beijing, China, e-mail: m.stynes@csrc.ac.cn

https://doi.org/10.1515/9783110571684-011

Observe that for almost all values of β, the solution (2) is not smooth at $x = 0$ (the exceptional values are $\beta = m-\alpha$ for $m = n, n+1, \dots$). This singularity at $x = 0$ is present even when β is an integer, i. e., even when the data of (1) lies in $C^\infty[0,1]$.

In particular, if we consider the very simple FDE ${}^C D^\alpha_{0+} u(x) = 1$ (i. e., $\beta = 0$) with $1 < \alpha < 2$, then the general solution contains a multiple of x^α, which lies in $C^1[0,1] \cap C^\infty(0,1]$ *but not in* $C^2[0,1]$; consequently, as the analysis of any numerical approximation of ${}^C D^\alpha_{0+} u$ will involve u'', this loss of regularity at $x = 0$ is an obstacle that must be overcome in the analysis.

Example 2 (Riemann–Liouville example). Let $n \in \mathbb{N}$ and $n-1 < \alpha < n$. Consider the differential equation

$$ {}^{\mathrm{RL}} D^\alpha_{0+} u(x) = x^\beta \quad \text{for } 0 < x < 1, \text{ where } \beta \geq 0 \text{ is constant.} \tag{3} $$

Here we shall assume that $\alpha - \beta \notin \mathbb{N}$; then by [7, Example 2.4] a particular solution is again $\Gamma(\beta+1)x^{\alpha+\beta}/\Gamma(\alpha+\beta+1)$ and by [7, p. 54] the general solution of (3) is

$$ u(x) = \frac{\Gamma(\beta+1)x^{\alpha+\beta}}{\Gamma(\alpha+\beta+1)} + \sum_{k=1}^{n} c_k x^{\alpha-k} \quad \text{for } 0 < x < 1, \tag{4} $$

where the constants c_k will be determined by initial or boundary conditions.

The Riemann–Liouville general solution (4) is more singular than the Caputo solution (2). While the $x^{\alpha+\beta}$ term is the same in each solution, all the terms $x^{\alpha-k}$ in (4) fail to be smooth at $x = 0$. Most strikingly, the worst-behaved term $x^{\alpha-n}$ is not even in $C[0,1]$!

Note that (4) is unable to accommodate any *inhomogeneous* initial/boundary condition $u^{(k)}(0) = \gamma$ with $k \in \{0,1,\dots,n-1\}$ and $\gamma \neq 0$, because each term from $\sum_{k=1}^{n} \cdots$ in $u^{(k)}(x)$ is either zero or blows up at $x = 0$. Riemann–Liouville FDEs in the research literature frequently assume *homogeneous* initial/boundary conditions

$$ u^{(k)}(0) = 0 \quad \text{for } k \in \{0,1,\dots,n-1\} \tag{5} $$

because these are the only conditions on $u^{(k)}$ that they can handle: they force each c_k in (4) to be zero. See [35] for some further discussion of boundary conditions for general Riemann–Liouville FDEs.

Thus to avoid having a singularity in the solution (4) of (3), one must choose $\beta = m - \alpha$ with $m \in \{n, n+1, \dots\}$ and impose the conditions (5). Observe how restrictive all these requirements are, compared with the generality of the original problem (3).

Example 3 (Two-sided FD example). In [9] Ervin et al. apply the two-sided FD operator

$$ {}^{\mathrm{EHR}} D^{\alpha,r} := r D I^{2-\alpha}_{0+} D + (1-r) D I^{2-\alpha}_{1-} D \tag{6} $$

to functions defined on $[0,1]$; here $1 < \alpha < 2$ and $r \in [0,1]$ are constants, and $D \equiv d/dx$. They discuss the general case of $r \in [0,1]$ but we confine our attention to $r = 1/2$.

It is shown in [9] that the kernel of ${}^{\mathrm{EHR}}D^{\alpha,1/2}$ is two-dimensional and spanned by the functions 1 and $\int_0^x s^{\alpha/2-1}(1-s)^{\alpha/2-1}\,ds$, so multiples of these functions will appear in the general solution to any problem ${}^{\mathrm{EHR}}D^{\alpha,1/2}u = f$. Consequently the solution will in general fail to be smooth at $x = 0$ and at $x = 1$ and will not lie in $C^1[0,1]$.

Example 4 (Space-time example). Consider the initial-boundary value problem

$$ {}^{C}D_{0+}^{\alpha,t}u(x,t) - p\frac{\partial^2 u(x,t)}{\partial x^2} + r(x)u(x,t) = f(x,t) \tag{7a}$$

for $(x,t) \in Q := (0,l)\times(0,T]$, with

$$u(0,t) = u(l,t) = 0 \quad \text{for } t \in (0,T], \tag{7b}$$

$$u(x,0) = \phi(x) \quad \text{for } x \in [0,l], \tag{7c}$$

where ${}^{C}D_{0+}^{\alpha,t}$ is a Caputo time derivative with $0 < \alpha < 1$, p is a positive constant, $r \in C[0,l]$ with $r \geq 0$, $f \in C(\bar{Q})$ where $\bar{Q} := [0,l]\times[0,T]$, and $\phi \in C[0,l]$. Assume also that $\phi(0) = \phi(l) = 0$ so that the initial and boundary conditions are zero-order compatible at the two corners $(0,0)$ and $(l,0)$ of the space-time domain Q.

It is shown in [64] that the main features of the solution of this problem are present also in the particular case when $p = 1$, $r \equiv 0$, $f \equiv 0$, $l = \pi$, and $\phi(x) = \sin x$, whose solution is

$$v(x,t) := E_{\alpha,1}(-t^{\alpha})\sin x. \tag{8}$$

Here $E_{\alpha,1}$ is the standard two-parameter Mittag-Leffler function. It is straightforward to deduce from well-known properties of $E_{\alpha,1}$ that all spatial derivatives of v are smooth, but for each fixed x the solution $v(x,t)$ behaves like a multiple of t^{α} near $t = 0$. Thus $v \in C(\bar{Q})$ but the partial derivative $v_t(x,t)$ blows up as $t \to 0+$ because $v_t(x,t) \sim t^{\alpha-1}$ for each fixed x. This loss of regularity in v_t at $t = 0$ complicates the analysis of any numerical method that is used to solve (7), as is demonstrated in [64].

Remark 1. In all the examples of Section 1.1, note that when the data of the problem are smooth, nevertheless the solution is not smooth. This is completely different from the situation with classical integer-order derivatives, where one has "shift theorems" that say that if the data and domain are smooth, then the solution is also smooth. The singularities appearing in the solutions of our examples are due entirely to the presence of a fractional-order derivative.

1.2 Theoretical bounds on solution singularities

The presence of singularities in the solution of FDEs is not a recent discovery. For example, Lubich [40] in 1983 investigated the singularities appearing in the solutions of

Volterra weakly singular integral equations of the second kind (see Section 3.2.4 for the close connection between this class of problems and FDEs).

Older works giving existence results for solutions of FDEs posed on the interval $[0,1]$ often content themselves with proving that solutions lie in $C[0,1]$; they do not discuss solutions in $C^m[0,1]$ for $m \geq 1$ because this may be much more restrictive (Example 1 hints that for Caputo problems one could hope to find solutions lying in $C^{\lfloor\alpha\rfloor}[0,1]$, but Example 2 suggests that for Riemann–Liouville FDEs one usually has only $u \in C^{\lfloor\alpha\rfloor-1}[0,1]$).

The question of smoothness of solutions is examined in Diethelm's elegant book [7, Section 6.4]. From this source we quote one result.

Theorem 1 ([7, Theorem 6.28]). *Let $n \in \mathbb{N}$ and $n-1 < \alpha < n$. Consider the non-linear initial value problem*

$$ {}^{C}D^{\alpha}_{0+}u(x) = f(x,u(x)) \quad \text{for } x > 0, \text{ with } u^{(k)}(0) = u_0^{(k)} \text{ for } k = 0,1,\dots,n-1. \tag{9}$$

Let $k \in \mathbb{N}$ with $k \geq n$. Assume that $f \in C^k(G)$ where $G = [-K,K] \times \mathbb{R}$ for some $K > 0$. Then there exists $h > 0$ such that a solution u of (9) exists in $C^k(0,h] \cap C^{n-1}[0,h]$, and for $\ell = n, n+1, \dots, k$ one has $u^{(\ell)}(x) = O(x^{\alpha-\ell})$ as $x \to 0$.

The conclusion of this theorem agrees precisely with our experience in Example 1. Increasing the smoothness of the data (i. e., increasing the value of k) will increase the smoothness of the solution on $(0,h]$ but not on $[0,h]$.

In Theorem 1 the statement that $u \in C^k(0,h] \cap C^{n-1}[0,h]$ is *qualitative*: it warns the researcher that in general the solution u fails to be smooth on $[0,h]$, but it does not give any precise information about the nature of the singularity in u at $x = 0$. To carry out a rigorous numerical analysis of some method for solving (9) approximately, we need *quantitative* information about the behavior of $u^{(n)}$ (and possibly some higher-order derivatives) as $x \to 0$; this is provided by the statement that for $\ell = n, n+1, \dots, k$ one has $u^{(\ell)}(x) = O(x^{\alpha-\ell})$ as $x \to 0$.

We continue by presenting theoretical bounds on the singularities that appear in various classes of FDEs that include (as special cases) some of the examples from Section 1.1.

1.2.1 Caputo two-point boundary value problem

In [55], Pedas and Tamme consider the general two-point linear boundary value problem

$$({}^{C}D^{\alpha_p}_{0+}u)(x) + \sum_{i=0}^{p-1} a_i(x)({}^{C}D^{\alpha_i}_{0+}u)(x) = f(x) \quad \text{for } 0 < x < b, \tag{10a}$$

subject to

$$\sum_{j=0}^{n_0} \alpha_{ij} u^{(j)}(0) + \sum_{j=0}^{n_1} \beta_{ij} u^{(j)}(b_1) = \gamma_i \quad \text{for } i = 0, \dots, n-1, \tag{10b}$$

where $0 \le \alpha_0 < \alpha_1 < \cdots < \alpha_p$, $n-1 < \alpha_p \le n$ for some $n \in \mathbb{N}$, with $n_0, n_1 \in \{0, \dots, n-1\}$, $b_1 \in (0, b]$, and $\alpha_{ij}, \beta_{ij}, \gamma_i \in \mathbb{R}$. The given functions a_i (for $i = 0, 1, \dots, p-1$) and f are assumed to lie in $C[0, b]$.

Example 1 is a special case of (10).

Definition 1. Set $\mathbb{N}_0 = \{0, 1, 2, \dots\}$. For $q \in \mathbb{N}_0$ and $-\infty < \eta < 1$ with η not an integer, let $C^{q,\eta}(0, b]$ denote the space of functions $y \in C[0, b] \cap C^q(0, b]$ such that

$$|y^{(k)}(x)| \le C[x^{(1-\eta)-k} + 1] \quad \text{for } k = 0, 1, \dots, q \text{ and } x \in (0, b].$$

In particular, $C^{0,\eta}(0, b] = C[0, b]$.

We can now describe quantitatively the structure of the singularity in the solution y of (10).

Theorem 2 ([55, Theorem 2.1]). *Assume that $a_i \in C^{q,\mu}(0, b]$ for $i = 0, \dots, p-1$ and $f \in C^{q,\mu}(0, b]$, where $q \in \mathbb{N}$ and $-\infty < \mu < 1$. Moreover, assume that the boundary value problem (10) with $f \equiv 0$ and $\gamma_i = 0$ for $i = 0, \dots, n-1$ has in $C[0, b]$ only the trivial solution $u \equiv 0$ and that from all polynomials y of degree $n-1$ only $u \equiv 0$ satisfies the boundary conditions (10b) with $\gamma_i = 0$ for $i = 0, \dots, n-1$.*

Then (10) has a unique solution $u \in C^{n-1}[0, b]$ such that ${}^C D_{0+}^{\alpha_p} u \in C[0, b]$. For this solution one has ${}^C D_{0+}^{\alpha_p} u \in C^{q,\nu}(0, b]$, where $\nu := \max\{\mu, \nu_1, \nu_2\}$ with

$$\nu_1 := \max\{1 - \alpha_p + \alpha_i : \alpha_p - \alpha_i \notin \mathbb{N},\ i = 0, \dots, p-1\},$$
$$\nu_2 := \max\{1 - \lceil \alpha_i \rceil + \alpha_i : \alpha_i < n-1, \alpha_i \notin \mathbb{N}_0,\ i = 0, \dots, p-1\}.$$

If for all indices $i = 0, \dots, p-1$ one has $\alpha_p - \alpha_i \in \mathbb{N}$, then we may set ν_1 equal to any number which is less than 1. Analogously, if one has $\alpha_i \in \mathbb{N}_0$ for all indices $i = 0, \dots, p-1$ such that $\alpha_i < n-1$, then we may set ν_2 equal to any number less than 1.

Theorem 2 says that, under hypotheses on its data that ensure existence and uniqueness of a solution u to (10), this solution has a certain singularity at $x = 0$ that is quantified in the theorem. In particular it shows that the highest-order derivative ${}^C D_{0+}^{\alpha_p} u$ cannot be smoother than the right-hand side f, which is obvious in Example 1 and demonstrates that the theorem is sharp in this case.

While Theorem 2 gives clear and detailed information about ${}^C D_{0+}^{\alpha_p} u$, the analysis of many numerical methods for solving (10) will require bounds on the integer-order derivatives of u. These bounds can be deduced from Theorem 2, as the next result shows.

Theorem 3 ([61, Theorem 3.4]). *Let $m \in \mathbb{N}$ and $m-1 < \sigma < m$. Assume that $r \in C^{m-1}[0, 1]$ and ${}^C D_{0+}^{\sigma} r \in C^{q,m-\sigma}(0, 1]$ for some $q \in \mathbb{N}$. Then $r \in C^{q+m-1}(0, 1]$ and, for* *all $x \in (0,1]$, there exists a constant C, which is independent of x, such that*

$$|r^{(i)}(x)| \le \begin{cases} C & \text{if } i = 0,1,\dots,m-1, \\ Cx^{\sigma-i} & \text{if } i = m, m+1,\dots,q+m-1. \end{cases} \tag{11}$$

In Theorem 3 the hypothesis that ${}^{C}D^{\sigma}_{0+}r \in C^{q,m-\sigma}(0,1]$, for which $m-\sigma$ must lie in $(0,1)$ because $m-1<\sigma<m$, is not restrictive since the definition of these spaces implies that $C^{q,\nu}(0,1] \subset C^{q,m-\sigma}(0,1]$ for any $\nu \le m-\sigma$.

Remark 2 (Other work on two-point boundary value problems). The Riemann–Liouville analog of the Caputo problem (10) – for which Example 2 is a special case – is discussed in [31, 35].

Sobolev spaces (possibly weighted) are used by several authors to describe precisely the regularity of solutions. In this setting Jin and Zhou [25] consider the problem

$$-{}^{\mathrm{RL}}D^{\alpha}_{0+}u + qu = f \quad \text{for } 0<x<1, \quad u(0)=u(1)=0; \tag{12}$$

in [1] a fractional Laplacian equation is investigated, while in [16] a problem resembling our Example 3 but with Riemann–Liouville derivatives is studied.

Remark 3 (Decompositions of the solution). In several papers the authors appeal to a decomposition of the unknown solution u into a sum of (roughly speaking) singular and regular terms to facilitate their analysis. Equations (2) and (4) are examples of such decompositions, but they are unusually simple because each term is known explicitly up to a positive multiplier. More typically, in [25, Section 3.1] the solution u of (12) is decomposed into two terms that lie in Sobolev spaces of different regularity; although these terms are not known explicitly, they can nevertheless be used – see [25] – to construct and analyze an effective numerical method for the solution of (12).

Further examples of this useful construction can be seen in [6, equation (8.2)], [15, Section 4.1], [31, Theorem 2], [55, equation (2.6)], and [70, equation (1)].

1.2.2 Space-time problem

Consider the space-time initial-boundary value problem (7) of Example 4. Using the classical technique of separation of variables, the solution of this problem is constructed as an infinite series in [44, 47, 57, 64]. Under certain regularity and compatibility conditions (at the two corners of the space-time domain) on the data of (7), one obtains the result of Theorem 4 below.

Let $\{(\lambda_i,\psi_i) : i = 1,2,\dots\}$ be the eigenvalues and eigenfunctions for the Sturm–Liouville two-point boundary value problem

$$\mathcal{L}\psi_i := -p\psi_i'' + c\psi_i = \lambda_i\psi_i \quad \text{on } (0,l), \quad \psi_i(0)=\psi_i(l)=0,$$

where the eigenfunctions are normalized by requiring $\|\psi_i\|_2 = 1$ for all i. From the theory of sectorial operators [17, 57], the fractional power $\mathcal{L}^\gamma$ of the operator $\mathcal{L}$ is defined for each $\gamma \in \mathbb{R}$ with domain

$$D(\mathcal{L}^\gamma) := \left\{ g \in L_2(0,l) : \sum_{i=1}^{\infty} \lambda_i^{2\gamma} |(g,\psi_i)|^2 < \infty \right\}.$$

For example, $D(\mathcal{L}^{1/2}) = H_0^1(0,l)$ in Sobolev space notation. Also set

$$\|g\|_{\mathcal{L}^\gamma} := \left(\sum_{i=1}^{\infty} \lambda_i^{2\gamma} |(g,\psi_i)|^2 \right)^{1/2}.$$

Theorem 4 ([64, Theorem 2.1]). *Assume that $\phi \in D(\mathcal{L}^{5/2})$, $f(\cdot,t) \in D(\mathcal{L}^{5/2})$, and $f_t(\cdot,t)$ and $f_{tt}(\cdot,t)$ are in $D(\mathcal{L}^{1/2})$ for each $t \in (0,T]$ with*

$$\|f(\cdot,t)\|_{\mathcal{L}^{5/2}} + \|f_t(\cdot,t)\|_{\mathcal{L}^{1/2}} + t^\rho \|f_{tt}(\cdot,t)\|_{\mathcal{L}^{1/2}} \le C_1$$

for all $t \in (0,T]$ and some constant $\rho < 1$, where C_1 is a constant independent of t. Then (7) *has a unique solution u that satisfies* (7a), (7b), *and* (7c) *point-wise, and there exists a constant C such that*

$$\left| \frac{\partial^k u}{\partial x^k}(x,t) \right| \le C \quad \text{for } k = 0,1,2,3,4, \tag{13a}$$

$$\left| \frac{\partial^\ell u}{\partial t^\ell}(x,t) \right| \le C(1 + t^{\alpha-\ell}) \quad \text{for } \ell = 0,1,2, \tag{13b}$$

for all $(x,t) \in [0,l] \times (0,T]$.

The particular example $u(x,t) = E_{\alpha,1}(-t^\alpha)\sin x$ that we met in (8) shows that the bounds of Theorem 4 are sharp. The time derivatives bound (13b) associated with the Caputo derivative ${}^C D_{0+}^{\alpha,t} u$ of (7a) resembles the bound (11) on the spatial derivatives for the Caputo two-point boundary value problem (10). Despite the apparent simplicity of (13b), it comes from the rather complicated term $E_{\alpha,1}(-t^\alpha)$, which any numerical method for (7) must approximate.

Sobolev-norm bounds on u and its derivatives are obtained in [47, 57].

When one has time-dependent coefficients in (7a), the above eigenfunction-based approach fails; an alternative analysis for this case is given in [32].

The regularity (in terms of Sobolev spaces) of a problem resembling problem (7), but with the Caputo time derivative replaced by a Riemann–Liouville derivative, is investigated in [33].

2 How much regularity can one assume?

In the numerical analysis of differential equations, simplifying assumptions are often needed if one is to make progress. For example, in (7a) we worked with $r(x)$ instead of

$r(x,t)$ because the separation of variables technique used to prove Theorem 4 will fail if r depends on t; despite this loss of generality, there are still infinitely many choices for the function $r(x)$, so the restrictive hypothesis that $r = r(x)$ is tolerable.

In this spirit one might ask the following. To make the numerical analysis of the space-time Example 4 more tractable, could one relax the bound (13b) for $\ell = 1$ to

$$\left|\frac{\partial u}{\partial t}(x,t)\right| \le C_1(1 + t^{\sigma-1}) \quad \text{for some constant } \sigma > \alpha? \tag{14}$$

Here C_1 is some fixed constant.

Surprisingly, this insignificant-looking weakening of (13b) has very serious consequences. To show this, we first prove a straightforward extension of a basic result [7, Lemma 3.11] for Caputo fractional derivatives.

Lemma 1 ([60, Lemma 2.1]). *Assume that* (14) *is valid. Then for each fixed x, one has*

$$\lim_{t\to 0^+} {}^C D_{0+}^{\alpha,t} u(x,t) = 0.$$

Proof. Fix x. For any $t \in (0,T]$,

$$\begin{aligned} |{}^C D_{0+}^{\alpha,t} u(x,t)| &= \left|\frac{1}{\Gamma(1-\alpha)} \int_{s=0}^{t} (t-s)^{-\alpha} \frac{\partial u}{\partial s}(x,s)\, ds\right| \\ &\le \frac{C_1}{\Gamma(1-\alpha)} \int_{s=0}^{t} (t-s)^{-\alpha}(1 + s^{\sigma-1})\, ds \\ &= \frac{C_1}{\Gamma(1-\alpha)} \left[\frac{t^{1-\alpha}}{1-\alpha} + \frac{\Gamma(1-\alpha)\Gamma(\sigma)}{\Gamma(1-\alpha+\sigma)} t^{\sigma-\alpha}\right], \end{aligned}$$

by splitting the integral into a sum of two terms and invoking a standard formula for the Euler Beta function [7, Theorem D.6]. The desired result now follows as $1-\alpha > 0$ and $\sigma - \alpha > 0$. □

The example $u(x,t) = t^\alpha$ shows that the conclusion of Lemma 1 does not hold if (14) is replaced by the stronger inequality (13b).

We can now present the main result of Section 2.

Theorem 5 ([60, Theorem 2.1]). *Suppose that the solution u of Example 4 satisfies* (14) *and that u and $\partial^2 u/\partial x^2$ are continuous on the closed domain $\bar{Q}$. Then for each f in* (7a)*, the initial value ϕ of u in* (7c) *is determined uniquely.*

Proof. For each fixed $x \in (0,l)$, consider the limit of equation (7a) as $t \to 0^+$. Invoking Lemma 1, we see that this limit is $-p\phi''(x) + r(x)\phi(x) = f(x,0)$. As well as this ordinary differential equation on the interval $(0,l)$ we have the boundary conditions $\phi(0) = \phi(l) = 0$ from Example 4. But the hypothesis $r \ge 0$ in Example 4 implies that the differential operator $w \mapsto -pw'' + rw$ satisfies a maximum principle [56, Corollary, p. 7]. Consequently for each f the function ϕ is unique. □

The conclusion of the theorem – that for each given f only one choice of initial condition ϕ is possible – is bizarre and extremely restrictive because typically in initial-boundary value problems the forcing function f and the initial condition ϕ are independent of each other except for some compatibility requirements at the corners $(0,0)$ and $(l,0)$ of $\bar{Q}$.

In Theorem 5 the hypotheses that u and $\partial^2 u/\partial x^2$ are in $C(\bar{Q})$ are reasonable; consider the example (8). The extraordinary conclusion of the theorem is caused by the unnatural assumption (14).

Remark 4. The condition (14) is automatically satisfied if one assumes that $\partial u/\partial t$ is bounded on $\bar{Q}$.

Example 5. Consider the fractional heat equation

$$D_t^\alpha v - \partial^2 v/\partial x^2 = 0 \quad \text{for } (x,t) \in Q := (0,\pi) \times (0,T]$$

with $0 < \alpha < 1$, boundary conditions $v(0,t) = v(\pi,t) = 0$, and the initial condition $v(x,0) = \phi(x)$, where $\phi \in C^2[0,\pi]$ is unspecified except that it satisfies the compatibility condition $\phi(0) = \phi(\pi) = 0$.

If we assume that for the solution v the partial derivatives v_{xx} and v_t lie in $C(\bar{Q})$, then the argument of Theorem 5 shows that ϕ must satisfy the conditions

$$-\phi''(x) = 0 \quad \text{on } (0,\pi), \quad \phi(0) = \phi(\pi) = 0.$$

It follows that $\phi \equiv 0$. But now all the data of this example are zero, so $v \equiv 0$. Thus the imposition of the arbitrary assumption that $v_t \in C(\bar{Q})$ forces $v \equiv 0$.

In [7, Section 6.4] the regularity of solutions to initial value problems for ordinary differential equations is examined; Theorems 6.26, 6.27, and 6.29 and Corollary 6.30 there, like our Theorem 5, describe how requiring smoothness of the solution on the closed domain places a restriction on the class of problems studied.

An analogous discussion for two-point boundary value problems with Caputo derivatives appears in [59].

3 Singularities and numerical methods

3.1 Effect of singularities on numerical methods

When an FDE is solved numerically, the presence or absence of a singularity in the true solution influences strongly the accuracy of most numerical methods.

Example 6. In [13] several numerical methods are tested on FD two-point boundary value problems. The FDE is $D_{0+}^\alpha u(x) = f(x)$ on $(0,1)$, where $1 < \alpha < 2$ and D_{0+}^α is a Caputo or a Riemann–Liouville derivative. Two fundamentally different test problems were used:

(i) $u(x) = 1 + 3x - 7x^2 + 4x^3 - 2x^4$ so $u(0) = 1$, $u(1) = -1$, with $f = D^{\alpha}_{0+}u$;
(ii) $f(x) \equiv 1$, $u(0) = 0$, $u(1) = 1$.

Thus in (i) the solution u is smooth but in (ii) it will exhibit a singularity at $x = 0$: the solutions for (ii) are (Examples 1 and 2)

$$\text{(Caputo case)}\ u(x) = \frac{x^{\alpha}}{\Gamma(1+\alpha)} + \left[1 - \frac{1}{\Gamma(1+\alpha)}\right]x,$$

$$\text{(RL case)}\ u(x) = \frac{x^{\alpha}}{\Gamma(1+\alpha)} + \left[1 - \frac{1}{\Gamma(1+\alpha)}\right]x^{\alpha-1}.$$

Finite difference methods were used in [13] on a uniform mesh of width h, and errors in computed solutions were measured in the discrete L^{∞} norm. We summarize the behavior of the schemes by the rates of convergence h^{rate} listed in Table 1, where our four test problems (Caputo (i), Caputo (ii), Riemann–Liouville (i), and Riemann–Liouville (ii)) form the columns and the schemes used form the rows.

Table 1: Rates of convergence from [13] for Example 6; "SGL" = shifted Grünwald–Letnikov.

	Caputo (i)	Caputo (ii)	RL (i)	RL (ii)
L2 scheme [54, 58]	1	1	1	$\leq \alpha - 1$
Spline Caputo scheme [58]	2	1	–	–
Spline RL scheme [58]	–	–	0	$\alpha - 1$
SGL scheme [50, 54]	–	–	0	$\alpha - 1$
Weighted SGL scheme [65]	–	–	0	$\alpha - 1$

At first sight of Table 1 it is surprising that the well-known Shifted Grünwald–Letnikov scheme fails to converge for the simple Riemann–Liouville Example 6(i), but an analysis [13, Theorem 4.1] proves that this scheme cannot achieve any positive order of convergence for polynomial solutions u that do not satisfy $u(0) = 0$. A similar argument shows that the spline Riemann–Liouville scheme and weighted shifted Grünwald–Letnikov (WSGL) scheme of Table 1 suffer from the same shortcoming. An extension of this argument in [63, Example 4.2] shows rigorously that these three schemes cannot achieve any order of convergence greater than $\alpha - 1$ for the Riemann–Liouville Example 6(ii), which corroborates the last three entries in the final column of Table 1. See also the analysis of the WSGL scheme in [70].

Remark 5. Extra assumptions such as $u(0) = 0$ or $f(0) = 0$ may affect the behavior of the above difference schemes. This aspect is discussed at length in [63, Section 5].

Example 7. In [9, Section 4], the authors apply a finite element method to solve the two-sided FDE of Example 3. Here the regularity of the solution depends on the value of the parameter r in (6), and the numerical results demonstrate clearly that the convergence rate of the method depends on r also.

Example 8. In [64] the initial-boundary value problem (7) of Example 4 is discretized on a uniform mesh whose widths are τ in time and h in space. The well-known L1 scheme is used to approximate the Caputo time derivative ${}^{C}D_{0+}^{\alpha,t}u$ of order $\alpha \in (0,1)$ and the spatial derivative u_{xx} is discretized by a standard second-order three-point finite difference. When this scheme is tested on a smooth solution, it is $O(\tau^{2-\alpha} + h^2)$ convergent in the discrete L^∞ norm. But when instead an example with a typical singularity at $t = 0$ is used, the scheme is only $O(\tau^\alpha + h^2)$ convergent. Here $O(h^2)$ is to be expected since typical solutions are smooth in space; it is the order of convergence in τ that is of interest.

Jin et al. [21, Tables 1–3] also present numerical results for the space-time problem (7), using several schemes that have been proposed in the literature. They measure errors in the $L^2(0,l)$ norm at a fixed time, and observe that these schemes attain high rates of convergence – $O(\tau^{2-\alpha} + h^2)$ or better – for the exceptional case of smooth solutions, but when they are tested on typical solutions with weak singularities at $t = 0$, these rates decline significantly.

3.2 Numerical methods designed to handle singularities

There are essentially four classes of methods that are designed to cope with the presence of typical singularities in solutions to FDEs:

1. Use a theoretical decomposition of the solution to construct a special scheme that can approximate accurately each term in the decomposition.
2. Use special basis functions to capture the essential singularity of the solution.
3. Use a special grid that is chosen a priori or adaptively so that extra grid points are clustered in a well-chosen way in the neighborhood of the singularity.
4. Reformulate the FD problem to facilitate its numerical solution.

Each of the next four subsections is devoted to one of the above classes of methods. As including full technical details would increase greatly the length of this survey, we merely outline the main features of each approach and cite relevant papers from the recent research literature. In particular, note that the theoretical convergence results in each of these papers require some regularity and compatibility assumptions on the problem's data that we do not state.

3.2.1 Methods based on solution decomposition

The numerical methods in this section are all based on exploiting some decomposition of the unknown solution (recall Remark 3). We shall focus mainly on the powerful technique known as *convolution quadrature*.

Lubich [39, 41, 42] analyzed the structure of typical solutions for various initial value and time-dependent problems and showed that convolution quadrature was a fruitful way of modifying standard difference schemes (which are inaccurate when applied to FD problems with weak singularities in their solutions) to obtain new and stable methods that are high-order accurate for these typical solutions. Without going into the details, his approach requires one to compute and introduce correction terms into each standard scheme for a fixed number of initial steps, with the aim of approximating exactly (or very accurately) the most singular terms in the decomposition of the unknown solution u while retaining the accuracy of the basic scheme for the remainder of u, which is of course a smoother function. See [5, 70] for a clear description of what this technique involves, its gains, and one potential drawback (the computation of the correction terms is an ill-posed problem if one tries to obtain a very high-order scheme).

Convolution quadrature is used in many interesting papers, e. g., [14, 21, 23, 24, 70, 71] and the references therein, which deal with various methods for time-dependent FD problems; see also [6, Section 8] and [43]. Its analysis is based on the theory of sectorial operators [17]. The papers above use uniform meshes; for convolution quadrature on non-uniform meshes, see [38] and its references.

Decompositions of the unknown solution play a role in some other numerical methods that are not based on convolution quadrature, and we now present a few of these.

In [25] Jin and Zhou consider the problem (12) and exhibit a theoretical decomposition of its solution u whose dominant singular term has an unknown constant coefficient. They design and analyze a finite element method that approximates separately this dominant term and the smooth remainder of u.

In [66] a spectral Galerkin method is developed for a class of Caputo two-point boundary value problems, based on a decomposition of the solution.

Finally, an extrapolation technique that can be applied to many standard schemes for various types of FD problems is developed and analyzed by Hao and Cao in [15]. The idea resembles Richardson extrapolation: if one knows the dominant singular terms in the solution decomposition, one can combine solutions on two (or more) grids to eliminate these terms and obtain improved accuracy in the computed solutions.

3.2.2 Special basis functions

The methods in this section aim to use basis functions that approximate well the dominant singular terms in the solution of the FD problem. This can be done explicitly if the structure of the solution is known a priori, or adaptively when one does not wish to construct special functions a priori or when the precise structure of the solution is unknown.

In [4, 9, 45, 69] the authors use spectral methods (finite element and collocation) based on appropriately modified Jacobi polynomials to solve various classes of FD two-point boundary value problems, including those of Examples 1–3. Theoretical and numerical results show that spectral accuracy can be attained in weighted Sobolev norms. Using similar techniques, problems posed in higher dimensions are considered in [46].

A discontinuous Petrov–Galerkin (DPG) finite element method with optimal test functions is used in [8] to solve a FD two-point boundary value problem and convergence is proven in an associated Sobolev-type norm.

In [26] the authors solve a variant of Example 4 using a Galerkin finite element in space and the L1 discretization in time. Then proper orthogonal decomposition (POD) is applied to the computed solutions to generate adaptively a reduced basis that is tailored to the FD problem and can be used to solve it efficiently and accurately. A related approach is used in [11], where special basis functions in time are constructed a priori and then combined with a standard finite difference method in space. The theoretical results in [11, 26] improve the error term $O(\tau^{\alpha})$ mentioned in Example 8 to $O(\tau^{2\alpha})$; while the discrete L^{∞} norm used in [11] appears stronger than the $L^{\infty}(L^2)$ norm used in [26], the latter paper requires less regularity of the data of the problem. An alternative technique that yields a similar improvement is described in Section 3.2.4.

3.2.3 Special grids

Suppose that the space-time problem of Example 4 is solved as described in Example 8. Then in the discrete L^{∞} norm, the L1 scheme on a uniform mesh in time achieves $O(\tau^{2-\alpha})$ for smooth solutions, but only $O(\tau^{\alpha})$ in the realistic case where a weak singularity is present at $t = 0$. If instead one uses the graded mesh $t_m = T(m/M)^r$ for $m = 0, 1, \ldots, M$, where $r \geq 1$ is a user-chosen parameter, then for typical solutions with a weak singularity it is shown in [64] that for $r \geq (2-\alpha)/\alpha$ the convergence rate of $O(M^{-(2-\alpha)})$ is restored. A closely related analysis, but in a more general setting, is presented in [36]. (Furthermore, in [36] it is shown how to remove the requirement that $r \geq 0$ in (7a) by means of a new discrete Grönwall inequality.) An alternative analysis to that of [64], which is also more general, is given in [29]. See also [53, Section 3].

Although it uses a uniform mesh in time, we mention here the “modified L1 scheme” of [67], which by changing the L1 scheme at a single time step achieves the same convergence order as the above graded-mesh papers.

A higher-order scheme (originally due to Alikhanov) is used in [3, 37] to solve the space-time problem on graded meshes.

Graded meshes of the type used in [3, 29, 36, 37, 64] put more points near the location of the weak singularity in an optimal way and are often used to solve weakly singular Volterra integral equations, as discussed in [2]; this class of integral equations is closely related to Example 4 (Section 3.2.4). Similar meshes were combined with finite element methods by McLean and Mustapha in several papers to solve fractional-diffusion problems (like Example 4) and fractional-wave problems, often in more than one space dimension; see [48, 51, 52] and their references. See also [18, 19].

In Section 3.2.4 other uses of this mesh are given.

Adaptive approaches to generating a suitable mesh for space-time problems with an initial singularity at $t = 0$ are discussed in [68] but the rigorous analysis of such a method, which would not be easy, remains an open problem.

3.2.4 Reformulation of the FD problem

The central idea of this subsection is that one begins by transforming the original FDE to a more amenable form before attempting to solve it numerically.

In [62] it is shown how knowledge of the structure of the solution of Example 4 enables a simple pre-processing of the problem that yields a new initial-boundary value problem with a smoother solution. Then applying the L1 scheme in time and standard finite differencing in space on a uniform space-time mesh, the discrete L^∞ norm bound $O(\tau^\alpha + h^2)$ of Example 8 is improved to $O(\tau^{2\alpha} + h^2)$. This pre-processing also allows one to use a less severely graded mesh to obtain the same accuracy as in the first paragraph of Section 3.2.3.

Two-point boundary value problems of arbitrary order can be re-formulated profitably as weakly singular Volterra integral equations of the second kind, for which accurate and efficient numerical methods are already known [2]. This approach is followed for Caputo problems in [30, 34, 55] and for Riemann–Liouville problems in [31, 35]; in these papers piece-wise polynomial collocation is used to solve the Volterra problem on a graded mesh like that of Section 3.2.3. A precursor of this approach is [10], where a similar method is used to solve initial value problems.

A variant of this re-formulation for Caputo two-point boundary value problems is considered by Pedas and Tamme and their coworkers in [28] and earlier papers: a change of independent variable is made to render the solution of the Volterra integral equation less singular before applying the collocation method.

A further variant is considered in [20], where the problem (12) is transformed to an integro-differential equation which is then solved via a finite element method.

3.2.5 Convergence away from the singularity

All the discussion up to this point has considered only global norms, i. e., norms that take into account the error behavior on the entire domain where the FDE is posed. But this ignores the following phenomenon: sometimes a numerical method achieves a certain rate of convergence in a global norm, yet when the norm is restricted to some fixed subdomain that excludes the location of the singularity, the rate of convergence is increased. (This behavior is well known in finite element solutions of classical parabolic problems where accuracy is often lost near the initial time $t = 0$.)

For example, for an initial-boundary value problem similar to Example 4 with $r \equiv 0$, in [21, Theorem 3.5] the solution U_h^n computed using a finite element method in space and finite differences in time is shown to satisfy the bound

$$\|u(t_n) - U_h^n\|_{L^2(\Omega)} \le C(t_n^{-1}\tau + t_n^{-\alpha}h^2)\|u(0)\|_{L^2(\Omega)}. \tag{15}$$

Here Ω is the spatial domain, h is the spatial mesh diameter, $t_n = n\tau$ where the temporal mesh is uniform of width τ, the true solution is u, and U_h^n is the solution computed at time $t = t_n$. The constant C is independent of the mesh. In particular, when $n = 1$ the right-hand side is at best $O(1)$, so we not have any positive rate of convergence in the global $L^\infty(L^2)$ norm, which is measured on $Q := \Omega \times [0, T]$ (here we use the standard norm notation for finite element methods applied to parabolic partial differential equations). But if we restrict our attention to the subdomain $\hat{Q} := \Omega \times [0.1, T]$, then (15) implies that

$$\|u(t_n) - U_h^n\|_{L^2(\Omega)} \le \hat{C}(\tau + h^2)\|u(0)\|_{L^2(\Omega)},$$

which yields $O(\tau + h^2)$ convergence in the restricted norm $L^\infty(L^2)|_{\hat{Q}}$. Here the constant $\hat{C}$ depends on the choice of $\hat{Q}$. In this choice the value 0.1 was arbitrary; any fixed positive value could be chosen.

Thus despite the failure of this method to achieve any positive order of convergence on Q it should not be dismissed as useless, because our final bound shows that it can compute an accurate solution on subdomains that are bounded away from $t = 0$.

Analogous results showing improved convergence behavior away from the location of the initial singularity appear in [12, 27, 49]. The convergence bounds for spectral methods obtained in [4, 9, 45] and some other papers use weighted Sobolev norms where the weight approaches zero as one approaches the singularity, so they imply convergence results in terms of standard unweighted Sobolev norms restricted to subdomains away from the singularity.

3.2.6 General reference

To close, we refer the reader to the useful survey article [22].

Bibliography

[1] G. Acosta, J. P. Borthagaray, O. Bruno, and M. Maas, Regularity theory and high order numerical methods for the (1D)-fractional Laplacian, *Math. Comput.*, **87**(312) (2018), 1821–1857.

[2] H. Brunner, *Collocation Methods for Volterra Integral and Related Functional Differential Equations*, Cambridge Monographs on Applied and Computational Mathematics, vol. 15, Cambridge University Press, Cambridge, 2004.
[3] H. Chen and M. Stynes, Error analysis of a second-order method on fitted meshes for a time-fractional diffusion problem, *J. Sci. Comput.*, (2018), 10.1007/s10915-018-0863-y.
[4] S. Chen, J. Shen, and L.-L. Wang, Generalized Jacobi functions and their applications to fractional differential equations, *Math. Comput.*, **85**(300) (2016), 1603–1638.
[5] X. Chen, F. Zeng, and G. E. Karniadakis, A tunable finite difference method for fractional differential equations with non-smooth solutions, *Comput. Methods Appl. Mech. Eng.*, **318** (2017), 193–214.
[6] E. Cuesta, C. Lubich, and C. Palencia, Convolution quadrature time discretization of fractional diffusion-wave equations, *Math. Comput.*, **75**(254) (2006), 673–696.
[7] K. Diethelm, *The Analysis of Fractional Differential Equations*, Lecture Notes in Mathematics, vol. 2004, Springer-Verlag, Berlin, 2010 An application-oriented exposition using differential operators of Caputo type.
[8] V. J. Ervin, T. Führer, N. Heuer, and M. Karkulik, DPG method with optimal test functions for a fractional advection diffusion equation, *J. Sci. Comput.*, **72**(2) (2017), 568–585.
[9] V. J. Ervin, N. Heuer, and J. P. Roop, Regularity of the solution to 1-D fractional order diffusion equations, *Math. Comput.*, **87**(313) (2018), 2273–2294.
[10] N. J. Ford, M. L. Morgado, and M. Rebelo, Nonpolynomial collocation approximation of solutions to fractional differential equations, *Fract. Calc. Appl. Anal.*, **16**(4) (2013), 874–891.
[11] J. L. Gracia, E. O'Riordan, and M. Stynes, A fitted scheme for a Caputo initial-boundary value problem, *J. Sci. Comput.*, **76**(1) (2018), 583–609.
[12] J. L. Gracia, E. O'Riordan, and M. Stynes, Convergence in positive time for a finite difference method applied to a fractional convection-diffusion problem, *Comput. Methods Appl. Math.*, **18**(1) (2018), 33–42.
[13] J. L. Gracia and M. Stynes, Formal consistency versus actual convergence rates of difference schemes for fractional-derivative boundary value problems, *Fract. Calc. Appl. Anal.*, **18**(2) (2015), 419–436.
[14] M. Gunzburger and J. Wang, A second-order Crank–Nicolson method for time-fractional PDEs, *Int. J. Numer. Anal. Model.*, **16**(2) (2019), 225–239.
[15] Z. Hao and W. Cao, An improved algorithm based on finite difference schemes for fractional boundary value problems with nonsmooth solution, *J. Sci. Comput.*, **73**(1) (2017), 395–415.
[16] Z.-P. Hao, G. Lin, and Z. Zhang, Regularity in weighted Sobolev spaces and spectral methods for a two-sided fractional reaction-diffusion equation, Preprint, Researchgate, 10.13140/RG.2.2.23591.04002, 2017.
[17] D. Henry, *Geometric Theory of Semilinear Parabolic Equations*, Lecture Notes in Mathematics, vol. 840, Springer-Verlag, Berlin, New York, 1981.
[18] C. Huang and M. Stynes, A direct discontinuous Galerkin method for a time-fractional diffusion equation with a Robin boundary condition, *Appl. Numer. Math.*, **135** (2019), 15–29.
[19] C. Huang, M. Stynes, and N. An, Optimal $L^{\infty}(L^2)$ error analysis of a direct discontinuous Galerkin method for a time-fractional reaction-diffusion equation, *BIT Numer. Math.*, **58** (2018), 661–690.
[20] B. Jin, R. Lazarov, X. Lu, and Z. Zhou, A simple finite element method for boundary value problems with a Riemann–Liouville derivative, *J. Comput. Appl. Math.*, **293** (2016), 94–111.
[21] B. Jin, R. Lazarov, and Z. Zhou, Two fully discrete schemes for fractional diffusion and diffusion-wave equations with nonsmooth data, *SIAM J. Sci. Comput.*, **38**(1) (2016), A146–A170.
[22] B. Jin, R. Lazarov, and Z. Zhou, Numerical methods for time-fractional evolution equations with nonsmooth data: a concise overview, *Comput. Methods Appl. Mech. Eng*, (2019), in press.

[23] B. Jin, B. Li, and Z. Zhou, Correction of high-order BDF convolution quadrature for fractional evolution equations, *SIAM J. Sci. Comput.*, **39**(6) (2017), A3129–A3152.
[24] B. Jin, B. Li, and Z. Zhou, An analysis of the Crank–Nicolson method for subdiffusion, *IMA J. Numer. Anal.*, **38**(1) (2018), 518–541.
[25] B. Jin and Z. Zhou, A finite element method with singularity reconstruction for fractional boundary value problems, *ESAIM: Math. Model. Numer. Anal.*, **49**(5) (2015), 1261–1283.
[26] B. Jin and Z. Zhou, An analysis of Galerkin proper orthogonal decomposition for subdiffusion, *ESAIM: Math. Model. Numer. Anal.*, **51**(1) (2017), 89–113.
[27] S. Karaa, K. Mustapha, and A. K. Pani, Optimal error analysis of a FEM for fractional diffusion problems by energy arguments, *J. Sci. Comput.*, **74**(1) (2018), 519–535.
[28] M. Kolk, A. Pedas, and E. Tamme, Smoothing transformation and spline collocation for linear fractional boundary value problems, *Appl. Math. Comput.*, **283** (2016), 234–250.
[29] N. Kopteva, Error analysis of the L1 method on graded and uniform meshes for a fractional-derivative problem in two and three dimensions, *Math. Comput.*, (2019), in press.
[30] N. Kopteva and M. Stynes, An efficient collocation method for a Caputo two-point boundary value problem, *BIT Numer. Math.*, **55**(4) (2015), 1105–1123.
[31] N. Kopteva and M. Stynes, Analysis and numerical solution of a Riemann–Liouville fractional derivative two-point boundary value problem, *Adv. Comput. Math.*, **43**(1) (2017), 77–99.
[32] A. Kubica and M. Yamamoto, Initial-boundary value problems for fractional diffusion equations with time-dependent coefficients, *Fract. Calc. Appl. Anal.*, **21**(2) (2018), 276–311.
[33] B. Li and X. Xie, Regularity of solutions to time fractional diffusion equations, *Discrete Contin. Dyn. Syst. Ser. B*, (2019), in press.
[34] H. Liang and M. Stynes, Collocation methods for general Caputo two-point boundary value problems, *J. Sci. Comput.*, **76**(1) (2018), 390–425.
[35] H. Liang and M. Stynes, Collocation methods for general Riemann–Liouville two-point boundary value problems, *Adv. Comput. Math.*, (2018), 10.1007/s10444-018-9645-1.
[36] H.-L. Liao, D. Li, and J. Zhang, Sharp error estimate of the nonuniform L1 formula for linear reaction-subdiffusion equations, *SIAM J. Numer. Anal.*, **56**(2) (2018), 1112–1133.
[37] H.-L. Liao, W. McLean, and J. Zhang, A second-order scheme with nonuniform time steps for a linear reaction-sudiffusion problem, arXiv:1803.09873, 2018.
[38] M. Lopez-Fernandez and S. Sauter, Generalized convolution quadrature based on Runge–Kutta methods, *Numer. Math.*, **133**(4) (2016), 743–779.
[39] C. Lubich, Convolution quadrature and discretized operational calculus. I, *Numer. Math.*, **52**(2) (1988), 129–145.
[40] Ch. Lubich, Runge–Kutta theory for Volterra and Abel integral equations of the second kind, *Math. Comput.*, **41**(163) (1983), 87–102.
[41] Ch. Lubich, Discretized fractional calculus, *SIAM J. Math. Anal.*, **17**(3) (1986), 704–719.
[42] Ch. Lubich, Convolution quadrature revisited, *BIT Numer. Math.*, **44**(3) (2004), 503–514.
[43] Ch. Lubich, I. H. Sloan, and V. Thomée, Nonsmooth data error estimates for approximations of an evolution equation with a positive-type memory term, *Math. Comput.*, **65**(213) (1996), 1–17.
[44] Y. Luchko, Initial-boundary-value problems for the one-dimensional time-fractional diffusion equation, *Fract. Calc. Appl. Anal.*, **15**(1) (2012), 141–160.
[45] Z. Mao, S. Chen, and J. Shen, Efficient and accurate spectral method using generalized Jacobi functions for solving Riesz fractional differential equations, *Appl. Numer. Math.*, **106** (2016), 165–181.
[46] Z. Mao and J. Shen, Efficient spectral-Galerkin methods for fractional partial differential equations with variable coefficients, *J. Comput. Phys.*, **307** (2016), 243–261.
[47] W. McLean, Regularity of solutions to a time-fractional diffusion equation, *ANZIAM J.*, **52**(2) (2010), 123–138.

[48] W. McLean and K. Mustapha, A second-order accurate numerical method for a fractional wave equation, *Numer. Math.*, **105**(3) (2007), 481–510.
[49] W. McLean and K. Mustapha, Time-stepping error bounds for fractional diffusion problems with non-smooth initial data, *J. Comput. Phys.*, **293** (2015), 201–217.
[50] M. M. Meerschaert and C. Tadjeran, Finite difference approximations for fractional advection-dispersion flow equations, *J. Comput. Appl. Math.*, **172**(1) (2004), 65–77.
[51] K. Mustapha, B. Abdallah, K. M. Furati, and M. Nour, A discontinuous Galerkin method for time fractional diffusion equations with variable coefficients, *Numer. Algorithms*, **73**(2) (2016), 517–534.
[52] K. Mustapha and W. McLean, Superconvergence of a discontinuous Galerkin method for fractional diffusion and wave equations, *SIAM J. Numer. Anal.*, **51**(1) (2013), 491–515.
[53] R. H. Nochetto, E. Otárola, and A. J. Salgado, A PDE approach to space-time fractional parabolic problems, *SIAM J. Numer. Anal.*, **54**(2) (2016), 848–873.
[54] K. B. Oldham and J. Spanier, *The Fractional Calculus*, Academic Press [A subsidiary of Harcourt Brace Jovanovich, Publishers], New York, London, 1974; Theory and applications of differentiation and integration to arbitrary order, With an annotated chronological bibliography by Bertram Ross, *Mathematics in Science and Engineering*, vol. 111.
[55] A. Pedas and E. Tamme, Piecewise polynomial collocation for linear boundary value problems of fractional differential equations, *J. Comput. Appl. Math.*, **236**(13) (2012), 3349–3359.
[56] M. H. Protter and H. F. Weinberger, *Maximum Principles in Differential Equations*, Springer-Verlag, New York, 1984, corrected reprint of the 1967 original.
[57] K. Sakamoto and M. Yamamoto, Initial value/boundary value problems for fractional diffusion-wave equations and applications to some inverse problems, *J. Math. Anal. Appl.*, **382**(1) (2011), 426–447.
[58] E. Sousa, How to approximate the fractional derivative of order $1 < \alpha \leq 2$, *Int. J. Bifurc. Chaos Appl. Sci. Eng.*, **22**(4) (2012), 1250075, 13.
[59] M. Stynes, A Caputo two-point boundary value problem: existence, uniqueness and regularity of a solution, *Model. Anal. Inform. Sist.*, **23**(3) (2016), 370–376.
[60] M. Stynes, Too much regularity may force too much uniqueness, *Fract. Calc. Appl. Anal.*, **19**(6) (2016), 1554–1562.
[61] M. Stynes and J. L. Gracia, A finite difference method for a two-point boundary value problem with a Caputo fractional derivative, *IMA J. Numer. Anal.*, **35**(2) (2015), 698–721.
[62] M. Stynes and J. L. Gracia, Preprocessing schemes for fractional-derivative problems to improve their convergence rates, *Appl. Math. Lett.*, **74** (2017), 187–192.
[63] M. Stynes, E. O'Riordan, and J. L. Gracia, Necessary conditions for convergence of difference schemes for fractional-derivative two-point boundary value problems, *BIT Numer. Math.*, **56**(4) (2016), 1455–1477.
[64] M. Stynes, E. O'Riordan, and J. L. Gracia, Error analysis of a finite difference method on graded meshes for a time-fractional diffusion equation, *SIAM J. Numer. Anal.*, **55**(2) (2017), 1057–1079.
[65] W. Y. Tian, H. Zhou, and W. Deng, A class of second order difference approximations for solving space fractional diffusion equations, *Math. Comput.*, **84**(294) (2015), 1703–1727.
[66] H. Wang and X. Zhang, A high-accuracy preserving spectral Galerkin method for the Dirichlet boundary-value problem of variable-coefficient conservative fractional diffusion equations, *J. Comput. Phys.*, **281** (2015), 67–81.
[67] Y. Yan, M. Khan, and N. J. Ford, An analysis of the modified L1 scheme for time-fractional partial differential equations with nonsmooth data, *SIAM J. Numer. Anal.*, **56**(1) (2018), 210–227.
[68] S. B. Yuste and J. Quintana-Murillo, Fast, accurate and robust adaptive finite difference methods for fractional diffusion equations, *Numer. Algorithms*, **71**(1) (2016), 207–228.

[69] F. Zeng, Z. Mao, and G. E. Karniadakis, A generalized spectral collocation method with tunable accuracy for fractional differential equations with end-point singularities, *SIAM J. Sci. Comput.*, **39**(1) (2017), A360–A383.
[70] F. Zeng, Z. Zhang, and G. E. Karniadakis, Second-order numerical methods for multi-term fractional differential equations: Smooth and non-smooth solutions, *Comput. Methods Appl. Mech. Eng.*, **327** (2017), 478–502.
[71] P. Zhu, S. Xie, and X. Wang, Nonsmooth data error estimates for FEM approximations of the time fractional cable equation, *Appl. Numer. Math.*, **121** (2017), 170–184.

Hong Wang

Fast numerical methods for space-fractional partial differential equations

Abstract: Because of the non-local nature of fractional differential operators, numerical methods for space-fractional partial differential equations (sFPDEs) often generate dense or full stiffness matrices with complex structures. The scenario is complicated further by the fact that linear elliptic and parabolic FPDEs with smooth data defined in smooth domains may generate solutions with boundary layers. Consequently, the numerical simulations of sFPDEs have significantly increased computational complexity and memory requirements, compared to their integer-order analogs. In this chapter we address the computational issues of sFPDEs, outline some of the recent developments of fast and accurate numerical methods for sFPDEs, and briefly discuss possible future directions in the field.

Keywords: Fractional partial differential equation, finite difference method, finite element method, finite volume method, Toeplitz matrix

PACS: 02.60.Lj, 02.60.Nm, 02.60.-x, 02.70.Bf, 02.70.Dh, 47.11.Df, 05.40.Fb

1 Fractional PDEs and their computational bottleneck

While they provide competitive tools for modeling challenging phenomena, including anomalous transport and long-range time memory or spatial interactions, compared to integer-order partial differential equations (PDEs) [5, 9, 31, 32, 39], fractional PDEs (FPDEs) present mathematical and numerical difficulties that are not encountered in the context of integer-order PDEs.

Because of the non-local nature of fractional differential operators, numerical methods for space-fractional PDEs (sFPDEs) generate dense or full stiffness matrices with complicated structures [27, 28, 30, 40]. These methods require $O(N^2)$ memory, where N is the number of spatial unknowns in the discretizations. A direct solver has an $O(N^3)$ computational complexity (per time step for time-dependent sFPDEs). A Krylov subspace iterative solver has an $O(N^2)$ computational complexity per matrix-

Acknowledgement: This work is funded by the OSD/ARO MURI Grant W911NF-15-1-0562 and by the National Science Foundation under Grant DMS-1620194.

Hong Wang, Department of Mathematics, University of South Carolina, Columbia, South Carolina 29208, USA, e-mail: hwang@math.sc.edu

https://doi.org/10.1515/9783110571684-012

vector multiplication and a large number of iterations is often needed, and may diverge even for very simple problems due to the impact of round-off errors (cf. Section 5.2). This is in sharp contrast to the numerical methods for integer-order PDEs, which yield sparse stiffness matrices that have a memory requirement of $O(N)$ and can be inverted in $O(N)$ or $O(N \log N)$ operations (per time step for time-dependent problems) [45, 57].

The significantly increased computational complexity and memory requirements of numerical methods for sFPDEs have a profound impact computationally. For example, if we use a finite difference method (FDM) or finite element method (FEM) with 1000 by 1000 by 1000 nodes to solve an integer-order PDE in three space dimensions, a multi-grid solver can solve the problem of the order of 10^9 operations. However, if we use a similar method with the same number of unknowns and time steps to solve a three-dimensional sFPDE, a direct solver would have a computational complexity of the order 10^{27} (per time step for a time-dependent problem). This will take a state-of-the-art supercomputer at least years of CPU time to finish the simulation. Here we have not taken into account the extra computational complexity due to the non-linearity and the evaluation and assembly of the stiffness matrices in the numerical simulation. The significantly increased computational complexity and memory requirements of FPDE models significantly limit their applications in reality, especially for problems in three space dimensions. Therefore, the development of fast and accurate numerical methods with efficient storage for FPDEs is of fundamental significance for their applications and requires rigorous mathematical, numerical, and computational studies.

In this chapter we discuss the recent developments of fast and accurate, matrix-free numerical methods for sFPDEs and we address some related issues. The rest of the chapter is organized as follows. In Section 2 we study some representative FDMs for a one-dimensional time-dependent sFPDE in detail and address related numerical, mathematical, and, especially, computational issues. Then we discuss the extensions to some other FDMs. In Sections 3 and 4 we outline the extension of the fast numerical methods in Section 2 to multi-dimensional sFPDEs in rectangular and general convex domains, respectively. In Section 5 we outline the development of fast FEMs and fast finite volume methods (FVMs) for sFPDEs and address the issue of pre-conditioners. In Section 6 we demonstrate that the solution to linear elliptic sFPDEs with smooth data in one space dimension may have boundary layer and low singularity, which is in sharp contrast to integer-order linear elliptic PDEs, and its impact on computational meshes. We then outline the development of fast numerical methods on the locally refined composite meshes. In Section 7 we provide concluding remarks and discuss possible future directions.

2 Fast FDMs for one-dimensional sFPDEs

We consider the following fractional diffusion equation (FDE) of order $1 < \alpha < 2$ [29, 30]:

$$\begin{aligned} &u_t - k_+(x,t)\,{}_a^G D_x^\alpha u - k_-(x,t)\,{}_x^G D_b^\alpha u = f(x,t), \quad (x,t) \in (a,b)\times(0,T], \\ &u(x,0) = u_0(x), \quad x \in [a,b], \quad u(a,t) = u(b,t) = 0, \quad t \in [0,T]. \end{aligned} \tag{1}$$

Here k_+ and k_- are the left and right fractional diffusivity, f is the source term, and the left and right Grünwald–Letnikov fractional derivatives ${}_a^G D_x^\alpha u$ and ${}_x^G D_b^\alpha u$ are defined by [39]

$$\begin{aligned} {}_a^G D_x^\alpha u(x,t) &:= \lim_{\varepsilon\to 0^+} \frac{1}{\varepsilon^\alpha} \sum_{k=0}^{\lfloor (x-a)/\varepsilon \rfloor} g_k^{(\alpha)} u(x-k\varepsilon, t), \\ {}_x^G D_b^\alpha u(x,t) &:= \lim_{\varepsilon\to 0^+} \frac{1}{\varepsilon^\alpha} \sum_{k=0}^{\lfloor (b-x)/\varepsilon \rfloor} g_k^{(\alpha)} u(x+k\varepsilon, t). \end{aligned} \tag{2}$$

Here $\lfloor x \rfloor$ is the floor of x and $g_k^{(\alpha)} = (-1)^k \binom{\alpha}{k}$ with

$$\binom{\alpha}{k} := \frac{\alpha(\alpha-1)\cdots(\alpha-k+1)}{k!}$$

being the fractional binomial coefficients.

2.1 An unconditionally stable shifted FDM

The variable-coefficient sFPDE (1) cannot be solved analytically. Hence, numerical means have to be used. Let M and N be positive integers. We define a uniform space-time partition

$$\begin{aligned} x_i &:= a + ih, \quad 0 \le i \le N+1, \quad h := \frac{b-a}{N+1}, \\ t^m &:= m\tau, \qquad 0 \le m \le M, \qquad \tau := \frac{T}{M}. \end{aligned} \tag{3}$$

It was proven in [29, 30] that the implicit FDM via a direct truncation of the series (2) is unconditionally unstable, which is in sharp contrast to its integer-order analog that is well known to be unconditionally stable. It was also proven in [29, 30] that the Meerschaert–Tadjeran FDM

$$\begin{aligned} &\frac{u_i^m - u_i^{m-1}}{\tau} - \frac{k_i^{+,m}}{h^\alpha} \sum_{l=0}^{i} g_l^{(\alpha)} u_{i-l+1}^m - \frac{k_i^{-,m}}{h^\alpha} \sum_{l=0}^{N-i+1} g_l^{(\alpha)} u_{i+l-1}^m = f_i^m, \\ &\quad 1 \le i \le N,\ 1 \le m \le M, \end{aligned} \tag{4}$$

which is obtained by a shifting of the unstable implicit FDM, with $k_i^{\pm,m} := k_\pm(x_i, t^m)$, $f_i^m := f(x_i, t^m)$, and u_i^m being the finite difference approximation to $u(x_i, t^m)$, is unconditionally stable and has first-order convergence rates in space and time.

The FDM (4) can be expressed in a matrix form

$$(I + \tau A^m)\boldsymbol{u}^m = \boldsymbol{u}^{m-1} + \tau \boldsymbol{f}^m, \quad 1 \le m \le M. \tag{5}$$

Here $\boldsymbol{u}^m := [u_1^m, \dots, u_N^m]^T$, $\boldsymbol{f}^m := [f_1^m, \dots, f_N^m]^T$, $\boldsymbol{I}$ is the identity matrix of order N, and $\boldsymbol{A}^m = [a_{i,j}^m]_{i,j=1}^N$ is the stiffness matrix of the form

$$a_{i,j}^m := \frac{1}{h^\alpha} \begin{cases} -(k_i^{+,m} + k_i^{-,m})g_1^{(\alpha)} > 0, & j = i, \\ -(k_i^{+,m} g_2^{(\alpha)} + k_i^{-,m} g_0^{(\alpha)}) < 0, & j = i-1, \\ -(k_i^{+,m} g_0^{(\alpha)} + k_i^{-,m} g_2^{(\alpha)}) < 0, & j = i+1, \\ -k_i^{+,m} g_{i-j+1}^{(\alpha)} < 0, & j < i-1, \\ -k_i^{-,m} g_{j-i+1}^{(\alpha)} < 0, & j > i+1. \end{cases} \tag{6}$$

It is clear from (6) that A^m is a full matrix, which has to be assembled in any traditional method, having an $O(N^2)$ memory requirement. A direct solver has $O(N^3)$ computational complexity (per time step) and a Krylov subspace iterative solve has $O(N^2)$ computational complexity per matrix-vector multiplication and a large number of iterations often has to be used.

It is clear that $g_k^{(\alpha)} := (-1)^k \binom{\alpha}{k}$ satisfy the following properties:

$$\begin{aligned} & g_1^{(\alpha)} = -\alpha < 0, \quad 1 = g_0^{(\alpha)} > g_2^{(\alpha)} > g_3^{(\alpha)} > \cdots > 0, \\ & \sum_{k=0}^{\infty} g_k^{(\alpha)} = 0, \quad \sum_{k=0}^{m} g_k^{(\alpha)} < 0 \quad (m \ge 1), \\ & g_k^{(\alpha)} = \frac{\Gamma(k-\alpha)}{\Gamma(-\alpha)\Gamma(k+1)} = \frac{1}{\Gamma(-\alpha)k^{\alpha+1}}\left(1 + O\left(\frac{1}{k}\right)\right). \end{aligned}$$

Hence, A^m is a strongly diagonally dominant M-matrix [30, 53], as

$$\begin{aligned} a_{i,i}^m - \sum_{j=1, j\neq i}^{N} |a_{i,j}^m| &= \frac{1}{h^\alpha}\left[-(k_i^{+,m} + k_i^{-,m})g_1^{(\alpha)} - k_i^{+,m} \sum_{l=0,l\neq 1}^{i} g_l^{(\alpha)} - k_i^{-,m} \sum_{l=0,l\neq 1}^{N-i} g_l^{(\alpha)}\right] \\ &> \frac{1}{h^\alpha}\left[-(k_i^{+,m} + k_i^{-,m})g_1^{(\alpha)} - (k_i^{+,m} + k_i^{-,m}) \sum_{l=0,l\neq 1}^{\infty} g_l^{(\alpha)}\right] = 0. \end{aligned}$$

Consequently, the FDM (4) satisfies the maximum principle and is monotone, yielding unconditional stability and an error estimate of the FDM in the discrete L^∞ norm under the assumption that the true solution of (1) is smooth. In contrast, the stiffness matrix of the unstable implicit FDM is not diagonally dominant, which partially explains its unconditional instability. In the following example, we present an alternative heuristic reasoning about the stability issue.

Example 1. Consider a one-sided steady-state version of (1) with $k_+ = 1$, $k_- = 0$, $f = 0$, and $(a, b) = (0, 1)$. Because the Grünwald–Letnikov fractional derivative ${}^G_aD^\alpha_x u$ coincides with the Riemann–Liouville fractional derivative ${}^R_aD^\alpha_x u$ [39], we rewrite the equation as

$$ {}^R_0D^\alpha_x u := D^2{}_0I^{2-\alpha}_x u = 0, \quad x \in (0,1), \qquad u(0) = 0, \quad u(1) = 1, \quad 1 < \alpha < 2. \tag{7} $$

Here ${}_aI^\beta_x$ and ${}_xI^\beta_b$, with $0 < \beta < 1$, are defined by [39]

$$ \begin{aligned} {}_aI^\beta_x w(x) &:= \frac{1}{\Gamma(\beta)} \int_a^x \frac{w(s)}{(x-s)^{1-\beta}} ds, \\ {}_xI^\beta_b w(x) &:= \frac{1}{\Gamma(\beta)} \int_x^b \frac{w(s)}{(s-x)^{1-\beta}} ds, \end{aligned} \tag{8} $$

where $\Gamma(\cdot)$ is the Gamma function.

Integrating the governing equation (7) twice yields

$$ {}_0I^{2-\alpha}_x u = C_1 x + C_0, \quad x \in (0,1). $$

We apply ${}_0I^{\alpha-1}_x$ to both sides of the equation and use the semi-group property of fractional integral operators and the formula

$$ {}_aI^\beta_x {}_aI^\gamma_x = {}_aI^{\beta+\gamma}_x, \quad {}_0I^\gamma_x x^\mu = \frac{\Gamma(\mu+1)}{\Gamma(\gamma+\mu+1)} x^{\gamma+\mu}, \quad 0 < \gamma < 1, \ \mu > -1, \tag{9} $$

with $\gamma = \alpha - 1$ and $\mu = 0, 1$, respectively, to get

$$ {}_0I_x u = {}_0I^{\alpha-1}_x (C_1 x + C_0) = \frac{C_1\, x^\alpha}{\Gamma(\alpha+1)} + \frac{C_0\, x^{\alpha-1}}{\Gamma(\alpha)}. $$

Differentiating the equation and enforcing both the boundary conditions in (7) yields the unique analytical solution to (7)

$$ u = x^{\alpha-1}, \quad x \in (0,1). $$

However, the directly truncated unstable implicit FDM yields a one-sided discretization, which is determined completely by the boundary condition at $x = 0$, yielding the trivial numerical solution $u_i = 0$ for $i = 1, 2, \ldots, N$. This is physically inconsistent with the problem (7) that needs to be determined by both boundary conditions. However, the shifted FDM (4) introduces at least one unknown in the other direction and so a two-way coupling, which has to be closed by both the boundary conditions. Hence, the shifted FDM is physically consistent with the continuous problem (7).

2.2 A lossless and matrix-free fast FDM with efficient storage

Theorem 2.1 ([53]). *The stiffness matrix A^m in (6) can be decomposed as*

$$
\begin{aligned}
A^m &= [K_+^m T^{\alpha,N} + K_-^m (T^{\alpha,N})^T]/h^\alpha, \\
K_+^m &:= \operatorname{diag}(\{k_i^{+,m}\}_{i=1}^N), \quad K_-^m := \operatorname{diag}(\{k_i^{-,m}\}_{i=1}^N), \\
T^{\alpha,N} &:= -\begin{bmatrix}
g_1^{(\alpha)} & g_0^{(\alpha)} & 0 & \dots & 0 & 0 \\
g_2^{(\alpha)} & g_1^{(\alpha)} & g_0^{(\alpha)} & \ddots & \ddots & 0 \\
\vdots & g_2^{(\alpha)} & g_1^{(\alpha)} & \ddots & \ddots & \vdots \\
\vdots & \ddots & \ddots & \ddots & \ddots & 0 \\
g_{N-1}^{(\alpha)} & \ddots & \ddots & \ddots & g_1^{(\alpha)} & g_0^{(\alpha)} \\
g_N^{(\alpha)} & g_{N-1}^{(\alpha)} & \dots & \dots & g_2^{(\alpha)} & g_1^{(\alpha)}
\end{bmatrix};
\end{aligned}
\tag{10}
$$

$A^m \boldsymbol{v}$ can be evaluated in $O(N \log N)$ operations in a lossless and matrix-free manner for any vector $\boldsymbol{v} \in \mathbb{R}^N$, and A^m can be stored in $O(N)$ memory.

Proof. By (10), to store the matrix A^m, we need only to store $k_i^{\pm,m}$ for $i = 1, \dots, N$ and $g_k^{(\alpha)}$ for $i = 0, 1, \dots, N$, i. e., $(3N + 1)$ parameters.

The matrix $T^{\alpha,N}$ can be embedded into a $2N \times 2N$ circulant matrix $C^{\alpha,2N}$, i. e.,

$$
C^{\alpha,2N} := \begin{bmatrix} T^{\alpha,N} & S^{\alpha,N} \\ S^{\alpha,N} & T^{\alpha,N} \end{bmatrix}, \quad S^{\alpha,N} := \begin{bmatrix}
0 & g_N^{(\alpha)} & \dots & \dots & g_3^{(\alpha)} & g_2^{(\alpha)} \\
0 & 0 & g_N^{(\alpha)} & \dots & \ddots & g_3^{(\alpha)} \\
0 & 0 & 0 & \ddots & \ddots & \vdots \\
\vdots & \ddots & \ddots & \ddots & \ddots & \vdots \\
0 & \dots & 0 & \ddots & 0 & g_N^{(\alpha)} \\
g_0^{(\alpha)} & 0 & \dots & 0 & 0 & 0
\end{bmatrix}.
$$

Let $\boldsymbol{c}^{\alpha,2N}$ be the first column vector of $C^{\alpha,2N}$. Then $C^{\alpha,2N}$ can be diagonalized by the $2N$-by-$2N$ discrete Fourier transform matrix as [6, 7, 33]

$$
C^{\alpha,2N} = F_{2N}^{-1} \operatorname{diag}(F_{2N} \boldsymbol{c}^{\alpha,2N})\, F_{2N}. \tag{11}
$$

For any $\boldsymbol{v} \in \mathbb{R}^N$, the matrix-vector multiplication $T^{\alpha,N}\boldsymbol{v}$ can be evaluated in a fast and lossless manner as follows. (i) Define $\boldsymbol{v}_{2N}$ by

$$
\boldsymbol{v}_{2N} = \begin{bmatrix} \boldsymbol{v} \\ 0 \end{bmatrix}, \quad C^{\alpha,2N} \boldsymbol{v}_{2N} = \begin{bmatrix} T^{\alpha,N} & S^{\alpha,N} \\ S^{\alpha,N} & T^{\alpha,N} \end{bmatrix} \begin{bmatrix} \boldsymbol{v} \\ 0 \end{bmatrix} = \begin{bmatrix} T^{\alpha,N}\boldsymbol{v} \\ S^{\alpha,N}\boldsymbol{v} \end{bmatrix}. \tag{12}
$$

(ii) Use (11) to evaluate $C^{\alpha,2N}\boldsymbol{v}_{2N}$ in $O(N \log N)$ operations, since $F_{2N}\boldsymbol{v}_{2N}$ can be carried out in $O(N \log N)$ operations via the fast Fourier transform (FFT). (iii) By (12), the first

N entries of $C^{\alpha,2N}\mathbf{v}_{2N}$ yield $T^{\alpha,N}\mathbf{v}$. (iv) Similarly, $(T^{\alpha,N})^T\mathbf{v}$ can be evaluated in $O(N\log N)$ operations. (v) By (10), $A^m\mathbf{v}$ can be evaluated in $O(N\log N)$ operations. We thus finish the proof of the theorem. □

Remark 2.1. The fundamental reason why the fast matrix-vector multiplication algorithm with efficient storage can be developed lies in the decomposition (10), which bridges the FPDE community and the numerical linear algebra community, especially those working on structured dense matrices [6, 7, 8, 33, 41]. The Toeplitz structure of $T^{\alpha,N}$ (and $(T^{\alpha,N})^T$) reflects the translation invariance property of the fractional finite difference operators in (4). The impact of the variable fractional diffusivity $\{k_i^{\pm}\}_{i=1}^N$, which are not translation-invariant, is reflected in the non-Toeplitz diagonal matrices $K_{\pm}^m$. Consequently, $A^m\mathbf{v}$ has the same computational complexity as $T^{\alpha,N}\mathbf{v}$ of $O(N\log N)$.

The Meerschaert–Tadjeran FDM (4) has only first-order accuracy in both space and time due to the use of the shifted approximation. High-order FDMs were developed subsequently in the literature. A Crank–Nicolson FDM [43] has second-order accuracy in time and uses a Richardson extrapolation of the shifted finite difference approximation in space to recover the second-order accuracy in space. The method requires the numerical solution at a finer mesh size of $h/2$ and so doubles the number of unknowns and thereby increases the computational cost and memory requirements, but retains the maximum principle and is monotone. Alternatively, Tian et al. used a weighted shifted finite difference approximation to recover second- (or higher-)order accuracy without halving the spatial mesh size [44]. It is not clear whether the scheme satisfies the maximum principle for any $0 \le \gamma \le 1$.

Since the fractional finite difference operators in these high-order FDMs are translation-invariant, fast solvers can be developed in parallel to Theorem 2.1 (e. g., refer to [3] for a fast Crank–Nicolson FDM).

3 Fast FDMs for multi-dimensional sFPDEs in cuboidal domains

We take the two-dimensional sFPDE in a rectangular domain $\Omega := (a,b)\times(c,d)$ [28, 42] to outline the development of the fast FDMs

$$\begin{aligned}
&u_t - k_{x,+}(x,y,t)\,{}_a^G D_x^{\alpha} u - k_{x,-}(x,y,t)\,{}_x^G D_b^{\alpha} u - k_{y,+}(x,y,t)\,{}_c^G D_y^{\beta} u - k_{y,-}(x,y,t)\,{}_y^G D_d^{\beta} u \\
&\quad = f(x,y,t), \quad (x,y,t) \in \Omega\times(0,T], \\
&u(x,y,0) = u_0(x,y), \quad (x,y)\in\overline{\Omega}, \\
&u(x,y,t) = 0, \quad (x,y,t)\in\partial\Omega\times[0,T], \quad 0<\alpha,\beta<1.
\end{aligned} \tag{13}$$

Here $k_{x,\pm}$ and $k_{y,\pm}$ are the left and right fractional diffusivity in the x and y directions, respectively. The fractional derivatives are defined as in (2).

Let N_1 and N_2 be two positive integers and $N := N_1N_2$. We define a uniform spatial partition

$$\begin{aligned} x_i &:= a + ih_1, \quad 0 \le i \le N_1 + 1, \quad h_1 := \frac{b-a}{N_1+1}, \\ y_j &:= c + ih_2, \quad 0 \le j \le N_2 + 1, \quad h_2 := \frac{d-c}{N_2+1}. \end{aligned} \tag{14}$$

The following two-dimensional Meerschaert–Tadjeran FDM was developed in [28]:

$$\begin{aligned} &\frac{u_{i,j}^m - u_{i,j}^{m-1}}{\tau} - \frac{k_{x,i,j}^{+,m}}{h^\alpha}\sum_{l=0}^{i} g_l^{(\alpha)} u_{i-l+1,j}^m - \frac{k_{x,i,j}^{-,m}}{h^\alpha}\sum_{l=0}^{N_1-i+1} g_l^{(\alpha)} u_{i+l-1,j}^m \\ &- \frac{k_{y,i,j}^{+,m}}{h^\beta}\sum_{l=0}^{j} g_l^{(\beta)} u_{i,j-l+1}^m - \frac{k_{y,i,j}^{-,m}}{h^\beta}\sum_{l=0}^{N_2-j+1} g_l^{(\alpha)} u_{i,j+l-1}^m = f_{i,j}^m, \\ &1 \le i \le N_1,\ 1 \le j \le N_2,\ 1 \le m \le M. \end{aligned} \tag{15}$$

Meerschaert et al. [28, 42] developed and analyzed an alternating direction implicit (ADI) FDM for problem (13). Starting from each time step t^{m-1} for $m = 1, 2, \dots, M$, the method splits the FDM (15) as a family of one-dimensional subsystems of the form (4) in terms of the unknowns $u_{i,j}^{m-1/2}$ in the x direction for $i = 1, 2, \dots, N_1$ while lagging the unknowns in the y direction at the time step t^{m-1} for each $j = 1, 2, \dots, N_2$. Next, the ADI method splits the FDM (15) as the other family of one-dimensional subsystems of the form (4) in terms of the unknowns $u_{i,j}^m$ in the y direction for $j = 1, 2, \dots, N_2$ while lagging the unknowns in the x direction at the time step $t^{m-1/2}$ for each $i = 1, 2, \dots, N_1$.

A direct solver of each subsystem in the first family, which has a full stiffness matrix of size N_1, has a computational complexity of $O(N_1^3)$ and memory requirement $O(N_1^2)$ for each $j = 1, 2, \dots, N_2$, leading to a computational complexity of $O(N_1^2N)$ and memory requirement of $O(N_1N)$ for this family of subsystems. By symmetry, a direct solve of the other family has a computational complexity of $O(N_2^2N)$ and a memory requirement of $O(N_2N)$. This leads to an overall computational complexity of $O(N(N_1^2 + N_2^2)) = O(N^2)$ and a memory requirement of $O(N(N_1 + N_2)) = O(N^{3/2})$, if N_1 and N_2 are of the same order [28]. A fast ADI method was developed in [52], in which the fast FDM in Section 2.2 was used to solve each of the subsystems in the two families, leading to an overall computational complexity of $O(N \log N_1) + O(N \log N_2) = O(N \log N)$. Numerical experiments show the strong potential of the fast ADI method.

Like their counterparts for integer-order PDEs [46], ADI methods often require the one-dimensional finite difference operators in the x direction and in the y direction to commute [28]. This condition is satisfied for (13), say, if $k_{x,\pm}(x, y, t) = k_x^\pm(x, t)$ and $k_{y,\pm}(x, y, t) = k_y^\pm(y, t)$, but does not hold for general variable fractional diffusivity $k_{x,\pm}(x, y, t)$ and $k_{y,\pm}(x, y, t)$. Moreover, because it is an approximation to the original

FDM (15), the ADI FDM does not retain certain (say, conservation) properties of the original FDM.

A fast solver was developed for the FDM (15) in [47]. Let $\boldsymbol{u}^m$ and $\boldsymbol{f}^m$ be the N-dimensional vectors defined by

$$\boldsymbol{u}^m := [u_{1,1}^m, \dots, u_{N_1,1}^m, u_{1,2}^m, \dots, u_{N_1,2}^m, \dots, u_{1,N_2}^m, \dots, u_{N_1,N_2}^m]^T,$$
$$\boldsymbol{f}^m := [f_{1,1}^m, \dots, f_{N_1,1}^m, f_{1,2}^m, \dots, f_{N_1,2}^m, \dots, f_{1,N_2}^m, \dots, f_{N_1,N_2}^m]^T.$$

The FDM (15) can be expressed in the matrix form (5). Due to the non-local coupling of the fractional finite difference operators in the x and y directions, the stiffness matrix A^m is dense. To analyze its structure, we split A^m as

$$A^m = A^{m,x} + A^{m,y}, \tag{16}$$

in which $A^{m,x}$ accounts for the coupling of all the nodes in the x direction and $A^{m,y}$ accounts for that in the y direction.

Note that the non-zero entries at each row of the matrix $A^{m,x}$ corresponding to the node (x_i, y_j) for $i = 1, \dots, N_1$ and $j = 1, \dots, N_2$ are indexed by (x_l, y_j) for $l = 1, \dots, N_1$. Hence, $A^{m,x}$ is a block-diagonal matrix of size N_2,

$$\boldsymbol{A}^{m,x} = \operatorname{diag}(\{A_j^{m,x}\}_{j=1}^{N_2}), \tag{17}$$

in which each diagonal matrix block $A_j^{m,x}$ is a full matrix of order N_1 of the form (6). By Theorem 2.1, $A^{m,x}$ can be stored in $N_2 O(N_1) = O(N)$ memory and $A^{m,x}\boldsymbol{v}$ can be evaluated in $N_2 O(N_1 \log N_1) = O(N \log N)$ operations in a lossless and matrix-free manner for each vector $\boldsymbol{v} \in \mathbb{R}^N$.

On the other hand, each row of the matrix $A^{m,y}$ has coupling to a node in each of the horizontal lines $y = y_j$ for $j = 1, 2, \dots, N_2$ with the same x coordinate. Namely, $\boldsymbol{A}^{m,y}$ is a full block matrix with sparse matrix blocks. In fact, $\boldsymbol{A}^{m,y}$ is a block-Toeplitz-circulant-block matrix of the form [47]

$$\begin{aligned}
A^{m,y} &= [K_+^{m,y}(T^{\beta,N_2} \otimes \boldsymbol{I}_{N_1}) + K_-^{m,y}((T^{\beta,N_2})^T \otimes \boldsymbol{I}_{N_1})]/h_2^\beta,\\
K_+^{m,y} &:= \operatorname{diag}(\{\operatorname{diag}(\{k_{y,i,j}^{+,m}\}_{i=1}^{N_1})\}_{j=1}^{N_2}),\\
K_-^{m,y} &:= \operatorname{diag}(\{\operatorname{diag}(\{k_{y,i,j}^{-,m}\}_{i=1}^{N_1})\}_{j=1}^{N_2}).
\end{aligned} \tag{18}$$

Consequently, $\boldsymbol{A}^{m,y}$ can be stored in $O(N)$ memory and $A^{m,y}\boldsymbol{v}$ can be evaluated in $O(N \log N)$ operations in a lossless and matrix-free manner for each vector $\boldsymbol{v} \in \mathbb{R}^N$ [47]. The fast FDMs have been extended to three-dimensional sFPDEs and sFPDEs with fractional derivative boundary conditions. We refer interested readers to, e. g., [19, 50, 51, 60] for details.

Table 1: Performance of the FDM and the fast FDM for Example 2.

$h = \tau$	# of nodes	CPU of the FDM	CPU of the fast FDM
2^{-4}	32,768	86 days	17 seconds
2^{-5}	262,144	N/A	5 minutes
2^{-7}	16,777,216	N/A	28 hours

Example 2. A three-dimensional extension of the Meerschaert–Tadjeran FDM (15) is used to solve a three-dimensional analog of (13) of order 1.8 with a fractional diffusivity 0.005 and $f = 0$ in the domain $\Omega = (-1,1)^3$ over the time period $[0,T] = [0,1]$. The true solution is given by the fundamental solution of the FPDE, which is expressed in terms of the inverse Fourier transform

$$u(x_1,x_2,x_3,t) := \frac{1}{\pi^3}\prod_{i=1}^{3}\int_0^\infty e^{-0.01|\cos(\frac{9\pi}{10})|(t+0.5)\xi_i^{1.8}}\cos(\xi_i x_i)d\xi_i.$$

In the numerical experiments, the Meerschaert–Tadjeran FDM was solved by a direct solver and the fast FDM was solved by the conjugate gradient squared (CGS) method. Both were implemented in Fortran 90 on a workstation of 120 GB of memory.

We present the numerical results in Table 1. The finest mesh size which the conventional FDM can use is $h = 1/16$ on that workstation that leads to a total of 32,768 unknowns per time step, as the method has to assemble the stiffness matrix. The method needed a calculation time of almost three months on that workstation. With the same mesh size and on the same workstation, the fast FDM consumed about 17 seconds. Moreover, the fast FDM uses a lossless matrix-free manner to solve the problem and hence has a linear memory requirement. We were able to use the fast FDM to solve the same problem on the same workstation at a much finer mesh size of $h = 2^{-7}$ to solve the problem in one day and four hours of calculation time. This would take the conventional FDM at least years of CPU time on a state-of-the-art supercomputer to finish the simulation. This simple example shows the strong potential of the fast FDM.

4 Fast FDMs for sFPDEs in convex domains

To date, most FDMs developed for multi-dimensional sFPDEs were restricted to rectangular domains, in which the stiffness matrices have certain tensor product forms of (block) Toeplitz-like structures. A fast FDM was developed in [21] for variable-coefficient multi-term sFPDEs or distributed-order sFPDEs in general convex domains in the plane. The stiffness matrices of these FDMs do not have a (block) Toeplitz-like structure any longer.

In this section we outline the development of the fast FDM for the time-dependent sFPDE on a two-dimensional convex domain. We have

$$\begin{aligned} &u_t - \left[k_{x,+}(x,y,t)\ {}_{a(y)}D_x^{\alpha}u(x,y,t) + k_{x,-}(x,y,t)\ {}_xD_{b(y)}^{\alpha}u(x,y,t)\right] \\ &\quad - \left[k_{y,+}(x,y,t)\ {}_{c(x)}D_y^{\beta}u(x,y,t) + k_{y,-}(x,y,t)\ {}_yD_{d(x)}^{\beta}u(x,y,t)\right] \\ &= f(x,y,t), \quad (x,y) \in \Omega, \quad t \times (0,T], \\ &u(x,y,0) = u_0(x,y), \quad (x,y) \in \Omega, \\ &u(x,y,t) = 0, \quad (x,y) \in \partial\Omega, \quad t \in [0,T]. \end{aligned}$$

Here Ω is a simply connected bounded convex domain in the plane. Let $a(y)$ and $b(y)$ represent the left and right boundary of Ω at given y and $a := \min a(y)$ and $b := \max a(y)$, and let $c(x)$ and $d(x)$ represent the lower and upper boundary of Ω at given x and $c := \min c(x)$ and $d := \max d(x)$. Then Ω is inscribed in the rectangular domain $(a,b) \times (c,d)$.

The left- and right-sided fractional derivatives are defined by

$$\begin{aligned} {}_{a(y)}D_x^{\alpha}u(x,y,t) &:= \lim_{\varepsilon\to0^+}\frac{1}{\varepsilon^{\alpha}}\sum_{k=0}^{\lfloor (x-a(y))/\varepsilon\rfloor} g_k^{(\alpha)}u(x-k\varepsilon,y,t), \\ {}_xD_{b(y)}^{\alpha}u(x,y,t) &:= \lim_{\varepsilon\to0^+}\frac{1}{\varepsilon^{\alpha}}\sum_{k=0}^{\lfloor (b(y)-x)/\varepsilon\rfloor} g_k^{(\alpha)}u(x+k\varepsilon,y,t), \end{aligned}$$

with the fractional derivatives in the y direction being defined by symmetry. Because Ω is a general convex domain, the lower limits of the left fractional derivatives and the upper limits of the right fractional derivatives in the x and y directions depend on y and x, respectively. This is in contrast to the case of a rectangular domain in which the lower and upper limits of the fractional derivatives are constant [28].

We still use the uniform spatial partition (14) and temporal partition (3). Since some of the nodes (x_i, y_j) defined by (14) may stay outside of the domain $\overline{\Omega}$, we let $\Omega_h := \Omega \cap \{(x_i, y_j)\}_{1\le i\le N_1;1\le j\le N_2}$ denote the set of the nodes that are contained in Ω. For $j = 1, \ldots, N_2$, let $(x_{i_1(j)}, y_j)$ and $(x_{i_2(j)}, y_j)$ defined by (14) denote the left-most and right-most nodes in $\overline{\Omega}$ in the line segment $y = y_j$. Because Ω is convex, the set Ω_h is connected, i. e., all the nodes between $(x_{i_1(j)}, y_j)$ and $(x_{i_2(j)}, y_j)$ are also in Ω. By symmetry, for $i = 1, \ldots, N_1$, let the nodes $(x_i, y_{j_1(i)})$ and $(x_i, y_{j_2(i)})$ denote the lower-most and upper-most nodes in $\overline{\Omega}$ on the line segment $x = x_i$. Then all the nodes in between are also in Ω_h. Let $\Omega_h^r \subset \Omega_h$ denote the subset of the regular interior nodes, i. e., the nodes of Ω_h such that all their four neighboring nodes are also in $\overline{\Omega}$. The index set Π_h of all the nodes in Ω_h^r and the total number of nodes in Π_h on which a finite difference equation needs to be set up are given by

$$\begin{aligned} \Pi_h &:= \{(i,j) : i_1(j) + 1 \le i \le i_2(j) - 1,\ 1 \le j \le N_2 - 1\}, \\ N &:= |\Pi_h| = \sum_{j=1}^{N_2-1} n_j, \quad n_j := i_2(j) - i_1(j) - 1. \end{aligned}$$

With these preparations, the FDM (15) can be formulated as follows:

$$\begin{aligned}&\frac{u_{i,j}^{m}-u_{i,j}^{m-1}}{\tau}-\frac{k_{x,i,j}^{+,m}}{h_1^{\alpha}}\sum_{k=0}^{i-i_1(j)+1} g_k^{(\alpha)}u_{i-k+1,j}^{m}-\frac{k_{x,i,j}^{-,m}}{h_1^{\alpha}}\sum_{k=0}^{i_2(j)-i+1} g_k^{(\alpha)}u_{i+k-1,j}^{m}\\&\quad-\frac{k_{y,i,j}^{+,m}}{h_2^{\beta}}\sum_{k=0}^{j-j_1(i)+1} g_k^{(\beta)}u_{i,j-k+1}^{m}-\frac{k_{y,i,j}^{-,m}}{h_2^{\beta}}\sum_{k=0}^{j_2(i)-j+1} g_k^{(\beta)}u_{i,j+k-1}^{m}=f_{i,j}^{m},\\&\quad\forall(i,j)\in\Pi_h.\end{aligned}\tag{19}$$

Unlike the case of a rectangular domain, the “boundary” nodes in the FDM (19) do not necessarily lie on the boundary $\partial\Omega$ but their distances from the boundary $\partial\Omega$ are less than h_1 or h_2, to which we enforce the homogeneous Dirichlet boundary condition. Then Taylor expansion shows that the values of u at these nodes are of order $O(h_1)$ or $O(h_2)$. In other words, the approximation of the Dirichlet boundary condition yields a first-order approximation, which is consistent with the accuracy of the FDM.

It is clear that the stiffness matrix A^m of the FDM (19) is a dense matrix due to the non-local coupling of the fractional finite difference operators in the x and y directions. However, a significant difference of the FDM (19) from its analog on a rectangular domain (15) is that A^m does not have a tensor product form of Toeplitz-like structure. Hence, the development in Section 3 does not apply.

To develop a fast solver for the FDM (19), we still split A^m as in (16) and $A^{m,x}$ is still of the form (17). However, each diagonal matrix block $A_j^{m,x}$ has a size $n_j := i_2(j)-i_1(j)-1$, which may vary with respect to j. Nevertheless, any $\boldsymbol{v}\in\mathbb{R}^N$ can be expressed in the form

$$\boldsymbol{v}=\left[\boldsymbol{v}_1^T,\boldsymbol{v}_2^T,\dots,\boldsymbol{v}_{N_2}^T\right]^T,\quad \boldsymbol{v}_j=[v_{i_1(j)+1,j},\dots,u_{i_2(j)-1,j}]^T,\quad 1\le j\le N_2.$$

Then $\boldsymbol{A}^{m,x}\boldsymbol{v}$ can be evaluated via the formula

$$\boldsymbol{A}_{\alpha}^{m,x}\boldsymbol{v}=\left[(A_1^{m,x}\boldsymbol{v}_1)^T,(A_2^{m,x}\boldsymbol{v}_2)^T,\dots,(A_{N_2}^{m,x}\boldsymbol{v}_{N_2})^T\right]^T,\tag{20}$$

in which $A_j^{m,x}\boldsymbol{v}_j$ can be evaluated in $O(n_j\log n_j)$ operations by Theorem 2.1, leading to an overall computational cost of $\sum_{j=1}^{N_2}O(n_j\log n_j)=O(N\log N)$ operations for $A^{m,x}\boldsymbol{v}$.

However, the tensor-product decomposition (18) of $A^{m,y}$ does not hold any longer. To develop a fast algorithm for evaluating $A^{m,y}\boldsymbol{v}$, we utilize the symmetry of the fractional differential operators in the x and y directions and borrow the idea of the relabeling in the ADI method in Section 3. Algorithmically, the fast evaluation of $A^{m,y}\boldsymbol{v}$ is carried out as follows. Let vector $\boldsymbol{w}$ denote the re-indexing of the vector $\boldsymbol{v}$ by labeling the nodes in the y direction first. Then

$$\boldsymbol{w}=\boldsymbol{P}\boldsymbol{v},\tag{21}$$

where $\boldsymbol{P}$ represents the permutation matrix that maps $\boldsymbol{v}$ to $\boldsymbol{w}$. Let $\boldsymbol{B}^{m,y}$ denote the analog of $\boldsymbol{A}^{m,y}$ that accounts for the spatial coupling in the finite difference schemes (19) by labeling the nodes in the y direction first. Then we have

$$\boldsymbol{A}^{m,y} = P^T \boldsymbol{B}^{m,y} P. \tag{22}$$

We combine (21) and (22) to obtain

$$\boldsymbol{A}^{m,y}\boldsymbol{v} = P^T \boldsymbol{B}^{m,y}\boldsymbol{w}.$$

The key points are as follows. (i) If we label the nodes in the y direction first, then the stiffness matrix $B^{m,y}$ is block diagonal-like (20). (ii) If we store $\boldsymbol{v}$ as a two-dimensional array corresponding to the indexing of the nodes (x_i, y_j), then the transform from $\boldsymbol{v}$ to $\boldsymbol{w}$ defined by (22) can be carried out simply by letting the index j go first in the two-dimensional array storing $\boldsymbol{v}$ and vice versa. In short, we can evaluate $A^{m,y}\boldsymbol{v}$ in $O(N \log N)$ operations in a lossless and matrix-free manner.

5 Pre-conditioned fast numerical methods for sFPDEs

The objectives of this section are twofold. (i) We address the issues in the development of FEMs and FVMs for sFPDEs. Because of the similarity in the development, we only outline the derivation for FVMs. (ii) We address the issues of pre-conditioning in the context of FVMs in parallel to FDMs and FEMs. For simplicity of presentation, we consider the following steady-state FDE of order $2-\beta$ to expose the idea:

$$\begin{aligned} &-D(k(x)(\gamma\, {}_a^C D_x^{1-\beta} u - (1-\gamma)\, {}_x^C D_b^{1-\beta} u)) = f(x), \quad x \in (a,b), \\ &u(a) = u(b) = 0, \quad 0 < \beta < 1, \quad 0 \le \gamma \le 1. \end{aligned} \tag{23}$$

We refer interested readers to [10, 11, 12, 40, 59] and the references therein for details.

The model problem (23) was obtained by incorporating a fractional Fick's law into a conventional local conservation law [9, 11, 58]. Here $0 \le \gamma \le 1$ represents the weight of forward versus backward transition probability. The left and right Caputo fractional derivatives are defined by

$${}_a^C D_x^{1-\beta} u := {}_a I_x^{\beta} Du, \quad {}_x^C D_b^{1-\beta} u := -{}_x I_b^{\beta} Du,$$

where the fractional integral operators ${}_a I_x^{\beta}$ and ${}_x I_b^{\beta}$ are defined in (8).

5.1 An FVM and its fast implementation

FEMs (and FVMs) are based on weak formulations and so apply directly to the conservative FDE (23). This is in contrast to the FDMs, which apply directly to the non-conservative FPDEs of the form (1) and (13). Although the conservative FDE (23) can be rewritten as an equivalent non-conservative FDE of the form (1) plus a fractional derivative term of order $0 < 1-\beta < 1$, the FDMs for the latter are not necessarily equivalent to the FEMs or FVMs for the former. Hence, it is not clear whether the FDMs applied to the latter retain the conservation property carried in the original conservative FDE (23). In other words, the direct applicability to conservative FDEs (23) is one of the key properties why FEMs and FVMs are desired in solving (23). Another reason why FEMs and FVMs are preferred to the FDMs based on the discretizations of the Grünwald–Letnikov fractional derivatives (2) will be addressed in Section 6.

We begin with the development of an FVM for problem (23), for which FVMs are desired due to their local conservation property that is crucial in many applications [4, 5, 58]. Let

$$a =: x_0 < x_1 < \cdots < x_i < \cdots < x_{N+1} := b$$

be a (not necessarily uniform) partition of $[a,b]$ and $x_{i-1/2} := (x_{i-1} + x_i)/2$ for $i = 1,2,\ldots,N+1$. Let $\{\phi_i\}_{i=1}^N$ be the piece-wise linear hat functions with $\phi_i(x_i) = 1$ and $\phi_i(x_j) = 0$ for $j \neq i$.

We integrate (23) over $[x_{i-1/2}, x_{i+1/2}]$ and replace the true solution u by its finite volume approximation

$$u_h = \sum_{j=1}^{N} u_j \phi_j$$

to obtain an FVM, which is expressed in a matrix form as follows:

$$\begin{aligned} &\boldsymbol{Au} = \boldsymbol{f}, \quad f_i := \int_{x_{i-1/2}}^{x_{i+1/2}} f(x)dx, \quad 1 \le i,j \le N, \\ &a_{i,j} := [k(x)(\gamma\, {}_a^C D_x^{1-\beta} \phi_j - (1-\gamma)\, {}_x^C D_b^{1-\beta} \phi_j)]_{x=x_{i+1/2}}^{x=x_{i-1/2}}. \end{aligned} \tag{24}$$

Although $\operatorname{supp}\{\phi_j\} = [x_{j-1}, x_{j+1}]$ has a compact support, ${}_a^C D_x^{1-\beta} \phi_j|_{x=x_{i+1/2}} \neq 0$ for $j \le i+1$ and ${}_x^C D_b^{1-\beta} \phi_j|_{x=x_{i-1/2}} \neq 0$ for $j \ge i-1$. Hence, the stiffness matrix A is full, which is a salient difference of the FVM from its integer-order analog.

It was proven in [48, 49] for a uniform spatial partition of the form (3) that the stiffness matrix A can be decomposed as

$$\begin{aligned} A &= \frac{1}{\Gamma(\beta+1)h^{1-\beta}}[K_L T_L^{\beta,N} + K_R T_R^{\alpha,N}], \\ K_L &:= \operatorname{diag}(\{k(x_{i-1/2})\}_{i=1}^N), \quad K_R := \operatorname{diag}(\{k(x_{i+1/2})\}_{i=1}^N), \\ T_L^{\beta,N} &= (l_{i-j}(\beta,\gamma)), \quad T_R^{\beta,N} = (r_{i-j}(\beta,\gamma)). \end{aligned} \tag{25}$$

Consequently, a fast FVM can be developed as the FDM in Section 2, which has a memory requirement of $O(N)$ and computational complexity of $O(N \log N)$ per Krylov subspace iteration.

5.2 Pre-conditioning of numerical discretizations to sFPDEs

When the coefficient matrix of a numerical discretization is ill-conditioned, the number of iterations can be large, which still leads to a high computational cost even if the computational complexity of each Krylov subspace iteration has been reduced significantly. For the steady-state FDE (23), the condition number of the stiffness matrix A of the FVM (24) on the uniform partition (3) is $\kappa(\boldsymbol{A}) = O(h^{-(2-\beta)})$. Although each Krylov subspace iteration has a computational complexity of $O(N \log N)$, the overall computational complexity is still of the order of $O(N^{2-\beta/2} \log N)$. Therefore, the development of an effective and efficient pre-conditioner for numerical methods to sFPDEs is of great importance, even if a fast matrix-vector multiplication algorithm has been developed.

For constant fractional diffusivity, the stiffness matrices A of the FDMs, FEMs, and FVMs for steady-state sFPDEs, for which circulant pre-conditioners are well developed, are Toeplitz and have an almost linear computational complexity [7, 8, 33, 41]. The discovery that the stiffness matrices of the numerical discretizations of sFPDEs are Toeplitz-like has motivated more development of circulant-like pre-conditioners [2, 26, 34, 35, 36, 37] as well as the development of the multi-grid method for sFPDEs [38]. In addition, Jiang and Xu [23, 24] extended the analysis of multi-grid and domain decomposition methods for integer-order symmetric and positive definite elliptic PDEs to the symmetric and positive definite sFPDEs in multiple space dimensions, which is a constant-coefficient multi-dimensional analog of (23) with $\gamma = 1/2$, in which the number of iterations were significantly reduced as for integer-order elliptic PDEs. However, they did address the issue of fast matrix-vector multiplications.

Here we outline the use of a superfast method, which is a direct solver for a symmetric and positive definite Toeplitz system with a proven $O(N \log^2 N)$ computational complexity [1], as a pre-conditioner for the fast FVM.

It was proven in [49] that $M := T_L^{\beta,N} + T_R^{\alpha,N}$ is a symmetric and positive definite Toeplitz matrix; hence M can be inverted in $O(N \log^2 N)$ operations. Let $K_M := \operatorname{diag}(\{k(x_i)\}_{i=1}^N)$. Then, we have

$$\begin{aligned}
K_M^{-1}[K_L T_L^{\beta,N} + K_R T_R^{\alpha,N}] &= K_M^{-1} K_L\, T_L^{\beta,N} + K_M^{-1} K_R\, T_R^{\beta,N} \\
&= K_M^{-1}[K_M + (K_L - K_M)]T_L^{\beta,N} + K_M^{-1}[K_M + (K_R - K_M)]T_L^{\beta,N} \\
&= M + K_M^{-1}[(K_L - K_M)T_L^{\beta,N} + (K_R - K_M)T_R^{\beta,N}] \\
&= M + O(h)(\|T_L^{\beta,N}\| + \|T_R^{\beta,N}\|).
\end{aligned}$$

We rewrite the matrix system (24) into the following equivalent form in which the stiffness matrix is just a perturbation of M:

$$(K_M^{-1} K_L\, T_L^{\beta,N} + K_M^{-1} K_R\, T_R^{\beta,N})\boldsymbol{u} = \Gamma(1+\beta) h^{1-\beta} \boldsymbol{f}. \tag{26}$$

Note that for variable fractional diffusivity, the FDE (23) is not self-adjoint in general, even if $\gamma = 1/2$. Hence, a non-symmetric Krylov subspace iterative method needs to be used to solve the linear system (26), in which the symmetric and positive definite Toeplitz matrix M is used as a pre-conditioner and is solved by the superfast direct solver [1] in $O(N \log^2 N)$ computations.

Example 3. We consider problem (23) with $\beta = 0.2$, $\gamma = 0.5$, $K(x) = \Gamma(1.2)(1+x)$, and $[a,b] = [0,1]$. The true solution $u(x) = x^2(1-x)^2$ and f is computed accordingly. We use the CGS method, the fast CGS (FCGS) in which we use the fast matrix-vector multiplication algorithm to evaluate $A\boldsymbol{v}$, and the superfast pre-conditioned FCGS (PFCGS) along with Gaussian elimination (Gauss) as a benchmark to solve (26). We present the numerical results in Table 2.

Table 2: Performance of Gauss, CGS, FCGS, and PFCGS for Example 3.

	Gauss			CGS		
N	$\lVert u-u_G\rVert_{L^\infty}$	CPU time		$\lVert u-u_C\rVert_{L^\infty}$	CPU time	Itr. #
2^7	1.294×10^{-5}	0.000 s		1.294×10^{-5}	0.016 s	128
2^9	7.893×10^{-7}	0.500 s		7.893×10^{-7}	3.359 s	599
2^{11}	4.030×10^{-8}	2 min 38 s		4.047×10^{-8}	21 min 13 s	2,624
2^{13}	5.783×10^{-9}	3 h 27 min		N/A	> 2 days	> 20,000
	FCGS			PFCGS		
	$\lVert u-u_F\rVert_{L^\infty}$	CPU time	Itr. #	$\lVert u-u_P\rVert_{L^\infty}$	CPU time	Itr. #
2^7	1.294×10^{-5}	0.031 s	128	1.294×10^{-5}	0.000 s	5
2^9	7.893×10^{-7}	0.578 s	576	7.893×10^{-7}	0.016 s	5
2^{11}	4.037×10^{-8}	9.953 s	1,997	4.038×10^{-8}	0.078 s	5
2^{13}	2.372×10^{-8}	2 min 52 s	7,410	4.345×10^{-9}	0.391 s	5

We make the following observations. (i) Although the CGS and FCGS should generate identical results in the ideal scenario of exact arithmetics, the numerical results show that the FCGS not only reduces the computational complexity and so the CPU time (from 3.5 h by Gauss to under 3 min) but also improves the convergence behavior of CGS, which diverges in this example due to the impact of significant round-off errors. (ii) The FCGS generates slightly less accurate results than Gauss at fine meshes due to round-off errors accumulated in the large number of iterations, even if the computational complexity of each iteration has been reduced significantly. (iii) The pre-conditioner M is optimal, so the PFCGS has an almost linear overall computational complexity of $O(N \log^2 N)$. It significantly reduces round-off errors and generates more accurate solutions than Gaussian elimination. It further reduces CPU time (from under 3 min for FCGS to under 0.4 s).

6 Singularity of solutions and local refinements

The development of [25, 54, 55, 56] linear elliptic FPDEs with smooth data on smooth domains yielding solutions with boundary layers and hence low regularity, which in turn affect the numerical discretizations to FPDEs, was not realized until recently.

Example 4. We consider problem (23) with $\gamma = 1$, $k = 1$, $f = -1$, and $(a,b) = (0,1)$ [55, 56]. We have

$$D({}_0^C D_x^{1-\beta} u) = 1, \quad x \in (0,1), \qquad u(0) = u(1) = 0.$$

This yields ${}_0 I_x^\beta Du = x + C_0$. We apply equation (9) with $\gamma = 1 - \beta$ to find

$$\int_0^x Du(s)ds = {}_0 I_x^{1-\beta}(x + C_0) = \frac{x^{2-\beta}}{\Gamma(3-\beta)} + \frac{C_0 x^{1-\beta}}{\Gamma(2-\beta)}.$$

This leads to

$$Du = \frac{x^{1-\beta}}{\Gamma(2-\beta)} + \frac{C_0}{\Gamma(1-\beta)x^\beta}, \qquad u = \frac{x^{2-\beta}}{\Gamma(3-\beta)} + \frac{C_0 x^{1-\beta}}{\Gamma(2-\beta)}.$$

We incorporate the boundary condition at $x = 1$ to get $C_0 = -1/(2-\beta)$ and

$$u(x) = \frac{x^{2-\beta} - x^{1-\beta}}{\Gamma(3-\beta)}.$$

It is clear that $u \notin W^{1,1/\beta}(0,1)$ for any $0 < \beta < 1$. In particular, $u \notin H^1(0,1)$ for any $1/2 \le \beta < 1$! This is in sharp contrast to its integer-order cousins [13, 14]. Because of the boundary layer of solutions to linear elliptic sFPDEs, their numerical approximations discretized on a uniform partition do not ensure an accurate approximation. Instead, a locally refined mesh should be used. FDMs are out of the question, as Grünwald–Letnikov fractional derivatives (2) are inherently defined on a uniform mesh. In contrast, the Riemann–Liouville and Caputo fractional derivatives offer such flexibilities.

A fast FVM discretized on a geometrically gridded mesh was derived in [18] for a one-sided version of problem (23) in which the stiffness matrix A was proven to be of the form

$$A = (K_- T_- + K_+ T_+) H_N + L, \quad H_N := \mathrm{diag}(\{h_i^{\beta-1}\}_{i=1}^N), \tag{27}$$

where T_- and T_+ are Toeplitz matrices and K_- and K_+ are diagonal matrices; L has rank one, which accounts for the fact that the last cell has the same size as its neighbor and hence does not follow the general pattern. The diagonal matrix H_N accounts for the scaling effect of different mesh sizes. It is clear that the stiffness matrix A can be stored in $O(N)$ memory and Av can be carried out in $O(N \log N)$ operations.

Remark 6.1. As discussed in Remark 2.1, the Toeplitz matrix reflects the translation invariance of the corresponding numerical discretization, which has been derived for numerical methods discretized on a uniform partition. The decomposition (27) implies that as long as the partition is translation-invariant under a certain "metric" (geometrically gridded) partition and "measured" under the diagonally scaled matrix in the current context, then one would expect a Toeplitz structure (multiplied by certain sparse matrices or Toeplitz matrices). This suggest that one may develop fast methods on more generally structured partitions.

A fast FVM was derived in [20] for problem (23) discretized on a locally refined composition mesh, in which a uniform spatial partition is introduced on the entire interval and then locally refined meshes are implemented within the cells where the true solution exhibits singularities. Such a composite mesh was used based on the following motivation. Ideally, a numerical method on an arbitrarily refined mesh offers great flexibility and effective approximation properties. However, such a method will destroy the structure of its stiffness matrix and hence efficiency for sFPDEs. Thus, such a composite mesh balances the flexibility of the meshes in the local refinement and the efficiency in the solution of the resulting linear algebraic system.

A representative scenario in which a geometrically gridded local refinement is introduced near the left boundary of the domain yields a stiffness matrix A of the form

$$\boldsymbol{A} = \begin{bmatrix} \boldsymbol{A}_{l,l} & \boldsymbol{A}_{l,r} \\ \boldsymbol{A}_{r,l} & \boldsymbol{A}_{r,r} \end{bmatrix}. \tag{28}$$

Here $A_{l,l}$ is the stiffness matrix of the FVM on the gridded mesh of size m and has the form (27), $A_{r,r}$ is the one on the uniform partition of size $N - m$ of the form (25). The matrices $A_{l,r}$ and $A_{r,l}$ are full rectangular matrices of sizes m-by-$(N-m)$ and $(N-m)$-by-m, respectively, which count for the coupling between the FVM on the gridded mesh and that on the uniform mesh, and can be decomposed as

$$\begin{aligned} A_{l,r} &= \frac{(1-\gamma)h^{\beta-1}}{\Gamma(\beta+1)}(K_l^- R_l^- - K_l^+ R_l^+), \\ A_{r,l} &= \frac{\gamma}{\Gamma(\beta+1)}(K_r^- R_r^- - K_r^+ R_r^+)H_m. \end{aligned} \tag{29}$$

Here $K_l^\pm$ and $K_r^\pm$ are diagonal matrices, $R_l^\pm$ and $R_r^\pm$ are full rectangular matrices but are not Toeplitz-like, and H_m is a diagonal matrix defined in (27). For example, the representative entries of R_l^+ are given by

$$(R_l^+)_{i,j} = 2(j + 1 - 3 \cdot 2^{i-m-1})^\beta - (j - 3 \cdot 2^{i-m-1})^\beta - (j + 2 - 3 \cdot 2^{i-m-1})^\beta.$$

It was proven in [20] that $R_l^\pm$ and $R_r^\pm$ have low-rank approximations, so A can still be stored in $O(N)$ memory and $A\boldsymbol{v}$ can be evaluated in $O(N \log N)$ operations.

7 Concluding remarks

In this chapter we have outlined some recent developments of fast numerical methods for sFPDEs discretized on certain structured meshes. These methods are derived by carefully analyzing the structure of the dense or even full stiffness matrices and decomposing these matrices as a finite combination of (block) Toeplitz-like matrices multiplied by certain sparse matrices. Consequently, FFT can be used to carry out the corresponding matrix-vector multiplication in $O(N \log N)$ operations in a lossless and matrix-free manner, and hence fast Krylov subspace iterative methods are obtained.

We note that there has been extensive research in the literature on the development of fast numerical methods to solve PDEs and integral equations, including (but not limited to) the fast multi-pole method (FMM) [15], the hierarchical matrix method [16], and the randomized matrix method [17]. Many of these methods have been applied in the development of fast numerical methods for FPDEs. Below we just briefly comment on some representative developments in the literature.

Zhao et al. [61] applied the H-matrix method to develop a fast numerical method for sFPDEs, in which the weakly singular translation-invariant kernel $(x-y)^{-\gamma}$ in the sFPDE is decomposed by a Taylor expansion as a finite combination of the products of univariant functions as long as the variable y is outside some neighborhood of the variable x, leading to a low-rank approximation of the stiffness matrix. Then a multi-grid method was developed to solve the resulting linear algebraic system. Jiang et al. [22] used the FMM to develop a fast numerical method for time-fractional PDEs, in which the weakly singular translation-invariant kernel in the time-fractional derivative was approximated by a finite combination of rational functions in the Laplace frequency space that in turn leads to an accurate approximation to the time-fractional derivatives. The key advantage of the method is that it only needs to store the numerical solutions at a fixed number of time steps instead of storing the numerical solutions at all the previous time steps. These methods can be applied to solve FPDEs in general domains with a general partition. It is not clear how the resulting fast numerical methods retain certain (e. g., conservation) properties of the underlying numerical discretizations as the methods use compressed approximations to the underlying numerical discretizations, while retaining such properties as conservation of the underlying numerical discretizations may be important, especially for problems with high uncertainty [4, 5, 58].

A hybrid numerical method that combines these powerful numerical methods with the lossless matrix-free methods discretized on structured meshes outlined in this chapter seems to provide a feasible approach. In the context of (28) this would suggest we use structured meshes on each subdomain to generate structured diagonal matrix blocks while using the FMM or H-matrix method to handle the coupling (say, $R_l^{\pm}$ and $R_r^{\pm}$ in (29)) by low-rank approximations with certain constraints to retain the conservation. The randomized matrix method [17] provides another powerful approach to provide a low-rank approximation to $R_l^{\pm}$ and $R_r^{\pm}$. It is not clear how to generate low-rank approximations without formulating the matrices $R_l^{\pm}$ and $R_r^{\pm}$, as the formation of these matrices is computationally very expensive and requires $O(N^2)$ memory in the context of FPDEs. These issues need to be investigated in the near future.

Bibliography

[1] G. S. Ammar and W. B. Gragg, Superfast solution of real positive definite Toeplitz systems, *SIAM J. Matrix Anal. Appl.*, **9** (1988), 61–76.

[2] Z. Bai, K. Lu, and J. Pan, Diagonal and Toeplitz splitting iteration methods for diagonal-plus-Toeplitz linear systems from spatial fractional diffusion equations, *Numer. Algorithms*, 10.1002/nla.2093.

[3] T. S. Basu and H. Wang, A fast second-order finite difference method for space-fractional diffusion equations, *Int. J. Numer. Anal. Model.*, **9** (2012), 658–666.

[4] J. Bear, *Dynamics of Fluids in Porous Materials*, American Elsevier, New York, 1972.

[5] D. Benson, R. Schumer, M. M. Meerschaert, and S. W. Wheatcraft, Fractional dispersion, Lévy motions, and the MADE tracer tests, *Transp. Porous Media*, **42** (2001), 211–240.

[6] R. Chan and X. Jin, *An Introduction to Iterative Toeplitz Solvers*, SIAM, Philadelphia, 2007.

[7] R. Chan and M. Ng, Conjugate gradient methods for Toeplitz systems, *SIAM Rev.*, **38** (1996), 427–482.

[8] T. Chan, An optimal circulant preconditioner for Toeplitz systems, *SIAM J. Sci. Stat. Comput.*, **9** (1988), 766–771.

[9] D. del-Castillo-Negrete, B. A. Carreras, and V. E. Lynch, Fractional diffusion in plasma turbulence, *Phys. Plasmas*, **11** (2004), 3854.

[10] N. Du and H. Wang, A fast finite element method for space-fractional dispersion equations on bounded domains in $\mathbb{R}^2$, *SIAM J. Sci. Comput.*, **37** (2015), A1614–A1635.

[11] V. J. Ervin and J. P. Roop, Variational formulation for the stationary fractional advection dispersion equation, *Numer. Methods Partial Differ. Equ.*, **22** (2005), 558–576.

[12] V. J. Ervin and J. P. Roop, Variational solution of fractional advection dispersion equations on bounded domains in R^d, *Numer. Methods Partial Differ. Equ.*, **23** (2007), 256–281.

[13] L. C. Evans, *Partial Differential Equations*, Graduate Studies in Mathematics, vol. 19, American Mathematical Society, Rhode Island, 1998.

[14] D. Gilbarg and N. Trudinger, *Elliptic Partial Differential Equations of Second Order*, 2nd edn., Springer, Berlin, 1983.

[15] L. Greengard and V. Rokhlin, A fast algorithm particle simulations, *J. Comput. Phys.*, **73** (1987), 325–348.

[16] W. Hackbusch, A sparse matrix arithmetic based on H-matrices, part I: Introduction to H-matrices, *Computing*, **62** (1999), 89–108.

[17] N. Halko, P.-G. Martinsson, and J. A. Tropp, Finding structure with randomness: Probabilistic algorithms for constructing approximate matrix decompositions, *SIAM Rev.*, **53** (2011), 217–288.

[18] J. Jia, C. Wang, and H. Wang, A fast locally refined method for a space-fractional diffusion equation, in *ICFDA'14 Catania, 23–25 June 2014*, IE0147, IEEE, 2014, ISBN 978-1-4799-2590-2.

[19] J. Jia and H. Wang, Fast finite difference methods for space-fractional diffusion equations with fractional derivative boundary conditions, *J. Comput. Phys.*, **293** (2015), 359–369.

[20] J. Jia and H. Wang, A preconditioned fast finite volume scheme for a fractional differential equation discretized on a locally refined composite mesh, *J. Comput. Phys.*, **299** (2015), 842–862.
[21] J. Jia and H. Wang, A fast finite difference method for space distributed-order fractional diffusion equations on convex domains, *Comput. Math. Appl.*, **75** (2018), 2031–2043.
[22] S. Jiang, J. Zhang, Q. Zhang, and Z. Zhang, Fast evaluation of the Caputo fractional derivative and its applications to fractional diffusion equations, *Commun. Comput. Phys.*, **21** (2017), 650–678.
[23] Y. Jiang and X. Xu, Multigrid methods for space fractional partial differential equations, *J. Comput. Phys.*, **302** (2015), 374–392.
[24] Y. Jiang and X. Xu, Domain decomposition methods for space fractional partial differential equations, *J. Comput. Phys.*, **350** (2017), 573–589.
[25] B. Jin, R. Lazarov, J. Pasciak, and W. Rundell, Variational formulation of problems involving fractional order differential operators, *Math. Comput.*, **84** (2015), 2665–2700.
[26] S. L. Lei and H. W. Sun, A circulant preconditioner for fractional diffusion equations, *J. Comput. Phys.*, **242** (2013), 715–725.
[27] F. Liu, V. Anh, and I. Turner, Numerical solution of the space fractional Fokker–Planck equation, *J. Comput. Appl. Math.*, **166** (2004), 209–219.
[28] M. M. Meerschaert, H. P. Scheffler, and C. Tadjeran, Finite difference methods for two-dimensional fractional dispersion equation, *J. Comput. Phys.*, **211** (2006), 249–261.
[29] M. M. Meerschaert and C. Tadjeran, Finite difference approximations for fractional advection-dispersion flow equations, *J. Comput. Appl. Math.*, **172** (2004), 65–77.
[30] M. M. Meerschaert and C. Tadjeran, Finite difference approximations for two-sided space-fractional partial differential equations, *Appl. Numer. Math.*, **56** (2006), 80–90.
[31] R. Metzler and J. Klafter, The random walk's guide to anomalous diffusion: A fractional dynamics approach, *Phys. Rep.*, **339** (2000), 1–77.
[32] R. Metzler and J. Klafter, The restaurant at the end of the random walk: recent developments in the description of anomalous transport by fractional dynamics, *J. Phys. A, Math. Gen.*, **37** (2004), R161–R208.
[33] M. K. Ng, *Iterative Methods for Toeplitz Systems*, Oxford University Press, New York, 2004.
[34] M. Ng and J. Pan, Approximate inverse circulant-plus-diagonal preconditioners for Toeplitz-plus-diagonal matrices, *SIAM J. Sci. Comput.*, **32** (2010), 1442–1464.
[35] J. Pan, R. Ke, M. Ng, and H. Sun, Preconditioning techniques for diagonal-times-Toeplitz matrices in fractional diffusion equations, *SIAM J. Sci. Comput.*, **36** (2014), A2698–A2719.
[36] J. Pan, M. Ng, and H. Wang, Fast iterative solvers for linear systems arising from time-dependent space-fractional diffusion equations, *SIAM J. Sci. Comput.*, **38** (2016), A2806–A2826.
[37] J. Pan, M. Ng, and H. Wang, Fast preconditioned iterative methods for finite volume discretization of steady-state space-fractional diffusion equations, *Numer. Algorithms*, **74** (2017), 153–173.
[38] H. Pang and H. Sun, Multigrid method for fractional diffusion equations, *J. Comput. Phys.*, **231** (2012), 693–703.
[39] I. Podlubny, *Fractional Differential Equations*, Academic Press, New York, 1999.
[40] J. P. Roop, Computational aspects of FEM approximation of fractional advection dispersion equations on bounded domains in R^2, *J. Comput. Appl. Math.*, **193** (2006), 243–268.
[41] G. Strang, A proposal for Toeplitz matrix calculations, *Stud. Appl. Math.*, **74** (1986), 171–176.
[42] C. Tadjeran and M. M. Meerschaert, A second-order accurate numerical method for the two-dimensional fractional diffusion equation, *J. Comput. Phys.*, **220** (2007), 813–823.

[43] C. Tadjeran, M. M. Meerschaert, and H. P. Scheffler, A second-order accurate numerical approximation for the fractional diffusion equation, *J. Comput. Phys.*, **213** (2006), 205–213.
[44] W. Tian, H. Zhou, and W. Deng, A class of second order difference approximations for solving space fractional diffusion equations, *Math. Comput.*, **84** (2015), 1703–1727.
[45] A. Toselli and O. B. Widlund, *Domain Decomposition Methods: Algorithms and Theory*, Springer-Verlag, 2005.
[46] R. S. Varga, *Matrix iterative Analysis*, 2nd edn., Springer-Verlag, Berlin, Heidelberg, 2000.
[47] H. Wang and T. S. Basu, A fast finite difference method for two-dimensional space-fractional diffusion equations, *SIAM J. Sci. Comput.*, **34** (2012), A2444–A2458.
[48] H. Wang, A. Cheng, and K. Wang, Fast finite volume methods for space-fractional diffusion equations, *Discrete Contin. Dyn. Syst., Ser. B*, **20** (2015), 1427–1441.
[49] H. Wang and N. Du, A superfast-preconditioned iterative method for steady-state space-fractional diffusion equations, *J. Comput. Phys.*, **240** (2013), 49–57.
[50] H. Wang and N. Du, A fast finite difference method for three-dimensional time-dependent space-fractional diffusion equations and its efficient implementation, *J. Comput. Phys.*, **253** (2013), 50–63.
[51] H. Wang and N. Du, Fast alternating-direction finite difference methods for three-dimensional space-fractional diffusion equations, *J. Comput. Phys.*, **258** (2013), 305–318.
[52] H. Wang and K. Wang, An $O(N \log^2 N)$ alternating-direction finite difference method for two-dimensional fractional diffusion equations, *J. Comput. Phys.*, **230** (2011), 7830–7839.
[53] H. Wang, K. Wang, and T. Sircar, A direct $O(N \log^2 N)$ finite difference method for fractional diffusion equations, *J. Comput. Phys.*, **229** (2010), 8095–8104.
[54] H. Wang, D. Yang, and S. Zhu, Inhomogeneous Dirichlet boundary-value problems of space-fractional diffusion equations and their finite element approximations, *SIAM J. Numer. Anal.*, **52** (2014), 1292–1310.
[55] H. Wang, D. Yang, and S. Zhu, Accuracy of finite element methods for boundary-value problems of steady-state fractional diffusion equations, *J. Sci. Comput.*, **70** (2017), 429–449.
[56] H. Wang and X. Zhang, A high-accuracy preserving spectral Galerkin method for the Dirichlet boundary-value problem of variable-coefficient conservative fractional diffusion equations, *J. Comput. Phys.*, **281** (2015), 67–81.
[57] J. Xin and L. Zikatanov, Algebraic multigrid methods, *Acta Numer.*, **26** (2017), 591–721.
[58] Y. Zhang, D. A. Benson, M. M. Meerschaert, and E. M. LaBolle, Space-fractional advection-dispersion equations with variable parameters: Diverse formulas, numerical solutions, and application to the MADE-site data, *Water Resour. Res.*, **43** (2007), W05439.
[59] M. Zhao, A. Cheng, and H. Wang, A preconditioned fast Hermite finite element method for space-fractional diffusion equations, *Discrete Contin. Dyn. Syst., Ser. B*, **22** (2017), 3529–3545.
[60] M. Zhao, A. Cheng, and H. Wang, A fast finite difference method for three-dimensional time-dependent space-fractional diffusion equations with fractional derivative boundary conditions, *J. Sci. Comput.*, (2017), 10.1007/s10915-017-0478-8.
[61] X. Zhao, X. Hu, W. Cai, and G. E. Karniadakis, Adaptive finite element method for fractional differential equations using hierarchical matrices, *Comput. Methods Appl. Mech. Eng.*, **325** (2017), 56–76.

Roberto Garrappa and Marina Popolizio

Fast methods for the computation of the Mittag-Leffler function

Abstract: In this chapter we discuss the problem of the fast and reliable computation of the Mittag-Leffler function. We start from the series representation, by which the function is commonly defined, to illustrate how it turns out to be not suitable for computation in most of the cases; then we present an alternative approach based on the numerical inversion of the Laplace transform. In particular, we describe a technique, known as the optimal parabolic contour, in which the inversion of the Laplace transform is performed by applying the trapezoidal rule along a parabolic contour in the complex plane; the contour and the integration parameters are chosen, on the basis of the error analysis and the location of the singularities of the Laplace transform, with the aim of achieving a target accuracy (which can be very close to machine precision) with a substantially low computational effort. Applications to the evaluation of derivatives of the Mittag-Leffler function and to matrix arguments are also discussed.

Keywords: Mittag-Leffler function, Laplace transform, quadrature rule, trapezoidal rule, complex contour, derivative of the Mittag-Leffler function, matrix function

MSC 2010: 33E12, 44A10, 65D32, 65F60

1 Introduction

In 1903 the Swedish mathematician Mittag-Leffler introduced the function [26]

$$E_{\alpha}(z) = \sum_{k=0}^{\infty} \frac{z^k}{\Gamma(\alpha k + 1)}, \quad \alpha \in \mathbb{C},\ \Re(\alpha) > 0,\ z \in \mathbb{C},$$

which is now universally recognized with his name and, only few years later, Wiman [38] defined the more general version depending on two (possibly) complex parameters

$$E_{\alpha,\beta}(z) = \sum_{k=0}^{\infty} \frac{z^k}{\Gamma(\alpha k + \beta)}, \quad \alpha, \beta \in \mathbb{C},\ \Re(\alpha) > 0,\ z \in \mathbb{C}. \tag{1}$$

Roberto Garrappa, Member of the INdAM Research group GNCS, Dipartimento di Matematica, Università degli Studi di Bari "Aldo Moro", Via E. Orabona 4, 70126 Bari, Italy, e-mail: roberto.garrappa@uniba.it, https://orcid.org/0000-0002-7881-0885
Marina Popolizio, Member of the INdAM Research group GNCS, Dipartimento di Ingegneria Elettrica e dell'Informazione, Politecnico di Bari, Via E. Orabona 4, 70126 Bari, Italy, e-mail: marina.popolizio@poliba.it, https://orcid.org/0000-0003-0474-2573

https://doi.org/10.1515/9783110571684-013

Actually, the work of Wiman did not focus on the investigation of the properties of this function, nor on its applications; it was only fifty years later, in 1953, that Agarwal and Humbert [22], and independently Djrbashyan [4] in 1954, deeply analyzed the two-parameter function (1).

At a later time, in 1971, the Indian mathematician Prabhakar [31] proposed a further generalization to three parameters, i. e.,

$$E_{\alpha,\beta}^{\gamma}(z) = \frac{1}{\Gamma(\gamma)} \sum_{j=0}^{\infty} \frac{\Gamma(j+\gamma)z^j}{j!\Gamma(\alpha j+\beta)}, \quad \alpha,\beta,\gamma \in \mathbb{C},\ \Re(\alpha) > 0,\ z \in \mathbb{C}, \tag{2}$$

and studied integral operators built on the basis of this function.

Nowadays, functions like (1) and (2) are all recognized as Mittag-Leffler (ML) functions.

Although the introduction of these functions was motivated by purely theoretical purposes, they are currently extremely important to describe and solve essential models in the fields of physics and engineering, like random walks, Lévy flights, superdiffusive transport, nerve conduction, dielectrical and mechanical relaxation, and, more generally, in complex systems theory. For a rich list of applications involving the ML functions we refer to [25, 29] and, in particular, to the recent book by Gorenflo and coauthors [17] completely devoted to ML functions. However, we have to stress the impulse given to the subject by Caputo and Mainardi at the beginning of the 1970s, when they proved the deep connection between the ML functions and the derivatives of fractional order in their pioneering works on linear viscoelastic bodies. Since then, ML functions have become popular above all for their key role in fractional calculus.

ML functions are indeed fundamental to describe the solution of Abel–Volterra-type equations and integral or differential equations of fractional-order. In this regard, the operational method is a well-established way to solve classes of these problems in closed form in which the solution is expressed in terms of ML functions (e. g., see [24]). Important contributions in this direction are due to Mainardi, Luchko, Gorenflo, and their coauthors (for a detailed list of references we refer to [17]).

Only recently the importance of the three-parameter ML function (2), also known as the Prabhakar function, has been re-discovered for defining operators involved in dielectric models and special cases of viscoelasticity (e. g., see [7, 8, 16]) but also because it provides an explicit representation of the derivatives of the two-parameter ML function, since (e. g., see [3, 5, 15, 23])

$$\frac{d^k}{dz^k} E_{\alpha,\beta}(z) = k! E_{\alpha,\alpha k+\beta}^{k+1}(z), \quad k \in \mathbb{N}.$$

The important role in fractional calculus has thus renewed the interest in ML functions and stimulated the investigation of efficient methods for their numerical evaluation.

In this chapter we address this issue, by showing the difficulties related to the computation of (1), thus to motivate the choice of alternative approaches with respect to the direct evaluation of the series representation (1). In particular, we present an approach based on the numerical inversion of the Laplace transform (LT) obtained by integration along parabolic contours in the complex plane by means of quadrature rules of trapezoidal type. The so-called “optimal parabolic contour” allows to select the contour and the quadrature parameters in an optimal way, on the basis of the location of the singularities of the LT and of the error analysis, in order to achieve accurate approximations of the ML function (with errors close to machine precision) with a reasonably small amount of computation. The present chapter covers scalar arguments, but also the case of matrix arguments is addressed at the end of the chapter, together with the computation of derivatives of the ML function.

2 The role of the ML functions

Hille and Tamarkin were probably the first authors, in their paper [21] dating back to 1930, to use the ML function to express the solution of Abel integral equations of the second kind, i. e.,

$$\Phi(x) - \frac{\lambda}{\Gamma(\alpha)} \int_0^x \frac{\Phi(t)}{(x-t)^{1-\alpha}} \mathrm{d}t = f(x), \quad 0 < \alpha < 1,\ 0 < x < 1.$$

Since then, the importance of the ML function in the study of different types of integral equations has been widely recognized. Later, in 1954, Barret used the ML function for the general solution of linear fractional differential equations (FDEs) with constant coefficients [1]. Indeed, given for instance an initial value problem for an FDE of order $0 < \alpha < 1$,

$${}^{\mathrm{C}}D_0^{\alpha} y(t) = \lambda y(t), \quad y(0) = y_0,$$

with ${}^{\mathrm{C}}D_0^{\alpha}$ denoting the fractional-order derivative according to the Caputo definition

$${}^{\mathrm{C}}D_0^{\alpha} y(t) = \frac{1}{\Gamma(m-\alpha)} \int_0^t (t-\tau)^{m-\alpha-1} y^{(m)}(\tau) \mathrm{d}\tau, \quad m = \lceil \alpha \rceil,$$

its solution can be expressed in terms of the ML function as

$$y(t) = E_{\alpha,1}(t^{\alpha}\lambda) y_0.$$

More generally, also for the solution of semi-linear systems of FDEs

$${}^{\mathrm{C}}D_0^{\alpha} Y(t) = AY(t) + F(t, Y(t)), \quad Y(0) = Y_0,$$

where $A \in \mathbb{R}^{N\times N}$ and Y and F are vector-valued functions, it is possible to draw a representation by means of convolution integrals with ML functions in the kernel, i. e.,

$$Y(t) = E_{\alpha,1}(t^\alpha A)Y_0 + \int_0^t (t-\tau)^{\alpha-1} E_{\alpha,\alpha}((t-\tau)^\alpha A)F(\tau, Y(\tau))\mathrm{d}\tau;$$

this representation has been also exploited for stable numerical computation [9, 10, 11, 12] once reliable procedures for the approximation of the ML function with matrix arguments are available [15, 27].

3 Is the series representation useful for computation?

At first glance, the series representation (1), which is very similar to the series representation of the exponential function, may appear as the natural tool for computing the ML function after truncation to a finite number of terms.

Anyway, it is essential to observe that (1) can be directly used for practical computation only when $|z|$ is sufficiently small (namely, less than 1 or, however, not much greater than 1). Indeed, an accurate approximation for larger $|z|$ requires to consider a large number of terms in (1), and so we have to observe that:

- in double precision arithmetic the Gamma function can be evaluated only for arguments not greater than 171.624 and, therefore, the maximum number of the series terms which can be actually computed is no larger than $(171.624 - \beta)/\alpha$; for arguments with large $|z|$, for which a greater number of terms in the series is necessary to get a satisfactory accuracy, no reliable approximation is possible by means of (1);
- when $|z| > 1$ the sequence of the terms in (1) does not decrease monotonically; indeed, in a first transient phase the terms of the series increase and may become very large in modulus before decreasing thanks to the supergeometric growth of the Gamma function; however, when $|\arg(z)| > \alpha\pi/2$ a small value $|E_{\alpha,\beta}(z)|$ of the final result must be expected which therefore results as the summation of large terms (hence with alternating signs), and this is a widely recognized source of heavy numerical cancelation [19].

In floating-point units the computation of elementary functions is often made by exploiting some characteristic properties of the function in order to map the argument to a suitable region where the accurate computation requires a reasonable effort. One of these techniques, known under the name of *range reduction*, when applied to compute $\exp(z)$ can be roughly summarized as

1. two values N and w are found such that $z = N\ln(2) + w$, with $|w| < 1$ and N an integer number;
2. $\exp(w)$ is in some way approximated, for instance by polynomial functions (also by truncating the series representation), by rational functions, or by some table-driven method;
3. $\exp(z)$ is hence reconstructed thanks to the formula $\exp(z) = 2^N \exp(w)$.

The application of the range reduction technique to the exponential is possible since the semi-group property (namely, $e^{a+b} = e^a \cdot e^b$) allows to reconstruct $e^z = 2^N e^w$. Unfortunately, although the ML function is a generalization of the exponential, methods for its computation cannot rely on similar strategies since the semi-group property does not apply to the ML function (unless the trivial case $\alpha = \beta = 1$ corresponding to the exponential).

It is also true that for arguments with large modulus, a different series expansion can be employed. For instance, whenever $0 < \alpha < 1$ the asymptotic expansion of the ML function for large arguments is given (see for instance [17]) by

$$E_{\alpha,\beta}(z) \sim \begin{cases} \frac{1}{\alpha} z^{(1-\beta)/\alpha} e^{z^{1/\alpha}} - \sum\limits_{k=1}^{\infty} \frac{z^{-k}}{\Gamma(\beta - \alpha k)} & |\arg z| < \pi\alpha, \\ -\sum\limits_{k=1}^{\infty} \frac{z^{-k}}{\Gamma(\beta - \alpha k)} & \pi\alpha < |\arg z| < \pi. \end{cases} \tag{3}$$

However, serious difficulties arise in the neighborhood of the Stokes lines $\arg z = \pm\pi\alpha$, where both algebraic and exponential terms are involved without a clear predominance of either of them [28], a problem which has been faced in [32].

Anyway, the two expansions (1) and (3) work out in restricted areas of the complex plane (when $|z|$ is small or large respectively) but in intermediate regions, where $|z|$ is only moderately small or moderately large, neither of them is suitable for computation and alternative approaches must be considered.

As discussed in the pioneering work by Gorenflo et al. [18], and successively in [32] and [6, 14], an alternative and reliable approach for the practical computation of special functions like the ML one can be outlined by representing the function by means of some integral formulation and then approximating the integral by accurate numerical procedures.

4 Numerical inversion of the Laplace transform

Many functions possess a LT whose analytical expression is easier to compute than the function itself. This feature often facilitates the numerical computation since accurate methods for an effective approximation of the inverse of the LT can be made up. The

ML function shares this quality and, indeed, it is well known that the LT of

$$t^{\beta-1}E_{\alpha,\beta}(t^{\alpha}\lambda), \quad t \in \mathbb{R}_{+}, \lambda \in \mathbb{C} \tag{4}$$

is given by [17, 29]

$$\mathcal{E}_{\alpha,\beta}(s;\lambda) = \frac{s^{\alpha-\beta}}{s^{\alpha}-\lambda}, \quad \Re(s) > 0 \quad \text{and} \quad |\lambda s^{-\alpha}| < 1.$$

Thus, after putting $t = 1$ in (4), the formula for the inversion of the LT leads to the Bromwich complex contour integral

$$E_{\alpha,\beta}(\lambda) = \frac{1}{2\pi \mathrm{i}} \int_{\sigma-\mathrm{i}\infty}^{\sigma+\mathrm{i}\infty} \mathrm{e}^{s}\mathcal{E}_{\alpha,\beta}(s;\lambda)\mathrm{d}s \tag{5}$$

with the abscissa $\sigma \in \mathbb{R}$ chosen in such a way that all the singularities of $\mathcal{E}_{\alpha,\beta}(s;\lambda)$ lie to the left of the Bromwich line $\Re(s) = \sigma$.

The integrand in (5) is a multi-valued function for the presence of non-integer powers; a branch cut is therefore mandatory to make the integrand single-valued and, for convenience, we fix the branch cut along the negative real semi-axis.

The numerical integration of (5) is however a task which requires some caution. Along the Bromwich line $s = \sigma + \mathrm{i}y$, $-\infty < y < \infty$, the exponential factor is highly oscillatory while $\mathcal{E}_{\alpha,\beta}(s;\lambda)$ slowly decays as $|y| \to \infty$, thus introducing an outstanding source of instability in the numerical computation.

In 1979 Talbot [33] proposed a clever solution for the numerical computation of integrals like (5) with functions $F(y)$ slowly decaying as $|y| \to \infty$. In practice the Bromwich line is deformed into a suitable contour $\mathcal{C}$ that begins and ends in the left half of the complex plane in order to force a rapid decay of the integrand in (5). After the pioneering work of Talbot, who proposed the use of cotangent contours, other contours were deeply investigated. For the three most used contours, namely cotangent, parabolas, and hyperbolas, the literature reports very fast convergence rates even with simple quadrature rules when the parameters are selected in optimal ways [35]. Trefethen and Weideman thus described an accurate procedure for selecting a set of optimal parameters by balancing the different components of discretization and truncation errors in the quadrature rule [37]; the procedure was further improved by Weideman [36] by incorporating also round-off errors (see [34] for an interesting review on this and related topics).

Among the possible contours $\mathcal{C}$ on which to perform the integration

$$E_{\alpha,\beta}(\lambda) = \frac{1}{2\pi \mathrm{i}} \int_{\mathcal{C}} \mathrm{e}^{s}\mathcal{E}_{\alpha,\beta}(s;\lambda)\mathrm{d}s, \tag{6}$$

parabolic contours have been observed to be more appropriate for the ML function [6, 14]. Indeed, despite the fact that the exponential decays slower on parabolas than on other contours [35], their very simple representation

$$z(u) = \mu(\mathrm{i}u + 1)^2, \quad -\infty < u < \infty, \tag{7}$$

represents a key factor. Actually the fact that the parabola depends on just one parameter $\mu > 0$ allows an easier and more specialized error analysis and hence a more precise tuning of the quadrature parameters in order to achieve a given target accuracy.

If the chosen parabola encompasses all the singularities of $\mathcal{E}_{\alpha,\beta}$, then the integral (6) can be reformulated as

$$E_{\alpha,\beta}(\lambda) = \frac{1}{2\pi \mathrm{i}} \int_{-\infty}^{\infty} \mathrm{e}^{z(u)} \mathcal{E}_{\alpha,\beta}(z(u);\lambda) z'(u) \mathrm{d}u, \tag{8}$$

where it is easy to compute $z'(u) = 2\mu(\mathrm{i} - u)$.

A suitable quadrature formula is then needed to numerically evaluate this integral and the trapezoidal rule is well adapted to tasks of this kind, as thoroughly analyzed in [34, 37]. As for the geometry of the contour, also the choice of the trapezoidal rule (among other possibly more efficient rules) is strongly forced by practical reasons: with the trapezoidal rule it is indeed possible to perform an accurate error analysis which drives the choice of the parameters.

The application of the trapezoidal rule leads to a discretized version of the integral (8)

$$E_{\alpha,\beta}^{[h]}(\lambda) = \frac{h}{2\pi \mathrm{i}} \sum_{k=-\infty}^{+\infty} \mathrm{e}^{z(u_k)} \mathcal{E}_{\alpha,\beta}(z(u_k);\lambda) z'(u_k), \tag{9}$$

where $u_k = kh$ and $h > 0$ is a selected step size. Finally the truncation to a finite number of nodes on the parabola allows to provide the computable approximation

$$E_{\alpha,\beta}^{[N,h]}(\lambda) = \frac{h}{2\pi \mathrm{i}} \sum_{k=-N}^{N} \mathrm{e}^{z(u_k)} \mathcal{E}_{\alpha,\beta}(z(u_k);\lambda) z'(u_k) \tag{10}$$

of the ML function $E_{\alpha,\beta}(\lambda)$; crucially, it can be evaluated with a reasonably small effort (and hence in a fast way) once the contour is chosen with the aim of selecting a small number N.

4.1 Selection of the contour and of the integration parameters

To proceed with the numerical evaluation of (10) three parameters need to be selected: the value μ, which defines the parabolic contour for the integration, and the parameters of the quadrature rule (namely, the distance h of the $2N + 1$ nodes and the parameter N, which actually determines the computational cost of the procedure). Their selection is made in such a way to achieve a prescribed accuracy $\varepsilon > 0$. In particular, values of ε very close to machine precision are usually considered, thus to assuming the evaluation of $E_{\alpha,\beta}(\lambda)$ to be virtually exact, i. e., within the tolerance of the double precision floating-point arithmetic of standard computers.

To drive the selection of the three parameters μ, N, and h, an accurate error analysis is therefore necessary which has to take into account all the possible sources of error introduced in the computation.

Denoting by $\hat{E}_{\alpha,\beta}^{[N,h]}(\lambda)$ the value actually evaluated in floating-point arithmetic by the quadrature rule (10), the following three types of error are considered in the error analysis:

1. the *round-off error* due to the finite precision of the computer arithmetic,
$$\mathbf{RE} = \left|E_{\alpha,\beta}^{[N,h]}(\lambda) - \hat{E}_{\alpha,\beta}^{[N,h]}(\lambda)\right|;$$
2. the *truncation error* introduced in (10) after truncating the quadrature rule (9) at a finite number of terms,
$$\mathbf{TE} = \left|E_{\alpha,\beta}^{[h]}(\lambda) - E_{\alpha,\beta}^{[N,h]}(\lambda)\right|;$$
3. the *discretization error* introduced by the quadrature rule (9) as a consequence of the approximation of the integrand
$$\mathbf{DE} = \left|E_{\alpha,\beta}(\lambda) - E_{\alpha,\beta}^{[h]}(\lambda)\right|.$$

The overall error $\mathbf{Err} = |E_{\alpha,\beta}(\lambda) - \hat{E}_{\alpha,\beta}^{[N,h]}(\lambda)|$ clearly satisfies

$$\mathbf{Err} \leq \mathbf{RE} + \mathbf{TE} + \mathbf{DE}.$$

4.2 Round-off error and residue subtraction

The need of encompassing all the singularities of $\mathcal{E}_{\alpha,\beta}(s;\lambda)$ may often lead to select contours extending very far in the right part of the complex plane and with a very broad shape; namely, these are parabolas with large values of the parameter μ. Due to the presence of a factor e^{μ} in the exponential term, the summation (10) presents, in these cases, terms with very large modulus but alternating signs; when summed up, the final result is in modulus much smaller than the various terms and, as is well known, this is a source of numerical cancelation [19].

The corresponding rounding errors, as also observed in [36], are therefore proportional to $e^{\mu}\epsilon$, with ϵ being the machine precision, and therefore, in order to ensure that these errors are under the desired target accuracy ε, it is necessary to impose the bound

$$\mu < \log \varepsilon - \log \epsilon$$

(obviously the target accuracy ε must be chosen such that $\varepsilon > \epsilon$).

Restricting μ in order to comply with the above requirement may not ensure that all the singularities of $\mathcal{E}_{\alpha,\beta}(s;\lambda)$ lie left of the contour. To preserve the invariance of the final result, a residue subtraction is necessary in (6) when using a contour having some of the singularities at the right. Then, if $\mathcal{C}$ is the selected contour and $S^\star_{\mathcal{C}}$ denotes the set of the singularities of $\mathcal{E}_{\alpha,\beta}(s;\lambda)$ laying in the rightmost part of the complex plane delimited by $\mathcal{C}$, the integral reads as

$$E_{\alpha,\beta}(\lambda) = \sum_{s^\star \in S^\star_{\mathcal{C}}} \text{Res}\left(\mathrm{e}^s \mathcal{E}_{\alpha,\beta}(s;\lambda), s^\star\right) + \frac{1}{2\pi i} \int_{\mathcal{C}} \mathrm{e}^s \mathcal{E}_{\alpha,\beta}(s;\lambda) \mathrm{d}s, \tag{11}$$

where $\text{Res}(f, s^\star)$ denotes the residue of the function f at $s^\star$.

From the computational point of view, the residue subtraction (11) does not represent a serious issue since an analytical and easy-to-compute formula for the residues of the ML function is known [17], i. e.,

$$\text{Res}(\mathrm{e}^s \mathcal{E}_{\alpha,\beta}(s;\lambda), s^\star) = \frac{1}{\alpha} (s^\star)^{1-\beta} \mathrm{e}^{s^\star}.$$

We have to keep in mind, however, that according to the parameter α and the argument λ, the function $\mathcal{E}_{\alpha,\beta}(s;\lambda)$ can have from none to several singularities. Indeed, the relevant poles (those in the main Riemann sheet) of $\mathcal{E}_{\alpha,\beta}(s;\lambda)$ are

$$\bar{s}^\star_j = \lambda^{1/\alpha} = |\lambda|^{1/\alpha} e^{i\frac{\theta+2j\pi}{\alpha}}, \quad j \in \left\{ j \in \mathbb{Z} \;\middle|\; -\frac{\alpha}{2} - \frac{\theta}{2\pi} < j \le \frac{\alpha}{2} - \frac{\theta}{2\pi} \right\}, \tag{12}$$

where $\theta = \arg(\lambda)$, $-\pi < \theta \le \pi$. To this set we have to add also the origin which, although it is a pole of $\mathcal{E}_{\alpha,\beta}(s;\lambda)$ only when $\beta > \alpha$, it is a branch cut singularity.

An additional source of round-off error arises when the numerical integration is performed in the proximity of some of the singularities.

As discussed in [2] for closely related problems, the conditioning of the numerical quadrature of Cauchy-type integrals depends on the extent of the analyticity region surrounding the integration contour. Thus, performing the integration on a contour very close to one of the singularities may lead to the amplification of possible round-off errors.

It is therefore necessary to introduce a further control in order to ensure that the integration is performed on “safe regions”, i. e., in regions sufficiently large, in order to keep the contour at a certain distance from the singularities. We will see later on that a control of this type can be made by keeping into account some algebraic factors in the discretization error.

4.3 Identifying regions of analyticity

As said above, the width of the analyticity region of $\mathcal{E}_{\alpha,\beta}(s;\lambda)$ over the integration contour influences the round-off error as well as the discretization error we are going to discuss later. It is thus important to identify regions in which the LT of the ML function is analytical.

The approach we propose consists in partitioning the complex plane in regions bounded by parabolas passing through the singularities of $\mathcal{E}_{\alpha,\beta}(s;\lambda)$. We will then choose, by means of the parameter estimates, the region in which to fix the parabolic contour for the numerical computation of (11).

To select these regions we need the function $\varphi : \mathbb{C} \to \mathbb{R}_+$ defined by

$$\varphi(s) = \frac{\mathbb{R}(s) + |s|}{2};$$

then, denoting by $\bar{s}_0^\star, \ldots, \bar{s}_J^\star$ all the possible singularities of $\mathcal{E}_{\alpha,\beta}(s;\lambda)$, the parabola $\mathcal{C}_j^\star$ passing through a given singularity $\bar{s}_j^\star$ is obtained by selecting the parameter μ in (7) as $\mu_j = \varphi(\bar{s}_j^\star)$.

For convenience we sort the singularities $\bar{s}_j^\star$ in ascending order with respect to the values of $\varphi(\bar{s}_j^\star)$ (such that the parameters μ_j define parabolas in the complex plane extending from left to right).

We then define $R_j^\star$ as the region bounded by the parabolas $\mathcal{C}_j^\star$ and $\mathcal{C}_{j+1}^\star$ corresponding to $\bar{s}_j^\star$ and $\bar{s}_{j+1}^\star$, with the exception of $R_J^\star$, which is unbounded to the right (since $\bar{s}_0^\star = 0$, the first contour $\mathcal{C}_0^\star$ collapses onto the negative real semi-axis).

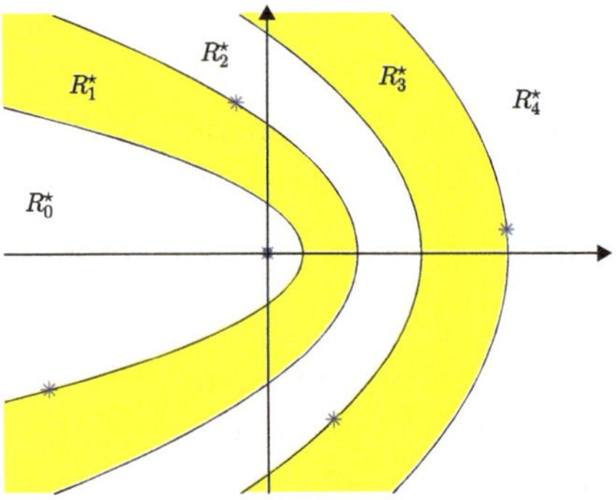

Figure 1: Partitioning of the complex plane into regions $R_j^\star$ by means of parabolic contours through the singularities $\bar{s}_j^\star$.

An example of the partitioning of the complex plane by means of parabolic contours through some possible singularities (denoted with the stars) is illustrated in Figure 1.

4.4 Discretization error

The width of each region $R_j^\star$ affects in a decisive way the discretization error $\mathbf{DE}_j$ resulting from the integration along a contour $\mathcal{C}_j$ in $R_j^\star$. By following the analysis in [6], $R_j^\star$ can be characterized as the strip

$$R_j^\star = \{z \in \mathbb{C} \,|\, z = \mu_j(iw+1)^2,\ w \in \mathbb{C} \text{ and } -d_j^\star < \Im(w) < c_j^\star\}, \tag{13}$$

with $c_j^\star > 0$ and $d_j^\star > 0$ defined as

$$c_j^\star = \begin{cases} 1 & j = 0, \\ 1 - \sqrt{\frac{\varphi(s_j^\star)}{\mu_j}} & j \geq 1, \end{cases} \qquad d_j^\star = \begin{cases} \sqrt{\frac{\varphi(s_{j+1}^\star)}{\mu_j}} - 1 & j \leq J, \\ +\infty & j = J. \end{cases} \tag{14}$$

In Figure 2 we observe the correspondence between the strip of analyticity in the w plane and the regions of analyticity of parabolic shape in the z plane.

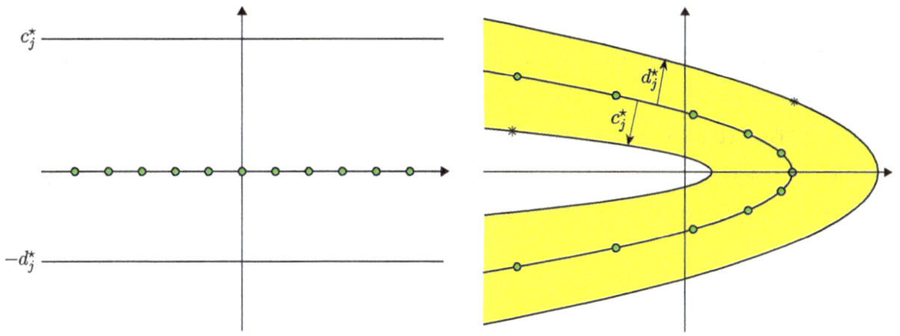

Figure 2: Correspondence between the representation of a region of analyticity $R_j^\star$ in the w plane (left plot) and in the z plane (right plot).

The analysis performed in [6, 14] on the basis of the results published in [34, 37] describes the discretization error $\mathbf{DE}_j$ as

$$\mathbf{DE}_j = \mathrm{DE}_j^+ + \mathrm{DE}_j^-,$$

where DE_j^+ (respectively DE_j^-) depends on the distance between the contour $\mathcal{C}_j$ and the boundary of the analyticity region at the left (respectively the right) of the contour.

The analysis of the discretization error on open regions, although possible, turns out to be useless for determining the desired parameters. We thus restrict our analysis to smaller regions by introducing two further coefficients $c_j < c_j^\star$ and $d_j < d_j^\star$, thus obtaining smaller regions $R_j \subset R_j^\star$ in which the integrand is bounded.

By restricting the analysis to the subregions of analyticity R_j it has been evaluated in [6] that when the step size $h_j \to 0$ we have

$$\mathrm{DE}_+(c_j) = \mathcal{O}\big((c_j^\star - c_j)^{-p_j} \mathrm{e}^{-2\pi c_j^\star/h_j}\big),$$
$$\mathrm{DE}_-(d_j) = \begin{cases} \mathcal{O}((d_j^\star - d_j)^{-q_j} \mathrm{e}^{-2\pi d_j^\star/h_j + \varphi(\bar{s}_{j+1}^\star)}) & j = 0, \dots, J-1, \\ \mathcal{O}(\mathrm{e}^{\pi^2/(\mu_J h_J^2) + 2\pi/h_J}) & j = J, \end{cases}$$

with

$$p_j = \begin{cases} \max\{0, -2(\alpha - \beta + 1)\} & j = 0, \\ 1 & j = 1, \dots, J \end{cases}$$

and $q_j = 1$. The algebraic terms $(c_j^\star - c_j)^{-p_j}$ and $(d_j^\star - d_j)^{-q_j}$ in $\mathrm{DE}_+(c_j)$ and $\mathrm{DE}_-(d_j)$ not only improve the description of the discretization error but allow to keep the round-off errors, which can become very large when the contour is chosen close to the singularity, under control.

4.5 Truncation error

Since the contour begins and ends in the left half of the complex plane, the integrand in (8) decays in a rapid way when $u \to \pm\infty$.

An estimation of the truncation error $\mathbf{TE}_j$ for each parabola $\mathcal{C}_j$ is hence provided by the last values of the integrand in the quadrature rule (10), i. e., by the outermost values for $k = N_j$ or $k = -N_j$, once the $2N_j + 1$ nodes $u_k = kh_j$, $k = -N, \dots, N$, are fixed on $\mathcal{C}_j$.

To simplify the analysis, and by the dominant presence of the exponential factor, for large N_j, this error can be approximated by

$$\mathbf{TE}_j = \mathcal{O}(|\mathrm{e}^{z(u_{N_j})}|), \quad N_j \to \infty,$$

which, after some elementary manipulation, can be expressed as

$$\mathbf{TE}_j = \mathcal{O}(|\mathrm{e}^{\mu_j(1-h_jN_j)}|), \quad N_j \to \infty.$$

4.6 Optimal parabolic contour

A mixed analytical and numerical approach allows to determine, in each region $R_j^\star$, appropriate parameters μ_j, h_j, and N_j in order to obtain an error close to the prescribed target tolerance ε by balancing the different sources of errors and imposing them to be close to ε (we refer to [6, 14] for more details concerning the ways in which optimal parameters are evaluated).

We have just to note that the parameter computation requires a computational effort which is negligible with respect to the computation of the quadrature rule (10). Thus, determining the quadrature parameters in each region of analyticity is not expensive and allows to choose in an optimal way the region in which to actually perform the inversion of the LT.

Indeed, among the different regions in which the numerical integration can be performed, the one requiring the minimum amount of computation, i. e., the one for which the balancing of the errors requires the minimum number N_j of nodes, is chosen.

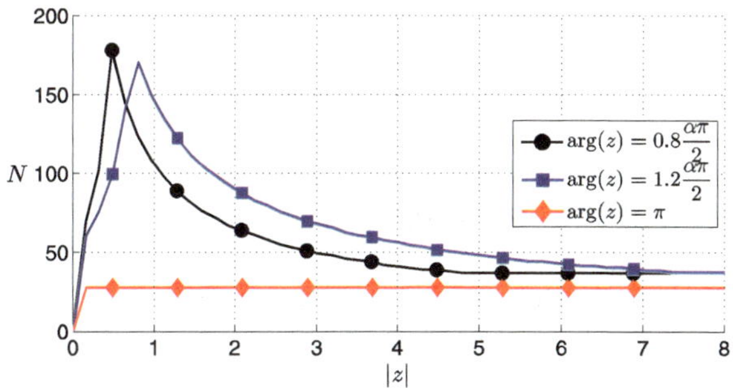

Figure 3: Number N of nodes necessary to compute the ML function when $\alpha = 0.7$ and $\beta = 1.0$.

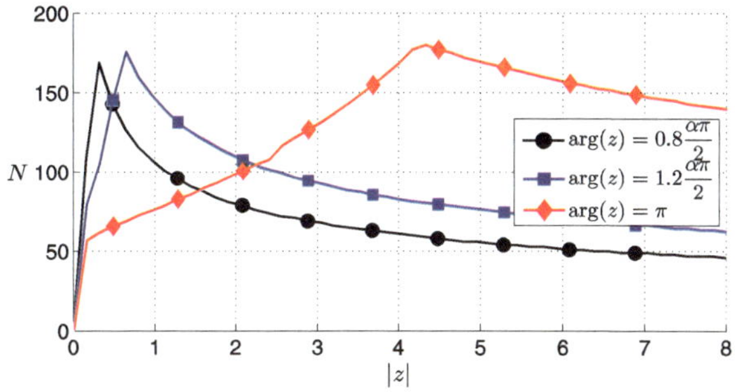

Figure 4: Number N of nodes necessary to compute the ML function when $\alpha = \beta = 1.2$.

A Matlab code implementing this algorithm is freely available in the file exchange service of the MathWorks website.[1]

In Figures 3 and 4 we show the number N of nodes which are actually employed, to achieve an accuracy $\varepsilon = 10^{-15}$, for some values of α and β and the argument z in different regions of the complex plane. As we can observe, the number of nodes, and hence the number of floating-point operations necessary to compute the ML function, is confined to some dozens and only in few cases it is proportional to a few hundred, a number of operations which on modern computers is executed in extremely short times.

5 Further applications

There are other interesting applications in which the technique discussed in the previous sections can be usefully employed or extended. We present in this section just

1 www.mathworks.com/matlabcentral/fileexchange/48154-the-mittag-leffler-function

two of them: the evaluation of derivatives of the ML function and the evaluation of the ML function with matrix arguments.

5.1 Evaluation of derivatives of the ML function

Gorenflo and coauthors were the first to face the numerical computation of the derivatives of the ML function, with results concerning just the first-order derivative [18]. Recently this topic has been addressed with a complete analysis resulting in a numerical method which turns out to be effective for any order derivative and for every argument [15]. We just describe here the main features of this approach, which is based on the numerical inversion of the LT. Indeed, the method for computing the ML function described in Section 4 can be extended to derivatives of the ML function of any order.

After a term-by-term derivation of the series (1), it is simple to observe that the derivatives of the ML function can be expressed as

$$\frac{\mathrm{d}^k}{\mathrm{d}z^k} E_{\alpha,\beta}(z) = \sum_{j=0}^{\infty} \frac{(j+k)_k z^j}{\Gamma(\alpha j + \alpha k + \beta)} = k! E^{k+1}_{\alpha,\alpha k+\beta}(z), \quad k \in \mathbb{N}, \tag{15}$$

where $E^{\gamma}_{\alpha,\beta}(z)$ is the Prabhakar function (2) and $(x)_n$ the falling factorial. Since also the Prabhakar function has a very simple LT, namely,

$$\mathcal{L}\big(t^{\beta-1} E^{\gamma}_{\alpha,\beta}(t^{\alpha} z)\,;\, s\big) = \frac{s^{\alpha\gamma-\beta}}{(s^{\alpha} - z)^{\gamma}}, \quad \mathbb{R}(s) > 0,\ |zs^{-\alpha}| < 1, \tag{16}$$

the computation of these derivatives can be performed by means of an algorithm similar to that outlined in Section 4.

Some changes are however necessary to adapt the whole procedure to the specific features of the derivatives of the ML function. In particular:

- the coefficients p_j and q_j in the formulas for the discretization errors $\mathrm{DE}_+(c_j)$ and $\mathrm{DE}_-(d_j)$ have to take into account the presence of the third parameter γ in (16), namely, $k+1$ for kth-order derivatives; on the basis of the analysis in [6] they are

$$p_j = \begin{cases} \max\{0, -2(\alpha - \beta + 1)\} & j = 0, \\ k+1 & j = 1, \dots, J, \end{cases}$$

$$q_j = k+1, \quad j = 0, 1, \dots, J-1;$$

- the representation of the residues in (11) is less immediate and requires some further manipulations.

We have first to observe that, when γ is not an integer, since a different branch cut must be imposed, it is not anymore possible to move the contour as in (11) and have the same freedom in the selection of the contour as in the case of the two-parameter ML function (namely, when $\gamma = 1$). And indeed, computing $E^{\gamma}_{\alpha,\beta}(z)$ by numerical inversion of the LT is a not simple task when $\gamma \in \mathbb{R} - \mathbb{N}$.

Anyway, for the computation of derivatives of the two-parameter ML function (1), just instances of the three-parameter ML function (2) having $\gamma \in \mathbb{N}$ are involved.

The problem of determining the residues at the singularities $\bar{s}_j^\star$ of

$$H_k(s;\lambda) = \frac{s^{\alpha-\beta}}{(s^\alpha - \lambda)^{k+1}}$$

has been addressed in [15], where it has been found that

$$\operatorname{Res}(e^s H_k(s;\lambda), s_\star) = \frac{1}{\alpha^{k+1}} e^{s_\star} s_\star^{1-\alpha k-\beta} P_k(s_\star),$$

where $P_k(x) = c_0^{(k)} + c_1^{(k)} x + \cdots + c_k^{(k)} x^k$ is a kth-degree polynomial with coefficients $c_j^{(k)}$ given by

$$c_j^{(k)} = \frac{1}{j!} \sum_{\ell=0}^{k-j} \frac{(\alpha-\beta)_\ell}{\ell!} H_{k-j-\ell}^{(k)}, \quad j = 0, 1, \ldots, k;$$

the terms $H_j^{(k)}$ are evaluated recursively as

$$H_0^{(k)} = 1, \quad H_j^{(k)} = -\frac{1}{\alpha} \sum_{\ell=1}^{j} \binom{\alpha}{\ell+1} \left(\frac{k\ell}{j} + 1 \right) H_{j-\ell}^{(k)}, \quad j = 1, 2, \ldots, k,$$

after defining, as usual, the generalized binomial coefficient as

$$\binom{\alpha}{j} = \frac{\alpha(\alpha-1)(\alpha-2)\cdots(\alpha-j+1)}{j!} = (-1)^j \frac{\Gamma(j-\alpha)}{\Gamma(-\alpha)\Gamma(j+1)}.$$

Once the residues at the singularities $\bar{s}_j^\star$ are known, the optimal parabolic contour method can be applied without particular difficulties.

We have to consider, however, that for derivatives of high order the strength of the singularities makes more difficult the task of finding regions of analyticity sufficiently large to get small errors. In these cases alternative techniques must be adopted, for instance the “derivatives balance” described in [15].

5.2 Application to matrix arguments

In many important applications the ML function with matrix arguments is involved. The presence of terms like $E_{\alpha,\beta}(A)$ is indeed spread out in fractional calculus, for example, for the efficient and stable solution of systems of FDEs [9, 10, 13, 37], to determine the solution of certain multi-term FDEs [30], in control theory, and in other related fields (see [15] for a review of these applications).

The formal definition of the matrix ML function can be given by simply substituting the scalar argument z in (1) with the desired matrix argument A, that is,

$$E_{\alpha,\beta}(A) = \sum_{j=0}^{\infty} \frac{A^j}{\Gamma(\alpha j + \beta)}.$$

Although theoretically unexceptionable, this series representation cannot be used for the actual computation; indeed, it presents the same issues raised in the scalar case (as discussed in Section 3), amplified by the computation increase due to the presence of the matrix argument. An alternative efficient numerical method to evaluate the matrix ML function is then necessary and this need has recently deserved much attention. Among all the available contributions we refer to our recent work [15], where an effective algorithm is discussed (and references to other strategies are extensively argued). The corresponding Matlab code is currently available in the file exchange service of the MathWorks website.[2]

The method proposed is based on the Schur–Parlett algorithm, which is currently the method of choice when dealing with general functions f (see, e. g., the comprehensive book by Higham [20]). It is essentially based on the Schur decomposition of the matrix argument A and the Parlett recurrence to evaluate f on the triangular factor. A key ingredient in this procedure is the use of the Taylor series expansion of f about a scalar argument (depending on the eigenvalues of A). This issue clearly requires the knowledge of the derivatives of f and the ability to numerically compute them. To this purpose, a thorough analysis for the case of the ML function has been carried out in [15], whose main steps are described in subsection 5.1.

The resulting method turns out to be very accurate and easily manageable, as shown by the different examples and numerical tests presented in [15].

Bibliography

[1] J. H. Barrett, Differential equations of non-integer order, *Can. J. Math.*, **6** (1954), 529–541.

[2] F. Bornemann, Accuracy and stability of computing high-order derivatives of analytic functions by Cauchy integrals, *Found. Comput. Math.*, **11**(1) (2011), 1–63.

[3] D. S. de Oliveira, E. Capelas de Oliveira, and S. Deif, On a sum with a three-parameter Mittag-Leffler function, *Integral Transforms Spec. Funct.*, **27**(8) (2016), 639–652.

[4] M. M. Džrbašjan, On the asymptotic behavior of a function of Mittag-Leffler type, *Akad. Nauk Armyan. SSR Dokl.*, **19** (1954), 65–72.

[5] R. Garra and R. Garrappa, The Prabhakar or three parameter Mittag-Leffler function: theory and application, *Commun. Nonlinear Sci. Numer. Simul.*, **56** (2018), 314–329.

2 www.mathworks.com/matlabcentral/fileexchange/66272-mittag-leffler-function-with-matrix-arguments

[6] R. Garrappa, Numerical evaluation of two and three parameter Mittag-Leffler functions, *SIAM J. Numer. Anal.*, **53**(3) (2015), 1350–1369.

[7] R. Garrappa, Grünwald–Letnikov operators for fractional relaxation in Havriliak–Negami models, *Commun. Nonlinear Sci. Numer. Simul.*, **38** (2016), 178–191.

[8] R. Garrappa, F. Mainardi, and G. Maione, Models of dielectric relaxation based on completely monotone functions, *Fract. Calc. Appl. Anal.*, **19**(5) (2016), 1105–1160.

[9] R. Garrappa, I. Moret, and M. Popolizio, Solving the time-fractional Schrödinger equation by Krylov projection methods, *J. Comput. Phys.*, **293** (2015), 115–134.

[10] R. Garrappa, I. Moret, and M. Popolizio, On the time-fractional Schrödinger equation: theoretical analysis and numerical solution by matrix Mittag-Leffler functions, *Comput. Math. Appl.*, **74**(5) (2017), 977–992.

[11] R. Garrappa and M. Popolizio, Generalized exponential time differencing methods for fractional order problems, *Comput. Math. Appl.*, **62**(3) (2011), 876–890.

[12] R. Garrappa and M. Popolizio, On accurate product integration rules for linear fractional differential equations, *J. Comput. Appl. Math.*, **235**(5) (2011), 1085–1097.

[13] R. Garrappa and M. Popolizio, On the use of matrix functions for fractional partial differential equations, *Math. Comput. Simul.*, **81**(5) (2011), 1045–1056.

[14] R. Garrappa and M. Popolizio, Evaluation of generalized Mittag-Leffler functions on the real line, *Adv. Comput. Math.*, **39**(1) (2013), 205–225.

[15] R. Garrappa and M. Popolizio, Computing the matrix Mittag-Leffler function with applications to fractional calculus, *J. Sci. Comput.*, **77**(1) (2018), 129–153.

[16] A. Giusti and I. Colombaro, Prabhakar-like fractional viscoelasticity, *Commun. Nonlinear Sci. Numer. Simul.*, **56** (2018), 138–143.

[17] R. Gorenflo, A. A. Kilbas, F. Mainardi, and S. Rogosin, *Mittag-Leffler Functions. Theory and Applications*, Springer Monographs in Mathematics, Springer, Berlin, 2014.

[18] R. Gorenflo, J. Loutchko, and Y. Luchko, Computation of the Mittag-Leffler function $E_{\alpha,\beta}(z)$ and its derivative, *Fract. Calc. Appl. Anal.*, **5**(4) (2002), 491–518.

[19] N. J. Higham, *Accuracy and Stability of Numerical Algorithms*, 2nd edn., Society for Industrial and Applied Mathematics (SIAM), Philadelphia, PA, 2002.

[20] N. J. Higham, *Functions of Matrices*, Society for Industrial and Applied Mathematics (SIAM), Philadelphia, PA, 2008.

[21] E. Hille and J. D. Tamarkin, On the theory of linear integral equations, *Ann. of Math. (2)*, **31**(3) (1930), 479–528.

[22] P. Humbert and R. P. Agarwal, Sur la fonction de Mittag-Leffler et quelques-unes des généralisations, *Bull. Sci. Math. (2)*, **77** (1953), 180–185.

[23] A. A. Kilbas, H. M. Srivastava, and J. J. Trujillo, *Theory and Applications of Fractional Differential Equations*, North-Holland Mathematics Studies, vol. 204, Elsevier Science B. V., Amsterdam, 2006.

[24] Y. Luchko and R. Gorenflo, An operational method for solving fractional differential equations with the Caputo derivatives, *Acta Math. Vietnam.*, **24**(2) (1999), 207–233.

[25] F. Mainardi, *Fractional Calculus and Waves in Linear Viscoelasticity*, Imperial College Press, London, 2010.

[26] M. G. Mittag-Leffler, Sur la nouvelle fonction $E_\alpha(x)$, *C. R. Acad. Sci. Paris*, **137** (1903), 554–558.

[27] I. Moret and P. Novati, On the convergence of Krylov subspace methods for matrix Mittag-Leffler functions, *SIAM J. Numer. Anal.*, **49**(5) (2011), 2144–2164.

[28] R. B. Paris, Exponential asymptotics of the Mittag-Leffler function, *R. Soc. Lond. Proc., Ser. A, Math. Phys. Eng. Sci.*, **458**(2028) (2002), 3041–3052.

[29] I. Podlubny, *Fractional Differential Equations*, Mathematics in Science and Engineering, vol. 198, Academic Press Inc., San Diego, CA, 1999.

[30] M. Popolizio, Numerical solution of multi-term fractional differential equations using the matrix Mittag-Leffler functions, *Mathematics*, **1**(6) (2018), 7.

[31] T. R. Prabhakar, A singular integral equation with a generalized Mittag-Leffler function in the kernel, *Yokohama Math. J.*, **19**(1) (1971), 7–15.

[32] H. Seybold and R. Hilfer, Numerical algorithm for calculating the generalized Mittag-Leffler function, *SIAM J. Numer. Anal.*, **47**(1) (2008/09), 69–88.

[33] A. Talbot, The accurate numerical inversion of Laplace transforms, *J. Inst. Math. Appl.*, **23**(1) (1979), 97–120.

[34] L. N. Trefethen and J. A. C. Weideman, The exponentially convergent trapezoidal rule, *SIAM Rev.*, **56**(3) (2014), 385–458.

[35] L. N. Trefethen, J. A. C. Weideman, and T. Schmelzer, Talbot quadratures and rational approximations, *BIT Numer. Math.*, **46**(3) (2006), 653–670.

[36] J. A. C. Weideman, Improved contour integral methods for parabolic PDEs, *IMA J. Numer. Anal.*, **30**(1) (2010), 334–350.

[37] J. A. C. Weideman and L. N. Trefethen, Parabolic and hyperbolic contours for computing the Bromwich integral, *Math. Comput.*, **76**(259) (2007), 1341–1356.

[38] A. Wiman, Über den Fundamentalsatz in der Teorie der Funktionen $E_{\alpha}(x)$, *Acta Math.*, **29**(1) (1905), 191–201.

Index

https://doi.org/10.1515/9783110571684-014